Cone Beam
Computed Tomography

IMAGING IN MEDICAL DIAGNOSIS AND THERAPY

William R. Hendee, Series Editor

Published titles

Quality and Safety in Radiotherapy
Todd Pawlicki, Peter B. Dunscombe,
Arno J. Mundt, and Pierre Scalliet, Editors
ISBN: 978-1-4398-0436-0

Adaptive Radiation Therapy
X. Allen Li, Editor
ISBN: 978-1-4398-1634-9

Quantitative MRI in Cancer
Thomas E. Yankeelov, David R. Pickens,
and Ronald R. Price, Editors
ISBN: 978-1-4398-2057-5

Informatics in Medical Imaging
George C. Kagadis and Steve G. Langer,
Editors
ISBN: 978-1-4398-3124-3

Adaptive Motion Compensation in Radiotherapy
Martin J. Murphy, Editor
ISBN: 978-1-4398-2193-0

Image-Guided Radiation Therapy
Daniel J. Bourland, Editor
ISBN: 978-1-4398-0273-1

Targeted Molecular Imaging
Michael J. Welch and William C. Eckelman,
Editors
ISBN: 978-1-4398-4195-0

Proton and Carbon Ion Therapy
C.-M. Charlie Ma and Tony Lomax, Editors
ISBN: 978-1-4398-1607-3

Comprehensive Brachytherapy: Physical and Clinical Aspects
Jack Venselaar, Dimos Baltas, Peter J. Hoskin,
and Ali Soleimani-Meigooni, Editors
ISBN: 978-1-4398-4498-4

Physics of Mammographic Imaging
Mia K. Markey, Editor
ISBN: 978-1-4398-7544-5

Physics of Thermal Therapy: Fundamentals and Clinical Applications
Eduardo Moros, Editor
ISBN: 978-1-4398-4890-6

Emerging Imaging Technologies in Medicine
Mark A. Anastasio and Patrick La Riviere,
Editors
ISBN: 978-1-4398-8041-8

Cancer Nanotechnology: Principles and Applications in Radiation Oncology
Sang Hyun Cho and Sunil Krishnan, Editors
ISBN: 978-1-4398-7875-0

Monte Carlo Techniques in Radiation Therapy
Joao Seco and Frank Verhaegen, Editors
ISBN: 978-1-4665-0792-0

Image Processing in Radiation Therapy
Kristy Kay Brock, Editor
ISBN: 978-1-4398-3017-8

Informatics in Radiation Oncology
George Starkschall and R. Alfredo C. Siochi,
Editors
ISBN: 978-1-4398-2582-2

Cone Beam Computed Tomography
Chris C. Shaw, Editor
ISBN: 978-1-4398-4626-1

Tomosynthesis Imaging
Ingrid Reiser and Stephen Glick, Editors
ISBN: 978-1-4398-7870-5

Stereotactic Radiosurgery and Radiotherapy
Stanley H. Benedict, Brian D. Kavanagh, and
David J. Schlesinger, Editors
ISBN: 978-1-4398-4197-6

IMAGING IN MEDICAL DIAGNOSIS AND THERAPY
William R. Hendee, Series Editor

Cone Beam
Computed Tomography

Edited by
Chris C. Shaw

CRC Press
Taylor & Francis Group
Boca Raton London New York

CRC Press is an imprint of the
Taylor & Francis Group, an **informa** business
A TAYLOR & FRANCIS BOOK

Taylor & Francis
Taylor & Francis Group
6000 Broken Sound Parkway NW, Suite 300
Boca Raton, FL 33487-2742

First issued in paperback 2020

© 2014 by Taylor & Francis Group, LLC
Taylor & Francis is an Informa business

No claim to original U.S. Government works

Version Date: 20131203

ISBN 13 : 978-0-367-57618-9 (pbk)
ISBN 13 : 978-1-4398-4626-1 (hbk)

Library of Congress Cataloging-in-Publication Data

Cone beam computed tomography (Shaw)
  Cone beam computed tomography / editor, Chris C. Shaw.
    p. ; cm. -- (Imaging in medical diagnosis and therapy)
  Includes bibliographical references and index.
  ISBN 978-1-4398-4626-1 (hardcover : alk. paper)
  I. Shaw, Chris C., editor of compilation. II. Title. III. Series: Imaging in medical diagnosis and therapy.
  [DNLM: 1. Cone-Beam Computed Tomography. WN 206]

RC78.7.T6
616.07'5722--dc23   2013047343

**Visit the Taylor & Francis Web site at
http://www.taylorandfrancis.com**

**and the CRC Press Web site at
http://www.crcpress.com**

Contents

Series preface

Since their inception more than a century ago, advances in the science and technology of medical imaging and radiation therapy are more profound and rapid than ever before. Furthermore, the disciplines are increasingly cross-linked as imaging methods become more widely used to plan, guide, monitor, and assess treatments in radiation therapy. Today, the technologies of medical imaging and radiation therapy are so complex and so computer-driven that it is difficult for the persons (physicians and technologists) responsible for their clinical use to know exactly what is happening at the point of care, when a patient is being examined or treated. The persons best equipped to understand the technologies and their applications are medical physicists, and these individuals are assuming greater responsibilities in the clinical arena to ensure that what is intended for the patient is actually delivered in a safe and effective manner.

However, the growing responsibilities of medical physicists in the clinical arenas of medical imaging and radiation therapy are not without their challenges. Most medical physicists are knowledgeable in either radiation therapy or medical imaging, and expert in one or a small number of areas within their discipline. They sustain their expertise in these areas by reading scientific articles and attending scientific talks at meetings. In contrast, their responsibilities increasingly extend beyond their specific areas of expertise. To meet these responsibilities, medical physicists periodically must refresh their knowledge of advances in medical imaging or radiation therapy, and they must be prepared to function at the intersection of these two fields. How to accomplish these objectives is a challenge.

At the 2007 annual meeting of the American Association of Physicists in Medicine in Minneapolis, this challenge was the topic of conversation during a lunch hosted by Taylor & Francis, involving a group of senior medical physicists (Arthur L. Boyer, Joseph O. Deasy, C.-M. Charlie Ma, Todd A. Pawlicki, Ervin B. Podgorsak, Elke Reitzel, Anthony B. Wolbarst, and Ellen D. Yorke). The conclusion of this discussion was that a book series should be launched under the Taylor & Francis banner, with each volume in the series addressing a rapidly advancing area of medical imaging or radiation therapy of importance to medical physicists. The aim would be for each volume to provide medical physicists with the information needed to understand technologies driving a rapid advance and their applications to safe and effective delivery of patient care.

Each volume in the series is edited by one or more individuals with recognized expertise in the technological area encompassed by the book. The editors are responsible for selecting the authors of individual chapters and ensuring that the chapters are comprehensive and intelligible to someone without such expertise. The enthusiasm of volume editors and chapter authors has been gratifying and reinforces the conclusion of the Minneapolis luncheon that this series of books addresses a major need of medical physicists.

Imaging in Medical Diagnosis and Therapy would not have been possible without the encouragement and support of the series manager, Luna Han of Taylor & Francis. The editors and authors, and most of all I, are indebted to her steady guidance of the entire project.

William Hendee, Series Editor
Rochester, MN

Preface

Cone beam computed tomography (CBCT) aroused interest early on in the development of CT technology. The motivation was probably the desire to complete a scan with one single rotation of the gantry. One striking example was the attempt to design and build the dynamic spatial reconstructor at the Mayo Clinic. The project did not lead to any commercialization for clinical use. The major obstacle was probably the lack of a large-area detector with sufficient dynamic range as well as a "flat" x-ray-absorbing layer. The image intensifier-video chain was the only large-area detector available for building CBCT systems back in the 1970s and 1980s. Although it is efficient and versatile for use in fluoroscopy, cardioangiography, and later on in digital subtraction angiography, it uses aperture control to match the limited dynamic range of a video camera to the exposures and a curvilinear cesium iodide layer for x-ray absorption. The former made it difficult or even impossible to properly measure x-ray attenuation information for image reconstruction. The latter made it difficult to properly incorporate the scanning geometry with the acquired images before reconstruction. Despite the unsuccessful attempt to develop CBCT systems' medical use, CBCT systems for small animal imaging as well as industrial testing have been developed and commercialized using an x-ray detector based on a charge-coupled device (CCD) for image acquisition. This situation lasted until large-area amorphous silicon-based x-ray detectors were developed and introduced for general radiography and mammography in 1999. These detectors have reasonably large dynamic range; ample resolution; and most importantly, a flat x-ray absorption layer. With this new type of detector, CBCT systems for various medical applications have been developed, investigated, and even commercialized. Most notable are the applications in radiation treatment, C-arm-based systems for interventional and head imaging, dental maxillofacial imaging, and breast imaging.

Along a totally separate path is the continuing and rigorous development of regular CT scanners. Soon after the introduction of CT, almost all CT scanners were designed as a fan beam CT system, including both the third- and fourth-generation scanners. A revolutionary development is the introduction of the idea of designing and using a continuously rotating gantry to perform a spiral or helical scan to cover a large section of the body in a much reduced time frame. Soon after and before the time the flat panel detectors were introduced, multiple detector row scanners were developed to increase the speed in covering a large section of body. What followed was a war of slice numbers, in which the number of detector rows quickly increased in an effort to widen the coverage and to further reduce the scanning time or alternatively to allow a large section of body to be continuously scanned at a high rate without longitudinal motion for dynamic imaging studies. Along with the use of a large number of detector rows is the change of the scanning geometry from fan beam CT to CBCT. Although the x-ray beam used with a multiple-slice scanner, with its small number of detector rows, could be treated as a fan beam, as the number of detector rows increased,

it became necessary to treat the x-ray beam as a cone beam. It is interesting to see that although CT technology initially evolved along two different paths with fan beam and cone beam scanning geometries, in the end they both converged back to the cone beam scanning geometry. However, although modern multiple detector row CT also uses cone beam x-rays, what people refer to as CBCT is usually based on flat panel (including amorphous silicon, CCD and CMOS) detectors. Thus, the term CBCT used in this book generally refers to techniques based on these detectors unless otherwise specified. Thus, to understand and work with either the flat panel–based cone beam CT systems or the multiple-slice CT scanners, professionals in the field of medical x-ray imaging should familiarize themselves with the principles, implementation, development, and application of CBCT. The purpose of this book is exactly that—to help these professionals achieve this goal.

The book has been divided into three parts. Part I covers fundamental principles and techniques. Chapter 1 provides a brief history of the development of the CT technology. Chapter 2 covers the image acquisition techniques and issues related to flat panel–based CBCT systems. Chapter 3 describes and discusses the major image reconstruction technique used in CBCT. Chapters 4 and 5 address the two most important issues of any x-ray imaging modality: image quality and radiation dose to the patient.

Part II covers advanced CBCT techniques. These topics may not reflect the newest development of CBCT technology. However, they depict the development of desirable or useful features beyond those provided by a basic CBCT scan. Chapter 6 does not address any particular CBCT technique but addresses the techniques of image simulation. Image simulation is widely used in the development of various new techniques and applications. This chapter, therefore, is included to help readers get started with image simulation should they be interested in working on further development of CBCT techniques or applications. Chapter 7 covers techniques used for processing, analysis, and visualization of the three-dimensional (3D) images generated by CBCT scans. Chapter 8 addresses the technique to scan a small volume of interest while minimizing exposures to outside regions. Chapter 9 addresses the ambitious task of obtaining dynamic 3D images, images that are essential in depicting respiratory and even cardiac motion. Chapter 10 addresses the various correction techniques that would help improve the accuracy of CT numbers; these correction techniques are essential for the accuracy of quantitative uses of CBCT images, for example, dose estimation with Monte Carlo simulation.

Part III provides an introduction to various applications of CBCT. Chapter 11 addresses multidetector row CT, the mainstream CT technology used in medicine. This technology may not be what people normally consider as CBCT but, like regular CBCT, the scanning techniques also are based on the use of cone beam x-rays. Chapter 12 addresses cone beam micro-CT for small-animal research. Cone beam micro-CT had been developed and continuously improved long before the flat panel–based CBCT appeared, although its use has been limited to industrial testing or small-animal research. Chapter 13 addresses the specific area of imaging the heart with multidetector row CT. Chapter 14 addresses the implementation and use of CBCT on a C-arm for interventional imaging applications. Chapter 15 addresses the integration and use of CBCT with a linear accelerator for verification and monitoring of radiation treatment. Chapter 16 addresses the implementation and application of CBCT to breast imaging.

The authors for the chapters have all worked intensively on various research topics related to CBCT and are experts in the field. In addition to a detailed introduction to the topics, they also have been asked to provide ample references that will allow readers to pursue a more thorough study. It is the hope of the editor and all authors that this book gives readers a detailed overview of CBCT technology and applications and helps them with their own research and development efforts or clinical applications.

Acknowledgments

The editor deeply thanks all contributing authors for their generosity in spending their precious time preparing the chapters and sharing their expertise, knowledge, and experiences in the field.

Editor

Chris C. Shaw, PhD, is a professor in the Department of Imaging Physics and the director of Digital Imaging Research at The University of Texas MD Anderson Cancer Center, Houston, Texas. He received his PhD in radiological sciences from the University of Wisconsin at Madison under the supervision of Charles A. Mistretta. As a graduate student, he participated in Dr. Mistretta's research efforts in developing digital subtraction angiography techniques that have been commercialized as a commonly used modality. After graduation, he held faculty positions at the Upstate Medical Center, the University of Rochester Medical Center, and the University of Pittsburgh Medical Center before joining the MD Anderson Cancer Center. His research interests have centered on the development and investigation of new x-ray imaging techniques for diagnostic applications. Early on, he worked on digital subtraction angiography and then switched to digital chest radiography and breast imaging. Since 2003, he has been working on the development and investigation of high-resolution cone beam breast CT, digital tomosynthesis imaging, and time-resolved CT and CBCT techniques.

Contributors

Amin Al-Ahmad
Department of Cardiology
Stanford University
Palo Alto, California, USA

Cem Altunbas
Department of Radiation Oncology
University of Colorado School of Medicine
Aurora, Colorado, USA

Zhiqiang Chen
Department of Engineering Physics
Tsinghua University
Beijing, China

Rebecca Fahrig
Department of Radiology
Stanford University
Palo Alto, California, USA

Martin Fiebich
Instituts für Medizinische Physik und Strahlenschutz
Technische Hochschule Mittelhessen
Munich, Germany

Elliot K. Fishman
Department of Radiology and Radiological Sciences
The Johns Hopkins University School of Medicine
Baltimore, Maryland, USA

Arundhuti Ganguly
Department of Radiology
Stanford University
Palo Alto, California, USA

Hewei Gao
Department of Radiology
Stanford University
Palo Alto, California, USA

Erin Girard
Department of Radiology
Stanford University
Palo Alto, California, USA

Stephen J. Glick
Department of Radiology
University of Massachusetts Medical Center
Worcester, Massachusetts, USA

Kenneth R. Hoffmann
Department of Neurosurgery
University at Buffalo, The State University of New York
Buffalo, New York, USA

Jiang Hsieh
GE Healthcare Technologies
Milwaukee, Wisconsin, USA

David A. Jaffray
Department of Radiation Oncology
University of Toronto
Toronto, Ontario, Canada

Nishita Kothary
Department of Radiology
Stanford University
Palo Alto, California, USA

Iacovos S. Kyprianou
Center for Devices and Radiological Health
Food and Drug Administration
Silver Spring, Maryland, USA

Liang Li
Department of Engineering Physics
Tsinghua University
Beijing, China

Peter B. Noël
Department of Diagnostic and International Radiology
Technische Universität München
Munich, Germany

Tinsu Pan
Department of Imaging Physics
The University of Texas MD Anderson Cancer Center
Houston, Texas, USA

Erik L. Ritman
Department of Physiology and Biomedical Engineering
Mayo Clinic College of Medicine
Rochester, Minnesota, USA

Ioannis Sechopoulos
Department of Radiology and Imaging Sciences
Emory University
Atlanta, Georgia, USA

Chris C. Shaw
Department of Imaging Physics
The University of Texas MD Anderson Cancer Center
Houston, Texas, USA

Jeffrey H. Siewerdsen
Department of Biomedical Engineering
Johns Hopkins University
Baltimore, Maryland, USA

Jared Starman
Department of Radiology
Stanford University
Palo Alto, California, USA

Katsuyuki (Ken) Taguchi
The Russell H. Morgan Department of Radiology and
    Radiological Sciences
The Johns Hopkins University School of Medicine
Baltimore, Maryland, USA

Xiangyang Tang
Department of Radiology and Imaging Sciences
Emory University
Atlanta, Georgia, USA

Ge Wang
Department of Biomedical Engineering
Rensselaer Polytechnic Institute
Troy, New York, USA

John W. Wong
Department of Radiation Oncology and Molecular Radiation
    Sciences
The Johns Hopkins University School of Medicine
Baltimore, Maryland, USA

Jennifer Xu
Department of Biomedical Engineering
Johns Hopkins University
Baltimore, Maryland, USA

Di Yan
Department of Radiation Oncology
William Beaumont Hospital
Royal Oak, Michigan, USA

Wojciech Zbijewski
Department of Biomedical Engineering
Johns Hopkins University
Baltimore, Maryland, USA

Wei Zhao
Department of Radiology
State University of New York at Stony Brook
Stony Brook, New York, USA

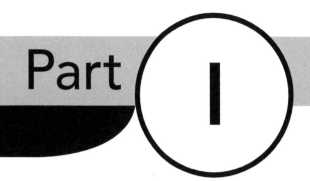

Part **1**

Fundamental principles and techniques

1

History of x-ray computed tomography

Jiang Hsieh

Contents

1.1   INTRODUCTION

Since the discovery of x-ray radiation by the German physicist Wilhelm C. Roentgen (Figure 1.1a) in 1895, x-ray radiation has been used extensively in a variety of applications and has generated numerous pictures, including the famous x-ray picture of the hand of Roentgen's wife (Figure 1.1b), the first x-ray picture of a human body ever taken. Although the power of x-rays rests on their ability to penetrate through dense materials and their material-dependent attenuation characteristics, it is not possible to visualize a particular structure along their paths without considering the attenuation caused by other structures along the same paths. This overlap is nicely illustrated in the picture of Mrs. Roentgen's hand, where the shadow of the wedding ring overlaps with the bony structure inside the ring (Figure 1.1b). The desire to remove the impact of the overlapping structures led to the development of conventional tomography.

One of the pioneers of conventional tomography was E. M. Bocage (1922). As early as 1921, Bocage described an apparatus to blur out structures above and below a plane of interest. The major components of his invention include an x-ray tube, an x-ray film, and a mechanical connection to ensure synchronous movement of the tube and the film (Figure 1.2). By properly controlling the speed of the x-ray tube and the film, the shadow generated by structures located on a particular plane parallel to the film does not change from different exposures, whereas the shadows generated by structures above or below the plane are blurred due to the positional shift in the shadow. Although conventional tomography offers a major step in reducing the impact of the overlaying structures in the x-ray radiographic images, it does not eliminate these overlaying structures. Often, when a dense object lies either above or below the object of interest, the blurring effect alone is insufficient to provide adequate visibility of the object of interest. To overcome some of these shortcomings, other scanning trajectories, such as pluridirectional tomography or transverse tomography, were proposed. However, these approaches do not fundamentally resolve the issue either.

The conversion from a set of measured images to a set of images that are free of overlapping structures is often called reconstruction. The mathematical formulation for reconstructing an object from multiple projections dates back to 1917 to the Austrian mathematician J. Radon who demonstrated mathematically that an object could be replicated from an infinite set of its projections. In 1956, R. N. Bracewell first applied the concept to reconstruct a map of solar microwave emission from a series of radiation measurements across the solar surface (Bracewell 1956). Between 1956 and 1958, several Russian papers accurately formulated the tomographic reconstruction problem as an inverse Radon transform (Tetel'baum 1956, 1957; Korenblyum et al. 1958). These papers discussed issues associated with implementation and proposed methodologies of performing reconstructions with television-based systems. Although these methodologies were somewhat inefficient, they offered satisfactory performance (Barrett et al. 1983).

The development of the medical x-ray computed tomography (CT) is generally credited to two physicists: Drs. G. N. Hounsfield (Figure 1.3a) and A. M. Cormack (Figure 1.3b). In 1963, Cormack reported the findings from investigations of perhaps the first CT scanner actually built (Cormack 1963). His work could be traced back to 1955 when he was asked to spend 1.5 days/wk at Groote Schuur Hospital (Cape Town, South Africa) to attend to the use of isotopes after the resignation of the hospital physicist. While observing the planning of radiotherapy treatments, Cormack came to realize the importance of knowing the x-ray attenuation coefficient distribution inside the body. During a sabbatical to Harvard University (Boston, MA) in late 1956, he derived a mathematical theory for image reconstruction and tested his theory with a laboratory simulation when he returned to South Africa in 1957. With a central cylinder of pure aluminum 20 cm in diameter surrounded by an aluminum alloy and oak annulus as phantom, a collimated ^{60}Co as the radiation source, and a Geiger counter as a detector, he calculated the attenuation coefficients for aluminum and wood by using his reconstruction technique.

In 1963, as a member of the physics department, he repeated his experiment at Tufts University (Medford, MA) with a circularly unsymmetrical phantom of aluminum and plastic. The phantom consists of an outer ring of aluminum simulating

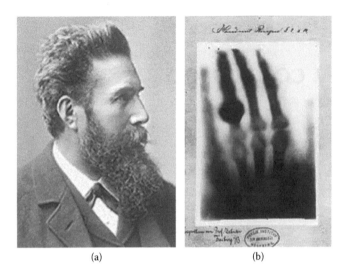

(a)   (b)

Figure 1.1 Wilhelm C. Roentgen (a) and the x-ray image of his wife's hand (b). (From http://wordinfo.info/unit/3151?letter=R&spage=4 http://cookit.e2bn.org/historycookbook/23-117-victorians-Health-facts.html. With permission.)

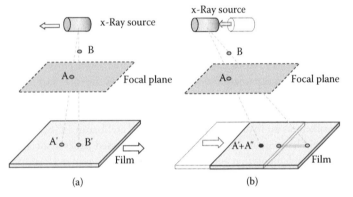

(a)   (b)

Figure 1.2 Illustration of the principle of conventional tomography. (a) At a particular time instant, object A (located on the focal plane) and object B (located off the focal plane) cast shadows A' and B' respectively on the film. (b) At a later time instant, both x-ray source and film move in opposite directions at specified speeds such that the shadow A" cast by object A overlaps with shadow A'. Shadow B" cast by object B does not overlap with B'. The signature of object A is therefore enhanced while signal associated with object B is blurred.

(a)   (b)

Figure 1.3 Inventors of x-ray CT: Godfrey N. Hounsfield (a) and Allan M. Cormack (b).

human skull, a polymethyl methacrylate (Lucite) filling representing soft tissue, and two aluminum disks within Lucite-mimicking tumors. Although the results of two experiments were published in 1963 and 1964, respectively, little attention was paid at the time to his work due to difficulties in performing the necessary calculation. Cormack remarked in his Nobel Lecture that, "There was virtually no response. The most interesting request for a reprint came from the Swiss Center for Avalanche Research. The method would work for deposits of snow on mountains if one could get either the detector or the source into the mountain under the snow!"

The development of the first clinical CT scanner began in 1967 with Hounsfield. While investigating pattern recognition techniques at the Central Research Laboratories of EMI, Ltd. in England, he deduced that x-ray measurements through a body taken from different directions would allow the reconstruction of its internal structure (Hounsfield 1976). Preliminary calculations by Hounsfield indicated that the accuracy in measuring the x-ray attenuation coefficients in a CT slice image could reach 0.5%, nearly an improvement of a factor of 100 over the conventional radiograph.

A laboratory scanner was built in 1967, and linear scan was performed on a rotating specimen in 1° steps; the specimen remained stationary during each linear scan (Figure 1.4). It took 9 days to complete the data acquisition, and an additional 2.5 hr were spent to solve 28,000 simultaneous equations and produce a picture. The use of modified interpolation method, higher intensity x-ray tube, and a crystal detector with a photomultiplier reduced the scan time to 9 hr and improved the accuracy from 4% to 0.5%.

The first clinically available CT device was installed at Atkinson Morley Hospital (London, England) in September 1971. After further refinement of the data acquisition and reconstruction techniques, images could be produced in 4.5 min. On October 4, 1971, the first patient with a large cyst was scanned, and the pathology was clearly visible in the image (Ambrose 1975). For their pioneer work in CT, Cormack and Hounsfield shared the Nobel Prize in Physiology and Medicine in 1979.

Figure 1.4 Laboratory scanner built by Hounsfield. (From http://miac.unibas.ch/BIA/06-Xray.html. With permission.)

The CT scanner developed by EMI is often called the first-generation scanner in which a single detector cell is used to collect the projection signals. After the detector and the x-ray source travel synchronously along a straight line to collect a set of parallel projection samples, the entire apparatus rotates 1° and the above-mentioned process is repeated to collect parallel projection samples along a slightly different orientation. Such data acquisition takes a substantial amount of time, so to speed up the data acquisition process, a second-generation scanner was built in which multiple detector cells were used to collect projection samples of multiple projection angles simultaneously. If six detector cells are used, for example, the angular increment for successive acquisition can increase from every 1° to every 6°, a factor-of-six increase in speed. Despite the improvement, the data acquisition speed was still fundamentally limited by the translation–rotation motion of the CT system, because the cross section of the entire object could not be covered by the small detector size. This limitation led to the development of a third-generation scanner in which a large number of detector cells were used to create a field of view large enough to cover the entire object of interest within the imaging plane. With this configuration, the linear translational motion of the x-ray source and the detector was no longer needed, and the entire data acquisition could be completed with the rotational motion of the gantry. The data acquisition speed was dramatically increased with this type of scanner. The third-generation design, however, came with a series of engineering issues. For example, because the x-ray focal spot and the detector cells were stationary with respect to each other, the projection samples violated the Nyquist sampling criteria that required the collection of two independent projection samples per detector cell. The fixed geometric relationship between the samples collected by each detector cell relative to the isocenter of the CT system also placed a stringent demand on the detector performance in terms of the fidelity of the measurement. Although these design issues have been resolved in later years of CT development, they paved the way to the development of the fourth-generation CT scanner in which the x-ray source and detector are no longer stationary to each other. In the fourth-generation scanner design, a ring of stationary detectors completely surrounds the patient while the x-ray tube rotates about the patient. A single projection view is formed with the readings of a single detector cell collected over time with the x-ray tube at different locations. Although this design overcomes some of the limitations of the third-generation scanners, it has issues of its own, such as the lack of ability to reject scattered radiation and a large increase in the number of detector cells. Figure 1.5 presents schematic diagrams of the four generations of scanners. Interestingly, the state-of-the-art CT scanners in the market today are all third-generation scanners.

Humans are not the sole beneficiaries of this wonderful invention. Over the years, CT scanners have been used to scan trees, animals, industrial parts, mummies, and just about everything that can fit inside a CT gantry. The technological advancement over the years also has pushed this medical imaging device from a backroom operation to a first-line defense in the emergency room. Two technologies stand out as the key drivers for such a role change: the development of helical or spiral scan mode and the introduction of multislice CT.

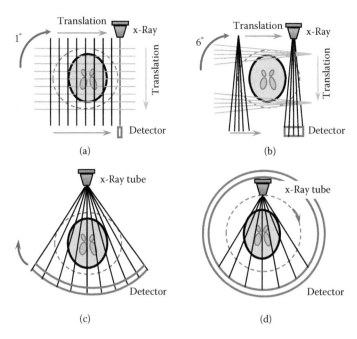

Figure 1.5 Illustration of the generations of CT scanners: first generation (a), second generation (b), third generation (c), and fourth generation (d).

Helical CT was introduced commercially in late 1980s and early 1990s. Previously, the only acquisition option was the step-and-shoot scanning mode containing both the data acquisition and nondata acquisition periods. During the data acquisition period, the patient remains stationary while the x-ray tube and detector rotate about the patient. Once a complete projection data set is acquired, the x-ray tube is turned off and the patient is indexed to the next scanning location. For typical CT scanners, the minimum nondata acquisition period is on the order of seconds due to mechanical and patient constraints. The mechanical constraint comes mainly from the fact that a patient table needs a finite amount of time to move a patient [weighing typically >45 kg (>100 lb)] from one location to another, and time to decelerate the gantry to stop before accelerating the gantry to rotate in the opposite direction. The patient constraint is due to the potential motion artifacts induced by a moving nonrigid human body (i.e., the internal organs can shift and deform) by the force of acceleration and deceleration (Mayo et al. 1987). As a result, a nonnegligible amount of time needs to elapse to minimize motion artifacts.

In helical or spiral CT, projections are continuously acquired while the patient is translated at a constant speed. (Although theoretically the patient can be translated with variable speed, most commercial scanners use constant speed for simplicity.) (Mori 1986; Nishimura and Miyazaki 1988; Kalender et al. 1989; Vock et al. 1989; Crawford and King 1990). Because there is no acceleration or deceleration of patients during the scan, the nondata acquisition period is eliminated, and a nearly 100% duty cycle is achieved. From a patient point of view, the x-ray source traverses a helical or spiral trajectory (Figure 1.6). Helical CT has expanded the traditional CT capability and enables the coverage of an entire organ in a single breath-hold. It is safe to state that helical CT is one of the key steps that move CT from a slice-oriented imaging modality to an organ-oriented modality.

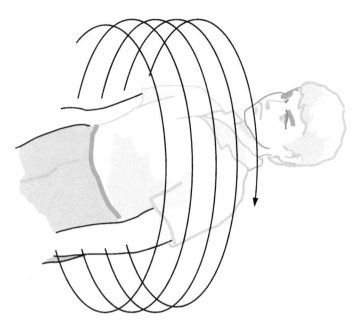

Figure 1.6 Illustration of helical scan mode.

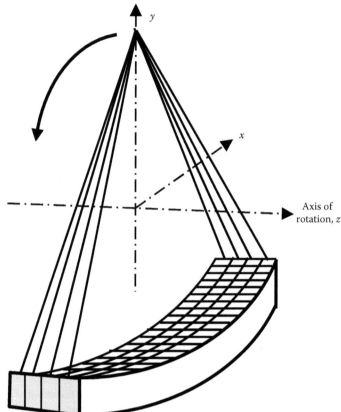

Figure 1.7 Geometrical relationship of a four-slice CT scanner.

To ensure optimal contrast enhancement and minimal patient motion, it is desirable to complete the entire study in as little time as possible, thereby demanding the patient table to travel at a faster speed. For example, a 60-cm coverage along the patient long axis in 30 s requires the patient table to travel at least 2 cm/s. At 0.5 s per gantry revolution, for example, it requires the table to increment 1 cm per gantry rotation and makes the reconstructed image slice thickness significantly larger than 5 mm. The unavoidable trade-off between slice thickness and volume coverage leads to the development of multislice scanner (Taguchi and Aradate 1998; Hu 1999; Bruder et al. 2000; Hsieh 2000; Proksa et al. 2000). Each detector cell of the single-slice scanner is divided into multiple smaller cells in the direction parallel to the axis of rotation (patient long axis). Each projection is therefore formed with a two-dimensional array of samples instead of the previous one-dimensional samples for the single-slice scanner. For illustration, Figure 1.7 depicts a four-slice scanner.

To illustrate the clinical benefits of multislice CT, we present two clinical examples. Figure 1.8a shows a volume-rendered image of a cardiac study. To visualize the fine vascular structures, the slice thickness has to be thin and the entire heart has to be covered in a short time to freeze cardiac motion. The level of image quality depicted in the figure is impossible to obtain with a single-slice scanner. Figure 1.8b shows a coronal image of an abdomen–pelvis study. Again, thin slice acquisition had to be used to depict fine anatomical details along the patient long axis while the entire study was completed well within a single breath-hold. Both studies were performed on a Discovery CT750HD high-definition CT system (GE Healthcare, Milwaukee, WI).

In recent years, the detector coverage along the patient long axis has increased steadily, from the original 1 cm for a single-slice scanner to 2, 4, 8, and 16 cm for multislice scanners, representing a gradual transition from a multislice scanner to a full cone beam scanner. This transition has brought a series of technical challenges ranging from engineering design obstacles and manufacturing difficulties to fundamental physics and

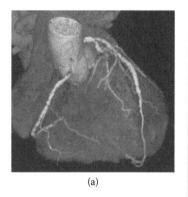

(a)

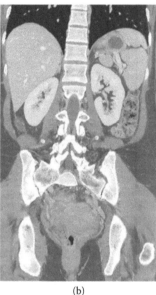

(b)

Figure 1.8 Examples of clinical images acquired on a Discovery CT750HD scanner. (a) a volume-rendered view of a cardiac coronary artery imaging study to illustrate visualization of fine vascular structures. (b) coronally reformatted image of an abdomen-pelvis study to demonstrate the isotropic spatial resolution offered by the state-of-the-art CT scanner. (From GE Healthcare Technologies. With permission.)

mathematical issues. Significant efforts have been made by both academic researchers and industry investigators to address and resolve these issues, and clinically acceptable image quality has been shown by commercially available CT scanners.

X-Ray CT is not the first and only medical imaging modality that tackles the cone beam tomography issues. Single-photon emission CT (SPECT), for example, explored various methodologies to handle the cone beam reconstruction issues. A micro-CT system for small animal imaging applications is another example of a cone beam CT scanner at a smaller scale. Although the detector technologies used for such scanners, charge-coupled device (CCD) based or flat-panel based, are somewhat different than detector technologies for clinical CT scanners, they often face similar technical challenges. Hybrid systems based on the flat-panel technologies, such as C-arm-based interventional or specialty scanners, also appeared to address particular clinical needs. Because of various physical constraints placed on these systems, they often face more technical challenges than the traditional cone beam CT.

With the increased detector coverage, the focus of CT research has moved beyond the pure three-dimensional anatomical information to include a fourth dimension: temporal. CT images are no longer limited to providing static anatomical details; they have expanded to provide dynamic information as well. Good examples of such studies are the investigation of cardiac wall motion over multiple cardiac phases and perfusion studies to understand the contrast uptake and washout of soft tissues. With the use of dual energy, the state-of-the-art CT capability has extended to the investigation of the functional aspect of an organ. These developments start to move x-ray CT from a pure anatomical modality to a physiological modality (Hsieh 2009). A new and exciting future for CT is underway.

REFERENCES

Ambrose, J. 1975. A brief review of the EMI scanner. *Proc Br Inst Radiol* 48: 605–6.

Barrett, H.H., Hawkins, W.G. and Joy, M.L.G. 1983. Historical note on computed tomography (letter). *Radiology* 147: 172.

Bocage, E.M. 1922. Procede et dispositifs de radiographie sur plaque en mouvement. French Patent No. 536464.

Bracewell, R.N. 1956. Strip integration in radiation astronomy. *Aust J Phys* 9: 198–217.

Bruder, H., Kachelrieb, M., Schaller, K., et al. 2000. Single-slice rebinning reconstruction in spiral cone-beam computed tomography. *IEEE Trans Med Imaging* 19: 873–87.

Cormack, A.M. 1963. Representation of a function by its line integrals, with some radiological applications. *J Appl Phys* 34: 2722–7.

Crawford, C.R. and King, K. 1990. Computed tomography scanning with simultaneous patient translation. *Med Phys* 17: 967–82.

Hounsfield, G.N. 1976. Historical notes on computerized axial tomography. *J Can Assoc Radiol* 27: 135–42.

Hsieh, J. 2000. CT image reconstruction. In: Goldman, L.W. and Fowlkes, J.B. (eds.). *Categorical Course in Diagnostic Radiology Physics: CT and US Cross-Sectional Imaging*. Oakbrook, IL: RSNA, 53–64.

Hsieh, J. 2009. *Computed Tomography: Principles, Design, Artifacts, and Recent Advances*. Bellingham, WA: SPIE and Wiley.

Hu, H. 1999. Multi-slice helical CT: scan and reconstruction. *Med Phys* 26: 1–18.

Kalender, W.K., Seissler, W. and Vock, P. 1989. Single-breath-hold spiral volumetric CT by continuous patient translation and scanner rotation. *Radiology* 173(P): 414.

Korenblyum, B.I., Tetel'baum, S.I. and Tyutin, A.A. 1958. About one scheme of tomography. *Bull Inst Higher Educ Radiophys* 1: 151–7.

Mayo, J.R., Muller, N.L. and Henkelman, R.M. 1987. The double-fissure sign: a motion artifact on thin section CT scans. *Radiology* 165: 580–1.

Mori, I. 1986. Computerized tomographic apparatus utilizing a radiation source. US Patent No. 4630202.

Nishimura, H. and Miyazaki, O. 1988. CT system for spirally scanning subject on a movable bed synchronized to x-ray tube revolution. US Patent No. 4,789,929.

Proksa, R., Kohler, T., Grass, M., et al. 2000. The n-PI-method for helical cone-beam CT. *IEEE Trans Med Imaging* 19: 848–63.

Taguchi, K. and Aradate, H. 1998. Algorithm for image reconstruction in multi-slice helical CT. *Med Phys* 25: 550–61.

Tetel'baum, S.I. 1956. About the problem of improvement of images obtained with the help of optical and analog instruments. *Bull Kiev Polytech Inst* 21: 222. [Russian]

Tetel'baum, S.I. 1957. About a method of obtaining volume images with the help of x-rays. *Bull Kiev Polytech Inst* 22: 154–60. [Russian]

Vock, P., Jung, H. and Kallender, W. 1989. Single-breath-hold spiral volumetric CT of the hepatobillary system. *Radiology* 173(P): 377.

2

Acquisition of projection images

Wei Zhao and Jeffrey H. Siewerdsen

Contents

2.1 INTRODUCTION

In this chapter, methods for the acquisition of projection images in a cone beam computed tomography (CBCT) scan are reviewed. Emphasis is placed on the detector technologies and the acquisition parameters adopted for several clinical applications.

2.2 IMAGING GEOMETRY AND GANTRY ROTATION

The x-ray beam used in CBCT for body imaging is a full cone with a solid angle defined by the size of the detector and the source-to-imager distance (SID). In specialized applications, for example, dedicated breast CBCT, a half-cone is used, where the central ray stays parallel to the chest wall for maximum coverage of breast tissue. Most clinical CBCT systems use circular gantry rotation with minimum angular coverage of 180 + cone angle. Typically, 200 to 300 images are acquired for each scan to avoid streak artifact. The gantry rotation speeds depend both on the mechanical stability of the gantry and the image acquisition speed of the detector.

It is advantageous to insert a bow-tie filter to the output of the x-ray tube to reduce radiation delivered to the periphery of the body (Mail et al. 2009); the bow-tie filter can reduce patient dose, reduce scattered radiation, and make more efficient use of the dynamic range of the detector.

2.3 X-RAY DETECTOR TECHNOLOGIES

The most widely used x-ray detectors for CBCT image acquisition are flat-panel imagers (FPIs) made with amorphous silicon (a-Si) thin-film transistor (TFT) technology (Rowlands and Yorkston 2000). They provide rapid image readout [30 frames per second (fps) with 1024 × 1024 detector matrix] and large-area image coverage [up to 43 cm × 43 cm (17 in. × 17 in.)]. In this section, an overview of the detector physics for FPIs is provided.

2.3.1 FPIs USING a-Si

FPIs are made using large-area a-Si semiconductor technology on glass substrates without tiling. This technology is enabled by the rapid advancement in flat-panel active matrix liquid crystal displays (AMLCDs) that use a two-dimensional array of a-Si TFTs to deliver the display signal to each pixel. Since the early 1990s, this technology has been incorporated into making large-area active matrix flat-panel imagers (AMFPIs). Depending on the x-ray detection materials, AMFPIs are divided into two main categories: direct and indirect detection. As shown in Figure 2.1a, direct detection AMFPIs use a uniform layer of x-ray-sensitive photoconductors, for example, amorphous selenium (a-Se) to convert incident x-rays directly to charge that is subsequently readout electronically by a TFT array (Zhao and Rowlands 1995); Figure 2.1b shows that indirect AMFPIs use an x-ray scintillator such as structured cesium iodide (CsI) to convert x-ray energy to optical photons that are then converted to charge by integrated photodiodes at each pixel of the TFT array (Antonuk et al. 1992). Both direct and indirect AMFPIs use the same readout scheme: the scanning control circuit turns on the TFTs one row at a time and transfers image charge from the pixel to external charge-sensitive amplifiers that are shared by all the pixels in the same column. The readout rate is dictated by the pixel RC time constant, where R is the on-resistance of the TFT and C is the capacitance for the pixel-sensing element. For complete readout of charge, the TFT needs to be ON for at least $T_{on} = 5RC$. With nominal values of $R = 4$ MΩ and $C = 1$ pF, each row of the AMFPI would require ~20 µs to readout. In addition to the time required for charge transfer, the parallel to serial conversion of the digitized signal necessitates an overhead. Hence, a detector with 1024 × 1024 pixels can be readout in real time (i.e., 30 fps). A faster readout rate, for example, 60 fps, may be possible by binning the pixels, that is, switching on more than one pixel at a time to reduce the image matrix. Both direct and indirect conversion AMFPIs have been commercialized for a wide variety of clinical x-ray imaging applications, including CBCT. The direct method has the advantages of higher image resolution and simpler TFT array structure that can be manufactured in a standard facility for AMLCDs. The indirect method has the advantage of higher x-ray quantum efficiency (QE) due to the higher atomic number of Cs and I compared with a-Se.

2.3.1.1 Indirect detectors

Indirect AMFPIs have been used by several major vendors in commercial CBCT systems for different clinical applications. They all use thallium (Tl)-doped CsI with columnar (needle-like) structure as the x-ray scintillator. As shown in Figure 2.2a, the columns in CsI help channel light photons that are generated by x-ray interaction in the forward direction. Although the light guidance is not as perfect as in fiber optics with smooth walls, columnar CsI provides much better imaging performance than powder phosphor screens (Rowlands and Yorkston 2000; Zhao et al. 2004). The CsI (Tl) used in AMFPIs is less hygroscopic than the CsI (Na) layers used in x-ray image intensifiers (XRIIs) and its optical emission spectrum (green) is a better match to the spectral response of a-Si photodiodes, as shown in Figure 2.3 (Rowlands and Yorkston 2000). The overall pixel x-ray sensitivity of an indirect AMFPI depends on four factors: (1) x-ray QE (η) of CsI; (2) inherent x-ray-to-optical photon conversion gain (g_c); that is, the number of optical photons emitted from the CsI for each absorbed x-ray; (3) optical QE of the a-Si photodiode; and (4) pixel fill factor, the fraction of pixel area occupied by the photodiode. The thickness of CsI (d_{CsI}) used in indirect AMFPIs varies depending on the clinical application. For the relatively high x-ray energy [>70 kilovoltage, peak (kVp)] used in clinical CBCT, d_{CsI} is typically 600 µm (Jaffray and Siewerdsen 2000). Columnar-structured CsI layers have lower density compared with single crystals. The packing density could vary depending on the deposition procedures; however, the widely quoted value is ~75%, resulting in a density, ρ, of 3.38 g/cm^3. Figure 2.4 shows the η of a 600-µm CsI layer as a function of x-ray photon energy. The k-edge of Cs (35 keV) and I (33 keV) creates a boost for η over the energy range typically used for CBCT. With an RQA5 spectrum (70-kVp tungsten spectrum with 21 mm of added Al filtration), η with the aforementioned detector parameters is ~0.84. As shown in Figure 2.3, the optical QE of a-Si photodiodes is ~0.7 for the green light emitted from CsI (Tl). The reported conversion

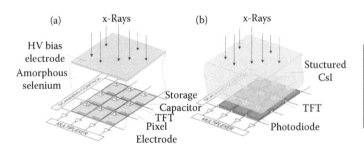

Figure 2.1 Diagram showing the concept of AMFPI with direct and indirect x-ray conversion. (a) Direct detector uses an x-ray photoconductor (e.g., a-Se) to convert x-rays directly charge. (b) Indirect detector uses a phosphor screen or structured scintillator to first convert x-rays to optical photons that are then converted to charge by an integrated photodiode at each pixel of the detector.

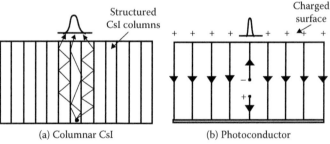

Figure 2.2 Image resolution of indirect and direct conversion x-ray detection materials. (a) Columnar structures of CsI (Tl) helps channeling light in the forward direction, thus providing better resolution than powder phosphor screens. (b) The applied electric field in photoconductors draws x-ray-generated image charge directly to surfaces without lateral spread.

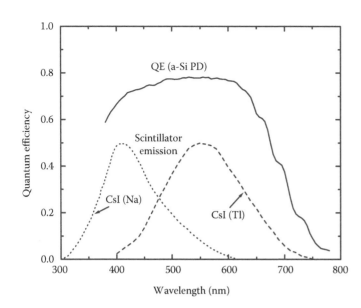

Figure 2.3 Optical quantum efficiency of a-Si photodiodes as a function of wavelength of light. Plotted in comparison are photon emission spectra for three types of x-ray scintillators: structured CsI (Na) and CsI (Tl). (Adapted from Rowlands, J.A. and Yorkston, J., *Medical Imaging: Volume 1. Physics and Psychophysics*, SPIE, Bellingham, WA, 2000.)

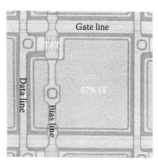 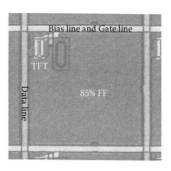

Figure 2.5 Micrograph showing the top view of a single pixel of two different indirect flat-panel designs: side-by-side TFT and photodiode (left) and photodiode on top of TFT (right). (Reproduced from Weisfield, R.L. et al., *Proc SPIE*, 5368, 338–48, 2004. With permission.)

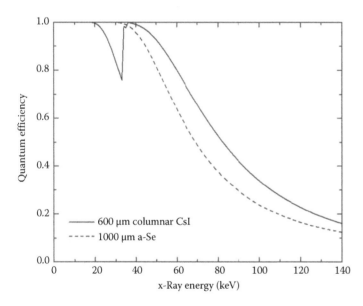

Figure 2.4 QE (η) as a function of x-ray photon energy for two materials, a-Se and columnar CsI, used in x-ray imaging detectors: Thickness of layer is 1000 μm for a-Se and 600 μm for CsI.

gain of CsI for indirect AMFPIs is ~25 eV per photon, resulting in g_c = 2000 for a 50-keV x-ray photon. The fill factor (f_p) depends on the pixel pitch (d) and the design rules of the a-Si photodiode–TFT array. Figure 2.5 shows the micrographs of two different pixel designs for an indirect AMFPI with 127-μm pixel size (Weisfield et al. 2004). The design on the left has the TFT at one corner of the pixel. The gate lines, data lines, and the photodiode occupy the rest of the space. This design is the typical design used in most commercial indirect AMFPIs. Because the space taken by the TFT and the lines does not

change as a function of d, f_p drops rapidly as d decreases (Rowlands and Yorkston 2000). The value of f_p is 0.57 for d = 127 μm. Shown on the right of Figure 2.5 is an advanced pixel design with boosted f_p, where the photodiode is built on top of the TFT and the gate and data lines, so that f_p for the same pixel size is increased to 0.85. With all factors considered, the overall gain of an indirect AMFPI is ~1000 electrons per 50 keV x-ray photon with f_p = 0.7.

The main disadvantage of AMFPIs compared with crystalline Si (c-Si) detector technologies, such as charge-coupled devices (CCDs), is its higher electronic noise that is dominated by the noise associated with the pixel reset and the charge amplifier (Weisfield and Bennett 2001). The nominal root-mean-square (rms) value for the pixel electronic noise is ~1500 electrons (e), a number that is higher than the number of light photons captured by the photodiode for each absorbed x-ray in CsI. The excessive electronic noise would lead to degradation in low-dose imaging performance, especially for high spatial frequency information that is already compromised by the image blur in CsI.

2.3.1.2 Direct detectors

The most highly developed x-ray photoconductor is a-Se that is currently used in all commercial direct AMFPIs. Because of its lower atomic number than CsI, the thickness (d_{Se}) used in the fluoroscopic detectors for CBCT applications is 1000 μm. The density of a-Se is 4.27 g/cm³, a value that is lower than that for crystalline selenium. For an RQA5 x-ray spectrum, η is 0.77 for d_{Se} = 1000 μm. The x-ray-to-charge conversion gain of a-Se depends on the electric field (E_{Se}). The nominal value for the energy required to generate an electron hole pair in a-Se W is 50 eV at E_{Se} = 10 V/μm (Rowlands et al. 1992). The geometric fill factor for direct AMFPIs is high because the pixel electrode is built on top of the TFT and the gate and data lines. In addition, the image charge collection in a-Se is governed by the electric field. Because the field lines in the gap between pixels bend toward the pixel electrodes, the image charge created in this region also can be collected (Pang et al. 1998). This leads to an effective fill factor of unity, as has been confirmed experimentally from direct AMFPIs (Zhao et al. 2003). Compared with indirect AMFPIs, a-Se direct detectors have approximately the same x-ray conversion gain (1000 e per incident 50 keV x-ray photon) and electronic noise; hence, they share the same advantages and limitations in low-dose imaging performance as the indirect AMFPIs. One of the advantages of the direct AMFPI compared with the indirect is the ability to make smaller pixels because of its simpler array structure (no need for the photodiodes) and the unity fill factor that is independent of pixel size.

2.3.2 FPIs IN CBCT

Both direct and indirect AMFPIs have been used in CBCT applications. Table 2.1 shows several examples of the detector design parameters and operating modes adopted for both types of AMFPIs in CBCT systems. The indirect AMFPI Paxscan 4030CB (Varian, Palo Alto, CA) has been used extensively by medical equipment manufacturers and research groups because of its ease of integration and excellent imaging performance (Colbeth et al. 2001; Suzuki et al. 2004). The detector is implemented with several image readout modes, which trade image resolution for frame rate. When 2 × 2 detector pixels are binned by turning on two rows of TFTs simultaneously, the frame rate can be doubled. Full resolution readout with 194-μm pixel pitch has been used for high-resolution imaging applications such as dedicated breast CBCT (Ning et al. 2007; Boone et al. 2010) and the binned mode in angiography and cardiac CBCT applications (Suzuki et al. 2004; Tognina et al. 2004). It is important to note that all CBCT system geometries use magnification to some extent; therefore, high frame rate with binning modes is often used to ensure rapid image acquisition without significant trade-off in reconstructed image resolution. In CBCT applications where the volume of interest (VOI) is moderate, such as for cardiac or mobile-C arm, FPIs with smaller active areas (e.g., 20 cm × 20 cm) can be used. Due to the shorter data line, the smaller detectors can have lower electronic noise and hence slightly better low-dose performance. Direct AMFPIs made by two manufacturers (Shimadzu and Anrad) also have been used in CBCT systems (Kakeda et al. 2007; Chen et al. 2009; Koyama et al. 2010). They have

virtually identical pixel size (150 μm) and a-Se layer thickness (1000 μm), as shown in Table 2.1. The detector size ranges from 22 cm × 22 cm up to 43 cm × 43 cm. Detector binning also has been implemented to enable rapid image acquisition in CBCT.

2.4 PROJECTION IMAGE QUALITY

Several international standards [International Electrotechnical Commission (IEC) and American Association of Physicists in Medicine (AAPM) Task Group] have adopted image quality metrics expressed in the spatial frequency domain to evaluate the image quality of projection x-ray images. These image quality metrics, including modulation transfer function (MTF), noise power spectrum (NPS), and detective quantum efficiency (DQE), are reviewed here. In addition to inherent detector properties, factors related to CBCT system operation and their impact on image quality are also reviewed.

Before quantitative evaluation of image quality can be performed, projection images acquired by FPIs need to be corrected for imperfection due to detector nonuniformity and defects. Defect pixels are unavoidable during fabrication of the active matrix. Due to the large number of pixels, even a 0.1% defect rate could result in 9000 bad pixels in a 3000 × 3000 pixel AMFPI. In addition, there is nonuniformity between pixels due to several reasons: (1) nonuniformity in the active matrix that results in variation in TFT characteristics, (2) variation in the thickness of the x-ray detector material, and (3) gain nonuniformity between different charge amplifier channels. This variation necessitates image correction through postprocessing. The standard method is an offset and gain nonuniformity correction followed by a defect pixel replacement. An offset (or dark) image is obtained without x-ray exposure and is subtracted from each x-ray image. To reduce the effect of electronic noise, the average of several dark images is usually used. Because there is temporal drift in offset due to device instability, offset images are constantly updated between x-ray examinations. The gain correction is performed by dividing the offset-subtracted image

Table 2.1 Examples of AMFPI detector parameters used in CBCT

DETECTOR TYPE	INDIRECT	INDIRECT	DIRECT
Maker/model	Varian Paxscan 4030CB	Varian Paxscan 2020	Shimadzu/Safire Anrad/FPD9 or 14
X-ray detection material thickness	CsI (Tl) 0.6 mm	CsI (Tl) 0.6 mm	a-Se 1.0 mm
Pixel pitch (μm)	194 × 194	194 × 194	150 × 150
Fill factor	0.7		1
Detector active area (cm × cm)	40 × 30	20 × 20	22 × 22
Detector matrix	2048 × 1536	1024 × 1024	1472 × 1472
Readout rate	60 fps (4 × 4 binning) 30 fps (2 × 2 binning) 7.5 fps (full resolution)	60 fps (2 × 2 binning) 30 fps (full resolution)	30 fps (1024 × 1024 ROI)
Maximum detector exposure (mR)	2.06 mR (2 × 2 binning) 3.55 mR (full resolution)		
X-ray quantum noise-limited exposure (μR)	1 μR (2 × 2 binning) 4 μR (1 × 1)		
Clinical application	Angiography, breast CT	Cardiac, mobile C-arm	Cardiac CBCT

by a gain table obtained during a calibration procedure. During calibration, the detector is exposed to uniform radiation. By averaging several x-ray images, the gain of each pixel can be determined. Defect pixels are identified by setting a lower threshold of x-ray sensitivity based on the pixel statistics, and the result is stored in a defect map. After gain correction, the bad pixels are replaced by the average values of neighboring good pixels. The gain table and bad pixel map are much more stable compared with offset; hence, in commercial detectors, the calibration procedure needs to be repeated only once a month or even less frequently.

2.4.1 SPATIAL FREQUENCY DOMAIN IMAGE QUALITY METRICS

2.4.1.1 Spatial resolution: Modulation transfer function

The spatial resolution of projection images is quantified by MTF. MTF is defined as the Fourier transform (FT) of the point spread function (PSF). This concept applies to a linear system that is shift invariant. In practice, MTF is usually measured in two orthogonal directions using FT of the line spread function (LSF).

For CBCT systems, the shift-invariance condition is violated due to several factors: (1) the digital detectors are undersampled, making the PSF position-dependent; (2) the projection image blur due to the finite focal-spot size of the x-ray tube varies with the location of the object plane; and (3) the focal-spot blur may be worsened by focal-spot motion during x-ray exposure of the CBCT scan. It is also important to note that the cone beam image reconstruction will result in further position dependence in the reconstructed image domain. For details, see Chapter 16, where nonstationarity of noise and spatial resolution are discussed. For a digital detector with pixel sensing element width a = 140 μm and pixel pitch d = 150 μm, the detector aperture response is a sinc function with the first zero at f = $1/a$ = 7.1 cycles per mm. The Nyquist frequency of pixel sampling is f_N = $1/(2d)$ = 3.3 cycles per mm. Thus, a digital detector is always undersampled except when the frequency response of the x-ray detection material is very poor. To apply MTF to a digital detector, the concept of presampling MTF is usually used. It describes the frequency response of the detector before sampling occurs. The standard experimental technique adopted by IEC for measuring the presampling MTF is the slanted edge method. Figure 2.6a shows the measured presampling MTF of an indirect AMFPI (Varian Paxscan 4030CB) with different detector pixel binning (Tognina et al. 2004). In full resolution with pixel pitch of 194 μm, the presampling MTF is dominated by the image blur in CsI, a blur that is 600 μm thick. It is important to note that even with such thick CsI layers, the detector is undersampled, with MTF = 0.2 at the Nyquist frequency of f_{NY} = 2.6 cycles per mm. With 2 × 2 and 4 × 4 pixel binning, which is often used in CBCT image acquisition to increase readout speed, the aperture function of the larger binned pixels becomes the dominant factor for spatial resolution. The presampling MTF of direct AMFPI, in contrast, is limited only by the pixel aperture function because there is essentially no image blur in the x-ray photoconductor. Figure 2.6b shows the measured MTF for an a-Se-based AMFPI with 150-μm pixel size (FPD14, Anrad) (Hunt et al. 2004). The

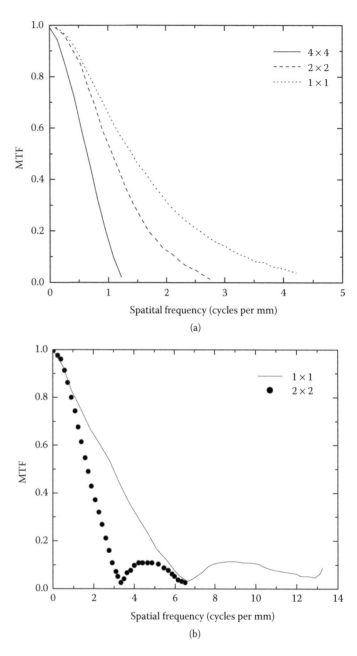

Figure 2.6 MTF of commercial AMFPI detectors. (a) Indirect FPI with 194 μm pixel size and 600-μm-thick columnar-structured CsI. (Adapted from Tognina, C.A. et al., *Proc SPIE*, 5368, 648–56, 2004.) (b) Direct FPI with 150-μm pixel size and 1000-μm-thick a-Se. (Adapted from Hunt, D.C. et al., *Med Phys*, 31, 1166–75, 2004.)

presampling MTF of the detector has its first zero at 6.1 cycles per mm, indicating an effective fill factor of unity for the 150-μm pixels. Image blur has been observed in some a-Se detectors due to charge trapping and recombination in the bulk of very thick a-Se layers or near the pixel–electrode interface, a source of presampling blur, and could contribute to ~10%–20% drop in presampling MTF at the Nyquist frequency depending on the material properties and thickness of a-Se (Zhao et al. 2003; Hunt et al. 2004).

When AMFPI are used in CBCT systems, other factors could contribute to the presampling MTF of projection images in addition to the inherent MTF of the detector. These factors include focal-spot blur and x-ray scatter that enter as a linear combination with the MTF of the detector in the system MTF. As discussed in Chapter 16, the reconstruction filters used in filtered backprojection also affect the (three-dimensional) MTF of the CBCT imaging system.

2.4.1.2 Noise power spectra

The noise in a projection image can be characterized in the spatial frequency domain using NPS, which is the FT of the autocorrelation of a flat-fielded x-ray image. The inherent stochastic (Poisson) noise of incident x-rays is white, that is, no spatial correlation. Image blur in an AMFPI detector could lead to spatial correlation of noise, resulting in a high-frequency drop of NPS. When a presampling NPS has frequency components above the Nyquist frequency of detector sampling, aliasing of NPS occurs, leading to an increase in NPS. Noise aliasing in direct conversion AMFPIs results in an NPS that is essentially white. The NPS of projection images can be measured experimentally using flat-fielded x-ray images under uniform x-ray exposures. For accurate measurement of NPS, the spatiotemporal behavior of detectors needs to be taken into account. Temporal performance of AMFPIs (to be discussed later), such as image lag, could lead to noise correlation between frames, resulting in a reduction in NPS if NPS analysis is performed using a temporal sequence of x-ray images. Two methods have been used to account for this factor: (1) measure the spatiotemporal NPS by adding time domain as a third-dimensional variable (in addition to the two spatial dimensions x and y) and determine the two-dimensional (2D) NPS after correction of lag effect (Friedman and Cunningham 2010); and (2) measure 2D spatial domain NPS by eliminating the temporal effect, that is, at low frame rate where temporal correlation is negligible. For accurate measurement of 2D NPS, fixed pattern in projection images must be removed through offset and gain correction. Then, a region of interest (ROI) $I(x,y)$ is selected from each flat-fielded image with its mean value subtracted before FT to obtain NPS: (IEC62220–1:2004):

$$NPS(u,v) = \frac{d_x d_y}{N_x N_y} < | FT[I(x,y) - \bar{I}(x,y)]|^2 > \quad (2.1)$$

where <> represents the ensemble average; N_x and N_y are the number of elements in the x and y directions, respectively; and d_x and d_y are the pixel pitch in each direction.

2.4.1.3 Detective quantum efficiency

The overall imaging performance of an x-ray detector is best represented by its DQE. It is defined as the ratio between the signal-to-noise ratio (SNR) squared at the output of the detector and that at the input, which is equal to the number of x-ray photons per unit area (q_0):

$$DQE = \frac{SNR_{out}^2}{q_0} \quad (2.2)$$

SNR_{out}^2 is also known as the number of noise equivalent quanta (NEQ). Hence, DQE describes the efficiency of the detector in using the incident x-rays, and its upper limit is the quantum efficiency (η) of the detection material. To describe the ability of the detector in transferring information with different frequency content, DQE is usually measured as a function of spatial frequency (f) using the following formula:

$$DQE(f) = \frac{MTF^2(f)}{q_0 NNPS(f)} \quad (2.3)$$

where $NNPS(f)$ is the NPS normalized by square of the pixel x-ray response of the detector at a given exposure. Any additional noise source in an imaging system (e.g., detector electronic noise) increases $NPS(f)$ from the x-ray quantum noise and degrades the DQE. Because $MTF(f)$ always decreases as a function of f, added noise, which is usually white, degrades $DQE(f)$ more severely at higher f. $DQE(f)$ is used as the standard for imaging performance comparison between different detectors.

Dose dependence of $DQE(f)$ is another important imaging performance criteria. The DQE of AMFPIs at very low exposures could be degraded due to the readout electronic noise, and it has been recognized as a major disadvantage compared with the more established x-ray detectors such as the XRII that has internal signal gain. Figure 2.7a shows the $DQE(f)$ of Varian Paxscan 2020 detector in both full resolution and 2 × 2 binning readout modes (Tognina et al. 2004). It shows that at detector entrance exposure of 5.3 μR (46 nGy), DQE(0) of the detector for an RQA5 spectrum is ~0.7, a value that is approaching the theoretical limit of $DQE(0) = \eta A_S$, where A_S is the Swank factor of CsI describing the added noise due to variation in x-ray to optical photon conversion gain (Swank 1973). However, as the exposure decreases to 0.6 μR (5.2 nGy), DQE(0) drops to below 0.6 due to the degradation effect of added electronic noise. The magnitude of this effect is more pronounced at high frequencies because the electronic noise is white; whereas, the x-ray quantum noise decreases with frequency. Figure 2.7b shows the measured DQE of direct conversion AMFPIs (FPD14, Anrad). A similar degradation in DQE at low exposures is observed. However, the magnitude of DQE drop at high spatial frequency is similar to that for DQE(0) because the x-ray quantum noise in a-Se is virtually white.

2.4.2 DYNAMIC RANGE

In CBCT, the detector received a wide range of x-ray exposures during each projection view, the highest being raw exposure without patient attenuation and the lowest behind bony structures, for example, mediastinum. AMFPIs were originally developed for radiography/fluoroscopy (RF) applications that require a dynamic range of 10^4–10^5. This range can be accommodated by a 14-bit digitizer without problem of detector saturation. For CBCT applications, however, a dynamic range of up to 10^7 may be required. This range posed two challenges for AMFPIs: (1) improve the low-dose performance that is limited by electronic noise; and (2) avoid detector saturation due to either the detection element itself (e.g., a-Si photodiode) or the charge amplifier (e.g., with high gain setting). One method implemented to overcome amplifier saturation is dynamic gain switching that effectively extends the dynamic range to 18 bits

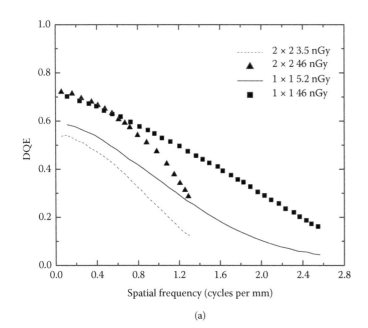

(a)

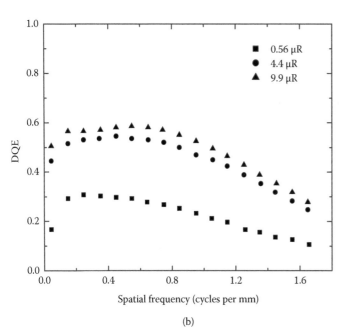

(b)

Figure 2.7 (a) DQE for an indirect AMFPI with pixel size of 194 μm in both full resolution and 2 × 2 binning operation. DQE degradation due to electronic noise of the AMFPI is evident at detector entrance exposure <5.2 nGy (0.6 μR). (Adapted from Tognina, C.A. et al., *Proc SPIE*, 5368, 648–56, 2004.) (b) DQE for a direct AMFPI with pixel size of 150 μm and 1000-μm-thick a-Se layer. The DQE in 2 × 2 binning mode shows degradation for exposures <4 μR. (Adapted from Hunt, D.C. et al., *Med Phys*, 31, 1166–75, 2004.)

(Roos et al. 2004). With dynamic gain readout, the signal from each detector pixel is sampled (nondestructively) during readout by a charge comparator, and the amplifier gain (i.e., feedback capacitor) is dynamically adjusted to either "high" or "low" gain according to a specified threshold. As a result, pixels receiving low exposure (i.e., low signal in the deepest shadow of the object) are automatically addressed in high-gain mode; whereas, pixels in the bare beam are automatically addressed in low-gain mode (to avoid sensor saturation).

In addition to detector performance (i.e., dynamic gain modes and reduced electronic noise), several aspects of the CBCT system design can be implemented to improve dynamic range: (1) bow-tie filter and (2) source modulation. As mentioned, a bow-tie filter tends to reduce patient dose, x-ray scatter, and the requirements on detector dynamic range. Bow-tie filter design typically assumes a cylindrical object (e.g., 32 cm of water) and is shaped such that the x-ray fluence transmitted to the detector is nearly uniform, that is, the detector signal in projection of the cylinder is nearly uniform. The usual means of accomplishing this is by adding a filter at the x-ray tube output varying in thickness from ~0 mm at center to ~2–10 mm in the region of the unattenuated beam (depending on material type) and of nearly constant thickness in the z direction. Of course, the assumption of a cylindrical object is grossly oversimplified in comparison to real patient anatomy, but the presence of a conservatively designed bow-tie (i.e., one that reduces the range of transmitted fluence to the detector) still tends to be beneficial. Variations on the basic bow-tie concept include a one-sided bow-tie filter for offset-detector CBCT acquisition (i.e., extended lateral field of view) and "dynamic" bow-tie filters in which a pair of opposing leaves, each of varying thickness, slide in or out laterally to modulate the transmitted fluence, thereby overcoming the simple assumption of a cylindrical patient.

Modulation of the x-ray source can have similar benefit to dynamic range and dose. The simplest form of source modulation is to vary the milliamperes (mAs) (and possibly kVp) in each projection during the source–detector orbit. This type of source modulation can be useful, for example, to increase signal (i.e., increase mAs and kVp) through thicker aspects of the patient (namely, lateral views) as well as to decrease dose, for example, by reducing mAs in anterior–posterior (AP) views. Alternative methods involve spatially varying source modulation within each view. An ROI filter is one such approach; for example, an aperture through which a higher fluence passes (increasing SNR in the center of reconstruction) surrounded by a more heavily attenuating regions (lower SNR in the periphery of reconstruction). The ROI modulation is therefore an extension of the bow-tie filter approach, where the goal is not necessarily to "flatten" the fluence at the detector but to provide higher SNR within an ROI (and allowing an increase in noise in surrounding regions). Methods in which the source is modulated both spatially and view to view are areas of ongoing work (Bartolac et al. 2011).

2.5 FRAME RATE AND TEMPORAL PERFORMANCE

2.5.1 FRAME RATE REQUIREMENT IN DIFFERENT APPLICATIONS

The frame rate of AMFPIs applied to CBCT is usually in the range of ~5–30 fps. Because CBCT typically requires several hundred projections for sufficient angular sampling, this requirement implies a fairly long scan time, certainly much slower than the 0.3 s per rotation typical of conventional diagnostic computed tomography (CT). For example, assuming one projection per degree over a 360° rotation, we have a total scan time of 12 s (for 30 fps readout) and upward of 72 s (for 5 fps readout). A faster scan can be obtained through a shorter arc (e.g., 180° + fan angle), reduced angular sampling, or both, recognizing that such approaches have implications for image quality (and dose) as well. Different CBCT applications present different capabilities and requirements in this regard. Bench-top specimen scanning applications may be completely tolerant to long acquisition times without concern for object motion; for example, 1000 projections over 360° at 5 fps, giving acquisition time of 200 s, a value that is still fast in comparison with some micro-CT scan protocols. In image-guided radiation therapy, the acquisition speed is actually limited by the fastest allowed rotation rate of the linear accelerator gantry, 360° rotation in 60 s, limited by regulations to allow for touch-guard interrupts on collision detection. A 60-s rotation at, for example, 10 fps readout yields 600 projections for a full orbit or ~320 projections for a half-scan. Mobile C-arms tend to operate somewhere in between, for example, ~200 projections acquired over an ~200° orbit, with the detector reading at ~5 fps, giving ~40-s acquisition time. Fixed-room C-arms (i.e., ceiling-mounted or floor-mounted) allow significantly faster rotation rates, particularly in "propeller" mode; for example, upward of 45°/s rotation speed. A detector reading at 30 fps gives ~130 projections acquired in a half-scan orbit covered in ~4.5 s. Clearly, the choice of AMFPI readout rate depends on a multitude of factors ranging from detector capabilities to the power of the x-ray source and, ultimately, the requirements of the clinical application.

2.5.2 TEMPORAL PERFORMANCE OF DIFFERENT X-RAY DETECTOR TECHNOLOGIES

Temporal imaging characteristics of AMFPIs can be separated into two categories: lag and ghosting. As shown in Figure 2.8, lag is the carryover of image charge generated by previous x-ray exposures into subsequent image frames. It is manifested as changes in dark images, that is, readout of the detector without an x-ray exposure. As shown in Figure 2.9, ghosting is the change of x-ray sensitivity, or gain, of the detector as a result of previous exposures to radiation. It can be seen only with subsequent x-ray

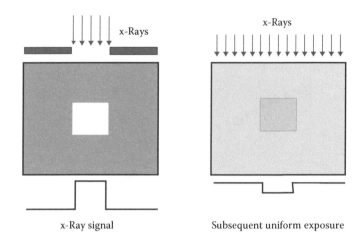

Figure 2.9 Conceptual images showing ghosting of an x-ray imaging detector. Ghosting is defined as the change in x-ray sensitivity as a result of the detector's exposure to radiation. It can be seen only with subsequent x-ray exposures.

exposures. Both lag and ghosting could lead to image artifacts in projection and reconstructed images in CBCT. An overview of the physical mechanism for and the measurement of lag and ghosting is provided here for both indirect and direct AMFPIs.

2.5.2.1 Temporal performance of indirect AMFPIs

The lag and ghosting of indirect AMFPIs can be attributed to three sources of mechanisms: (1) charge trapping and release in a-Si photodiode, (2) after-glow from the CsI scintillator, and (3) incomplete readout of charge from the pixel to the charge amplifiers (when $T_{on} < 5RC$) (Overdick et al. 2001). During x-ray exposure, the a-Si photodiode is biased with an electric field for image charge to be collected efficiently. Electrons in a-Si have better transport properties; therefore, most a-Si photodiodes are negatively biased at the light-entrance side. When electrons move toward pixel electrodes, they could be captured by localized state (traps) in the a-Si material and then released at a later time, for example, during the subsequent image frames. Lag has been investigated extensively under different imaging conditions, for example, detector exposure and frame rate (Siewerdsen and Jaffray 1999), with first-frame lag typically in the range ~1%–10%. Figure 2.10a is the relative signal intensity measured from an indirect AMFPI (Varian 2020) with the x-ray exposure delivered to frame zero, whose signal is set as the reference level (100%) (Tognina et al. 2004). It shows that the first frame lag depends on the frame rate, and ranged between 2% and 10% depending on operational conditions and entrance exposure. The time required for trapped charge to be released depends on the energy depth of the traps. Shallow traps are responsible for short-term lag and deep traps for long-term residual signal, which could be visible tens of minutes after exposures, and the magnitude of long-term lag depends on the degree of pixel saturation and frame time (Siewerdsen and Jaffray 1999). Usually, lag is more severe at higher exposures when the electric field across a-Si photodiodes nearly collapses due to pixel saturation because charge is more likely to be trapped under low electric field.

Ghosting of indirect AMFPIs has been observed as an increase in x-ray sensitivity after the detector is exposed to

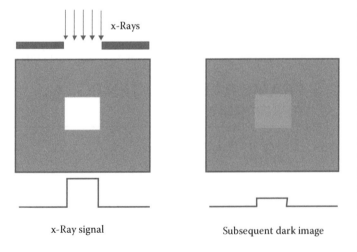

Figure 2.8 Conceptual images showing lag of an x-ray imaging system. Lag is defined as the residual signal from the detector's previous exposure to radiation. It is manifested as an enhanced signal in a subsequent dark image (acquired without x-rays).

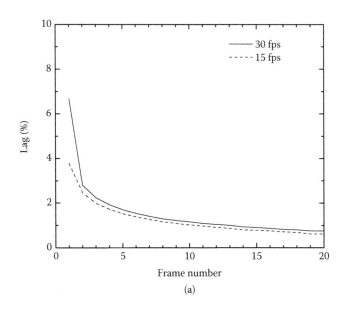

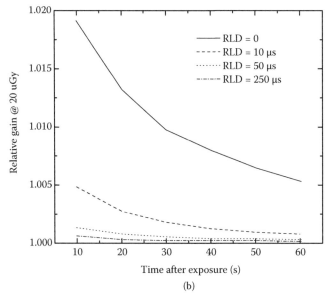

Figure 2.10 (a) Measured lag for indirect AMFPI (Varian 2020) with frame rates of 15 and 30 fps. (Adapted from Tognina, C.A. et al., *Proc SPIE*, 5368, 648–56, 2004.) (b) Measured lag for ghosting of indirect AMFPI (Trixel dynamic FPI). (Adapted from Fig. 13 of Overdick, M. et al., *Proc SPIE*, 4320, 47–58, 2001.)

radiation (Overdick 2001). Figure 2.10b shows the relative gain (x-ray sensitivity) of indirect AMFPIs as a function of exposure, exhibiting a 2% increase (without reset light) in x-ray sensitivity even at 10 s after x-ray exposure of 20 μGy. This increase is because charge trapped in a-Si due to previous radiation exposure fills the traps and reduces the probability of further charge trapping in subsequent exposures, whereas a "rested" (i.e., no recent history of radiation exposure) detector experiences reduction in x-ray sensitivity due to charge trapping. To alleviate ghosting due to this mechanism, reset light exposure has been implemented, where short pulses (~100 μs) of light delivered between x-ray exposures would generate charge to fill the traps in a-Si photodiode, thereby

minimizing the probability of charge trapping during x-ray exposure (Overdick 2001). Figure 2.10b shows that the longer the reset light duration (RLD), the lower the sensitivity ghost. This reset also was found to improve the x-ray sensitivity of AMFPIs, compared with that without reset light, by ~8% for RLD > 100 μs. An alternative method developed to overcome lag and ghosting caused by charging trapping is to put a-Si photodiodes in the forward-bias condition for a short period between two subsequent exposures. Forward-bias causes a large number of charge carriers injected from the bias electrodes of a-Si photodiode, and fill the traps before the next x-ray exposure (Mollov et al. 2008).

2.5.2.2 Temporal performance of direct AMFPIs

Lag and ghosting in a-Se AMFPIs are due to charge trapping in the bulk of the a-Se layer or at the interface between a-Se and the bias electrodes (Zhao et al. 2002, 2005a). Comparison of temporal performance of complete AMFPIs and a-Se samples (without pixelated TFT readout circuit) showed that the dominant factors are the charge trapping and recombination in the a-Se layer (Tousignant et al. 2005). The drift mobility of electrons in a-Se is only ~1/50 of that of holes, and there are a large number of deep electron traps that could capture electrons for up to several hours. Trapped electrons enhance the electric field near the positive bias electrode and increase injection of holes that is manifested as lag, that is, elevated dark signal, after radiation exposure. Figure 2.11a shows lag measurements from a real-time a-Se AMFPI (FPD14, Anrad), as well as an a-Se layer identical to that used in an AMFPI but without TFT readout (Tousignant et al. 2005). The first frame lag (30 fps) of the AMFPI depends on the radiation exposure, and the value increases from 1.7% at 48 μR to 3.9% at 384 μR. Ghosting in a-Se detectors is manifested as a reduction in x-ray sensitivity due to the recombination between previously trapped electrons in the bulk of a-Se and the x-ray-generated free holes (Fogal et al. 2004). It increases as a function of radiation dose and decreases with increasing electric field (E_{Se}). Figure 2.11b shows the quantitative measurements of ghosting in the same a-Se AMFPI (FPD14), where the relative x-ray sensitivity is measured as a function of time after x-ray has been delivered at a rate of 33 mR/min. It shows that the x-ray sensitivity continues to decrease with accumulation of exposure. Sensitivity recovery, or ghosting erasure, can be achieved through charge recombination technique. It has been shown previously that the injection of holes into the bulk of a-Se between subsequent exposures provides recovery of x-ray sensitivity because the trapped electrons are neutralized through recombination with holes (Zhao and Zhao 2005; Zhao et al. 2005a). This approach is different from the ghost erasure method used for indirect AMFPIs, where saturation of electron traps through injection of charge carriers was shown to be the effective mechanism.

2.5.3 IMPACT OF DETECTOR TEMPORAL PERFORMANCE ON ARTIFACT IN CBCT

The image persistence resulting from lag and ghosting can lead to artifacts in the reconstructed images. Residual signal frame-to-frame in the course of CBCT acquisition leads to a

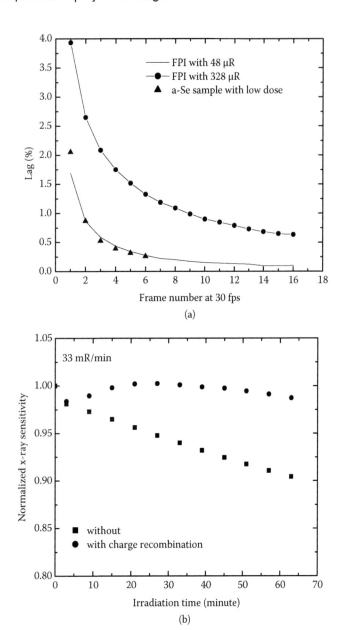

Figure 2.11 Temporal performance of a-Se FPI. (Adapted from Tousignant, O. et al., *Proc SPIE*, 5745, 207–15, 2005.) (a) Lag for FPI at two different exposures, as well as for a matching a-Se layer (without TFT readout) at the lower exposure. (b) Ghosting, that is, measurement of relative x-ray sensitivity, as a function of radiation exposure.

so-called "comet" artifact, that is, azimuthal blur of objects, and this blur tends to be greater for higher contrast objects (e.g., metal) at greater distance from the center of rotation (i.e., higher angular velocity). Such artifacts are similarly observed at the periphery of large, noncircular objects (e.g., a pelvis) in which the detector quickly experiences a drop in fluence as the shadow of the pelvis covers the detector, resulting in an azimuthal shading of the reconstruction. Similarly, inconsistency between the first few projections (in which the detector is first readout) and subsequent frames (in which the detector is closer to frame-to-frame signal equilibrium) can result in streaks along the direction of the first few views. Such effects can be mitigated somewhat by correction algorithms that subtract an (weighted) estimate

of the temporal response function applied to previous frames from the current frame (Mail et al. 2008).

2.6 EMERGING DETECTOR TECHNOLOGY FOR CBCT

2.6.1 ADVANCED AMFPIs

As discussed in Section 2.4, one of the major challenges for AMFPIs in CBCT is the degradation of DQE at low exposures due to electronic noise. Recently, many new AMFPI detector concepts have been proposed to overcome this limitation. The strategies can be divided into two categories: (1) increase the x-ray-to-image charge conversion gain so that the x-ray quantum noise can overcome the electronic noise (Street et al. 2002; Zhao et al. 2005b) and (2) decrease the electronic noise by incorporating amplification using two or more TFTs at each pixel, also referred to as active pixel sensor (APS) (Karim et al. 2003; El-Mohri et al. 2009). In the first case, the challenge is to increase the gain of the converter without sacrificing dynamic range or adding new sources of image degradation (e.g., blur or noise). For example, increasing the gain by a factor of ~10 may allow AMFPIs to maintain high DQE (x-ray quantum noise-limited) at the lowest exposure used in CBCT [with a single x-ray absorbed per pixel, even with the current level of electronic readout noise (~1500 e rms)] (Zhao et al. 2005b). In the second case, the noise due to the readout electronics can be reduced through the incorporation of pixel amplification with two or more a-Si or poly-Si TFTs. Noise reduction to ~500 e has been demonstrated (El-Mohri et al. 2009) that would allow the x-ray quantum noise-limited DQE performance to extend to lower exposures than in existing AMFPIs.

2.6.2 CMOS ACTIVE PIXEL SENSORS

With the advancement in very large–scale integrated circuit, there has been a steady increase in the effort of making wafer-scale c-Si complementary metal oxide semiconductor (CMOS) image sensors for x-ray imaging (Scheffer 2007; Heo et al. 2011). Due to the limited wafer size [mostly 20 cm (8 in.) in diameter due to cost considerations], the largest monolithic "tile" dimension is approximately 12 cm × 12 cm (or rectangular tiles with comparable total surface area) (Bohndiek et al. 2009). The majority of wafer-scale CMOS sensors are APSs that have an amplification circuit at each pixel using three or more transistors (Farrier et al. 2009). They have the following advantages over a-Si FPIs: (1) Pixel amplification permitting nondestructive readout and lower electronic noise (100–300 e rms), (2) faster readout speed, and (3) smaller pixel size (~30–40 μm for breast imaging and micro-CT). To make large-area detectors, several CMOS tiles may be butted side by side with minimal dead zone (e.g., less than one pixel wide) between them. With each CMOS tile, three-sided buttable, tiled CMOS detectors with sizes up to 29 cm × 23 cm have been made (Naday et al. 2010). They have potential applications in digital breast tomosynthesis and dental CBCT. With future cost reduction and increase in yield, tiled wafer-scale CMOS APSs are expected to expand their applications in CBCT.

REFERENCES

Antonuk, L.E., Boudry, J., Huang, W., et al. 1992. Demonstration of megavoltage and diagnostic x-ray imaging with hydrogenated amorphous silicon arrays. *Med Phys* 19: 1455–66.

Bartolac, S., Graham, S., Siewerdsen, J., et al. 2011. Fluence field optimization for noise and dose objectives in CT. *Med Phys* 38: S2–17.

Bohndiek, S.E., Blue, A., Cabello, J., et al. 2009. Characterization and testing of LAS: a prototype 'large area sensor' with performance characteristics suitable for medical imaging applications. *IEEE Trans Nucl Sci* 56: 2938–46.

Boone, J.M., Yang, K., Burkett, G.W., et al. 2010. An x-ray computed tomography/positron emission tomography system designed specifically for breast imaging. *Tech Canc Res Treat* 9: 29–44.

Chen, L., Shen, Y., Lai, C.J., et al. 2009. Dual resolution cone beam breast CT: a feasibility study. *Med Phys* 36: 4007–14.

Colbeth, R.E., Boyce, S.J., Fong, R., et al. 2001. 40 × 30 cm flat-panel imager for angiography, R&F, and cone-beam CT applications. *Proc SPIE* 4320: 94–102.

El-Mohri, Y., Antonuk, L.E., Koniczek, M., et al. 2009. Active pixel imagers incorporating pixel-level amplifiers based on polycrystalline-silicon thin-film transistors. *Med Phys* 36: 3340–55.

Farrier, M., Graeve Achterkirchen, T., Weckler, G.P., et al. 2009. Very large area CMOS active-pixel sensor for digital radiography. *IEEE Trans Electron Dev* 56: 2623–31.

Fogal, B., Kabir, M.Z., O'Leary, S.K., et al. 2004. X-ray-induced recombination effects in a-Se-based x-ray photoconductors used in direct conversion x-ray sensors. *J Vac Sci Technol A* 22: 1005–9.

Friedman, S.N. and Cunningham, I.A. 2010. A spatio-temporal detective quantum efficiency and its application to fluoroscopic systems. *Med Phys* 37: 6061–9.

Heo, S.K., Kosonen, J., Hwang, S.H., et al. 2011. 12-inch-wafer-scale CMOS active-pixel sensor for digital mammography. *Proc SPIE* 7961, 79610-O.

Hunt, D.C., Tousignant, O. and Rowlands, J.A. 2004. Evaluation of the imaging properties of an amorphous selenium-based flat panel detector for digital fluoroscopy. *Med Phys* 31: 1166–75.

IEC62220-1:2004. *Medical Electrical Equipment—Characteristics of Digital X-Ray Imaging Devices—Part 1: Determination of the Detective Quantum Efficiency.*

Jaffray, D.A. and Siewerdsen, J.H. 2000. Cone-beam computed tomography with a flat-panel imager: initial performance characterization. *Med Phys* 27: 1311–23.

Kakeda, S., Korogi, Y., Miyaguni, Y., et al. 2007. A cone-beam volume CT using a 3D angiography system with a flat panel detector of direct conversion type: usefulness for superselective intra-arterial chemotherapy for head and neck tumors. *AJNR Am J Neuroradiol* 28: 1783–8.

Karim, K.S., Nathan, A. and Rowlands, J.A. 2003. Amorphous silicon active pixel sensor readout circuit for digital imaging. *IEEE Trans Electron Dev* 50: 200–8.

Koyama, S., Aoyama, T., Oda, N. and Yamauchi-Kawaura, C. 2010. Radiation dose evaluation in tomosynthesis and C-arm cone-beam CT examinations with an anthropomorphic phantom. *Med Phys* 37: 4298–306.

Mail, N., Moseley, D.J., Siewerdsen, J.H., et al. 2008. An empirical method for lag correction in cone-beam CT. *Med Phys* 35: 5187–96.

Mail, N., Moseley, D.J., Siewerdsen, J.H., et al. 2009. The influence of bowtie filtration on cone-beam CT image quality. *Med Phys* 36: 22–32.

Mollov, I., Tognina, C. and Colbeth, R. 2008. Photodiode forward bias to reduce temporal effects in a-Si based flat panel detectors. *Proc SPIE* 6913: 69133S.

Naday, S., Bullard, E.F., Gunn, S., et al. 2010. Optimised breast tomosynthesis with a novel CMOS flat panel detector. *10th International Workshop on Digital Mammography*, June 16–18, 2010. Vol. 6136. *LNCS.* 428–35.

Ning, R., Conover, D., Yu, Y., et al. 2007. A novel cone beam breast CT scanner: system evaluation. *Proc SPIE* 6510: 651030.

Overdick, M., Solf, T. and Wischmann, H. 2001. Temporal artifacts in flat-dynamic x-ray detectors. *Proc SPIE* 4320: 47–58.

Pang, G., Zhao, W. and Rowlands, J.A. 1998. Digital radiology using active matrix readout of amorphous selenium: geometrical and effective fill factors. *Med Phys* 25: 1636–46.

Roos, P.G., Colbeth, R.E., Mollov, I., et al. 2004. Multiple-gain-ranging readout method to extend the dynamic range of amorphous silicon flat-panel imagers. *Proc SPIE* 5368: 139–49.

Rowlands, J.A., Decrescenzo, G. and Araj, N. 1992. X-ray imaging using amorphous selenium: determination of x-ray sensitivity by pulse height spectroscopy. *Med Phys* 19: 1065–9.

Rowlands, J.A. and Yorkston, J. 2000. Flat panel detectors for digital radiography. In: Beutel, J., Kundel, H. and Van Metter, R. (eds.). *Medical Imaging: Volume 1. Physics and Psychophysics.* Bellingham, WA: SPIE, 223–328.

Scheffer, D. 2007. A wafer scale active pixel CMOS image sensor for generic X-ray radiology. *Proc SPIE* 6510, 65100-O.

Siewerdsen, J.H. and Jaffray, D.A. 1999. A ghost story: spatio-temporal response characteristics of an indirect-detection flat-panel imager. *Med Phys* 26: 1624–41.

Street, R.A., Ready, S.E., Van Schuylenbergh, K., et al. 2002. Comparison of Pbl2 and Hgl2 for direct detection active matrix x-ray image sensors. *J Appl Phys* 91: 3345.

Suzuki, K., Ikeda, S., Ueda, K., et al. 2004. Development of angiography system with cone-beam reconstruction using large-area flat-panel detector. *Proc SPIE* 5368: 488–98.

Swank, R.K. 1973. Absorption and noise in x-ray phosphors. *J Appl Phys* 44: 4199–203.

Tognina, C.A., Mollov, I., Yu, J.M., et al. 2004. Design and performance of a new a-Si flat-panel imager for use in cardiovascular and mobile C-arm imaging systems. *Proc SPIE* 5368: 648–56.

Tousignant, O., Demers, Y., Laperriere, L., et al. 2005. Spatial and temporal image characteristics of a real-time large area a-Se x-ray detector. *Proc SPIE* 5745: 207–15.

Weisfield, R.L. and Bennett, N.R. 2001. Electronic noise analysis of a 127-mu m pixel TFT/photodiode array. *Proc SPIE* 4320: 209–18.

Weisfield, R.L., Yao, W., Speaker, T., et al. 2004. Performance analysis of a 127-micron pixel large-area TFT/photodiode array with boosted fill factor. *Proc SPIE* 5368: 338–48.

Zhao, B. and Zhao, W. 2005. Temporal performance of amorphous selenium mammography detectors. *Med Phys* 32: 128–36.

Zhao, W., Decrescenzo, G., Kasap, S.O., et al. 2005a. Ghosting caused by bulk charge trapping in direct conversion flat-panel detectors using amorphous selenium. *Med Phys* 32: 488–500.

Zhao, W., Decrescenzo, G. and Rowlands, J.A. 2002. Investigation of lag and ghosting in amorphous selenium flat-panel x-ray detectors. *Proc SPIE* 4682: 9–20.

Zhao, W., Ji, W.G., Debrie, A., et al. 2003. Imaging performance of amorphous selenium based flat-panel detectors for digital mammography: characterization of a small area prototype detector. *Med Phys* 30: 254–63.

Zhao, W., Li, D., Reznik, A., et al. 2005b. Indirect flat-panel detector with avalanche gain: fundamental feasibility investigation for SHARP-AMFPI (scintillator HARP active matrix flat panel imager). *Med Phys* 32: 2954–66.

Zhao, W., Ristic, G. and Rowlands, J.A. 2004. X-ray imaging performance of structured cesium iodide scintillators. *Med Phys* 31: 2594–605.

Zhao, W. and Rowlands, J.A. 1995. X-ray imaging using amorphous selenium: feasibility of a flat panel self-scanned detector for digital radiology. *Med Phys* 22: 1595–604.

Fundamental principles and techniques

3 Reconstruction algorithms

Liang Li, Zhiqiang Chen, and Ge Wang

Contents

3.1 INTRODUCTION

The objective of computed tomography (CT) is to reconstruct two-dimensional (2D) or three-dimensional (3D) images of internal structures from collected signals through an object. In x-ray CT, a reconstructed image represents a distribution of radiation-ray linear attenuation coefficients. As discussed in Chapter 2, data recorded on an x-ray detector array are actually x-ray intensity values after an x-ray beam penetrates an object. The attenuation of the x-ray intensity follows the Lambert–Beer's law. After applying the negative logarithmic operation of the ratio between the output x-ray intensity and the input x-ray intensity, we obtain the line integral of the attenuation coefficient distribution along an x-ray path. The presentation of line integrals is typically associated with x-ray projections. Projection data are the immediate input to an image reconstruction algorithm. Mathematically, CT image reconstruction is a linear inverse problem.

CT image reconstruction is a very interesting and challenging topic and an active research area. Novel algorithms are being continually developed. In this chapter, we first briefly review the history of reconstruction algorithms that can be traced back to as early as 1917, when J. Radon, an Austrian mathematician, first presented a mathematics solution for reconstruction of a function from these line integrals. However, his work did not attract much attention, and no progress was made until the late 1950s, when the development of CT scanners gradually gained more attention in the medical community. Allan M. Cormack (1963) solved the problem of how to reconstruct images by using a finite number of projections, an important contribution. In the same year, William H. Oldendorf (1963) developed a direct backprojection method. Later, the idea of filtered backprojection was first proposed by Bracewell and Riddle (1967), probably the most influential development in this area. Gordon et al. (1970) proposed the algebraic reconstruction technique (ART), which may produce a good reconstruction when projections are not uniformly distributed or limited. During this period, a breakthrough was made by Godfrey N. Hounsfield at the Central Research Laboratory of EMI, Ltd., in England. During 1968–1972, he built the first CT scanner and obtained the first image of a patient's head using an algebraic algorithm. For their pioneer work, Cormack and Hounsfield shared the Nobel Prize in Physiology or Medicine in 1979.

From the late 1970s to early 1980s, tremendous progress was made in CT technology. There are roughly five generations of CT scanners. By the late 1990s, multislice helical CT had become the predominant mode for medical applications. Correlated to the evolution of CT scanners, image reconstruction algorithms have been intensively developed for clinical and preclinical applications. Smith (1983) and Tuy (1983) independently studied the sufficient conditions and reconstruction theory for cone beam geometry based on the seminal work of A. Kirillov and obtained the so-called Smith–Tuy sufficiency condition for exact reconstruction. Furthermore, Feldkamp et al. (1984) derived a stable approximate formula, the Feldkamp–Davis–Kress (FDK) algorithm, for circular cone beam reconstruction that was then generalized by Wang et al. to obtain spiral cone beam scanning and other trajectories (Kudo and Saito 1991; Wang and Yu 1991, 1993). As a result, the approximate algorithms form an important class of CT reconstruction algorithms. Grangeat (1991) established a geometrically attractive connection between cone beam data and radial derivatives of the 3D Radon transform that turned out to be particularly important for exact reconstruction in cone beam geometry. Finally, Katsevich (2002) proposed an exact filtered backprojection type algorithm for cone beam helical reconstruction that greatly advanced the research on analytic reconstruction.

As mentioned, CT reconstruction is a mature yet relatively new subject. Numerous research papers, proceedings, and books have been published in this field. Covering all the aspects of CT reconstruction in this chapter is impossible. Therefore, we introduce only fundamentals of image reconstruction and some ground-breaking algorithms without rigorous derivations. We hope that this chapter helps readers to grasp this domain. Those who want to learn more are referred to the literature (Herman 1980; Natterer 1986; Kak and Slaney 1988; Censor et al. 2010; Zeng 2010).

3.2 FOURIER SLICE THEOREM

CT reconstruction is used to recover a cross-sectional image of an object from projections consisting of line integrals. To solve this problem, we need to establish the relationship between the image and its projections. For this purpose, we introduce the Fourier slice theorem, also known as the central slice theorem.

In the Cartesian coordinates defined in Figure 3.1, the object is represented by a 2D function $f(x,y)$. Its parallel-beam projection $p(\theta,s)$ is defined as

$$p(\theta, s) = \int_{-\infty}^{+\infty} \int_{-\infty}^{+\infty} f(x, y)\delta(y\cos\theta - x\sin\theta - s)\,dxdy. \qquad (3.1)$$

Taking the one-dimensional (1D) Fourier transform of the above-mentioned parallel-beam projection with respect to s at angle θ, we obtain $P_\theta(\omega)$, where ω denotes the frequency component. The 2D Fourier transform of the object function in the polar coordinates can be written as

$$F(\theta, \omega) = \int_{-\infty}^{+\infty} \int_{-\infty}^{+\infty} f(x, y)e^{-j2\pi\omega(x\cos\theta + y\sin\theta)}\,dxdy. \qquad (3.2)$$

The Fourier slice theorem can be expressed in the following equation:

$$P_\theta(\omega) = F(\theta, \omega). \qquad (3.3)$$

The Fourier slice theorem can be stated as the 1D Fourier transform of a parallel projection of an object $f(x,y)$ obtained at

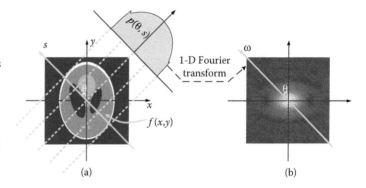

Figure 3.1 Fourier slice theorem relating a projection to the 2D Fourier transform along a radial line. (a) Shepp–Logan phantom in the spatial domain and its projection at angle θ. (b) Shepp–Logan phantom in the Fourier domain.

angle θ and equals the profile along the radial line at the same angle in the 2D Fourier space. Figure 3.1a shows the Shepp–Logan phantom and its parallel projection $p_\theta(s)$ at angle θ, and Figure 3.1b shows the image $F(\theta,\omega)$ of the phantom in the Fourier domain. Here, we omit the proof, which is quite straightforward. Interested readers should refer to the book by Kak and Slaney (1988).

From the Fourier slice theorem, it is easy to see that every radial profile in the 2D Fourier transform of the object can be obtained by performing the 1D Fourier transform on the parallel projection at the corresponding projection angle. If a sufficient number of projections are taken over the range from 0 to π, $F(\theta,\omega)$ would be known at all points in the frequency domain. Then, the object can be exactly reconstructed by the 2D inverse Fourier transform

$$f(x, y) = \int_{0}^{2\pi} \int_{0}^{+\infty} F(\theta, \omega) \cdot \omega \cdot e^{j2\pi\omega(x\cos\theta + y\sin\theta)}\,d\omega\,d\theta. \qquad (3.4)$$

Although the afore mentioned Fourier approach is straightforward, there is a problem in practical implementation: the Fourier slice theorem gives us samples on a polar grid, whereas the standard fast Fourier transform (FFT) requires data on a Cartesian grid. To perform a 2D inverse FFT, the samples have to be interpolated from a polar grid to a Cartesian grid. However, interpolation in the frequency domain is not as straightforward as in the spatial domain. Inappropriate interpolation may cause severe image artifacts; for example, bilinear interpolation or nearest neighbor interpolation is not optimal in this context. An interpolation error analysis was given by Natterer (1986, pp. 120–5). There are several methods to improve the standard direct Fourier reconstruction, such as the gridding method (O'Sullivan 1985; Schomberg and Timmer 1995) and the linogram method (Edholm and Herman 1987, 1988). However, these techniques are application-specific and protocol- or machine-dependent. Therefore, although the complexity of the Fourier reconstruction is low $O(N^2 \log N)$, it is not widely used in practice. Thus, it is desirable to find a different approach for implementation of the Fourier slice theorem. Currently, the method of choice is the filtered backprojection algorithm (see Section 3.3).

Note that the Fourier slice theorem was derived in the case of parallel beam imaging. When the x-ray beam is divergent like widely used fan beam and cone beam geometry, the one-to-one correspondence does not hold between the Fourier transform of a divergent beam projection and a subset of information in the Fourier space of the image. More complex relationships occur between divergent projection data and Fourier information; see discussions concerning the cases of cone beam (Tuy 1983) and fan beam geometry (Chen et al. 2005).

3.3 FILTERED BACKPROJECTION ALGORITHM

The filtered backprojection (FBP) algorithm derived from the Fourier slice theorem is widely used in many applications because of its high accuracy and efficiency. We first present the FBP algorithm for parallel beam projections. Starting with the 2D inverse Fourier transform, the object function $f(x,y)$ can be expressed by Equation 3.4 in the polar coordinates. Because $f(x,y)$ is a real function, its Fourier transform has the following symmetry property in the frequency domain:

$$F(\theta,\omega) = F(\theta+\pi, -\omega). \tag{3.5}$$

Substituting Equation 3.5 into Equation 3.4 and changing the integral domain, we have

$$f(x,y) = \int_0^\pi \int_{-\infty}^{+\infty} F(\theta,\omega) \cdot |\omega| \cdot e^{j2\pi\omega(x\cos\theta+y\sin\theta)} \, d\omega \, d\theta. \tag{3.6}$$

Recalling the Fourier slice theorem and substituting Equation 3.3 into Equation 3.5, we arrive at the following equation:

$$f(x,y) = \int_0^\pi \int_{-\infty}^{+\infty} P(\theta,\omega) \cdot |\omega| \cdot e^{j2\pi\omega(x\cos\theta+y\sin\theta)} \, d\omega \, d\theta, \tag{3.7}$$

where $P(\theta,\omega)$ is the 1D Fourier transform of the projection at angle θ. The inside integral represents a filtering operation on $P(\theta,\omega)$, where the frequency response of the filter is $|\omega|$. Therefore, in the spatial domain, we call this integral a "filtered projection," which may be rewritten as a new function $p^F(\theta,s)$:

$$p^F(\theta,s) = \int_{-\infty}^{+\infty} P(\theta,\omega) \cdot |\omega| \cdot e^{j2\pi\omega(x\cos\theta+y\sin\theta)} \, d\omega. \tag{3.8}$$

Then, Equation 3.7 can be expressed as

$$f(x,y) = \int_0^\pi p^F(\theta, x\cos\theta + y\sin\theta) \, d\theta. \tag{3.9}$$

Equation 3.9 indicates that the function $f(x,y)$ can be exactly reconstructed by adding the filtered projection data at

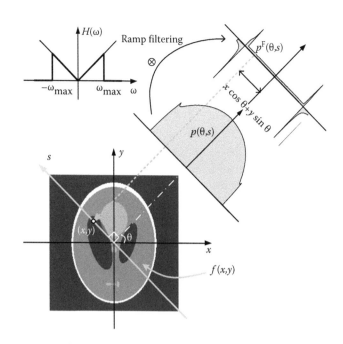

Figure 3.2 FBP process. Taking the Shepp–Logan phantom as an example of an underlying image. $p(\theta,s)$ is the parallel projection, $p^F(\theta,s)$ is the filtered projection, and $H(\omega)$ is the frequency representation of the band-limited ramp filter.

location $s = x\cos\theta + y\sin\theta$ for all the viewing angles. The value of s depends on both the image point (x,y) and the projection angle θ. It is the distance from the origin of the coordinates to the x-ray line through the point (x,y) at the angle θ as shown in Figure 3.2. Note that an x-ray line can be determined by a given θ and s. Therefore, the filtered projection datum $p^F(\theta,s)$ associated with this line makes the same contribution to the reconstruction at all of the points on this line. This process looks like smearing $p^F(\theta,s)$ back equally to all the points on the x-ray path, so it is called *backprojection*. Combining Equations 3.8 and 3.9, we obtain the final image reconstruction algorithm called FBP.

Now, let us consider in detail Equation 3.8), which performs projection filtration with the $|\omega|$ function in the frequency domain. As we know, this filtration in the frequency domain is often transformed to a convolution in the spatial domain in the following form:

$$p^F(\theta,s) = \int_{-\infty}^{+\infty} p(\theta,s') \cdot h(s-s') \, ds', \tag{3.10}$$

where $p(\theta,s)$ is the measured parallel projection and $h(s)$ the 1D inverse Fourier transform of the function $|\omega|$ in the frequency domain as shown in Figure 3.2:

$$h(s) = \int_{-\infty}^{+\infty} |\omega| \cdot e^{j2\pi\omega s} \, d\omega. \tag{3.11}$$

Because of its shape in the frequency domain, $h(s)$ is called a ramp filter. However, Equation 3.11 is not an integrable function; that is, $h(s)$ cannot be directly calculated. Usually, we assume that a projection is a band-limited signal in the frequency domain. Thus, Equation 3.11 can be performed over a range of $(-\omega_{max}, \omega_{max})$ as follows:

$$h(s) = \int_{-\omega_{max}}^{+\omega_{max}} |\omega| \cdot e^{j2\pi\omega s} \, d\omega, \qquad (3.12)$$

where ω_{max} is the cut off frequency of the projection. Based on the Nyquist sampling theorem, ω_{max} equals $1/2\,\delta$, where δ is the projection sampling interval. Then, Equation 3.12 can be simplified to the following equation:

$$h(s) = \frac{1}{2\delta^2} \frac{\sin 2\pi s/2\delta}{2\pi s/2\delta} - \frac{1}{4\delta^2} \left(\frac{\sin \pi s/2\delta}{\pi s/2\delta} \right)^2. \qquad (3.13)$$

Finally, combining Equations 3.9, 3.10, and 3.13, we obtain the complete FBP algorithm, also called the convolution backprojection algorithm (CBP). Note that the above-mentioned equations are derived in continuous form and need to be discretized for computer implementation (Kak and Slaney 1988).

3.4 FAN BEAM RECONSTRUCTION

The image reconstruction algorithms discussed earlier are for parallel beam data. Actually, almost all the current commercial CT scanners use point sources such as an x-ray tube and produce divergent x-ray beam projections in either the fan beam (2D) or cone beam (3D) geometry. As mentioned in Section 3.2, the Fourier slice theorem is very complex in the divergent beam geometry and hard to use to derive reconstruction algorithms. Thus, we use an alternative strategy that modifies the parallel beam algorithms for fan beam reconstruction through an appropriate coordinate transformation. There are two popular fan beam geometries: equiangular and equispatial fan beam setups (Hsieh 2003). The main difference lies in the detector arrangement as shown in Figure 3.3. In this chapter, the algorithms are derived for equispatial geometry. The reconstruction formulas for

equiangular geometry also are presented, but no derivation is given.

3.4.1 REBINNING ALGORITHM

For fan beam reconstruction, the previous parallel beam algorithms may be applied after we reformat fan beam projections into parallel beam projections. This reformatting process is called *rebinning* and is used in many applications. As shown in Figure 3.4a, each x-ray fan beam projection datum may be rearranged to form parallel beam data. We assume that the x-ray source moves in the $x-y$ plane along a circle $R(-\sin\beta, \cos\beta)$. Here, we place a virtual detector on the axis of rotation. The fan beam projection is written as $g(\beta, t)$ in the equispatial geometry. The relationship between the parallel beam projections $p(\theta, s)$ and fan beam projections $g(\beta, t)$, shown in Figure 3.4b, can be expressed as

$$\begin{cases} \theta = \beta + \arctan \dfrac{t}{R} \\ s = \dfrac{tR}{\sqrt{R^2 + t^2}} \end{cases}. \qquad (3.14)$$

Using Equation 3.14, a parallel projection set can be obtained from the measured fan beam data as

$$p(\theta, s) = g\left(\theta - \arcsin \frac{s}{R}, \frac{sR}{\sqrt{R^2 - s^2}} \right). \qquad (3.15)$$

In the equiangular geometry, the fan beam projection is written as $g(\beta, \gamma)$, where γ is the fan angle from an arbitrary ray to the central ray as shown in Figure 3.3b. Similarly, parallel projections can be obtained from the measured data $g(\beta, \gamma)$ as follows:

$$p(\theta, s) = g\left(\theta - \arcsin \frac{s}{R}, \arcsin \frac{s}{R} \right). \qquad (3.16)$$

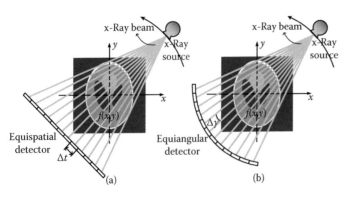

Figure 3.3 Fan beam geometry. (a) Equispatial sampling. (b) Equiangular sampling.

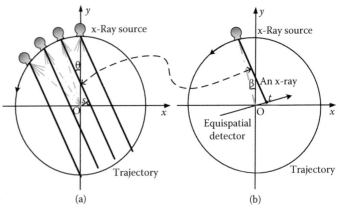

Figure 3.4 Fan beam to parallel beam rebinning. (a) Fan beam projections (gray lines) reformatted into parallel beam projections (black lines). (b) x-Ray projection sample in both fan beam and parallel beam geometries.

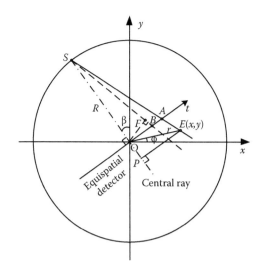

Figure 3.5 Illustration of equispatial fan beam reconstruction. *R*, radius of the scanning circle; *E*, pixel to be reconstructed, and the *t*-axis the virtual detector.

After data rebinning, the reconstruction can be done using a parallel FBP algorithm, described by Equations 3.9 and 3.10, over the interval $\theta \in [0,\pi]$. Using Equation 3.15, we may find redundant data in a fan beam full scan for $\beta \in [0,2\pi]$. Short-scan data are enough to obtain all the parallel data within $\theta \in [0,\pi]$ (Parker 1982). With a weighting function, a rebinning algorithm can easily integrate the redundant information in the input data.

Note that the rebinning algorithm is not simply a rearrangement of projection data, because an interpolation process is always necessary. We typically use the bilinear interpolation with respect to the projection angle and the detector position. Another important advantage is that the rebinning algorithm can be easily extended to other trajectories. However, the interpolation process compromises the spatial resolution of the reconstructed image. Hence, it is desirable to derive a reconstruction algorithm that uses fan beam data directly. Doing so leads to the fan beam FBP algorithm.

3.4.2 FILTERED BACKPROJECTION ALGORITHM

This section derives the FBP algorithm for equispatial fan beam data. We begin with the parallel FBP algorithm, Equation 3.9. Substituting Equation 3.10 into Equation 3.9, we get

$$f(x,y) = \int_0^\pi \int_{-\infty}^{+\infty} p(\theta,s) \cdot h(x\cos\theta + y\sin\theta - s)\,ds\,d\theta. \quad (3.17)$$

Given the symmetry of parallel projections, Equation 3.17 can be modified to include all projections over 2π,

$$f(x,y) = \frac{1}{2}\int_0^{2\pi}\int_{-\infty}^{+\infty} p(\theta,s)\cdot h(x\cos\theta + y\sin\theta - s)\,ds\,d\theta. \quad (3.18)$$

Using polar coordinates (r,φ) instead of Cartesian coordinates (x,y), we have

$$f(r,\varphi) = \frac{1}{2}\int_0^{2\pi}\int_{-\infty}^{+\infty} p(\theta,s)\cdot h\big[r\cos(\theta-\varphi)-s\big]\,ds\,d\theta. \quad (3.19)$$

For fan beam reconstruction, we need to convert the integrals over (θ,s) to (β,t). Using the relationships in Equation 3.14, we have

$$ds\,d\theta = \frac{R^3}{(R^2+t^2)^{3/2}}\,dt\,d\beta. \quad (3.20)$$

Substituting Equations 3.14 and 3.20 into Equation 3.19, we have

$$f(r,\varphi) = \frac{1}{2}\int_{-\arctan(t/R)}^{2\pi-\arctan(t/R)}\int_{-\infty}^{+\infty} p(\theta,s)\cdot h\left[r\cos\left(\beta+\arctan\frac{t}{R}-\varphi\right)\right.$$

$$\left. -\frac{tR}{\sqrt{R^2+t^2}}\right]\frac{R^3}{(R^2+t^2)^{3/2}}\,dt\,d\beta. \quad (3.21)$$

As shown in Figure 3.5, $p(\theta,s)$ in parallel data is simply $g(\beta,t)$ in fan beam data. Moreover, because all the functions of β have a period of 2π, the integral limits may be replaced by 0 and 2π. Introducing these changes into Equation 3.21, we have

$$f(r,\varphi) = \frac{1}{2}\int_0^{2\pi}\int_{-\infty}^{+\infty} g(\beta,t)\cdot h\left[r\cos\left(\beta+\arctan\frac{t}{R}-\varphi\right)\right.$$

$$\left. -\frac{tR}{\sqrt{R^2+t^2}}\right]\frac{R^3}{(R^2+t^2)^{3/2}}\,dt\,d\beta. \quad (3.22)$$

As shown in Figure 3.5, t is a variable along the detector axis, and we can randomly select a point B on the t-axis where $\overline{OB}=t$. Let A be the intersection point of a ray SE and the t-axis where $\overline{OA}=t'$. Now, let us examine the argument of h. We may find that $\beta + \arctan(t/R) - \varphi$ denotes the angle $\angle EOF$, where E is the pixel to be reconstructed and F the projection point of the origin on the line SB. Furthermore, $tR/\sqrt{R^2+t^2}$ is the length of OF. We define SP as the projection of the source to pixel distance SE on the central ray. We have $\overline{SP}=R+r\sin(\beta-\varphi)$. Thus, the argument of h may be rewritten as

$$r\cos\left(\beta+\arctan\frac{t}{R}-\varphi\right)-\frac{tR}{\sqrt{R^2+t^2}}$$

$$= \overline{SP}\cdot\big(\tan\angle ESO-\tan\angle BSO\big)\cdot\cos\angle BSO$$

$$= \overline{SP}\cdot\frac{t'-t}{\sqrt{R^2+t^2}}. \quad (3.23)$$

Recalling h according to its definition in Equation 3.11, it is easy to prove that

$$h\left(\overline{SP} \cdot \frac{t'-t}{\sqrt{R^2+t^2}}\right) = \frac{R^2+t^2}{\overline{SP}^2} h(t'-t). \qquad (3.24)$$

Substituting Equation 3.24 and $\overline{SP}$ into Equation 3.22 and replacing (r,φ) by (x,y), we have the equispatial fan beam FBP algorithm

$$f(x,y) = \frac{1}{2}\int_0^{2\pi} \frac{R^2}{(R+x\sin\beta-y\cos\beta)^2}$$

$$\times \int_{-\infty}^{+\infty} g(\beta,t)\frac{R}{\sqrt{R^2+t^2}} \cdot h(t'-t)\,dt\,d\beta, \qquad (3.25)$$

where

$$t' = \overline{OA} = R\frac{y\sin\beta+x\cos\beta}{R+x\sin\beta-y\cos\beta}. \qquad (3.26)$$

Compared with the parallel beam FBP algorithm Equation 3.19, Equation 3.25 is a bit more complex, but these two formulas are quite similar in the computational structure for the FBP format. There are two weighting factors in the convolution and backprojection processes of Equation 3.25. Thus, the computational expense for fan beam reconstruction is more than it is for parallel beam reconstruction.

As mentioned in Section 3.4.1, short-scan fan beam data are enough to reconstruct the whole image exactly. Parker (1982) proposed the first algorithm that uses half-scan fan beam data with a smooth windowing function before the ramp filtering. Another important result is the super-short-scan fan beam algorithm proposed by Noo et al. (2002), which may reconstruct exactly a region-of-interest (ROI) image from less data than the half scan. Their work uses the link between the Hilbert transform of a 2D object density and projection data (Li et al. 2010), being a special case of the general theory developed by Gel'fand and Graev (1991). The interested reader is referred to relevant papers (Gel'fand et al. 1966a, 1966b; Gel'fand and Gindikin 1977; Rullgard 2004; Zou and Pan 2004b).

3.5 CONE BEAM RECONSTRUCTION

Thus far, we have discussed 2D image reconstruction. In many practical cases, 3D imaging is highly desirable or even necessary. A 3D image, a series of slice-by-slice 2D images or a 3D volume-rendered image, is particularly helpful for screening, diagnosis, and image-guided intervention. In some cases, 3D image reconstruction can be decomposed into a series of 2D image reconstructions when projections can be divided into groups satisfying the data-sufficiency condition for 2D image reconstruction. However, as shown in Figure 3.6, for cone beam CT, the x-rays run through multiple slices and the 2D reconstruction approach will not work. Thus, we need to derive cone beam reconstruction algorithms.

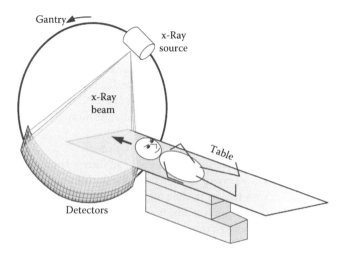

Figure 3.6 Cone beam helical CT.

Because there are many advantages to cone beam CT, it is widely used in both medical and industrial applications. Numerous research papers and books exist on this subject. It is impossible to cover all of the cone beam reconstruction algorithms in a single chapter. We present here several of the most important algorithms in the two categories: exact and approximate.

3.5.1 EXACT RECONSTRUCTION

An exact algorithm is mathematically correct, meaning that it is able to reconstruct an object function as closely as desired if a sufficient amount of noise-free projection data are provided. For 2D parallel beam CT, we have the Fourier slice theorem as a foundation for the design of algorithms. For cone beam CT, we have to establish a similar relationship to link cone beam projections to a 3D image.

3.5.1.1 3D Radon transform and Grangeat's formula

Different from the x-ray transform or line integral, also referred to as the 2D Radon transform, the 3D Radon transform is defined as the plane integral

$$Rf(\rho\vec{n}) = \int_{-\infty}^{+\infty}\int_{-\infty}^{+\infty}\int_{-\infty}^{+\infty} f(\vec{x}) \cdot \delta(\vec{x}\cdot\vec{n}-\rho)\,d\vec{x}, \qquad (3.27)$$

where $\vec{n}$ is a unit vector. Equation 3.27 indicates that the Radon transform value at a point $\rho\vec{n}$ is the integral of a 3D function $f(\vec{x})$ over the plane through $\rho\vec{n}$ with a normal $\vec{n}$. It is well known that $f(\vec{x})$ can be reconstructed using the 3D inverse Radon transform if $Rf(\rho\vec{n})$ is available for all planes through the vicinity of a point $\vec{x}$ (Deans 1983). The inverse Radon formula is given by

$$f(\vec{x}) = -\frac{1}{8\pi^2}\int_{-\pi/2}^{+\pi/2}\int_0^{2\pi} \frac{\partial^2}{\partial\rho^2} Rf\left[(\vec{x}\cdot\vec{n})\vec{n}\right]\big|\sin\theta\big|\,d\varphi\,d\theta. \qquad (3.28)$$

Equation 3.28 is a simple algorithm for 3D image reconstruction. However, cone beam projections measured in medical imaging are line integrals along divergent x-rays, quite different from the 3D Radon data. Therefore, we should build a bridge between cone beam data and Radon data.

The integral of a function over a plane that yields a 3D Radon transform datum can be rewritten as

$$Rf(\rho\vec{n}) = \iint f(\rho, r, \theta) r\, dr\, d\theta, \qquad (3.29)$$

where (r, θ) are polar coordinates in the plane through $\rho\vec{n}$ with a normal $\vec{n}$. In the cone beam projection, each individual x-ray measurement is of the form $\int f(\rho, r, \theta) dr$, from which we may compute the integral of the projections over θ:

$$\iint f(\rho, r, \theta)\, dr\, d\theta \quad \text{or} \quad \iint \frac{1}{r} f(\rho, r, \theta) r\, dr\, d\theta. \qquad (3.30)$$

Equation 3.30 shows that the line integral of cone beam projections gives a weighted Radon transform with a nonuniform weighting function $1/r$. If the undesired $1/r$ is somehow removed, cone beam data will be linked to 3D Radon data.

Grangeat (1991) derived a fundamental relationship between a derivative of a line integral of cone beam data and the first derivative of the 3D Radon transform. As shown in Figure 3.7, the relationship can be expressed as

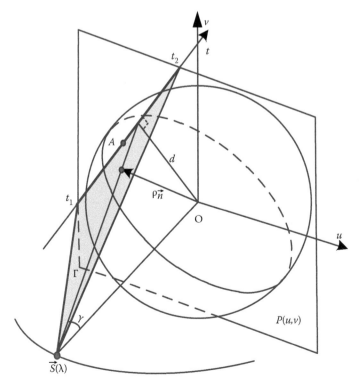

Figure 3.7 Relationship between the cone beam projection and the 3D Radon transform.

$$\frac{\partial}{\partial \rho} Rf(\rho\vec{n}) = \frac{1}{\cos^2 \gamma} \frac{\partial}{\partial d} \int_{t_1}^{t_2} \frac{SO}{SA} p\left[\lambda(\rho\vec{n}), d(\rho\vec{n}), t\right] dt, \quad (3.31)$$

where p denotes a cone beam projection on the central virtual detector plane orthogonal to $\vec{S}(\lambda)$, $\vec{S}(\lambda)$ is the x-ray source trajectory, O is the detector center, and a point $\rho\vec{n}$ defines a plane Γ perpendicular to $\vec{n}$ and through that point. The fan beam projection Γ associated with $\vec{S}(\lambda)$ makes a line t on the virtual detector plane. Let d denote the distance from the center O to the line t, SA the distance between the source and an arbitrary point A along t, and γ the angle between SO and the plane Γ.

Grangeat's algorithm can be applied to any trajectories. It consists of two parts. First, derivatives of Radon data are computed by Equation 3.31. The results are distributed on "Radon shells" determined by the x-ray source trajectory. Then, the data are inverted by Equation 3.28, whose computational complexity is $O(N^5)$. Based on Grangeat's fundamental work, various exact cone beam reconstruction algorithms were developed (Danielsson 1992; Defrise and Clack 1994; Kudo et al. 1998). Using the linogram method, the complexity of the Grangeat's framework can be reduced to $O(N^3 \log N)$ (Axelsson and Danielsson 1994). However, Grangeat's algorithm is not in an FBP-type format and is often too complicated to be implemented for real-world applications.

3.5.1.2 Katsevich's algorithm

Earlier FBP algorithms for helical cone beam CT (Kudo and Saito 1991; Wang et al. 1991, 1993) were only approximate. In 2002, Katsevich (2002) first proposed an FBP formula for theoretically exact helical cone beam image reconstruction, considered a breakthrough (Chen 2003). Katsevich's algorithm was initially developed for helical cone beam geometry and was later extended to general trajectories (Katsevich 2003; Pack et al. 2005; Ye et al. 2005; Lu et al. 2010). We denote a helical trajectory in the Cartesian coordinates by

$$\vec{S}(\lambda) = \left(R\cos\lambda, R\sin\lambda, \frac{l}{2\pi}\lambda \right), \qquad (3.32)$$

where R is the radius and l the helix pitch. Let $p\left(\vec{S}(\lambda), \vec{\beta}\right)$ be a cone beam projection, where $\vec{\beta}$ is a unit vector defined as an x-ray direction from a focal spot. Katsevich's algorithm can be expressed as follows:

$$f(\vec{x}) = -\frac{1}{2\pi^2} \int_{I_{PI}(\vec{x})} \frac{1}{\left|\vec{x} - \vec{S}(\lambda)\right|}$$

$$\times \int_{-\pi/2}^{+\pi/2} \frac{\partial}{\partial q} p\left(\vec{S}(\lambda), \cos\gamma\vec{\beta}(\lambda, \vec{x}) + \sin\gamma\vec{\alpha}\right)\bigg|_{q=\lambda} \frac{d\gamma}{\sin\gamma} d\lambda, \quad (3.33)$$

where $\vec{\beta}(\lambda, \vec{x})$ is defined as a unit vector from $\vec{S}(\lambda)$ to $\vec{x}$, $\vec{\alpha}$ is a unit vector perpendicular to $\vec{\beta}(\lambda, \vec{x})$ denoting the filtering

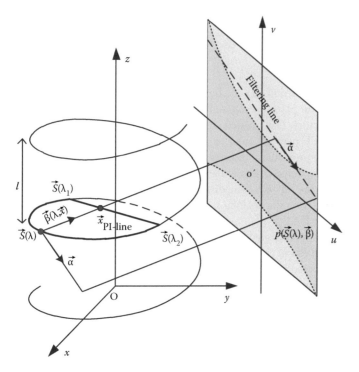

Figure 3.8 PI-line and the coordinate system for Katsevich's helical cone beam algorithm. The dark area in the detector shows the Tam–Danielsson window that contains the minimum amount of data for exact and stable cone beam reconstruction. (From Danielsson, P.E. et al., Towards exact reconstruction for helical cone beam scanning of long object. A new detector arrangement and a new completeness condition. *Proceedings of the 1997 Meeting on Fully 3D Image Reconstruction in Radiology and Nuclear Medicine*, Pittsburgh, PA, 141–4, 1997.)

direction on the detector plane as shown in Figure 3.8, and the limit of the integration over λ, $I_{\mathrm{PI}}(\vec{x}) = [\lambda_1, \lambda_2]$ is determined by the PI-line of a point $\vec{x}$ to be reconstructed. A PI-line is a line segment with endpoints located on the helix and separated by less than one pitch. It was proved that, for any point $\vec{x}$ strictly inside the helix, there is one and only one PI-line through it (Danielsson et al. 1997). Note that the choice of the filtering direction $\vec{\alpha}$ is not unique. Different selections of $\vec{\alpha}$ lead to different algorithms.

The structure of Equation 3.33 is of FBP-type and is highly efficient. The implementation of Katsevich's algorithm consists of the following steps:

1. Compute derivatives of the cone beam data with respect to the trajectory parameter λ.
2. For each point $\vec{x}$ to be reconstructed, determine its PI-line and then the limit of the integration $I_{\mathrm{PI}}(\vec{x})$.
3. Perform the 1D shift-invariant Hilbert filtering operations along the slant directions, which are specially selected.
4. Perform a cone beam backprojection with a weighting function that is inversely proportional to the distance from $\vec{x}$ to the source $\vec{S}(\lambda)$.

3.5.1.3 Backprojection filtration algorithm

Inspired by Katsevich's work, an alternative approach for theoretically exact image reconstruction from the minimum amount of cone beam data was proposed in the backprojection

filtration (BPF) format (Zou and Pan 2004a,b). This approach allows cone beam data to be transversely truncated to a certain degree. Different from Katsevich's algorithm, the BPF reconstruction does the 1D Hilbert filtering along a PI-line in the image domain after differential cone beam data are backprojected over the PI-line. Hence, their algorithm is often called the *BPF algorithm*. As mentioned, the relationship between the Hilbert transform of an object density and projection data was first established by Gel'fand and Graev (1991), whose general inversion formula covers the fan beam and cone beam BPF schemes as its special cases (Noo et al. 2004; Zhuang and Leng 2004; Ye et al. 2005; Zeng 2007).

For helical cone beam CT, the BPF algorithm is given as follows (Zou and Pan 2004c):

$$f(x_{\mathrm{PI}}, \lambda_1, \lambda_2) = \frac{1}{2\pi} \frac{1}{\sqrt{(x_2 - x_{\mathrm{PI}})(x_{\mathrm{PI}} - x_1)}}$$

$$\times \left[\int_{x_1}^{x_2} \frac{\sqrt{(x_2 - x'_{\mathrm{PI}})(x'_{\mathrm{PI}} - x_1)}}{\pi(x_{\mathrm{PI}} - x'_{\mathrm{PI}})} g(x'_{\mathrm{PI}}, \lambda_1, \lambda_2) dx'_{\mathrm{PI}} + C \right],$$
(3.34)

$$g(\vec{x}) = \int_{\lambda_1}^{\lambda_2} \frac{d\lambda}{|\vec{x} - \vec{S}(\lambda)|} \frac{\partial}{\partial q} p\left(\vec{S}(\lambda), \vec{\beta}(\lambda, \vec{x})\right)\Big|_{q=\lambda},$$
(3.35)

$$C = 2 \cdot p\left(\vec{S}(\lambda_1), \vec{\beta}(\lambda_1, (x_{\mathrm{PI}}, \lambda_1, \lambda_2))\right) \text{ or}$$

$$C = 2 \cdot p\left(\vec{S}(\lambda_2), \vec{\beta}(\lambda_2, (x_{\mathrm{PI}}, \lambda_1, \lambda_2))\right),$$
(3.36)

where $(x_{\mathrm{PI}}, \lambda_1, \lambda_2)$ is a point to be reconstructed that is expressed in the local coordinate system associated with the PI-line specified by λ_1 and λ_2. The above-mentioned BPF algorithm indicates that the reconstruction needs to be done PI-line by PI-line. To obtain the CT value at a point $\vec{x}$, we can simultaneously reconstruct all the points on the PI-line through $\vec{x}$ because the Hilbert filtration is not a local operator. In addition, because Equation 3.34 can reconstruct an image only on a PI-line, a resampling process is required to obtain the final 3D image in the Cartesian coordinates. In other words, the BPF algorithm consists of the following main steps:

1. Compute derivatives of cone beam data with respect to the trajectory parameter λ.
2. For each point $\vec{x}$, determine its PI-line and then the limit of the integration $I_{\mathrm{PI}}(\vec{x})$.
3. Perform a cone beam backprojection with a weighting function that is inversely proportional to the distance from $\vec{x}$ to $\vec{S}(\lambda)$.
4. Perform the 1D finite Hilbert inverse transform along the PI-line.
5. Map the reconstructed image from the PI-line-based coordinates onto the Cartesian coordinates by using an appropriate interpolation method.

Because Hilbert filtering does not allow data truncation, Katsevich's algorithm requires more data than that needed in the Tam–Danielsson window (Danielsson et al. 1997). By exchanging the order of backprojection and filtration, the BPF algorithm can handle the data truncation problem better, relying only on the data in the Tam–Danielsson window, as shown in Figure 3.8. Most importantly, in the BPF framework, a series of ROI reconstruction algorithms ware developed (Defrise et al. 2006; Ye et al. 2007; Kudo et al. 2008; Li et al. 2009, 2010) and are further discussed in the following.

3.5.2 APPROXIMATE RECONSTRUCTION

As an important class of reconstruction algorithms, in approximate cone beam algorithms, the reconstruction result will deviate somewhat from the true object function, regardless of the measurement quality. In this section, we present the most popular approximate reconstruction algorithm, called the FDK algorithm, and several important variants for circular or helical cone beam CT.

3.5.2.1 FDK and its variants for circular cone beam CT

As an extension of the fan beam FBP algorithm, the FDK (also called Feldkamp) algorithm was derived in 1984 to reconstruct a 3D object from cone beam data measured with a circular trajectory (Feldkamp et al. 1984). Here, we denote the circular trajectory as $R(-\sin\beta,\cos\beta)$ in the Cartesian coordinate system shown in Figure 3.9. The cone beam data are usually collected on either a planar or cylindrical detector surface. We present only the formula for the planar detector geometry, a formula that also can be modified to use data from a cylindrical detector. Similar to the process in Section 3.4, a virtual planar detector (u,v) is placed on the axis of rotation so that the v-axis of this detector coincides with the z-axis, as shown in Figure 3.9. The cone beam data are written as $p(\beta,u,v)$. The FDK formula is given as follows:

$$f(x,y,z) = \frac{1}{2}\int_0^{2\pi} \frac{R^2}{(R + x\sin\beta - y\cos\beta)^2}$$

$$\times \int_{-\infty}^{+\infty} p(\beta,u,v')\frac{R}{\sqrt{R^2+u^2+v'^2}}\cdot h(u'-u)\,du\,d\beta, \quad (3.37)$$

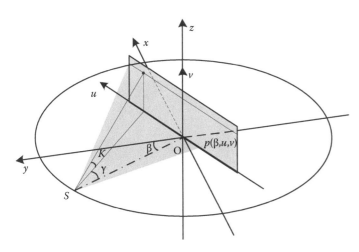

Figure 3.9 Circular cone beam geometry with a virtual planar detector.

$$v' = \frac{zR}{R + x\sin\beta - y\cos\beta}, \quad (3.38)$$

where u' is defined by Equation 3.26 and $h(\bullet)$ denotes the ramp filter.

Comparing Equations 3.25 and 3.37, we find that the structures of these two formulas are exactly the same. They both include three steps: preweighting, 1D filtration, and backprojection. The differences exist in the implementation details. The preweighting factor in Equation 3.37 equals the preweighting factor in Equation 3.25, multiplied by cosine of the cone angle, rewritten as

$$\frac{R}{\sqrt{R^2+u^2+v'^2}} = \frac{\sqrt{R^2+u^2}}{\sqrt{R^2+u^2+v'^2}}\cdot\frac{R}{\sqrt{R^2+u^2}} = \cos\kappa\cdot\cos\gamma, \quad (3.39)$$

where γ is the fan-angle and κ the cone angle for a special ray, as shown in Figure 3.9.

Then, the 1D ramp filtration and backprojection are implemented independently for each tilted fan beam projection. The weighting factor for backprojection in Equation 3.37 is identical to the factor in Equation 3.25. That is, it is independent of the z-coordinate of the voxel to be reconstructed. The difference lies in that the backprojection in Equation 3.37 is a process in 3D instead of 2D. Both the u- and v-coordinates need to be identified for the ray through the voxel. Thus, a 2D interpolation operation of the filtered data is required on the detector plane.

The computational structure of the FDK algorithm is straightforward, highly parallel, and hardware-supported. It is very robust and practical in various applications. Moreover, despite its approximate nature, the FDK algorithm preserves four exactness properties (Feldkamp et al. 1984; Wang et al. 1999; Rodet et al. 2004; Li et al. 2006a):

1. The reconstruction is exact on the circular trajectory plane.
2. The reconstruction is exact for any object homogeneous in the z-direction.
3. The longitudinal integral of the reconstruction is exact.
4. The integral of the reconstruction is exact along a set of oblique lines.

The first three properties were proved in the original paper (Feldkamp et al. 1984). The fourth property was reported in three publications (Wang et al. 1999; Rodet et al. 2004; Li et al. 2006a). Although the reconstruction domain is expanded to the whole 3D space, it was proved that the FDK reconstruction preserves the integrals along oblique lines with the angle from the z-axis less than arccos (R_{object} / R), where R_{object} and R are the radii of the spherical object support and the circular trajectory, respectively (Li et al. 2006a). This property provides new insight into the accuracy of this widely used algorithm.

Although the FDK algorithm uses all available data, some information is backprojected outside the actual object support. If we can draw the information outside the object, the FDK reconstruction could be improved (Li et al. 2006b).

However, as an approximate algorithm, FDK has a well-known shortcoming that, with an increasingly large cone angle, the cone beam artifacts will be more prominent.

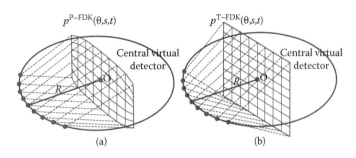

$p^{\text{P-FDK}}(\theta,s,t)$ $p^{\text{T-FDK}}(\theta,s,t)$

Central virtual detector

Central virtual detector

(a)   (b)

Figure 3.10 Different data rebinning schemes in the central virtual detector for P-FDK (a) and T-FDK (b). In both cases, the preweighting factors are the cosine of the angle between a ray path and the midplane.

As a result, the reconstructed value would have a clear drop in a slice far away from the midplane. Consequently, improving the accuracy of circular cone beam reconstruction has been one of the most interesting and challenging tasks for 3D cone beam CT.

Using the helical rebinning method, the P-FDK algorithm was proposed, which transforms native cone beam data into cone parallel beam data (Turbell 2001). The resulting beam geometry is parallel in the fan direction but divergent in the cone direction, as shown in Figure 3.10a. The rebinned data are obtained by

$$p^{\text{P-FDK}}\left(\theta,s,v\right)=p\left(\theta-\arcsin\frac{s}{R},\frac{sR}{\sqrt{R^2-s^2}},v\right). \quad (3.40)$$

Comparing Equations 3.15 and 3.40, the P-FDK algorithm rebins only cone beam data in the u-axis. The v-axis is left unchanged. After the data rebinning, an image is reconstructed by preweighting, ramp filtration, and backprojection according to the following:

$$\tilde{p}^{\text{P-FDK}}\left(\theta,s,v\right)=\left(\frac{R^2}{\sqrt{R^4+R^2v^2-v^2s^2}}\cdot p^{\text{P-FDK}}\left(\theta,s,v\right)\right)*h(s), \quad (3.41)$$

$$f^{\text{P-FDK}}(x,y,z)=\frac{1}{2}\int_0^{2\pi}\tilde{p}^{\text{P-FDK}}\left(\theta,s(x,y,\theta),v(x,y,z,\theta)\right)d\theta, \quad (3.42)$$

$$s(x,y,\theta)=x\cos\theta+y\sin\theta, \quad (3.43)$$

$$v(x,y,z,\theta)=\frac{zR^2}{\left(x\cos\theta+y\sin\theta\right)\sqrt{R^2-s(x,y,\theta)^2}+R^2-s(x,y,\theta)^2}. \quad (3.44)$$

As shown in Figure 3.10a, the filtration in P-FDK is performed along bulgy curves in the central virtual detector plane because of the variable distance between the cone parallel source position and the virtual detector plane. The P-FDK algorithm improves noise uniformity and computational efficiency, but it does

not improve the reconstruction quality outside the midplane. Considering the P-FDK algorithm, Grass et al. (2000) proposed the T-FDK algorithm, which has an additional rebinning operation along the vertical lines on the virtual detector. The rebinned data are obtained as follows:

$$p^{\text{T-FDK}}\left(\theta,s,t\right)=p\left(\theta-\arcsin\frac{s}{R},\frac{sR}{\sqrt{R^2-s^2}},\frac{tR^2}{R^2-s^2}\right). \quad (3.45)$$

After the data rebinning, the T-FDK reconstruction can be accomplished as follows:

$$\tilde{p}^{\text{T-FDK}}\left(\theta,s,t\right)=\left(\frac{\sqrt{R^2-s^2}}{\sqrt{R^2-s^2+t^2}}\cdot p^{\text{T-FDK}}\left(\theta,s,t\right)\right)*h(s), \quad (3.46)$$

$$f^{\text{T-FDK}}(x,y,z)=\frac{1}{2}\int_0^{2\pi}\tilde{p}^{\text{T-FDK}}\left(\theta,s(x,y,\theta),t(x,y,z,\theta)\right)d\theta, \quad (3.47)$$

$$t(x,y,z,\theta)=\frac{z\sqrt{R^2-s(x,y,\theta)^2}}{\sqrt{R^2-s(x,y,\theta)^2}+x\cos\theta+y\sin\theta}, \quad (3.48)$$

where $s(x,y,\theta)$ is given by Equation 3.43. Different from the P-FDK algorithm, the T-FDK algorithm filters projection data along horizontal lines on the rectangular virtual detector plane. T-FDK improves the image quality in the planes far from the midplane.

Figure 3.11 shows the FDK and T-FDK reconstruction results. In this numerical simulation, we used a 3D Shepp–Logan phantom (Turbell 2001) within a spherical support of radius 100 mm. Data acquisition was performed on a circular trajectory of radius 400 mm. The distance from the focal spot to the detector plane was 800 mm. A flat-panel detector of 256 × 256 elements was used. The physical size of this detector was 400 mm × 400 mm. Therefore, the cone angle was about ±14°. The number of projections simulated per rotation was 360. Noise-free cone beam projections were analytically calculated. Images were reconstructed in 256 × 256 × 256 matrices with cubic voxels of 0.78 mm per side. Clearly, FDK has a distinct drop in image

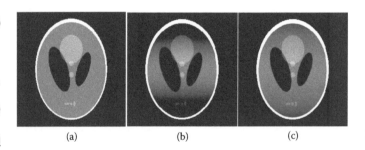

(a)   (b)   (c)

Figure 3.11 Actual and reconstructed images of the 3D Shepp–Logan phantom at $y = -25$ mm with a display window [1, 1.05]. (a) An ideal slice. (b) The slice reconstructed with FDK. (c) The slice reconstructed with T-FDK.

intensity for voxels far from the midslice. T-FDK reconstruction is in significantly better agreement with the ideal phantom.

To reduce the cone beam artifacts, other effective methods are available. The reader is referred to the related articles (Hu 1996; Tang et al. 2005; Mori et al. 2006; Nett et al. 2007; Zhu et al. 2008; Grimmer et al. 2009).

3.5.2.2 Helical FDK and generalized FDK

As mentioned, the original FDK algorithm was developed for circular cone beam scanning. For volume imaging of a long object, the most effective and widely used CT geometry is helical or spiral CT. Fan beam helical CT was first suggested in 1986 by a Japanese engineer, Mori (1986). Cone beam helical CT was first proposed in 1991 by Wang et al. (1991).

For fan beam helical CT, a typical reconstruction approach is to perform 2D fan beam or parallel beam reconstruction slice by slice. Because a helical trajectory has only a single intersection point with any plane perpendicular to the z-axis, there is only one projection measured for any slice. Therefore, various linear interpolation (LI) methods were developed to estimate sufficient data for 2D image reconstruction from measured data. Two commonly used LI methods are the 360°LI and 180°LI (Crawford and King 1990; Kalender et al. 1990). The 360°LI algorithm uses helical CT projections from the two adjacent full turns (360°) of the helix to estimate one set of projections at a prescribed location. In contrast, the 180°LI algorithm uses only the two half-scan data sets to estimate the required "planar" data set. Because it uses only data 180° apart, the 180°LI algorithm produces a narrower slice than the 360°LI algorithm, given the same helical pitch. As a trade-off, the image noise is increased with 180°LI compared with 360°LI.

The above-mentioned LI methods for synthesis of a single-slice data set may be naturally extended for multislice (multisection) computed tomography (MSCT), also referred to as multidetector-row CT. The linear interpolation along the z-axis is performed between the two measured rays closest to a desired ray. These two rays may belong to different rows of a detector array. Figure 3.12 shows how this linear interpolation is performed for a single-slice and a dual-slice CT.

The LI-based methods described earlier are approximate 2D FBP algorithms and work well for commercial MSCT scanners when the cone angle is small enough (typically less than a few degrees). However, with the rapid development of the MSCT technology, the cone angle has become quite large, spanned by more detector rows. For example, the cone angle of a Toshiba

320-slice CT scanner is approximately 15.2° (Siebert et al. 2009). With such a large cone angle, the cone beam artifacts associated with 2D reconstruction methods are clear because most backprojected values are accumulated to voxels out of the assumed imaging plane. Hence, truly 3D reconstruction algorithms must be developed.

Based on the FDK algorithm, initial efforts were made in 1991 to derive a helical FDK and solve the long object problem with spiral cone beam scanning (Kudo and Saito 1991; Wang et al. 1991, 1993, 2007). Here, we denote the helical trajectory as $(-R \sin\beta, R\cos\beta, l\beta/2\pi)$ in the Cartesian coordinate system, where l is the helical pitch. A virtual planar detector (u,v) is placed on the axis of rotation so that the v-axis of this detector coincides with this z-axis. The cone beam data are written as $p(\beta,u,v)$. We summarize the helical FDK algorithm as follows:

$$f(x,y,z) = \frac{1}{2} \int_{\beta_z-\pi}^{\beta_z+\pi} \frac{R^2}{(R + x\sin\beta - y\cos\beta)^2}$$

$$\times \int_{-\infty}^{+\infty} p(\beta,u,v') \frac{R}{\sqrt{R^2+u^2+v'^2}} \cdot h(u'-u)\,dud\beta, \quad (3.49)$$

$$v' = \frac{R(z - l\beta/2\pi)}{R + x\sin\beta - y\cos\beta}, \quad (3.50)$$

where u' is given by Equation 3.26 and $h(\bullet)$ is the ramp filter. For a plane that contains a point (x,y,z) and is perpendicular to the z-axis, we can easily find β_z as the β-coordinate of the intersection point of that plane and the helical trajectory, $\beta_z = 2\pi z/l$.

Comparing Equation 3.49 and the original FDK formula, Equation 3.37, we find that these two formulas are identical except that the preweighting, filtration, and backprojection are all performed with respect to the helix. In other words, the helical FDK algorithm also includes the three steps: preweighting, 1D filtration along each detector row, and cone beam backprojection from the closest full turn of the helix. The reader may modify Equation 3.49 to work on data from a cylindrical detector.

Furthermore, the FDK algorithm can be developed for cone beam image reconstruction from not only a helical trajectory but also quite general scanning trajectories (Wang et al. 1993). We denote a general trajectory as $(-R(\beta)\sin\beta, R(\beta)\cos\beta, l(\beta))$ in the Cartesian coordinate system. A virtual planar detector (u,v) is placed on the z-axis of rotation. The cone beam data are again written as $p(\beta,u,v)$. A generalized FDK algorithm is expressed as follows:

$$f(x,y,z) = \frac{1}{2} \int_{\beta_z-\pi}^{\beta_z+\pi} \frac{R(s)^2}{(R(s) + x\sin\beta - y\cos\beta)^2}$$

$$\times \int_{-\infty}^{+\infty} p(\beta,u,v') \frac{R(s)}{\sqrt{R^2(s)+u^2+v'^2}} \cdot h(u'-u)\,dud\beta, \quad (3.51)$$

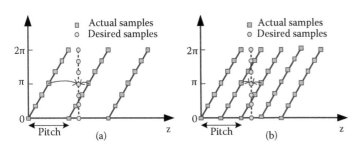

Figure 3.12 Linear interpolation along the z-axis for single-slice (a) and dual-slice (b) CT, where the solid squares are for actual samples and circular dots for desired samples.

$$v' = \frac{R(\beta) \cdot \left(z - l(\beta)\right)}{R(\beta) + x\sin\beta - y\cos\beta}, \quad (3.52)$$

where u' is given by Equation 3.26 and β_z is the β-coordinate of the intersection point of the helical trajectory and the plane through a point (x,y,z) and perpendicular to the z-axis, $\beta_z = l^{-1}(z)$.

The merits of these approximate algorithms are several. First, the approximate algorithms may produce comparable or even better spatial/contrast resolution than the exact algorithms when data are incomplete or contaminated by noise. Second, the algorithms have higher computational efficiency. Third, the approximate algorithms can be more easily modified to work with various scanning geometries. However, when the cone angle becomes too large, an FDK-type algorithm will produce significant cone beam artifacts. In general, the larger the cone angle becomes, the worse an approximate algorithm performs. Currently, the cone angle used for commercial MSCT is still moderate. Therefore, the approximate algorithms are still widely used in medical and industrial applications, especially the FDK-type algorithms, either in original form or variants for circular, helical, or other trajectories.

It is worth considering why commercial CT scanners still use approximate algorithms instead of exact algorithms (Wang et al. 1999, 2000, 2007; Pan et al. 2009). To answer this question, we need to know what kinds of algorithms are most desirable for commercial CT. The ultimate goal of commercial CT is to obtain high-quality images to meet practical requirements at a minimized total cost. From this perspective, commercial CT is not particularly focused on whether a reconstruction algorithm is theoretically exact or not.

According to the definition of exact reconstruction algorithms in Section 3.5.1, exact algorithms would have an outstanding, possibly even perfect, performance in ideal-world applications with a sufficient amount of noise-free data. However, this mathematical exactness cannot be completely implemented for practical imaging applications, primarily because we do not have an imaging model that reflects a real-world imaging process perfectly. Any model is only an approximation of the true imaging process. Therefore, reconstruction with an exact algorithm would be compromised. Moreover, in real applications, the collected data and algorithm implementations contain unavoidable approximations, such as discretization and noise. As a result, in many cases, exact algorithms have not yielded images with better quality than those reconstructed with approximate algorithms. Therefore, it remains the case that all major CT manufacturers still use approximate cone beam algorithms as the working horse.

The development of theoretically exact algorithms is significant and necessary. First, they may help optimize approximate algorithms. In contrast, for important applications, exact algorithms may eventually demonstrate their superiority over approximate algorithms when various practical considerations—such as noise, uniformity, and minimization—are taken into account. Ultimately, exact algorithms are based on rigorous analysis and have built-in benefits, so their potential should be further developed and fully realized. Because the ultimate criterion for image reconstruction is practical use, the fusion of exact and approximate algorithms is certainly welcome.

3.5.3 DATA SUFFICIENCY CONDITION

The discussions thus far have been limited to exact and approximate reconstruction algorithms. Essentially, exact or approximate reconstruction depends on the available data. Various algorithms provide different ways to recover images from collected data. Therefore, we need to know how much data is sufficient for exact reconstruction, a classic topic in the CT reconstruction field. In practice, the study of data sufficiency is closely correlated to the development of reconstruction algorithms. Especially, new data-sufficiency conditions are always accompanied by new reconstruction algorithms.

Orlov (1975a,b) proposed a data-sufficiency condition based on a set of parallel beam projections. His condition states that, to reconstruct an object exactly, a scanning curve on a unit sphere of directions for parallel beam projections must "have points in common with any arc of a great circle." Subsequently, Tuy (1983) and Smith (1985) independently derived a cone beam data-sufficiency condition for exact cone beam reconstruction. Their condition states that if, on every plane that intersects an object, there exists at least one cone beam source point, then one can reconstruct the object (Tuy 1983; Smith 1985; Li et al. 2008). It can be shown that Grangeat's framework is equivalent to the Tuy–Smith condition (Grangeat 1991). Because they are fundamental, these sufficiency conditions are often used in theoretical analysis on cone beam reconstruction. Note that the Orlov condition, the Tuy–Smith condition, and the Grangeat framework all assume that the projection of the object is not truncated at any source position.

We know that a circular trajectory does not satisfy the Tuy–Smith condition because a plane parallel to the trajectory has no intersection with the focal spot. This incompleteness of circular cone beam acquisition limits image quality regardless of the algorithms used. None of the above-mentioned data-sufficiency conditions permit truncation of projection data. The latest investigations of data sufficiency are consequences of the PI-line-based reconstruction research, with the PI-line being interpreted as an arbitrary line segment with two endpoints on a general trajectory.

The original BPF algorithms lead to the following sufficient condition (Gel'fand et al. 1991; Noo et al. 2004; Zou and Pan 2004a,b): An ROI can be exactly reconstructed from truncated data if (1) all the line integrals are known through the ROI and (2) there is a set of PI-line segments that cover the whole ROI with two endpoints located outside the object support. Then, Defrise et al. (2006) developed an advancement for exact ROI reconstruction. They proposed a more flexible data-sufficiency condition enlarging the ROI exactly reconstructable from a given data set. Their condition may be stated as follows: An ROI can be exactly reconstructed from truncated data if (1) all the line integrals are known through the ROI and (2) there is a set of PI-line segments that covers the whole ROI with at least one endpoint outside the object support.

The above-mentioned two conditions both require an ROI located on an edge of the object. When an ROI is inside the object, it becomes the classic interior problem. Mathematically, the interior problem was already proven to have no unique solution even when exact information on the object support is known. Nevertheless, in 2007, the interior problem was revisited, showing that it can be exactly and stably solved if *a priori* knowledge is known in the form of a known subregion in the ROI (Ye et al. 2007; Courdurier et al. 2008; Kudo et al. 2008). The new data-sufficiency results can be summarized as follows: An interior ROI can be exactly reconstructed from purely local data if (1) all the line integrals are known through the ROI and (2) the ROI contains a subregion for which the image is precisely known.

It has been further proved that *a priori* knowledge of a known subregion is not necessary (Li et al. 2009). This line of research gives the new data-sufficiency condition: An interior ROI can be exactly reconstructed from purely local data if (1) all the line integrals are known through the ROI and a small subregion B outside the ROI and (2) the subregion B is known or is on an edge of the object.

Figure 3.13 shows the differences among the above-mentioned four data-sufficiency conditions. Figure 3.13a through d corresponds to these four conditions, respectively. The support of the object is bounded by the solid curve. The x-ray projections are measured only along the lines that intersect the dashed circular field of view (FOV). Thus, exact and stable reconstruction can be achieved within the light-shaded ROI.

Figure 3.14 shows three ROI images reconstructed from simulated data according to the different data-sufficiency conditions. In this simulation, the 2D Shepp–Logan phantom was used. The results indicate the validity of exact local reconstruction in reference to known subregions.

Most interestingly, in light of compressed sensing theory, an interior ROI can be exactly reconstructed if the ROI can be assumed to be piecewise constant or piecewise

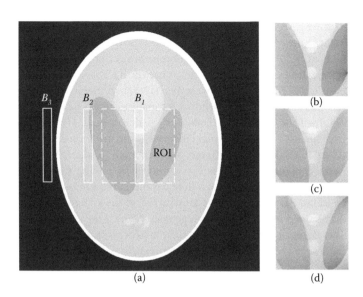

(b)

(c)

(a)

(d)

Figure 3.14 Differences among the four data-sufficiency conditions. (a) Dotted square is an ROI with three known subregions B_1, B_2, and B_3. (b) Reconstruction only from the truncated data through the ROI with *a priori* knowledge that the image is known on B_1. (c) Reconstruction from the truncated data through the ROI and B_2, with a known B_2. (d) Reconstruction from the truncated data through the ROI and B_3.

polynomial (Yu and Wang 2009; Wang et al. 2010, Yang et al. 2010). It provides a new approach to solve the interior problem. Finally, because of the Grangeat's framework and results of interior tomography, a new data-sufficiency condition in cone beam geometry may indicate that any plane intersecting a volume of interest (VOI) must contain at least one focal spot from which a narrow cone beam of x-rays goes through the ROI. This condition is the same as the Tuy–Smith condition except that truncated projection data are assumed.

As discussed, all the existing data-sufficiency conditions are sufficient for exact reconstruction, but a unifying theory is still under investigation. This development indicates exciting new research opportunities ahead. Collaboration with mathematicians is crucial in future success in the CT field.

3.6 DISCUSSION AND CONCLUSIONS

Over the past decades, numerous reconstruction algorithms have been proposed for various situations in which projection data can be longitudinally and transversely truncated and a source scanning trajectory can be not only a circle or a standard helix but also a fairly general class of trajectories such as a saddle curve, a nonstandard helix, or a circle-plus-line combination (Wang et al. 2008). In medical and industrial applications, it is crucial to select an appropriate reconstruction algorithm that depends on a specific application. In practice, projection data are always corrupted by various noise and misalignment errors. An ideal reconstruction algorithm needs to produce the best possible image resolution and noise characteristic while being implemented efficiently.

In this chapter, we have divided the algorithms into two categories—exact and approximate—that can be grouped into

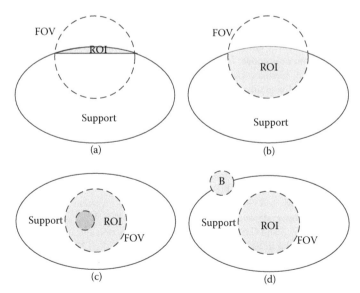

(a)

(b)

(c)

(d)

Figure 3.13 Differences among the four data-sufficiency conditions illustrated in a, b, c, and d, respectively. (c) Dark-shaded known subregion is inside the object support. (d) Line integrals are collected through a small subregion B outside the ROI, which is either known or on an edge of the object.

a larger category of analytic algorithms. However, iterative algorithms represent another popular type of algorithm. Generally, analytic algorithms use the projection data only once, whereas iterative algorithms use the data several or even many times until convergence is reached. Iterative reconstruction has been studied for many years. When he obtained the first clinical image of a patient's head, Hounsfield used an ART that is a typical iterative reconstruction method (Gordon et al. 1970). Different from an analytic algorithm, iterative algorithms reconstruct an image by solving a system of linear equations $p = Af + e$, where p is measured projection data, f a 2D or 3D image to be reconstructed, A the system matrix determined by the CT system geometry, and e the error vector. Therefore, the iterative reconstruction process is to estimate the image vector f based on a measurement vector p corrupted by the error vector e. Iterative algorithms can be divided into two categories: algebraic iterative (Kak and Slaney 1988) and statistical iterative algorithms (Shepp and Vardi 1982; Lange and Carson 1984; Manglos et al. 1995; Fessler 2000; Leahy and Byrne 2000; Zeng and Gullberg 2000; Zeng 2001). In algebraic iterative algorithms, we ignore the statistical nature of the imaging process and assume e to be constant. In statistical iterative algorithms, a statistical model is incorporated into the reconstruction so that the likelihood or a properly-defined probability is maximized. Often, we have information about an image to be reconstructed in addition to measured data. This information can be put into the objective function as a penalty term for iterative reconstruction (Candes et al. 2006). An example is the use of roughness penalty. Compared to analytic algorithms, iterative algorithms are generally advantageous for low-dose CT. Therefore, iterative algorithms deserve more efforts in the future.

The major drawback of iterative algorithms is their high computational cost. Usually, one iteration requires more computational time than does the FBP algorithm. Thus, we must consider how to improve the efficiency of iterative algorithms. An important approach is to accelerate iterative algorithms using hardware, such as an application-specific integrated circuit (ASIC), a field programmable gate array (FPGA), and graphics processing units (GPUs). Traditionally, GPUs are installed for accelerating graphically intensive applications like animations and games. Because they were designed to execute tasks in parallel, GPUs contain many programmable pipelines that enable parallel data processing. As a result, GPUs are an excellent platform for CT reconstruction and have been used in both medical and industrial applications (HPIR 2009). With the rapid development of high-speed processor and parallel computing technology, we believe that iterative algorithms will receive increasing attention in the future.

In conclusion, we have reviewed the CT principles and reconstruction algorithms with an emphasis on analytic algorithms. The recent advancements in this field have been impressive, especially in exact cone beam reconstruction, interior tomography, compressed sensing (CS)–based reconstruction, and iterative reconstruction. It is anticipated that CT will play an even more important role in biomedical, industrial, and other applications.

REFERENCES

Axelsson, C. and Danielsson, P.E. 1994. Three-dimensional reconstruction from cone-beam data in $O(N^3 \log N)$ time. *Phys Med Biol* 39: 477–91.

Bracewell, R.N. and Riddle, A.C. 1967. Inversion of fan beam scans in radio astronomy. *Astrophys J* 150: 427–34.

Candes, E., Romberg, J. and Tao, T. 2006. Robust uncertainty principles: exact signal reconstruction from highly incomplete frequency information. *IEEE Trans Inform Theor* 52: 489–509.

Censor, Y., Jiang, M. and Wang, G. 2010. *Biomedical Mathematics: Promising Directions in Imaging, Therapy Planning, and Inverse Problems.* Madison, WI: Medical Physics Publishing.

Chen, G.H. 2003. An alternative derivation of Katsevich's cone-beam reconstruction formula. *Med Phys* 30(12): 3217–26.

Chen, G.H., Leng, S. and Mistretta, C.A. 2005. A novel extension of the parallel-beam projection-slice theorem to divergent fan-beam and cone-beam projections. *Med Phys* 32(3): 654–65.

Cormack, A.M. 1963. Representation of a function by its line integrals, with some radiological applications I. *J Appl Phys* 34: 2722–7.

Courdurier, M., Noo, F, Kudo, H., et al. 2008. Solving the interior problem of computed tomography using a priori knowledge. *Inverse Probl* 24: 65001.

Crawford, C.R. and King, K.F. 1990. Computed tomography scanning with simultaneous patient translation. *Med Phys* 17(6): 967–82.

Danielsson, P.E. 1992. From cone-beam projections to 3D Radon data in $O(N^3 \log N)$ time. *Proc of IEEE Med Imaging Conf,* 1992., Conference Record of the 1992 IEEE: 1135–7.

Danielsson, P.E., Edholm, P., Eriksson, J., et al. 1997. Towards exact reconstruction for helical cone-beam scanning of long object. A new detector arrangement and a new completeness condition. *Proceedings of the 1997 Meeting on Fully 3D Image Reconstruction in Radiology and Nuclear Medicine.* Pittsburgh, PA, 141–4.

Deans, S.R. 1983. *The Radon Transform and Some of Its Applications.* New York: Wiley-Interscience.

Defrise, M. and Clack, R. 1994. A cone-beam reconstruction algorithm using shift-variant filtering and cone-beam backprojection. *IEEE Trans Med Imaging* 13: 186–95.

Defrise, M., Noo, F., Clack, R., et al. 2006. Truncated Hilbert transform and image reconstruction from limited tomographic data. *Inverse Probl* 22: 1037–53.

Edholm, P. and Herman, G. 1987. Linograms in image reconstruction from projections. *IEEE Trans Med Imaging* 6(4): 301–7.

Edholm, P., Herman, G. and Roberts, D. 1988. Image reconstruction from linograms: implementation and evaluation. *IEEE Trans Med Imaging* 7(3): 239–46.

Feldkamp, L.A., Davis, L.C. and Kress, J.W. 1984. Practical cone-beam reconstruction. *J Opt Soc Am A* 1: 612–19.

Fessler, J. 2000. Statistical image reconstruction methods for transmission tomography. In: Sonka, M. and Fitzpatrick, J. (eds.) *Handbook of Medical Imaging, Volume 2, Medical Image Processing and Analysis.* Bellingham, WA: SPIE, 1–70.

Gel'fand, I.M. and Gindikin, S.G. 1977. Nonlocal inversion formulas in real integral geometry. *Funct Anal Appl* 11: 173–9.

Gel'fand, I.M. and Graev, M.I. 1991. Crofton function and inversion formulas in real integral geometry. *Funct Anal Appl* 25(1): 1–5.

Gel'fand, I.M., Graev, M.I. and Shapiro, Z.Y. 1966a. Integral geometry on manifolds of k-dimensional planes. *Sov Math Dokl* 7: 801–4.

Gel'fand, I.M., Graev, M.I. and Vilenkin, N.Y. 1966b. *Generalized Functions: Integral Geometry and Representation Theory.* New York: Academic Press.

Gordon, R., Bender, R. and Herman, G.T. 1970. Algebraic reconstruction technique (ART) for three-dimensional electron microscopy and x-ray photography. *J Theor Biol* 29: 471–81.

Grangeat, P. 1991. Mathematical framework of cone beam 3D reconstruction via the first derivative of the Radon transform. *Proceedings of Lecture Notes in Mathematics* 1497. New York: Springer.

Grass, M., Kohler, T. and Proksa, R. 2000. 3D cone-beam CT reconstruction for circular trajectories. *Phys Med Biol* 45(2): 329–47.

Grimmer, R., Oelhafen, M., Elstrom, U., et al. 2009. Cone-beam CT image reconstruction with extended z range. *Med Phys* 36(7): 3363–70.

Herman, G.T. 1980. *Image Reconstruction from Projections: The Fundamentals of Computerized Tomography*. New York: Academic Press.

Hsieh, J. 2003. *Computed Tomography: Principles, Design, Artifacts, and Recent Advances*. Bellingham, WA: SPIE.

Hu, H. 1996. An improved cone-beam reconstruction algorithm for the circular orbit. *Scanning* 18: 572–81.

Kalender, W., Seissler, E. and Klotz, E. et al. 1990. Spiral volumetric CT with single-breath-hold technique, continuous transport, and continuous scanner rotation. *Radiology* 176: 181–3.

Kak, A.C. and Slaney, M. 1988. *Principles of Computerized Tomographic Imaging*. New York: IEEE Press.

Katsevich, A. 2002. Theoretically exact filtered backprojection-type inversion algorithm for spiral CT. *SIAM J Appl Math* 62(6): 2012–26.

Katsevich, A. 2003. A general scheme for constructing inversion algorithms for cone-beam CT. *Int J Math Math Sci* 21: 1305–21.

Kudo, H. and Saito, T. 1991. Helical-scan computed tomography using cone-beam projections. *Rec Nuclear Science Symposium and Medical Imaging Conference*, November 1991., Conference Record of the 1991 IEEE: 1958–62.

Kudo, H., Courdurier, M., Noo, F., et al. 2008. Tiny a priori knowledge solves the interior problem in computed tomography. *Phys Med Biol* 53(9): 2207–31.

Kudo, H., Noo, F. and Defrise, M. 1998. Cone-beam filtered-backprojection algorithm for truncated helical data. *Phys Med Biol* 43(10): 2885–909.

Lange, K. and Carson, R. 1984. EM reconstruction algorithm for emission and transmission tomography. *J Comput Assisted Tomogr* 8(2): 306–16.

Leahy, R. and Byrne, C. 2000. Editorial: recent development in iterative image reconstruction for PET and SPECT. *IEEE Trans Med Imaging* 19: 257–60.

Li, L., Chen, Z.Q., Kang, K.J., et al. 2010. Recent advance in exact ROI/VOI image reconstruction. *Curr Med Imag Rev* 6(2): 112–18.

Li, L., Chen, Z.Q., Xing, Y.X., et al. 2006a. A general exact method for synthesizing parallel-beam projections from cone-beam projections via filtered backprojection. *Phys Med Biol* 51: 5643–54.

Li, L., Kang, K.J., Chen, Z.Q., et al. 2006b. The FDK algorithm with an expanded definition domain for cone-beam reconstruction preserving oblique line integrals. *J X Ray Sci Tech* 14: 217–33.

Li, L., Kang, K.J., Chen, Z.Q., et al. 2008. An alternative derivation and description of Smith's data sufficiency condition for exact cone-beam reconstruction. *J X Ray Sci Tech* 16: 43–9.

Li, L., Kang, K.J., Chen, Z.Q. et al. 2009. A General region-of-interest image reconstruction approach with truncated Hilbert transform. *J X Ray Sci Tech* 17: 135–52.

Lu, Y., Katsevich, A., Zhao, J., et al. 2010. Fast exact/quasi-exact FBP algorithms for triple-source helical cone-beam CT. *IEEE Trans Med Imaging* 29: 756–70.

Manglos, S., Gange, G., Krol, A., et al. 1995. Transmission maximum-likelihood reconstruction with ordered subsets for cone beam CT. *Phys Med Biol* 40: 1225–41.

Mori, I. 1986. Computerized tomographic apparatus utilizing a radiation source. US Patent No. 4630202.

Mori, S., Endo, M., Komatsu, S., et al. 2006. A combination-weighted Feldkamp-based reconstruction algorithm for cone-beam CT. *Phys Med Biol* 51: 3953–65.

Natterer, F. 1986. *The Mathematics of Computerized Tomography*. New York: John Wiley & Sons.

Nett, B.E., Zhuang, T.L., Leng, S., et al. 2007. Arc based cone-beam reconstruction algorithm using an equal weighting scheme. *J X Ray Sci Tech* 15: 19–48.

Noo, F., Defrise, M., Clackdoyle, R., et al. 2002. Image reconstruction from fan-beam projections on less than a short scan. *Phys Med Biol* 47: 2525–46.

Noo, F., Clackdoyle, R. and Pack, J.D. 2004. A two-step Hilbert transform method for 2D image reconstruction. *Phys Med Biol* 49: 3903–23.

Oldendorf, W.H. 1963. Radiant energy apparatus for investigating selected areas of the interior of objects obscured by dense material. US Patent No. 3106640.

Orlov, S. 1975a. Theory of three dimensional reconstruction: I. Conditions for a complete set of projections. *Sov Phys Crystallogr* 20: 312–14.

Orlov, S. 1975b. Theory of three dimensional reconstruction: II. The recovery operator. *Sov Phys Crystallogr* 20: 429–33.

O'Sullivan, J. 1985. A fast sinc function gridding algorithm for Fourier inversion in computer tomography. *IEEE Trans Med Imaging* 4(4): 200–7.

Pack, J.D., Noo, F. and Clackdoyle, R. 2005. Cone-beam reconstruction using the backprojection of locally filtered projections. *IEEE Trans Med Imaging* 24: 70–85.

Pan, X., Sidky, E. and Vannier, M. 2009. Why do commercial CT scanners still employ traditional, filtered back-projection for image reconstruction? *Inverse Probl* 25: 123009.

Parker, D.L. 1982. Optimal short scan convolution reconstruction for fan-beam CT. *Med Phys* 9(2): 254–7.

Proceedings of 2nd Workshop on High Performance Image Reconstruction (HPIR). *10th International Meeting on Fully Three-Dimensional Image Reconstruction in Radiology and Nuclear Medicine*. Beijing, China, 5–10 September 2009, Available from: http://www.fully3d.org/2009/index09.htm.

Rodet, T., Noo, F. and Defrise, M. 2004. The cone-beam algorithm of Feldkamp, Davis, and Kress preserves oblique line integrals. *Med Phys* 31: 1972–5.

Rullgard, H. 2004. An explicit inversion formula for the exponential Radon transform using data from 180°. *Ark Mat* 42: 353–62.

Schomberg, H. and Timmer, J. 1995. The gridding method for image reconstruction by Fourier transformation. *IEEE Trans Med Imaging* 14(3): 596–607.

Shepp, L. and Vardi, Y. 1982. Maximum likelihood reconstruction for emission tomography. *IEEE Trans Med Imaging* 1(2): 113–22.

Siebert, E., Bohner, G., Dewey, M., et al. 2009. 320-slice CT neuroimaging: initial clinical experience and image quality evaluation. *Br J Radiol* 82: 561–70.

Smith, B.D. 1983. Cone-beam convolution formula. *Comput Biol Med* 13: 81–7.

Smith, B.D. 1985. Image reconstruction from cone-beam projections: necessary and sufficient conditions and reconstruction methods. *IEEE Trans Med Imaging* 4: 14–25.

Tang, X.Y., Hsieh, J., Hagiwara, A., et al. 2005. A three-dimensional weighted cone beam filtered backprojection (CB-FBP) algorithm for image reconstruction in volumetric CT under a circular source trajectory. *Phys Med Biol* 50: 3889–905.

Turbell, H. 2001. *Cone-Beam Reconstruction Using Filtered Backprojection, Linkoping Studies in Science and Technology*, Dissertation No. 672. Linkopings Universitet, Sweden.

Tuy, H. 1983. An inversion formula for cone-beam reconstruction. *SIAM J Appl Math* 43: 546–52.

Wang, G., Lin, T.H., Cheng, P.C., et al. 1991. Scanning cone-beam reconstruction algorithms for x-ray microtomography. *Proc SPIE* 1556: 99–112.

Wang, G., et al. 1993. A general cone-beam reconstruction algorithm. *IEEE Trans Med Imaging* 12: 486–96.

Wang, G., Crawford, C.R. and Kalender, W.A. 2000. Multirow detector and cone-beam spiral/helical CT. *IEEE Trans Med Imaging* 19(9): 817–21.

Wang, G., Ye, Y. and Yu, H. 2007. Approximate and exact cone-beam reconstruction with standard and non-standard spiral scanning. *Phys Med Biol* 52: R1–13.

Wang, G. and Yu, H. 2010. Can interior tomography outperform lambda tomography? *Proc Natl Acad Sci U S A* 107(22): E92–3.

Wang, G., Yu, H. and Deman, B. 2008. An outlook on x-ray CT research and development. *Med Phys* 35(3): 1051–64.

Wang, G., Zhao, S.Y. and Cheng, P.C. 1999. Exact and approximate cone-beam x-ray microtomography. In: Cheng, P.C., et al. (eds.) *Focus on Multidimensional Microscopy (I)*. Singapore: World Scientific, 233–61.

Yang, J., et al. 2010. High order total variation minimization for interior tomography. *Inverse Probl* 26: 35013.

Ye, Y., et al. 2005. A general exact reconstruction for cone-beam CT via backprojection-filtration. *IEEE Trans Med Imaging* 24: 1190–8.

Ye, Y., Yu, H., Wei, Y., et al. 2007. A general local reconstruction approach based on a truncated Hilbert transform. *Int J Biomed Imaging*, Article No.: ID 63634.

Yu, H. and Wang, G. 2009. Compressed sensing based Interior tomography. *Phys Med Biol* 54: 2791–805.

Zeng, G. 2001. Image reconstruction—a tutorial. *Comput Med Imag Graph* 25(2): 97–103.

Zeng, G. 2007. Image reconstruction via the finite Hilbert transform of the derivative of the backprojection. *Med Phys* 34: 2837–43.

Zeng, G. 2010. *Medical Image Reconstruction: A Conceptual Tutorial*. New York: Springer.

Zeng, G. and Gullberg, G. 2000. Unmatched projector/backprojector pairs in an iterative reconstruction algorithm. *IEEE Trans Med Imaging* 19(5): 548–55.

Zhu, L., Starman, J. and Fahrig, R. 2008. An efficient estimation method for reducing the axial intensity drop in circular cone-beam CT. *Int J Biomed Imaging*, 2008, doi:10.1155/2008/242841

Zhuang, T., Leng, S., Nett, B.E., et al. 2004. Fan-beam and cone-beam image reconstruction via filtering the backprojection image of differentiated projection data. *Phys Med Biol* 49: 5489–503.

Zou, Y. and Pan, X. 2004a. Exact image reconstruction on PI-line from minimum data in helical cone-beam CT. *Phys Med Biol* 49: 941–59.

Zou, Y. and Pan, X. 2004b. An extended data function and its backprojection onto PI-lines in helical cone-beam CT. *Phys Med Biol* 49: N383–7.

Zou, Y. and Pan, X. 2004c. Image reconstruction on PI-lines by use of filtered backprojection in helical cone-beam CT. *Phys Med Biol* 49: 2717–31.

4 Cone-beam CT image quality

Jeffrey H. Siewerdsen, Wojciech Zbijewski, and Jennifer Xu

Contents

4.1 INTRODUCTION

This chapter approaches the broad question of cone-beam computed tomography (CBCT) image quality in two broad strokes: **(1) a review of the *metrics* by which CBCT imaging performance may be assessed and (2) the physical and mathematical *factors* that affect CBCT imaging performance.** Discussion throughout is limited to CBCT imaging by three-dimensional (3D) filtered backprojection (FBP), assuming a circular source–detector orbit (Feldkamp et al. 1984), and the applicability of the various performance metrics to other reconstruction methods is a subject of other work.

The metrics of CBCT imaging performance include basic measures of contrast resolution and spatial resolution extended to spatial-frequency-dependent measures and performance of a particular imaging task. Discussion of image artifacts also is

summarized in a manner that attempts to distinguish artifacts as deterministic effects (e.g., shading and streaks that are reproducible in a given source-object-detector configuration) from other sources of image degradation arising from stochastic effects (e.g., quantum noise).

The physical and mathematical factors affecting CBCT imaging performance present a broad space of parameters, including factors of CBCT system design (e.g., system geometry, focal spot size, and detector size), acquisition "technique factors" [e.g., peak kilovoltage (kVp), milliamperes-seconds (mAs), and number of projection views], and 3D image reconstruction parameters (e.g., selection of voxel size and reconstruction filter).

Treatment of such a broad question within the pages of a single chapter is limited to a somewhat simplified view of the major considerations. The reader is directed to a wealth of scientific literature in the bibliography and other textbooks on specific aspects of CBCT performance metrology, image artifacts, and the myriad effects on image quality. The broad perspective provided by this chapter is intended to provide a basic foundation in the physics of CBCT image quality from which readers may, first, set out to evaluate imaging performance in quantifiable terms suitable to rigorous system design, assessment, and technique optimization and, second, consider for themselves the factors at play in the performance measured for any particular CBCT system.

4.2 METRICS OF IMAGING PERFORMANCE

4.2.1 IMAGE NOISE

Image "noise" refers to stochastic variation in CBCT voxel values. For example, in a CBCT image of a uniform water phantom, noise refers to the fluctuations in voxel values about the mean. More generally, for example, within a heterogeneous object, noise refers to voxel value fluctuations that are not associated with the actual variation in attenuation coefficient of the underlying physical structures. For example, Figure 4.1 illustrates a CBCT image of a cylinder containing a variety of plastic rods within a uniform background. In a broader sense, noise could be considered any fluctuation in the image that is not associated with a structure of interest, and thereby invites interpretation of deterministic effects (i.e., artifacts) and background anatomy (i.e., anatomy not associated with the structure of interest) within the context of noise. We touch on such interpretation below with regard to so-called generalized noise models. For now, discussion is focused on purely stochastic

fluctuations, the most common source of which is quantum noise owing to the random distribution of individual x-ray photons incident on each detector element. The fluctuations illustrated in Figure 4.1 demonstrate variations that are characteristic of stochastic quantum noise as well as correlations (streaks) arising from other structures in the phantom.

Noise is characterized in terms of its magnitude and correlation. The simplest description of noise magnitude is the standard deviation of voxel values within an otherwise uniform region (Barrett et al. 1976; Chesler et al. 1977). This is the classic metric of image noise common to computed tomography (CT) quality assurance and is the value entering the denominator of the contrast-to-noise ratio, below. The variance is simply the standard deviation squared. Noise correlation refers to the extent to which fluctuation in one voxel correlates with fluctuation in its neighbors. As detailed below, a common metric of noise correlation (and magnitude) is the noise-power spectrum (NPS).

Noise in CBCT is nonstationary—i.e., dependent on position in the image. The spatial dependence (nonstationarity) of the noise owes to a multitude of factors, such as the variation in the mean number of x-rays (and therefore the variance in the number of x-rays) across the detector. For example, the mean (and variance) in fluence at the center of the detector (i.e., in the deepest shadow of the patient) is different from that at the edge of the detector (in the unattenuated beam). This particular source of nonstationarity results in noise that is typically greater at the center of the image than at the edge (Riederer et al. 1978; Hanson 1979; Pineda et al. 2008; Baek and Pelc 2011b). For example, the image in Figure 4.2 illustrates that both the magnitude and correlation of noise vary from center to edge.

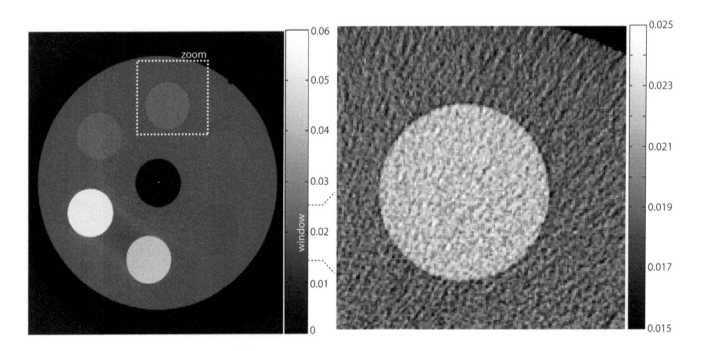

Figure 4.1 Axial CBCT image of a cylindrical phantom containing a variety of tissue-simulating plastic rods. The zoomed in image illustrates fluctuations associated with stochastic quantum noise as well as correlations (in this case, streaks) associated with other structures in the phantom (namely, the high-contrast rod in the 8 o'clock position).

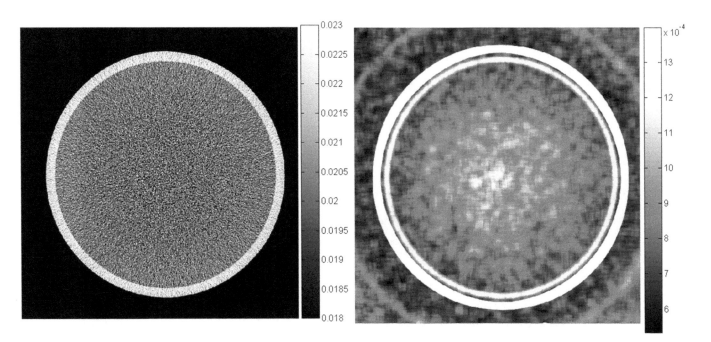

Figure 4.2 (See color insert.) (Left) Axial CBCT image of a uniform water phantom. (Right) Map of the noise (standard deviation in voxel values). Note the variation in noise magnitude and correlation within the cylindrical enclosure between the center and edge of the phantom. In this case, the phantom was imaged without a bow-tie filter, so the nonstationarity in quantum noise (i.e., higher noise at the center of the image) is attributed to variation in fluence incident on the detector between the center and edge. The colorbar shows voxel value (or noise) in units of attenuation coefficient (mm⁻¹).

Noise correlation is quantified by the image NPS, which may be analyzed from the 3D discrete Fourier transform (DFT_{3D}) of "noise-only" images:

$$\text{NPS}(f_x, f_y, f_z) = \frac{a_x}{N_x} \frac{a_y}{N_y} \frac{a_z}{N_z} \left\langle \left| \text{DFT}_{3D} \left[\Delta\mu (x, y, z) \right] \right|^2 \right\rangle . \quad (4.1)$$

The noise-only image $\Delta\mu(x,y,z)$ in Equation 4.1 implies subtraction of the mean voxel value and "detrending" of the data to minimize any slowly varying background trends (e.g., shading artifacts). A common means of detrending is to acquire a pair of CBCT images of the same object in succession, subtract the two, and correct Equation 4.1 by a factor of one-half (because subtraction of independent distributions doubles the variance). The angle brackets denote an ensemble average—i.e., the mean DFT_{3D} (modulus squared) computed over a number of realizations. Such realizations can be drawn from a large number of independent CBCT images acquired in succession or, more commonly, from smaller regions of interest (ROIs) within a given CBCT image. The terms a_x, a_y, and a_z are the voxel size in the (x,y,z) spatial domain and N_x, N_y, and N_z are the corresponding side-lengths (number of voxels) of the ROIs forming the ensemble. The terms f_x, f_y, and f_z are the Fourier domain coordinates, with (f_x, f_y) denoting the "axial," (f_y, f_z) the "sagittal," and (f_x, f_z) the "coronal" planes of the 3D Fourier domain. The NPS carries units (signal²)(mm³), and for CBCT reconstruction of attenuation coefficient the units are (mm⁻¹)²(mm³). To avoid confusion in canceling (mm) terms, this may be denoted (µ²)(mm³) or alternatively as (HU²)(mm³). At the time of writing, there is no established standard for NPS measurement or analysis

methodology, but typical methods include detrending by subtraction of CBCT images and an ensemble consisting of at least ~100 ROIs of side-length at least ~64 voxels.

The NPS is a 3D quantity. The DFT in Equation 4.1 is a 3D transform operating on 3D image data $\Delta\mu(x,y,z)$, and it is *incorrect* to apply a two-dimensional (2D) Fourier transform to 2D slices of the image, or more specifically, if one should choose do so, the result is *not* the image NPS, is *not* the term relevant to the noise-equivalent quanta (NEQ, see below), and does *not* carry meaningful units (signal²mm³) of a 3D image NPS (Siewerdsen et al. 2002). Analysis on slices extracted from the volume ignores correlation in the orthogonal direction and is subject to bias (unless corrected by scale factors associated with the bandwidth integral in the orthogonal direction). One may be quickly convinced of this by considering the analogy in projection radiography, for which the NPS is a 2D function analyzed from 2D realizations, and extracting individual rows from the image for one-dimensional (1D) NPS analysis ignores correlation orthogonal to the row and yields a biased result that is not the image NPS. An example 3D NPS measured from a CBCT image of a water cylinder is shown in Figure 4.3, illustrating that the NPS is nonisotropic in three dimensions—specifically, a ramp + low-pass filter characteristic in the f_x and f_y directions (i.e., the axial domain) compared with a low-pass characteristic in the f_z direction. The axial domain NPS is therefore (approximately) rotationally symmetric, whereas the sagittal and coronal NPS exhibit a distinctly asymmetric noise characteristic (Siewerdsen and Jaffray 2000b, 2003; Siewerdsen et al. 2002; Siewerdsen et al. 2004; Tward and Siewerdsen 2008, 2009). Note also that the 3D integral of the NPS over the Nyquist region of the Fourier domain is equal to the variance, that is, $\sigma^2 = \iiint \text{NPS} \, df_x \, df_y \, df_z$.

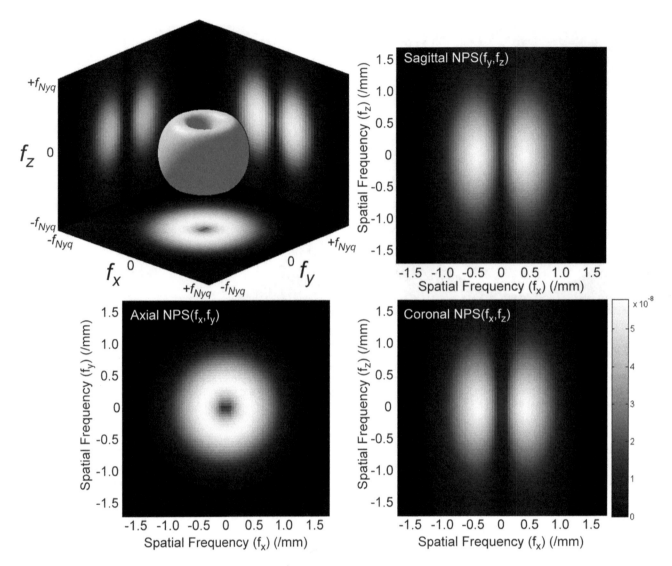

Figure 4.3 3D NPS analyzed from a CBCT image of a water phantom.

Because image noise is nonstationary, the NPS requires a "local" approximation. Analysis of the NPS includes an assumption of "weak" stationarity, meaning that the first and second moments of the distribution of voxel values are independent of position ("strong" or "wide-sense" stationarity implies that all moments are spatially invariant; "cyclo-" stationarity suggests invariance under periodic displacements). Just as the magnitude of the noise (i.e., the second moment of the distribution) is seen to vary spatially in Figure 4.2, its correlation varies as well. Therefore, the ROI realizations forming the ensemble in Equation 4.1 should be drawn from a region carrying the same noise characteristic. A reasonable choice is to draw ROIs from an annulus at fixed distance from the center of reconstruction. Factors governing the stationarity of the noise include use of a bow-tie filter, size and shape of the phantom, variation in beam quality and intensity reaching the detector, number of projection views (view sampling), and detector lag (Pineda et al. 2008).

The NPS is a meaningful noise metric, but alternative metrology can be envisioned. As a descriptor of the variance of the Fourier coefficients of the noise, the NPS captures the magnitude and correlation of noise and is an increasingly prevalent metric of CBCT system evaluation and quality assurance. Equivalent quantitation is given in the spatial domain by the autocovariance function (ACF). A fuller description of image fluctuations is conveyed by the full covariance matrix of the (3D) discrete Fourier transform of noise-only images, of which the NPS is simply the diagonal. Such analysis can avoid the strong assumptions intrinsic to NPS analysis (i.e., that the covariance matrix is diagonal), and although calculation of the full covariance matrix can be a computationally daunting task, the off-diagonal elements may convey a richer description of higher order correlation and the degree of noise stationarity.

4.2.2 CONTRAST RESOLUTION

Contrast refers to the (mean) signal difference between two regions of an image, typically two large-area, uniform regions such as those of the plastic rods in Figure 4.1. By this definition, contrast is simply a signal difference that carries the same units as the image signal [i.e., attenuation coefficient or Hounsfield units (HU)]. Alternatively, one may encounter definitions of contrast normalized by the average signal value, but herein, we refer to contrast simply as a mean signal difference. The contrast in CBCT images is therefore given ideally by the difference in (mean) attenuation coefficient between two materials of interest: $C = \triangle\mu = \overline{\mu_1} - \overline{\mu_2}$, where the μ values refer to the effective linear attenuation coefficient for a given x-ray spectrum. In practice, however, contrast is reduced below this theoretical maximum by factors such as x-ray scatter (Siewerdsen and Jaffray 2001). Furthermore, because attenuation coefficient is a function of the energy of x-ray photons forming the image (i.e., kVp), so, too, is contrast dependent on beam energy.

Contrast resolution refers to the smallest detectable difference in attenuation coefficient and depends on the (mean) signal difference presented in the CBCT image ($C \le \triangle\mu$) in proportion to the image noise. Characterized classically in terms of the contrast-to-noise ratio (CNR) or more specifically as the signal-difference-to-noise ratio (SDNR), such simple metrics of contrast resolution provide a basic description of imaging performance with respect to large, low-contrast structures (e.g., muscle–fat discrimination). SDNR is a "large-area" transfer characteristic (i.e., meaningful in reference to structures with size much greater than the correlation length associated with the transfer characteristics of the imaging system) and is applicable to "small-signal-differences" (i.e., contrast well within an order of magnitude of the noise such that SDNR is in the range ~0.5–5.0). Seminal work (Rose 1957) in signal detection theory invoked the CNR as a measure of detectability, spawning the often-quoted "Rose criterion," suggesting visibility of structures for which SDNR $\ge$ 3–5. Depending on factors such as the degree of noise correlation, background clutter, and the imaging task, observer studies in CBCT show that large-area, low-contrast structures may be detectable at considerably lower levels of SDNR, for example, $\ge$~1.5 as illustrated below (Tward et al. 2007). Such rules should be recognized as coarse approximations and exercised with careful recognition of the complexities of visual signal detection. Still, SDNR provides a useful metric of large-area, low-contrast imaging performance that may be tested simply

in phantoms and measured as a function of the multitude of factors governing image quality (see Section 4.3).

4.2.3 SPATIAL RESOLUTION

Spatial resolution refers to the smallest detectable feature size in an image and describes the extent to which fine detail structures may be discriminated. The term typically pertains to high-contrast features in the image or, conversely, images in the limit of zero noise. Simple scalar metrics of spatial resolution are sometimes invoked in terms of the correlation length (width of the point-spread function) or the highest spatial frequency (line pairs per millimeter, lp/mm) that can be readily distinguished in an image. Subjective assessment of spatial resolution in such terms can be obtained from the image of a line-pair pattern presenting high-contrast structures at various lp/mm as in Figure 4.4. More complete functional metrics of spatial resolution include the impulse response function [e.g., the point spread function (PSF), line spread function (LSF), or edge spread function (ESF)] or its Fourier counterpart, the modulation transfer function (MTF).

Spatial resolution can be quantified in terms of the response to a known input. Under the assumption of a linear, shift-invariant (LSI) system, spatial resolution may be described in terms of input–output relationships common to digital signal processing—i.e., the transfer function derived from Fourier transform of the impulse response function. In CBCT, common "impulse" functions include a small, high-contrast sphere [the image of which is an approximation of the PSF(x,y,z)]; a long, thin wire oriented at a slight angle to the z axis [from which LSF(x,y) may be derived]; a thin foil (a slice of which also yields the LSF within a given plane); a flat edge oriented at a slight angle with respect to a particular plane (the image of which gives the ESF); or a large sphere (from which the ESF may be analyzed in any direction perpendicular to the surface of the sphere). Any of these impulse response functions allows calculation of the transfer function (i.e., MTF) of the imaging system. The simplest and most prevalent test object for such is a

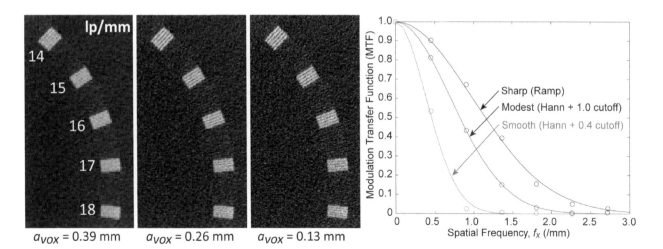

Figure 4.4 (Left) CBCT images of a line pair pattern common to quality assurance tests and subjective assessment of spatial resolution. The examples show CBCT images reconstructed at various voxel sizes, with finer voxels seen to improve the visibility of the higher lp/mm groups. (Right) MTF measured in the axial plane using a simple wire phantom. The various reconstruction filters are the same as used in the CBCT images of Figure 4.13.

phantom incorporating a long thin wire, in which case the LSF is given by Radon transform of the axial image (i.e., summation along either the x or y direction), and the MTF is given by Fourier transform. Numerous experimental details affect the accuracy of such MTF measurement, including the angle of the wire, interleaving (oversampling) of data from multiple slices, fitting of the LSF tails, and normalization of the LSF to unity area. This simple approach provides measurement of MTF(f_x) and/or MTF(f_y), and robust experimental methods for fuller evaluation of MTF(f_x,f_y,f_z) are under development. Example MTF measurements obtained from an angled wire phantom are shown in Figure 4.4.

Spatial resolution in CBCT is nonstationary. The assumption of shift-invariance is violated to a measurable degree within the axial plane and may be observed as degradation in MTF at greater distance from the center of reconstruction (i.e., center of rotation). More generally, the fully 3D MTF(f_x,f_y,f_z) depends on position within the 3D volume, most notably with respect to the z direction: the transfer function is null within the region of a cone about the f_z axis, with the angle of the null cone described by the angle between the x-ray source and the point of measurement [and tilt of the null cone determined by the (x,y) position] (Bartolac et al. 2009). The fully 3D transfer characteristic is therefore a fairly complex function; aside from slight variation in the central axial plane, it exhibits a null cone about the f_z axis that increases with distance $|z|>0$ and precesses about the f_z axis depending on the (x,y) position. The often quoted "cone-beam" artifact is, in fact, the manifestation of the null transfer characteristic of the imaging system in response to objects with signal power concentrated near $f_z \sim 0$ (the best example of which is a flat plate parallel to the axial plane; see Section 4.2.6).

Spatial resolution in CBCT is nearly isotropic. CBCT [and for that matter multidetector computed tomography (MDCT)] is often ascribed to provide "isotropic" spatial resolution, owing to the fact that the detector pixels are square, thereby overcoming the limitations of (single-slice) axial CT in which spatial resolution in the z direction was significantly less (~10 mm slice thickness) than that in the axial plane. Such ascription as "isotropic" also refers to the fact that CBCT voxels are typically cubic with equal side-length (e.g., 0.5 × 0.5 × 0.5 mm³), and "slice averaging" is an option in front-end visualization software, rather than a fundamental limitation of the detector/acquisition system. The evolution to (nearly) "isotropic" spatial resolution was a tremendous advance in helical MDCT and is similarly a strength of CBCT, offering multiplanar/oblique views and volumetric renderings without the coarse "stacked slice" appearance of classic axial CT. However, CBCT is only *nearly* isotropic, and a variety of factors impart distinct transfer characteristics in the (x,y) and z directions. First and simplest is the "smoothing" filter (also termed the apodization filter) that is optionally applied to counter the amplification of high-frequency noise by the ramp filter; because the ramp filter is applied only in the u direction of the detector (i.e., along detector rows), so, too, are the corresponding smoothing filters. As a result, the system MTF includes the transfer characteristic of the smoothing filter only in the axial (x,y) plane of CBCT reconstructions, and unless a symmetric 2D smoothing filter is used, such may be absent in the z direction.

Thus, the projection data are often filtered nonisotropically, and the resulting 3D MTF is therefore nonisotropic. More fundamentally, the MTF may vary with direction (and position) owing to the finite number and distribution of projection views acquired, as discussed earlier with respect to nonstationarity. Finally, and most fundamentally in the context of 3D FBP from a circular source–detector orbit, the MTF in the f_z direction is technically zero, and the fully 3D MTF(f_x,f_y,f_z) exhibits a null cone about the f_z axis as described earlier with respect to nonstationarity and the cone-beam artifact.

4.2.4 NOISE-EQUIVALENT QUANTA

The NEQ combines spatial-frequency-dependent measures of spatial resolution and noise. It is therefore a measure of the "fidelity" of signal transfer at various spatial frequencies in the imaging system. Carrying units of fluence, it can be interpreted as the effective number of photons (per unit area) that are used (at any particular spatial frequency) in forming the image. In the limit of high dose or zero electronic noise, NEQ is proportional to dose. Normalizing the NEQ by the total fluence yields the detective quantum efficiency (DQE), or the fraction of photons that were used (at any particular spatial frequency) in forming the image. The NEQ is widespread in application to 2D projection x-ray imaging systems and was applied to CT in early work by Hanson and others (Hanson 1979). For CBCT, the 3D NEQ is given by the ratio of the MTF and NPS:

$$\mathrm{NEQ}(f_x,f_y,f_z) = \Theta_{arc}\, f\, \frac{\mathrm{MTF}^2(f_x,f_y,f_z)}{\mathrm{NPS}(f_x,f_y,f_z)}, \qquad (4.2)$$

where Θ_{arc} is the extent of the source–detector orbit ($\Theta_{arc} = \pi$ for CBCT; $\Theta_{arc} < \pi$ for tomosynthesis) (Tward and Siewerdsen 2008, 2009; Gang et al. 2010, 2011). The MTF in the numerator includes that of the detector as well as the transfer characteristics of any interpolation or smoothing filters applied in 3D reconstruction. The NPS in the denominator includes the quantum noise (typically ramp-filtered random noise normalized per unit fluence) as well as the additive electronic noise of the detector. The proportionality of NEQ to dose is evident in that the (normalized) NPS in the denominator is inversely proportional to dose. "Generalized" forms of the NEQ (denoted GNEQ) also include a power-spectral density term in the denominator associated with background anatomical clutter (Bochud et al. 2000; Metheany et al. 2008; Engstrom et al. 2009; Gang et al. 2010; Reiser and Nishikawa 2010), typically of power-law form K/f^β, along with focal spot and scatter terms in the MTF (Kyprianou et al. 2004, 2005a, 2005b):

$$\mathrm{GNEQ}(f_x,f_y,f_z)$$

$$= \Theta_{arc}\, f\, \frac{\mathrm{MTF}^2_{detector}\mathrm{MTF}^2_{spot}\mathrm{MTF}^2_{scatter}}{\mathrm{NPS}_{quantum} + \mathrm{NPS}_{electronics} + \mathrm{MTF}^2\mathrm{NPS}_{background}}, \qquad (4.3)$$

where MTF = MTF$_{detector}$MTF$_{spot}$MTF$_{scatter}$. As noted earlier, the NPS has units (signal²)(mm³), denoted (μ²)(mm³). The NEQ therefore has units (mm⁻²), interpreted as (photons/mm²).

4.2.5 DETECTABILITY INDEX

Metrics of noise, contrast, and spatial resolution can be combined in simple models of detectability, and a key element to such simple models is quantitation of the imaging task. The simplest approach is derived from signal detection theory in which the imaging task is modeled as a difference of two hypotheses (International Commission on Radiation Units and Measurements 1996) - i.e., the task is for the observer to decide to which of two classes an image belongs. The task of classification therefore amounts to deciding if the image more closely resembles hypothesis 1 ($h_1(x)$) or hypothesis 2 ($h_2(x)$), and the difference ($h_1(x) - h_2(x)$) conveys the structures relevant to the task. In the Fourier domain, the difference presents a spectrum denoted $W_{task}(f) = FT[h_1(x) - h_2(x)] = H_1(f) - H_2(f)$. The term W_{task} is termed the "task function" and describes the spatial frequencies of interest in accomplishing a hypothesis-testing task. It also carries the contrast (magnitude of signal difference) associated with the structure of interest: $W_{task} = C_{task}F_{task}(f)$. Example tasks include the detection of an object against a uniform background, for which $h_1(x)$ is the object function and $h_2(x) =$ constant, and discrimination of two objects (International Commission on Radiation Units and Measurements 1996). The task function therefore weighs low frequencies for the detection of a large object against a uniform background, or weighs middle and high frequencies for discrimination of fine detail structures. As described below, the task function is applied as a weighting function on the spatial-frequency-dependent factors of NPS and MTF to quantify the spatial frequencies required for accomplishing a task versus the spatial frequencies that the imaging system can actually transfer.

The detectability index integrates the noise and spatial resolution of an imaging system with the imaging task. The simplest such model weighs the MTF and NPS (i.e., the NEQ) by the task function and integrates over the 3D Nyquist regime:

$$d'^2 = \iiint_{Nyquist} \frac{MTF^2(f_x, f_y, f_z)W_{task}^2(f_x, f_y, f_z)}{NPS(f_x, f_y, f_z)} df_x\, df_y\, df_z. \qquad (4.4)$$

The form in Equation 4.4 amounts to the so-called prewhitening (PW) ideal observer model, and a number of straightforward variations on such a model can be considered, including the non-prewhitening (NPW) model and variations that include a model of the human eye (denoted PWE and NPWE) in both its spatial-frequency-dependent transfer characteristic and internal noise. The 3D integral in Equation 4.4 suggests an observer that is somehow able to fully appreciate ("grok") volumetric 3D data—a model that may bear limited correspondence to a real observer scrolling 2D slices and relying on memory to form a 3D mental image. Display mechanisms and volumetric renderings that aim to capture the full dimensionality of the data similarly introduce departures from the simple, ideal 3D form of Equation 4.4. Accordingly, the model can be extended to a 2D form of "slice" detectability corresponding to performance in viewing a single slice extracted from the 3D volume, namely, by integration of the numerator and denominator over the

spatial frequency direction orthogonal to the extracted slice. Extracting an axial slice, for example, convolves over the orthogonal (f_z) direction, such that detectability in a single axial slice is:

$$d'^2_{axial} = \iint \frac{\int MTF^2(f_x, f_y, f_z)W_{task}^2(f_x, f_y, f_z)\, df_z}{\int NPS(f_x, f_y, f_z)\, df_z} df_x\, df_y. \qquad (4.5)$$

Note that the fully 3D signal and noise transfer characteristics of the imaging system are still intrinsic to the "slice" detectability, and correlations orthogonal to the extracted slice are handled by the (single) integral over the orthogonal direction.

Models of detectability can be "generalized" to include factors other than quantum noise. Factors that may impart significant effect on detectability include, for example, x-ray focal spot size (focal spot blur); x-ray scatter (loss of contrast and addition of additive quantum noise); and anatomical background clutter, which may diminish conspicuity of the structure of interest. Among the earliest and more widely investigated forms of generalization is the "generalized NEQ" (Barrett 1990) as mentioned earlier, in which anatomical background clutter is characterized in terms of its power spectral characteristics and included in sum with the quantum NPS in the denominator of the NEQ. Application of the generalized NEQ to detectability index provides a fairly comprehensive objective function for system optimization (Prakash et al. 2011):

$$d'^2 = \iiint_{Nyquist} \frac{MTF_{detector}^2 MTF_{spot}^2 MTF_{scatter}^2 W_{task}^2}{NPS_{quantum} + NPS_{electronics} + MTF^2 NPS_{background}} df_x\, df_y\, df_z. $$

$$(4.6)$$

Note that this generalized form can be similarly extended to PWE, NPW, NPWE, and "slice" variations. In the following section, the broad spectrum of physical and mathematical factors that bear on CBCT image quality are surveyed. In each case, readers are encouraged to consider these factors with respect to Equation 4.6 as a general touchstone for understanding how each factor weighs in—often in more than one term and imparting trade-offs among, for example, noise, spatial resolution, and contrast.

Detectability index provides a metric for system optimization that demonstrates basic agreement with real observer performance for simple imaging tasks. It has been used in applications such as analysis of optimal scintillator thickness, system geometry, detector selection, tomosynthesis angular range, and overall system design. Although such a simple model does not pretend to capture the complexities of real human observer response, nor the complexities of diagnostic decision making, which clearly go well beyond a simple binary hypothesis test of A versus B, such modeling does provides an abstract, idealized starting point, and, as Chomsky reminds us, is the basis of intellectual pursuit. It is furthermore encouraging that such models actually bear correspondence to real observer performance for simple imaging tasks, for example, in the analyzing effect of tomosynthesis angular range on detectability

(Gang et al. 2011) and therefore provide a reasonable basis for analyzing performance trade-offs in imaging system design and optimization.

4.2.6 ARTIFACTS

CBCT exhibits every form of image artifact as MDCT—and more. The scientific literature and textbooks are replete with excellent descriptions of the physical/mathematical basis of such artifacts, as well as the methods by which such artifacts may be mitigated by a variety of image processing and correction methods (Glover and Pelc 1980; Hsieh 1998; De Man et al. 1999; Siewerdsen and Jaffray 1999; Barrett and Keat 2004). It is worth noting that because CBCT is relatively new in clinical applications in comparison with MDCT, it has not enjoyed the decades of effort invested in artifact correction as intrinsic to any diagnostic MDCT scanner. CBCT images in many areas of new application are correspondingly "raw," and direct comparison with MDCT should be mindful of more sophisticated (sometimes proprietary) correction algorithms that may be exercised in the latter. Some of the more prevalent and significant artifacts that can limit image quality in CBCT are summarized later, with examples shown in Figure 4.5.

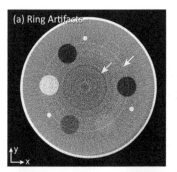

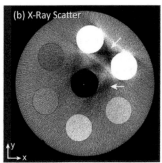

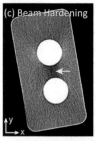

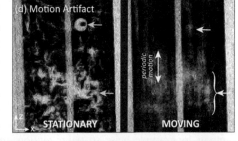

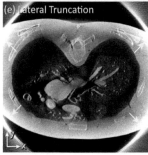

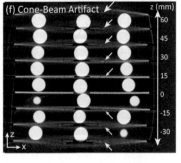

Figure 4.5 Example CBCT images illustrating various forms of image artifact. (a) Ring artifacts. (b) X-Ray scatter. (c) Beam hardening. (d) Motion artifact. (e) Lateral truncation. (f) Cone-beam artifact. Here and throughout this chapter, the (x,y) domain is the axial plane, and the (y,z) or (x,z) domains are the sagittal or coronal planes, respectively.

Ring artifacts arise from uncorrected variations in detector dark current, gain, linearity, and defects. Because the erroneous signal is from a fixed location on the detector, it presents a fixed (i.e., position independent of angle) defect in the sinogram and reinforces in backprojection at a fixed radius in the reconstruction, that is, on a circle. As shown in Figure 4.5a, ring artifacts can be bright or dark, depending on the direction (polarity) of the erroneous signal. The most common method to minimize ring artifacts is the application of robust offset-gain-linearity corrections based on previously acquired dark-field calibrations, flood-field calibrations, measurement of detector linearity on an element-by-element basis, and mapping of detector elements that exhibit defective signal (e.g., dead or noisy). Postprocessing methods also can be applied to selectively identify and filter circularly correlated structure in the reconstruction.

X-ray scatter is a significant source of artifact due to high scatter fractions in a volumetric beam. The scatter-to-primary ratio (SPR) at the detector depends on system geometry (air gap), object size, and field of view (FOV), with SPR exceeding 100% not uncommon in CBCT - i.e., the majority of photons arriving at the detector have undergone at least one scattering event. The resulting artifacts are of two forms: cupping or low-frequency shading across the image and streaks, particularly between heavily attenuating structures. As illustrated in Figure 4.5b, the dark shading and streaks associated with x-ray scatter (i.e., an underestimate of the true attenuation coefficient) are the result of extraneous fluence at the detector (increased detector signal due to scatter). Bright streaks also are observed in Figure 4.5b about the dark streak artifacts and are attributed to error arising from the ramp filter applied to projection data that are inconsistent along views with low or high x-ray scatter. Both coherent and incoherent scatter contribute to such artifacts in varying degrees and exhibiting low-frequency as well as high-frequency structure, depending on the system geometry. Physical methods to minimize x-ray scatter artifacts include a large air gap, minimizing the FOV, and incorporation of an antiscatter grid. Scatter correction methods exist in many forms, most seeking to correct the measured (scatter+primary) fluence by estimation (and subtraction) of the scatter-only fluence. Scatter estimates may be based on direct measurement (e.g., placement of beam blockers within the FOV), parameterization based on prior experimentation/calibration, analytical models, and Monte Carlo simulation (Boone and Sibert 1988a,b; Siewerdsen and Jaffray 2001; Kwan et al. 2005; Wiegert et al. 2005; Jarry et al. 2006; Lei et al. 2006; Siewerdsen et al. 2006; Zbijewski and Beekman 2006; Kyriakou and Kalender 2007; Maltz et al. 2008).

Beam hardening artifacts result from changes in the x-ray spectrum transmitted through different materials and thicknesses within the patient. Such artifacts are evident as shading (cupping) throughout the center of large objects (long ray paths), brightening at the periphery of the object, and streaks between highly attenuating structures within the object (De Man et al. 2001). They are therefore difficult to differentiate from x-ray scatter artifacts that appear with similar characteristics. Beam hardening artifacts, however, arise from the shift in energy of a polychromatic x-ray spectrum transmitted through the object. Longer path-lengths impart a greater degree of hardening (i.e., shift to higher mean energy and therefore lower attenuation coefficient) and appear correspondingly darker in the reconstruction. For example, Figure 4.5c exhibits a beam-hardening artifact between two regions of dense material along the long axis of the object. As mentioned earlier, bright streaks associated with ramp filtering of inconsistent projection data also may be evident. Methods to mitigate beam-hardening artifacts include physical and algorithmic approaches: use of a bow-tie filter; added filtration (e.g., Cu) to "preharden" the beam; calibration of energy-dependent response using cylindrical/wedge phantoms; and model-based reconstruction methods that incorporate the x-ray spectrum in the system model.

Motion artifacts arise from organ movement during the CBCT scan. Because most current CBCT systems involve a fairly long acquisition time (e.g., ~10–60 s for a single rotation compared with ~0.3 s for MDCT), there is significant potential for involuntary organ motion during the scan. Such motion can be approximately periodic (e.g., respiratory or cardiac) or aperiodic (e.g., peristalsis) and rigid or nonrigid. The result is an apparent blur of structures in the direction of motion, as illustrated in Figure 4.5d where a phantom containing lung-simulating tissue and a plastic sphere were imaged in a stationary state (left) and during ~10-mm-amplitude periodic motion during the scan (right). Motion correction methods include (Sonke et al. 2005; Li et al. 2006a,b, 2007; Rit et al. 2008, 2009; Tang et al. 2010): prospective gating in which projection data are acquired only within periodic intervals at which the phase of motion is consistent, for example, derived from a ventilator or electrocardiogram (ECG); retrospective gating in which long, slow projection sequences are retrospectively sorted according to the phase of motion, for example, reconstruction at multiple phases of breathing motion in 4D CBCT; and motion-compensated reconstruction in which a model for 4D motion is estimated and applied in 3D backprojection along "curved" rays (Sonke et al. 2005; Li et al. 2006a, 2006b, 2007; Rit et al. 2008, 2009; Tang et al. 2010).

Truncation artifacts result from a FOV smaller than the lateral extent of the patient. The artifact is evident as a bright ring about the periphery of reconstruction and a dark cupping within as illustrated in Figure 4.5e. A variety of methods exist for correcting truncation artifacts, including extrapolation of the projection data at truncated edges by a cylindrically symmetric or roll-off function (e.g., exponential) tending to zero at some distance outside the lateral FOV. More sophisticated methods in the context of "local tomography" aim to reconstruct the interior problem exactly (Yu and Wang 2009; Wang and Yu 2010; Yang et al. 2010; Xu et al. 2011; Yu et al. 2011).

Image lag artifacts are a result of frame-to-frame residual signal in the detector. Flat-panel detectors are known to exhibit some degree of residual signal frame to frame, for example, at the level of a few percent. In fluoroscopy, the effect is visible in subsequent frames as image lag, and in CBCT, it is responsible for an azimuthal "comet" artifact as well as possible streaks along the first few projection angles (Siewerdsen and Jaffray 1999). For prevalent designs of flat-panel detectors in which each detector element includes an amorphous silicon photodiode and thin-film transistor (TFT), the source of such residual signal is primarily charge trapping and release in metastable states of the amorphous silicon semiconductor in the

photodiode and, to a lesser extent, the TFT. Other potential sources of residual signal include incomplete charge transfer (e.g., reading out each line of the detector for less than some multiple of RC time constants) and afterglow in the scintillator. Although lag artifacts are often fairly subtle (associated primarily with high contrast structures at greater distance from the center of reconstruction), correction of lag artifacts has been demonstrated through subtraction of previous frame signal weighted by the temporal response function. An effect distinct from image lag is "ghosting", which refers to a change in detector gain depending on the history of exposure, for example, reduction in gain after prolonged exposure to high signal. Ghosting artifacts appear as bright- or dark-shaded regions primarily about the periphery of the patient where the detector experiences large swings in incident exposure.

The cone-beam artifact is a result of incomplete sampling of a volume from a circular source–detector orbit. It appears as bright or dark smearing of signal from edges that are parallel to the plane of source–detector orbit and worsens at increasing distance from the central axial plane. The classic example is a stack of disks, where coronal or sagittal images show a faithful representation of disks near the central plane but degrade significantly at increasing z. Figure 4.5f shows an example coronal image of a phantom incorporating a stack of disks separated by plastic spheres. Note the faithful reconstruction of the disk near the central plane, the degradation in the disk image at increasing z, and the seeming lack of such artifact on the spheres. Why then do flat disks suffer such a significant effect, whereas spheres or other structures (rich in 3D spatial frequency content) exhibit little artifact? The answer resides in incomplete sampling of the full 3D Fourier domain, specifically in a "null cone" (of angular extent determined by the cone angle) about the f_z axis. The effect is due to incomplete sampling of the volume from a circular source–detector trajectory, namely, Tuy's condition for complete reconstruction that every plane passing through a given point in the reconstruction must intersect the source–detector orbit. For CBCT from a circular source–detector orbit, every location above or below the central axial plane fails this condition and is subject to the cone-beam artifact. The incomplete sampling is evident in a "null cone" about the f_z axis in the Fourier domain, and as mentioned earlier in relation to the 3D MTF, the angle subtended by the null cone is equal to its angle above the central axial plane. A variety of noncircular source–detector trajectories have been proposed to mitigate the cone-beam artifact, including circle-and-line, double-arc, sinusoid-on-a-cylinder, and others. For anatomical structure that tends to be rich in 3D spatial frequency content (e.g., trabecular detail), the conspicuity of the cone-beam artifact and its influence on image quality can be (perhaps surprisingly) minor, with the exception being flat, high-contrast structures such as the superior and inferior surfaces of vertebrae and the superior aspect of the tibial plateau.

4.2.7 A FEW REMARKS ON TERMINOLOGY

A few brief remarks on terminology deserve clarification and consistency for the remainder of the chapter, if not for daily life in interaction among imaging physicists, radiologists, computer scientists, and other professionals for whom medical imaging is becoming a common basis of communication.

Images are numbers, not anatomy. Just as Magritte's painting is not a pipe, neither is a CBCT image of a prostate a prostate. An image is a matrix of numbers, and for CBCT it is a 3D matrix. In this respect, the purpose of this chapter is to describe how these numbers may be subject to fluctuation and correlation and to what extent a signal or stimulus (i.e., a specific pattern of numbers) may be detected within the matrix.

The element of a 3D CBCT reconstruction matrix is the "voxel." It is a 3D construct with width (a_x), length (a_y, with $a_x = a_y$ typically), and depth (a_z, commonly referred to as slice thickness). Even when a single 2D slice is extracted from a 3D volume, its elements are properly termed "voxels," because they carry an implicit (out-of-plane) depth and three-dimensionality. **An element on a 2D detector is the "dexel."** The term *dexel* refers to a physical 2D detector element with width (a_u) and height (a_v). **An element of a 2D picture (image) portrayed on a 2D display is the "pixel."** It has width and height characteristic of the display system. A CBCT system involves all three elements, and it is worthwhile keeping terminology distinct and consistent.

Cuiusmodi resolution? Often the word "resolution" is used casually in reference to "spatial resolution." However, given that imaging systems present the ability to "resolve" a variety of measures beyond space, it is valuable to carry the complete term consistently in each instance, for example, spatial resolution, contrast resolution, temporal resolution, and energy resolution.

Noise is stochastic. Artifacts are deterministic. The former term is best reserved for effects that are truly random and unreproducible, whereas the latter is better applied to systematic, reproducible effects that arise as a systematic consequence of a given imaging system applied to a given object. The best example of noise is quantum noise, for example, Poisson-distributed photons impinging on detector elements in a random process that in turn imparts a random distribution in the image signal. Examples of artifact abound and include shading associated with a constant offset in the projection data and streaks associated with finite sampling of a sharp edge in the object. "Streaks," for example, may arise from noise (e.g., strongly correlated noise associated with photon starvation through the lateral direction of an obese patient) or may be properly considered artifact (e.g., beam-hardening effects from a small, high-z object and a polyenergetic beam). Although it can be challenging to maintain precise terminology, consistent use has long-term benefit to the rigor of the field, particularly in regard to image interpretation where ambiguity is a potential liability, and an unambiguous, quantitative perspective on terminology is a strength.

4.3 FACTORS AFFECTING IMAGING PERFORMANCE

The previous section outlined the *metrics* of CBCT imaging performance; this section surveys the *factors* that govern it. These factors include both physical and mathematical factors, ranging from the geometry of the system to the acquisition and reconstruction technique. A broad summary of trends is provided in Figure 4.6 (effect of system configuration on imaging performance) and Figure 4.7 (effect of acquisition and reconstruction parameters on imaging performance). For each factor, the effect on metrics of contrast, noise, spatial resolution,

detectability, and possible artifacts is summarized. In subsequent sections, these trends are illustrated in images of simple and anthropomorphic phantoms imaged on an adjustable CBCT bench allowing free variation of, for example, system geometry, kVp, and dose.

4.3.1 SYSTEM GEOMETRY

System geometry affects every factor of imaging performance. Factors of system geometry include the source–detector distance (SDD, alternatively called source–image distance, SID), source–axis distance (SAD, marking the distance to the center of rotation or "isocenter"), and magnification (ratio of SDD to SAD). System geometry is governed foremost by clinical and logistical factors of the application (e.g., size of the patient and FOV) and in turn imparts significant effect on all aspects of image quality—contrast, noise, spatial resolution, and artifacts as summarized in Figure 4.6 (top row). For example, with all other system parameters held constant, increasing the system magnification tends to increase contrast (by increasing the object-to-detector air gap and thereby reducing x-ray scatter), increase noise (due to inverse-square-law reduction in fluence at the detector), improve spatial resolution up to a point (due to geometric magnification of the voxel size, $a_{xy} = a_u/M$) beyond which spatial resolution degrades due to focal spot blur, and reduce artifacts associated with x-ray scatter (due to a larger air gap). For a particular imaging task, the optimal system geometry can be analyzed by multivariate analysis of detectability index. For example, in development of CBCT for image-guided radiation therapy, it was shown that a magnification of M ~ 1.55 was optimal for relevant soft-tissue detection tasks in the presence of high x-ray scatter (e.g., pelvis) (Siewerdsen and Jaffray 2000b). In the development of a compact CBCT scanner for orthopaedics, in contrast, a magnification of M ~ 1.3 was shown optimal for bone and soft-tissue visualization in the context of musculoskeletal extremities (Prakash et al. 2011). The strong effect of system geometry on imaging performance is illustrated in Figure 4.8, where the same objects [a 16-cm water cylinder (top) and a 16-cm cylinder containing a variety of tissue-simulating inserts (bottom)] were imaged on an adjustable CBCT imaging bench in both a "compact" geometry (corresponding to an extremity scanner; Zbijewski et al. 2011) and in an "extended" geometry (corresponding to a mobile C-arm; Siewerdsen et al. 2005). In neither case was an antiscatter grid used; see Section 4.3.5. The compact configuration shows a marked increase in x-ray scatter artifacts (cupping and streaks) due to the smaller air gap (distance from object to detector), but it also demonstrates reduced noise due to higher fluence at the detector (shorter SDD). Despite the increase in artifacts, therefore, the compact configuration actually demonstrates increased SDNR, that is, the reduction in contrast (due to scatter) is outweighed by the reduction in noise (due to shorter SDD). By comparison, the extended geometry demonstrates significantly reduced cupping artifact, but the SDNR is lower due to decreased fluence at the detector (longer SDD). These results illustrate the multitude of image quality trade-offs associated with system geometry, with x-ray scatter among the most significant factors; therefore, the role of antiscatter grids (see Section 4.3.5) and scatter correction algorithms deserve careful consideration, and their

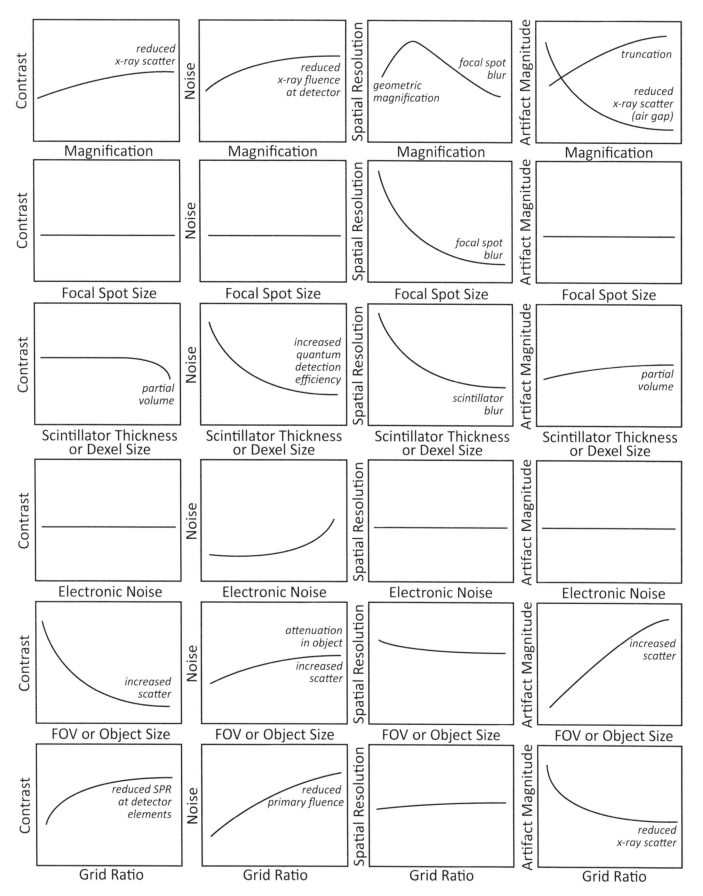

Figure 4.6 Effect of various factors of system configuration on imaging performance. Each plot gives a descriptive representation of the effect of a given parameter (where all other parameters are held constant).

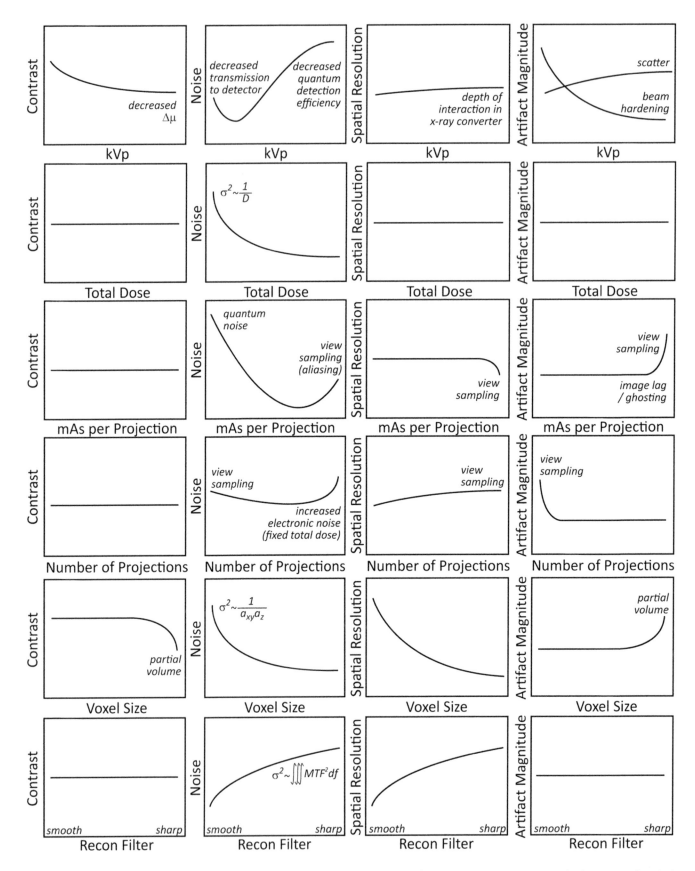

Figure 4.7 Effect of various acquisition and reconstruction parameters on imaging performance. As in Figure 4.6, each plot gives a descriptive representation of the effect of a given parameter (where all other parameters are held constant).

Fundamental principles and techniques

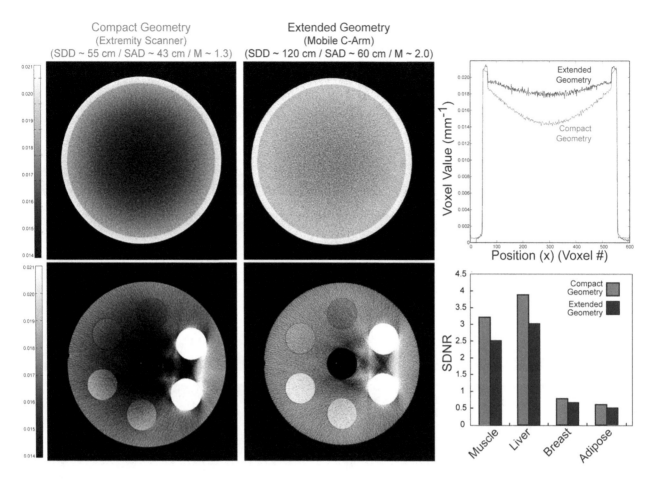

Figure 4.8 Effect of system geometry on CBCT image quality.

importance will, in turn, depend on the system geometry—e.g., more important for compact geometries, potentially reducing the scatter artifacts while maintaining high SDNR.

4.3.2 FOCAL SPOT SIZE

Focal spot size affects spatial resolution. Particularly for systems with higher magnification, focal spot blur can be a significant factor in system spatial resolution. Simple analytic forms can be used to relate the size of the focal spot to the MTF as a function of magnification—for example, a 2D rect of size a_{spot} giving a separable sinc in the 2D Fourier domain:

$$\text{MTF}_{spot}(f_u, f_v) = |\text{sinc}((M-1)a_{spot}f_u)||\text{sinc}((M-1)a_{spot}f_v)|. \quad (4.7)$$

For typical CBCT geometries ($1.3 < M < 2.0$) focal spot size of ~0.3–0.6 mm is required such that MTF is not dominated by focal spot blur. Note that although MTF_{spot} is a factor in the linear combination of MTFs comprising the system MTF, and it similarly appears in the numerator of the detectability index, it does *not* enter the (quantum noise) NPS. That is, although the quantum NPS is (approximately) proportional to other factors of system MTF (squared), it is independent of MTF_{spot}. The implication: unlike other sources of blur, focal spot blur does not reduce quantum noise. The reason is that x-ray photons produced from a distributed focal spot are still statistically uncorrelated (e.g., Poisson distributed)—just as with a point source; although the distributed source correlates "signal" (i.e., blurs the projection of some structure within the object), it does not correlate the quantum noise.

4.3.3 FIELD OF VIEW

The FOV affects contrast and artifacts. The benefit of larger FOV is obvious—a greater extent of the patient captured in the image. However, as the FOV is increased, either laterally (x,y) or longitudinally (z), the amount of x-ray scatter generated in the patient and reaching the detector increases. The primary effects are a loss in contrast and an increase in scatter artifacts (cupping and streaks) as illustrated in the fourth row of Figure 4.6. The lateral FOV (denoted FOV_{xy}) also relates to lateral truncation and corresponding truncation artifacts, although several correction algorithms exist that can mitigate the effect, for example, by "lateral extrapolation" of the projection data or more sophisticated solution of the "interior problem." The longitudinal FOV (denoted FOV_z) is defined by the cone angle, and it has been shown that the SPR at the detector increases in proportion to the cone angle such that, depending on the size of the object, the SPR at the detector can exceed 100%. Because a smaller FOV implies lower SPR [as well as dose-area product (DAP) and dose-length product (DLP)], the FOV should be minimized such that structures of interest are still captured

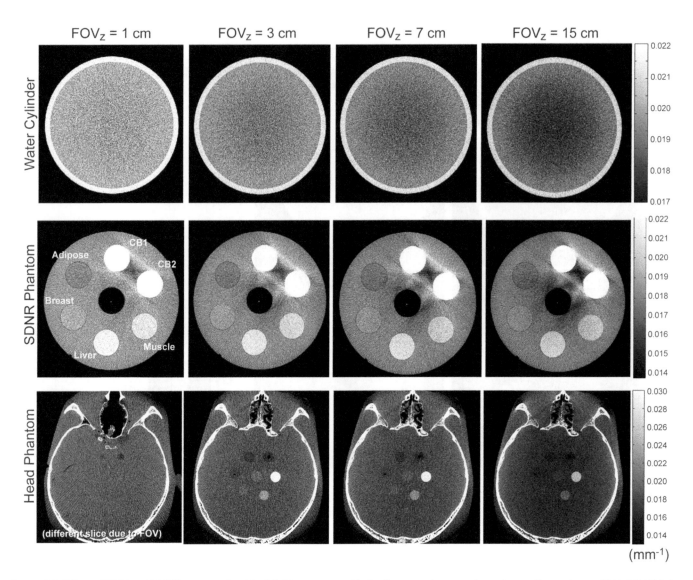

Figure 4.9 Effect of FOV on CBCT image quality. (Top) A 16-cm water cylinder illustrates increased cupping artifact at larger FOV. (Middle) An SDNR phantom containing a variety of tissue-simulating rods illustrates increased streak artifact and reduced contrast at larger FOV. (Bottom) An anthropomorphic head phantom illustrates both the increase in artifact and reduction in contrast within a semirealistic context.

within the image. High-quality lateral truncation correction and adjustable (or knowledgeably chosen) collimation are therefore important to a judiciously designed CBCT system. Example images acquired as a function of FOV_z are shown in Figure 4.9. Note the reduction in contrast and increase in scatter-related artifacts for larger FOV_z.

4.3.4 OBJECT SIZE

The object size affects noise, contrast, and artifacts. Increased object size (for fixed FOV and system geometry) results in a greater attenuation of the beam and therefore reduced detector signal and increased image noise. Similarly, a larger object increases the amount of x-ray scatter and therefore reduces contrast and increases the magnitude of scatter-related artifacts. It also results in a greater degree of beam hardening and potential for truncation artifacts. For soft-tissue imaging tasks, these are among the most challenging factors in CBCT image quality and are major challenges in applications such as CBCT-guided radiation therapy of tumor targets in the pelvis (e.g., prostate) and abdomen

(e.g., liver). The overall trends are similar to those associated with larger FOV, as illustrated in the fourth row of Figure 4.6.

4.3.5 ANTISCATTER GRID

An antiscatter grid affects contrast, noise, and artifacts. Antiscatter grids improve contrast (due to reduction in scatter fluence reaching the detector); increase noise (due to attenuation of the primary beam); and reduce scatter artifacts (shading and streaks), thereby imparting a trade-off in SDNR, dose, and artifacts. With proper flood-field correction and geometric calibration, subtle, residual gridlines in the projections have been shown to have fairly minor effect on CBCT reconstructions (Siewerdsen et al. 2004; Schafer et al. 2012). Whether an antiscatter grid has overall benefit on image quality depends on the system geometry (air gap), object size (scatter magnitude), and imaging task (soft-tissue visibility or high-contrast detail). For CBCT systems with a more extended geometry (e.g., M ~1.5–2.0 as with a mobile C-arm or linear accelerator gantry), the trade-off between improved contrast and increased noise associated with

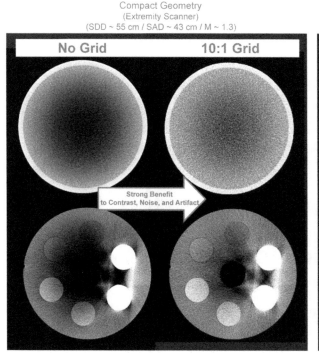
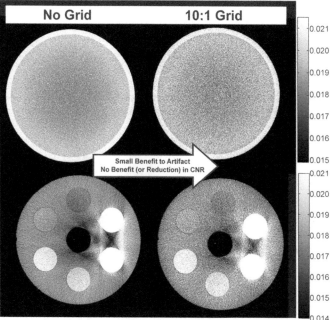

Figure 4.10 Effect of antiscatter grid on CBCT image quality.

grids has been shown to favor gridless operation, even for large objects (at fixed total dose). For more compact CBCT geometries, however (e.g., $M \sim 1.3$ as with a miniscanner dedicated to extremity imaging), the "race" between contrast and noise tends to favor the use of a grid. The effect of system geometry on the question of whether or not an antiscatter grid improves image quality is illustrated in Figure 4.10. For the extended geometry (at right), scatter artifacts are already fairly small (due to the larger air gap), and the grid imparts little or no improvement in SDNR—actually, a reduction in SDNR [and SDNR per unit sqrt(dose)]. In the extended geometry, therefore, the large air gap is the best first defense against x-ray scatter. For the compact geometry (at left), in contrast, the grid improves SDNR (without increase in dose) and significantly reduces the magnitude of scatter artifacts. In either case, there is still room for improvement in image quality through incorporation of a scatter correction algorithm, and given the large SPR characteristic of CBCT, such algorithms can be considered a requirement for high-quality imaging (with or without grids), particularly for soft-tissue imaging tasks.

4.3.6 BEAM ENERGY (kVp AND ADDED FILTRATION)

Beam energy affects contrast, noise, and artifacts. The energy of the x-ray beam is characterized not only by its peak energy (kVp) but also the shape (quality) of the spectrum (hardened by addition of added filtration, typically Al, Cu, Sn, or a combination). A higher energy beam will tend to reduce contrast, because the difference in attenuation coefficient, $\mu(E)$, between various soft tissues tends to decrease with energy. (Exceptions include lung tissues and materials with strong K-edge absorption.) A higher beam energy also tends to increase noise due to reduced quantum detection efficiency of the x-ray converter. A higher energy beam also exhibits a greater proportion of incoherent

scatter interactions and reduced photoelectric effect. All of these effects would tend to argue for a lower energy beam. However, a lower energy beam can result in increased noise due to reduced fluence transmitted to the detector. A softer beam (or more specifically a broader x-ray spectrum) also suffers increased beam-hardening artifacts, because the x-ray spectrum transmitted to the detector is significantly different from that in the unattenuated beam (I_0). The trade-offs in contrast, noise, and artifact at various kVp are illustrated in the first row of Figure 4.7 and in the example images of Figure 4.11. Optimization of kVp furthermore depends on the object size and the imaging task.

4.3.7 DOSE

The dose relates directly to image noise. The subject of CBCT dosimetry is an area of broad interest and ongoing standardization, and other chapters provide a more complete description of dose metrics and methodology. For present purposes, we can consider the dose in terms of a simple estimate derived from measurement of exposure at the detector and also in terms of absolute dose measured at the center of a uniform (water) cylinder. To estimate the dose (to the center of a water cylinder of radius R) from the exposure measured at the detector, we write

$$D(\text{mGy}) = N_{\text{projection}}\, mAs_{\text{projection}} \left(\frac{X}{\text{mAs}} \right) \left(\frac{\text{SDD}}{\text{SAD}} \right)^2 e^{\mu R}_{\text{water}}\ f_{\text{water}} \text{BSF},$$

(4.8)

where $N_{\text{projection}}$ is the number of projections, $mAs_{\text{projection}}$ is the mAs per projection, and (X/mAs) is the exposure per unit mAs (measured in air at the detector after attenuation by the object of radius R). The first three terms are therefore simply the

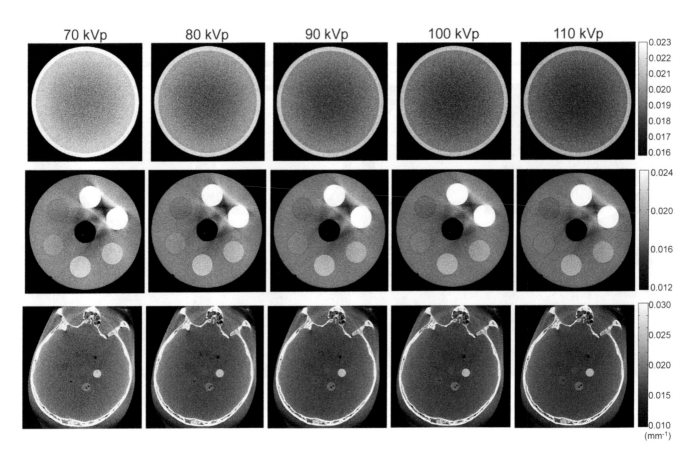

Figure 4.11 Effect of kVp on CBCT image quality. (Top) Images of a 16-cm water cylinder illustrate the change in attenuation coefficient with increasing kVp. (Middle and bottom) The SDNR and anthropomorphic head phantoms show trade-offs in contrast, noise, and artifact with kVp. All images are at approximately equivalent dose [absolute dose to the center of a 16-cm water cylinder ~0.4 milligray (mGy)].

total exposure at the detector in the shadow of the object. The (SDD/SAD)^2 terms scales to isocenter by the inverse square law, and the exp($\mu_{water}R$) term scales ("unattenuates") to the center of the water cylinder. The f_{water} term is the exposure-to-dose conversion, f-factor (cGy/R) for water, and BSF is the backscatter factor (i.e., scale factor for dose due to scatter in a volumetric beam). More direct measurement of the dose can be made using a small ionization chamber (e.g., 0.6-cm^3 Farmer chamber) at the center of a plastic (nearly water-equivalent) cylinder. In terms of CBCT image quality, the image noise is related to dose by the inverse-square-root ($\sigma^2 \propto 1/D$) that, in turn, depends on factors entering Equation 4.8, such as beam energy (kVp), FOV, object size, mAs/projection, and total number of projections. Similarly, the NPS scales in inverse proportion to dose. Increasing dose reduces quantum noise [and thereby improves contrast resolution (SDNR)], but it has no effect on spatial resolution (MTF) or contrast ($\Delta\mu$) as defined in this chapter. Presenting a brute force lever in image quality optimization, dose should be knowledgably selected (and minimized) with respect to the imaging task, clinically acceptable dose levels, and with recognition of the multitude of other factors that can be adjusted to maximize image quality.

4.3.8 mAs PER PROJECTION

The mAs per projection relates to image noise and artifacts. As discussed earlier, it is also a direct multiplier in the total dose. For a fixed number of projections, the dose scales with

mAs per projection, and the image noise is therefore simply related by the inverse-square-root. For a fixed total dose (i.e., fixed total mAs given by the product of mAs per projection and $N_{projection}$), the situation is more complicated, and the inverse-square-root relation of dose and noise can be dominated by effects of view sampling (in the limit of few projections) and electronic noise (in the limit of many projections). For example, at fixed total dose, decreasing the mAs per projection implies increasing the number of projections: view sampling effects may be correspondingly low, but the exposure to the detector in each projection may become so low that electronic noise arises as a significant contributor to the total noise. Conversely, increasing the mAs per projection implies reducing the number of projections, in which case electronic noise may be negligible, but view sampling effects increase when the number of projections is so low that the volume is angularly undersampled. The scenarios are illustrated in the third and fourth rows of Figure 4.7.

For systems employing a pulsed x-ray tube, mAs per projection is the product of tube current (mA) and exposure time (t_X), whereas for systems using a continuous fluoroscopy tube, mAs per projection is the product of mA and detector frame time (t_{frame}). Increasing the mAs per projection by increasing either t_X or t_{frame} can introduce angular (azimuthal) blur due to motion of the source–detector during the exposure (Siewerdsen and Jaffray 1999; Yang et al. 2007; Baek and Pelc 2011a). Similarly, technique selection that prolongs the total scan time raises the potential for artifacts associated with object motion.

Finally, the mAs per projection should be considered in relation to the sensitive range of the detector. As discussed earlier, a very low mAs per projection at the "toe" of detector response will increase the relative contribution of electronic noise. It also tends to increase the prominence of ring artifacts associated with imperfect offset-gain corrections. At the other extreme, a very high mAs per projection at the "shoulder" of detector response tends to artifacts associated with increase image lag (azimuthal "comet" artifact) and detector nonlinearity (ring artifacts). For very large body sites (e.g., imaging the prostate in an obese patient), there is sometimes the need to increase the mAs per projection to achieve a reasonably high detector signal in the deepest shadow of the patient (i.e., "drill for SNR"). In the absence of a strong bow-tie filter or other ROI attenuator, this can cause the detector to saturate in less heavily attenuated regions of the beam, causing a host of potentially significant artifacts, typically azimuthal shading ("radar" artifact) or complete loss of information (saturation) about the periphery of reconstruction analogous to truncation.

4.3.9 NUMBER OF PROJECTIONS

The number of projections affects noise, spatial resolution, and artifact. Many of the effects associated with the number of projections are interlinked with the total dose and mAs per projection, as discussed earlier. For example, at fixed mAs per projection, the number of projections is a direct multiplier in the total dose, and noise scales in an inverse-square-root manner. In the limit of few projections distributed over the source–detector orbit however, view sampling effects (angular undersampling) arise in the form of increased noise (undersampling), a loss in spatial resolution (particularly at greater distance from the center of reconstruction), and radial streak artifacts (spokes).

4.3.10 X-RAY CONVERTER/SCINTILLATOR

The x-ray converter affects noise and spatial resolution. Flat-panel detectors include some form of material (the converter) to convert x-rays to electronic charge. This conversion can be accomplished "directly" by way of a semiconductor [e.g., amorphous selenium (a-Se)] or "indirectly" by way of a scintillator [e.g., thallium-activated cesium iodide (CsI:Tl)]. In the former case, x-rays are converted directly to electron-hole pairs in the semiconductor, and the resulting charge is collected by means of an electric field applied across the converter. Because charge tends to move along the electric field lines (with little or no lateral spread), the converter can be made fairly thick without significant loss in spatial resolution. The quantum detection efficiency (i.e., the fraction of incident x-rays that interact in the converter) can therefore be increased with little or no loss in MTF. In the latter case, however, x-rays are converted to optical photons within the scintillator that, in turn, are collected by an optically sensitive photodiode. The propagation of optical photons can be preferentially forward-directed (e.g., using a needle-like structured scintillator that constrains lateral spread by reflection at crystal interfaces), but there is still significant lateral spread that increases with the thickness of the converter. Indirect detectors therefore tend to exhibit a trade-off between quantum detection efficiency and spatial resolution. Note that in each case, the spatial spreading of secondary quanta in the converter occurs at a "presampling" stage, that is, before sampling of the signal at discrete locations on the detector. The corresponding "presampling NPS" (proportional to $MTF^2_{converter}$) is subject to aliasing, giving a constant ("white") NPS for the direct converter ($MTF_{converter}$ ~ sinc) and a band-limited NPS for the indirect converter ($MTF_{converter}$ ~ low-pass). Therefore, although direct converters tend to provide higher spatial resolution, they suffer noise aliasing to a greater extent, and the trade-offs between quantum detection efficiency, spatial resolution, and NEQ are nontrivial and task-dependent (Siewerdsen et al. 2004, Tward and Siewerdsen 2008). As summarized in Figure 4.6 (third row), increasing converter thickness decreases noise but (for indirect detectors) reduces spatial resolution. For a given imaging task, the "optimal" thickness can be determined by analysis (maximization) of the detectability index (Siewerdsen and Antonuk 1998).

4.3.11 DETECTOR ELEMENT ("DEXEL") SIZE

The size of detector elements affects noise and spatial resolution. In a similar manner as described earlier for the x-ray converter, increasing the size (aperture) of detector elements reduces noise and reduces spatial resolution. For square dexels, the 2D detector MTF is given by a sinc function (MTF_{dex} ~ $|sinc(a_u f_u)||sinc(a_v f_v)|$) and the quantum NPS is (approximately) proportional to the MTF squared. Similarly, the detector MTF enters the 3D system MTF in a linear combination with MTF_{spot}, $MTF_{converter}$, and so on. The trade-off between spatial resolution and noise should therefore be clear, as illustrated in the third row of Figure 4.6. A task-dependent "optimal" dexel size can be analyzed from Equation 4.6 and suggests the need for knowledgeable consideration of detector binning (e.g., 1 × 1, 2 × 2, and so on) in balancing noise and spatial resolution trade-offs that are, in turn, tied to trade-offs in, for example, dose and readout speed.

Dexel size can be controlled through adjustment of detector element binning and imparts various effects on spatial and contrast resolution, as illustrated in Figure 4.12. For example, the detector may be read at "full resolution" (1 × 1 dexel binning), "half resolution" (2 × 2 dexel binning), or in nonisotropic binning modes (e.g., 1 × 2 binning for thicker slices). In the example in Figure 4.12, the detector has a native dexel size of $a_u = a_v = 0.194$ mm, may be read with 1 × 1 or 2 × 2 dexel readout and may be operated in either high-gain or low-gain mode (Roos et al. 2004). These readout options must be weighed against considerations of dose, acquisition speed, and of course the effect on spatial and contrast resolution. For simplicity, we define a "natural" voxel size $a_{vox} = a_{dex}/M$, where a_{dex} is the "full resolution" dexel size, and the notation ($n × n × n$) denotes reconstruction at voxel size of ($n × a_{vox}$). For example, the first two columns of Figure 4.12 show that for the conditions considered (fixed low dose) with 1 × 1 readout (and 1 × 1 × 1 reconstruction), the high-gain mode is advantageous, because the signal in such fine (1 × 1) dexels is low; therefore, low-gain operation suffers a comparatively large contribution of electronic noise. Binning the projections to 2 × 2

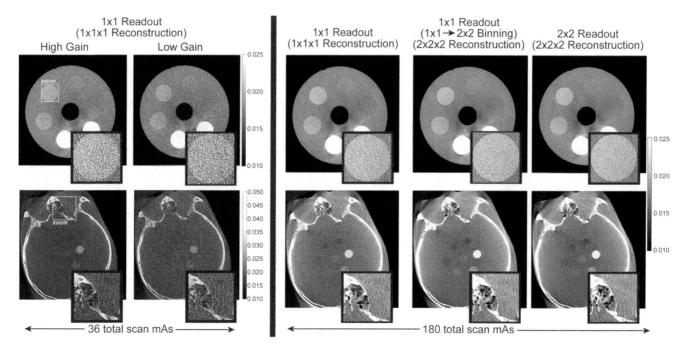

Figure 4.12 Effect of detector element binning on CBCT image quality.

reduces the noise (improves contrast resolution) and reduces spatial resolution, as should now be understood given all that has been discussed earlier.

A question therefore arises, *If we desire both high spatial resolution and high contrast resolution, can we acquire with 1 × 1 readout and then reconstruct twice*—once at "full resolution" (1 × 1 readout → 1 × 1 × 1 reconstruction) for visualization of fine details (third column of Figure 4.12) and once at "half resolution" (1 × 1 readout → 2 × 2 binning and downsampling → 2 × 2 × 2 reconstruction) for visualization of soft tissues (fourth column of Figure 4.12)? The answer depends on whether 1 × 1 readout subsequently binned and downsampled to 2 × 2 (fourth column) yields an equivalent "half resolution" (2 × 2 × 2) reconstruction as 2 × 2 readout (last column). They are not equivalent: the 1 × 1 readout method (followed by 2 × 2 binning and downsampling) has superior sampling (reduced aliasing) in the original projection data, whereas the 2 × 2 readout method (without postreadout binning and downsampling) has a smaller electronics noise contribution. The two methods may have nearly identical spatial resolution, but they differ in the magnitude and correlation of the noise associated with sampling versus electronic noise. For a broad range of conditions, however, the differences can be subtle, as illustrated in Figure 4.12, where 1 × 1 readout followed by 2 × 2 binning and downsampling (fourth column) is shown to be a good choice for soft-tissue visualization— indistinguishable from the 2 × 2 readout case (last column) and with equivalent SDNR in the various soft-tissue inserts therein. The advantage of the former method, however, is the availability of a "full resolution" (1 × 1 × 1) reconstruction (third column) for visualization of fine details. The relative advantages of such readout and binning scenarios, as well as the conditions where one method or another is distinctly superior, may be more fully appreciated by analysis of sampling in the 3D NEQ

as detailed by Tward and Siewerdsen (2009). In the end, of course, the readout mode also must satisfy overriding logistical requirements such as acquisition speed (typically slower for 1 × 1 readout) and detector saturation (lower for 2 × 2 readout).

4.3.12 VOLUME ELEMENT ("VOXEL") SIZE

The voxel size affects noise, spatial resolution, and (partial volume) artifact. Recognizing that voxel size is a free parameter of reconstruction that may be selected independent of any other parameter in the system, we define the "natural" voxel size to be simply given by the detector element (dexel) aperture divided by the geometric magnification: $a_{xy} = a_u/M$ and $a_z = a_v/M$, where a_{xy} is the voxel size in the axial plane and a_z is that in the longitudinal direction (slice thickness). The voxel size may be chosen arbitrarily finer (effectively oversampling the reconstruction) or arbitrarily coarser in which case it is advisable to first bin dexels to a corresponding natural dexel size ($a_{dex} = a_{vox}M$) to avoid undersampling the data. Voxel size presents a pair of independent levers that may be independently adjusted in balancing trade-offs in noise and spatial resolution: the axial voxel size (a_{xy}), to which noise is related by the inverse-cube-root ($\sigma^2 \propto 1/a_{xy}^3$) and longitudinal voxel size (a_z), to which noise is related by the inverse-square-root ($\sigma^2 \propto 1/a_z$). The effects are illustrated in the fifth row of Figure 4.7.

In addition to the strong effect on noise and spatial resolution, voxel size is associated with partial volume artifact—i.e., an erroneous estimate of attenuation coefficient owing to integration of fine detail $\mu(x,y,z)$ over the volume of the voxel (Glover and Pelc 1980). The effect is pronounced for larger voxels and invites semantics of what may be properly described as a reduction in spatial resolution versus what is more accurately considered an artifact. Such may be evident as a loss in spatial resolution at the edges of structures (i.e., the edge passes through the voxel, and the resulting voxel value is averaged over the edge and the background) or in cases where a small feature (e.g., a microcalcification) is entirely contained within a voxel, and the resulting voxel value is again an average of the bulk attenuation coefficient within the voxel. A prevalent example of the partial volume effect is in imaging of the lung, where it may often be stated that the attenuation coefficient of "lung" is approximately −700 HU. In fact, there is no such tissue; rather, the voxel value of −700 HU represents a volume average of −1000 HU (air) and ~0 HU (soft tissue) averaged over the volume of the voxel.

4.3.13 RECONSTRUCTION FILTER

The choice of reconstruction filter affects noise and spatial resolution. Another freely selected parameter of reconstruction, the filters applied to counter high-frequency noise amplified by the ramp filter are variously called "smoothing" or "apodization" filters. Common choices include various low-pass linear filters, such as a cosine window (e.g., Hann, Hamming), Shepp–Logan, Butterworth, and a variety of other filters derived largely from digital signal processing and spectral estimation methods. In addition to attenuating the high-frequency noise otherwise amplified by the ramp filter, apodization filters can be constructed to enhance ("sharpen") spectral response within a given frequency range—i.e., the transfer characteristic of the filter exceeding unity within a given frequency range. The effect of linear apodization filters on noise and spatial resolution is fairly straightforward: the variance is proportional to the bandwidth integral ($\sigma^2 \propto \iiint \mathrm{MTF}^2\, df$), and the transfer characteristic of

the filter enters as a linear combination with other factors in the system MTF. For a given imaging task, therefore, an "optimal" filter can be derived that maximizes Equation 4.6 in balancing the trade-offs of noise and spatial resolution. The trends are illustrated in the bottom row of Figure 4.7 and the example images of Figure 4.13.

4.4 IMAGING TASK

The performance of an imaging system depends on the imaging task. Given that the purpose of this chapter is to describe the factors governing CBCT image quality, this point bears reiteration. A system providing low noise may be well suited to the task of soft-tissue visualization but may fail entirely in discrimination of fine details. Conversely, a system with high image noise may fail to delineate soft-tissue boundaries but be perfectly adequate for visualization of high-contrast bone details. And if the task is quantitative measurement of attenuation coefficient, then metrics of noise and spatial resolution may be of secondary importance to artifacts that affect the accuracy of the reconstructed voxel value. As noted by Barrett et al. (1995),

> A medical image is always produced for some specific purpose or task, and the only meaningful measure of its quality is how well it fulfills that purpose. An objective approach to assessment of image quality must therefore start with a specification of the task and then determine quantitatively how well the task is performed.

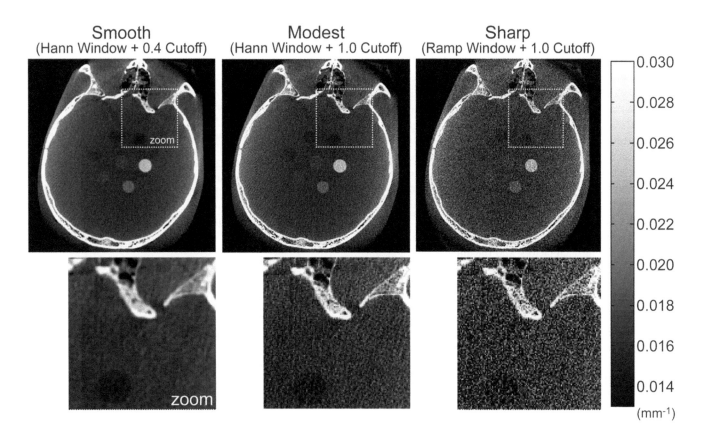

Figure 4.13 Effect of reconstruction filter on CBCT image quality. (Left to right) Sharper filters improve spatial resolution (e.g., high-contrast bone details) at the cost of increased noise (reduced soft-tissue visibility).

The individual metrics of contrast, noise, SDNR, NPS, MTF, and NEQ each quantify an important dimension of what may (or may not) relate to the utility or performance of an imaging system. The simplest first step beyond such individual metrics is realized in the quantitation of imaging task in terms of the spatial frequencies of importance in binary hypothesis testing, for example, detection of a stimulus within a known background, discrimination of two stimuli, etc., as described in Section 4.2.4. Such idealized task functions weight the signal and noise transfer characteristics (i.e., the "fidelity") of the imaging system in proportion to the frequencies of interest, yielding a simple objective function for task-based performance evaluation and optimization. Although such idealized models have demonstrated correspondence with real observers for simple imaging tasks and have provided a basis for the design of novel CBCT prototypes in new applications, extension beyond the idealized task models is an area of active ongoing research, particularly with regard to the flourishing space of nonlinear 3D reconstruction techniques, spectral tomography, and advanced models of human (or machine) observers.

4.5 CONCLUSIONS

This chapter has sought to accomplish two basic goals: (1) summarize the *metrics* by which CBCT imaging performance may be quantified (Section 4.2) and (2) survey the *factors*, both physical and mathematical, that govern CBCT imaging performance (Section 4.3). The various metrics of contrast, noise, and spatial resolution may be conveyed in terms of Fourier-domain descriptors such as NPS, MTF, and NEQ, recognizing that such LSI metrics face limitations in application to systems that are strictly neither linear nor shift-invariant. "Local" description of such metrics is therefore *de rigeur*, that is, analysis of "local" NPS, MTF, and NEQ over ranges in signal and space for which the system is approximately LSI, accompanied by a (quantitative) check on such assumptions and recognizing the role of more complete (but computationally intense) metrics of noise and correlation. The spectrum of factors that govern imaging performance includes broad consideration of system geometry, the specification of individual system components (e.g., x-ray tube and detector), the acquisition technique, and the reconstruction method. The result is a deep, multiparametric space that can be quantitatively considered in maximizing CBCT image quality. Rigorous design and application of CBCT systems can benefit tremendously from recognition of the manifold trade-offs that exist among such factors, and the understanding arising from the last several decades of CBCT image science has yielded a fairly comprehensive arsenal of theoretical and experimental tools to enable imaging physicists to design new, higher performance systems, optimize imaging techniques, minimize radiation dose, and maximize imaging performance with respect to the imaging task.

REFERENCES

Baek, J. and Pelc, N.J. 2011a. Effect of detector lag on CT noise power spectra. *Med Phys* 38(6): 2995–3005.

Baek, J. and Pelc, N.J. 2011b. Local and global 3D noise power spectrum in cone-beam CT system with FDK reconstruction. *Med Phys* 38(4): 2122–31.

Barrett, H.H. 1990. Objective assessment of image quality: effects of quantum noise and object variability. *J Opt Soc Am A* 7(7): 1266–78.

Barrett, H.H., Denny, J.L., Wagner, R.F. and Myers, K.J. 1995. Objective assessment of image quality. II. Fisher information, fourier crosstalk, and figures of merit for task performance. *J Opt Soc Am A* 12(5): 834–52.

Barrett, H.H., Gordon, S.K. and Hershel, R.S. 1976. Statistical limitations in transaxial tomography. *Comput Biol Med* 6(4): 307–23.

Barrett, J.F. and Keat, N. 2004. Artifacts in CT: recognition and avoidance. *Radiographics* 24(6): 1679–91.

Bartolac, S., Clackdoyle, R., Noo, F., Siewerdsen, J., Moseley, D. and Jaffray, D. 2009. A local shift-variant fourier model and experimental validation of circular cone-beam computed tomography artifacts. *Med Phys* 36(2): 500–12.

Bochud, F.O., Abbey, C.K. and Eckstein, M.P. 2000. Visual signal detection in structured backgrounds. III. Calculation of figures of merit for model observers in statistically nonstationary backgrounds. *J Opt Soc Am A Opt Image Sci Vis* 17(2): 193–205.

Boone, J.M. and Seibert, J.A. 1988a. An analytical model of the scattered radiation distribution in diagnostic radiology. *Med Phys* 15(5): 721–5.

Boone, J.M. and Seibert, J.A. 1988b. Monte Carlo simulation of the scattered radiation distribution in diagnostic radiology. *Med Phys* 15(5): 713–20.

Chesler, D.A., Riederer, S.J. and Pelc, N.J. 1977. Noise due to photon counting statistics in computed X-ray tomography. *J Comput Assist Tomogr* 1(1): 64–74.

De Man, B., Nuyts, J., Dupont, P., Marchal, G. and Suetens, P. 1999. Metal streak artifacts in x-ray computed tomography: a simulation study. *IEEE Trans Nucl Sci* 46(3): 691–6.

De Man, B., Nuyts, J., Dupont, P., Marchal, G. and Suetens, P. 2001. An iterative maximum-likelihood polychromatic algorithm for CT. *IEEE Trans Med Imaging* 20(10): 999–1008.

Engstrom, E., Reiser, I. and Nishikawa, R. 2009. Comparison of power spectra for tomosynthesis projections and reconstructed images. *Med Phys* 36(5): 1753–8.

Feldkamp, L.A., Davis, L.C. and Kress, J.W. 1984. Practical cone-beam algorithm. *J Opt Soc Am A* 1(6): 612–19.

Gang, G.J., Lee, J., Stayman, J.W., Tward, D.J., Zbijewski, W., Prince, J.L. and Siewerdsen, J.H. 2010. The generalized NEQ and detectability index for tomosynthesis and cone-beam CT: from cascaded systems analysis to human observers. *Proc SPIE Phys Med Imaging* 7622(1): 76220Y.

Gang, G.J., Lee, J., Stayman, J.W., Tward, D.J., Zbijewski, W., Prince, J.L. and Siewerdsen, J.H. 2011. Analysis of fourier-domain task-based detectability index in tomosynthesis and cone-beam CT in relation to human observer performance. *Med Phys* 38(4): 1754–68.

Gang, G.J., Tward, D.J., Lee, J. and Siewerdsen, J. H. 2010. Anatomical background and generalized detectability in tomosynthesis and cone-beam CT. *Med Phys* 37(5): 1948–65.

Glover, G.H. and Pelc, N.J. 1980. Nonlinear partial volume artifacts in x-ray computed tomography. *Med Phys* 7(3): 238–48.

Hanson, K.M. 1979. Detectability in computed tomographic images. *Med Phys* 6(5): 441–51.

Hsieh, J. 1998. Adaptive streak artifact reduction in computed tomography resulting from excessive x-ray photon noise. *Med Phys* 25(11): 2139–47.

International Commission on Radiation Units and Measurements. *Report No. 54, Medical Imaging – The Assessment of Image Quality*. 1996.

Jarry, G., Graham, S.A., Moseley, D.J., Jaffray, D.J., Siewerdsen, J.H. and Verhaegen, F. 2006. Characterization of scattered radiation in kV CBCT images using Monte Carlo simulations. *Med Phys* 33(11): 4320–9.

Kwan, A.L., Boone, J.M. and Shah, N. 2005. Evaluation of x-ray scatter properties in a dedicated cone-beam breast CT scanner. *Med Phys* 32(9): 2967–75.

Kyprianou, I.S., Ganguly, A., Rudin, S., Bednarek, D.R., Gallas, B.D. and Myers, K.J. 2005a. Efficiency of the human observer compared to an ideal observer based on a generalized NEQ which incorporates scatter and geometric unsharpness: evaluation with a 2AFC experiment. *Proc Soc Photo Opt Instrum Eng* 5749: 251–62.

Kyprianou, I.S., Rudin, S., Bednarek, D.R. and Hoffmann, K.R. 2004. Study of the generalized MTF and DQE for a new microangiographic system. *Proc Soc Photo Opt Instrum Eng* 5368: 349–60.

Kyprianou, I.S., Rudin, S., Bednarek, D.R. and Hoffmann, K.R. 2005b. Generalizing the MTF and DQE to include x-ray scatter and focal spot unsharpness: application to a new microangiographic system. *Med Phys* 32(2): 613–26.

Kyriakou, Y. and Kalender, W.A. 2007. X-ray scatter data for flat-panel detector CT. *Phys Med* 23(1): 3–15.

Lei, Z., Bennett, N.R. and Fahrig, R. 2006. Scatter correction method for x-ray CT using primary modulation: theory and preliminary results. *IEEE Trans Med Imaging* 25(12): 1573–87.

Li, T., Koong, A. and Xing, L. 2007. Enhanced 4D cone-beam CT with inter-phase motion model. *Med Phys* 34(9): 3688–95.

Li, T., Schreibmann, E., Yang, Y. and Xing, L. 2006a. Motion correction for improved target localization with on-board cone-beam computed tomography. *Phys Med Biol* 51(2): 253–67.

Li, T., Xing, L., Munro, P., McGuinness, C., Chao, M., Yang, Y., Loo, B. and Koong, A. 2006b. Four-dimensional cone-beam computed tomography using an on-board imager. *Med Phys* 33(10): 3825–33.

Maltz, J.S., Gangadharan, B., Bose, S., Hristov, D.H., Faddegon, B.A., Paidi, A. and Bani-Hashemi, A.R. 2008. Algorithm for x-ray scatter, beam-hardening, and beam profile correction in diagnostic (kilovoltage) and treatment (megavoltage) cone beam CT. *IEEE Trans Med Imaging* 27(12): 1791–810.

Metheany, K.G., Abbey, C.K., Packard, N. and Boone, J.M. 2008. Characterizing anatomical variability in breast CT images. *Med Phys* 35(10): 4685–94.

Pineda, A.R., Siewerdsen, J.H. and Tward, D.J. 2008. Analysis of image noise in 3D cone-beam CT: spatial and fourier domain approaches under conditions of varying stationarity. *Proc SPIE Phys Med Imaging* 6913(1): 69131Q.

Prakash, P., Zbijewski, W., Gang, G.J. Ding, Y., Stayman, J.W., Yorkston, J., Carrino, J.A. and Siewerdsen, J.H. 2011. Task-based modeling and optimization of a cone-beam CT scanner for musculoskeletal imaging. *Med Phys* 38(10): 5612–29.

Reiser, I. and Nishikawa, R.M. 2010. Task-based assessment of breast tomosynthesis: effect of acquisition parameters and quantum noise. *Med Phys* 37(4): 1591–600.

Riederer, S.J., Pelc, N.J. and Chesler, D.A. 1978. The noise power spectrum in computed x-ray tomography. *Phys Med Biol* 23(3): 446–54.

Rit, S., Wolthaus, J., van Herk, M. and Sonke, J.J. 2008. On-the-fly motion-compensated cone-beam CT using an a priori motion model. *Med Image Comput Comput Assist Interv* 11(Pt 1): 729–36.

Rit, S., Wolthaus, J.W., van Herk, M. and Sonke, J.J. 2009. On-the-fly motion-compensated cone-beam CT using an a priori model of the respiratory motion. *Med Phys* 36(6): 2283–96.

Roos, P.G., Colbeth, R.E., Mollov, I., Munro, P., Pavkovich, J., Seppi, E.J., Shapiro, E.G., Tognina, C.A., Virshup, G.F., Yu, J.M., Zentai, G., Kaissl, W., Matsinos, E., Richters, J. and Riem, H. 2004. 139–49.

Rose, A. 1957. Quantum effects in human vision. *Adv Biol Med Phys* 5: 211–42.

Schafer, S., Stayman, J., Zbijewski, W., Schmidgunst, C., Kleinszig, G. and Siewerdsen, J. 2012. Antiscatter grids in mobile C-arm cone-beam CT: effect on image quality and dose. *Med Phys* 39(1): 153–9.

Siewerdsen, J.H. and Antonuk, L.E. 1998. DQE and system optimization for indirect-detection flat-panel imagers in diagnostic radiology. *SPIE Physics of Medical Imaging* 3336: 546–55.

Siewerdsen, J.H., Cunningham, I.A. and Jaffray, D.A. 2002. A framework for noise-power spectrum analysis of multidimensional images. *Med Phys* 29(11): 2655–71.

Siewerdsen, J.H., Daly, M.J., Bakhtiar, B., Moseley, D.J., Richard, S., Keller, H. and Jaffray, D.A. 2006. A simple, direct method for x-ray scatter estimation and correction in digital radiography and cone-beam CT. *Med Phys* 33(1): 187–97.

Siewerdsen, J.H. and Jaffray, D.A. 1999. Cone-beam computed tomography with a flat-panel imager: effects of image lag. *Med Phys* 26(12): 2635–47.

Siewerdsen, J.H. and Jaffray, D.A. 2000a. Cone-beam CT with a flat-panel imager: noise considerations for fully 3D computed tomography. *SPIE Physics of Medical Imaging* 3977: 408–16.

Siewerdsen, J.H. and Jaffray, D.A. 2000b. Optimization of x-ray imaging geometry (with specific application to flat-panel cone-beam computed tomography). *Med Phys* 27(8): 1903–14.

Siewerdsen, J.H. and Jaffray, D.A. 2001. Cone-beam computed tomography with a flat-panel imager: magnitude and effects of x-ray scatter. *Med Phys* 28(2): 220–31.

Siewerdsen, J.H. and Jaffray, D.A. 2003. Three-dimensional NEQ transfer characteristics of volume CT using direct- and indirect-detection flat-panel imagers. *SPIE Physics of Medical Imaging* 5030: 92–102.

Siewerdsen, J.H., Moseley, D.J., Bakhtiar, B., Richard, S. and Jaffray, D.A. 2004. The influence of antiscatter grids on soft-tissue detectability in cone-beam computed tomography with flat-panel detectors. *Med Phys* 31(12): 3506–20.

Siewerdsen, J.H., Moseley, D.J., Burch, S., Bisland, S.K., Bogaards, A., Wilson, B.C. and Jaffray, D.A. 2005. Volume CT with a flat-panel detector on a mobile, isocentric C-arm: pre-clinical investigation in guidance of minimally invasive surgery. *Med Phys* 32(1): 241–54.

Siewerdsen, J.H., Moseley, D.J. and Jaffray, D.A. 2004. Incorporation of task in 3D imaging performance evaluation: the impact of asymmetric NPS on detectability. *SPIE Physics of Medical Imaging* 5368: 89–97.

Sonke, J.J., Zijp, L., Remeijer, P. and van Herk, M. 2005. Respiratory correlated cone beam CT. *Med Phys* 32(4): 1176–86.

Tang, J., Hsieh, J. and Chen, G.H. 2010. Temporal resolution improvement in cardiac CT using PICCS (TRI-PICCS): performance studies. *Med Phys* 37(8): 4377–88.

Tward, D.J. and Siewerdsen, J.H. 2008. Cascaded systems analysis of the 3D noise transfer characteristics of flat-panel cone-beam CT. *Med Phys* 35(12): 5510–29.

Tward, D.J. and Siewerdsen, J.H. 2009. Noise aliasing and the 3D NEQ of flat-panel cone-beam CT: effect of 2D/3D apertures and sampling. *Med Phys* 36(8): 3830–43.

Tward, D.J., Siewerdsen, J.H., Daly, M.J., Richard, S., Moseley, D.J., Jaffray, D.A. and Paul, N.S. 2007. Soft-tissue detectability in cone-beam CT: evaluation by 2AFC tests in relation to physical performance metrics. *Med Phys* 34(11): 4459–71.

Wang, G. and Yu, H. 2010. Can interior tomography outperform lambda tomography? *Proc Natl Acad Sci USA* 107(22): E92–93.

Wiegert, J., Bertram, M., Rose, G. and Aach, T. 2005. Model based scatter correction for cone-beam computed tomography. *Proc SPIE Phys Med Imaging* 5745: 271–82.

Xu, Q., Mou, X., Wang, G., Sieren, J., Hoffman, E.A. and Yu, H. 2011. Statistical interior tomography. *IEEE Trans Med Imaging* 30(5): 1116–28.

Yang, J., Yu, H., Jiang, M. and Wang, G. 2010. High order total variation minimization for interior tomography. *Inverse Probl* 26(3): 350131–29.

Yang, K., Kwan, A.L. and Boone, J.M. 2007. Computer modeling of the spatial resolution properties of a dedicated breast CT system. *Med Phys* 34(6): 2059–69.

Yu, H. and Wang, G. 2009. Compressed sensing based interior tomography. *Phys Med Biol* 54(9): 2791–805.

Yu, H., Wang, G., Hsieh, J., Entrikin, D.W., Ellis, S., Liu, B. and Carr, J.J. 2011. Compressive sensing-based interior tomography: preliminary clinical application. *J Comput Assist Tomogr* 35(6): 762–4.

Zbijewski, W. and Beekman, F.J. 2006. Efficient Monte Carlo based scatter artifact reduction in cone-beam micro-CT. *IEEE Trans Med Imaging* 25(7): 817–27.

Zbijewski, W., De Jean, P., Prakash, P., Ding, Y., Stayman, J.W., Packard, N., Senn, R., Yang, D., Yorkston, J., Machado, A., Carrino, J.A. and Siewerdsen, J.H. 2011. A dedicated cone-beam CT system for musculoskeletal extremities imaging: design, optimization, and initial performance characterization. *Med Phys* 38(8): 4700–13.

5 Radiation dose

Ioannis Sechopoulos

Contents

5.1 INTRODUCTION

Since its introduction in 1972 by Hounsfield and Ambrose (Ambrose 1973; Hounsfield 1973), computed tomography (CT) has evolved into a key element of radiological clinical practice. However, along with the increase in its clinical use, CT's contribution to medical and total population dose has also increased substantially. In the United States, from the early 1980s to 2006, the portion attributed to CT of the total effective dose per capita due to all sources increased from less than 0.5% to 24% (NCRP 2009). Of the effective per capita dose due to medical procedures only, in the United States, CT's contribution has increased from 3% to 49% over the same period, even though CT scans accounted for only 17% of the total medical procedures performed during 2006 (NCRP 2009). Similar trends have been experienced worldwide, with, for example, CT reports contributing more than half of the effective dose due to medical procedures in Germany (Nekolla et al. 2009) and Australia (Heggie et al. 2006).

Given the levels of radiation involved in CT, it is important to understand how CT doses are measured, communicated, and ultimately optimized during clinical use. This chapter introduces the metrics used in radiation dosimetry applicable to all x-ray-based radiologic applications and the methods to estimate these metrics, followed by a discussion of CT-specific dosimetry and its limitations, in addition to a recent proposal for new CT dose metrics aimed to overcome these limitations. Also, CT image acquisition parameters and how these parameters affect patient dose are reviewed. Finally, two cone beam computed tomography (CBCT) applications that involve application-specific dosimetry metrics are discussed.

5.2 GENERAL RADIATION DOSE UNITS

Many metrics (e.g., exposure, dose, absorbed dose, effective dose) are usually associated with measuring the potential damage to tissues caused by radiation, and often these metrics are, incorrectly, used interchangeably. However, each of these metrics refers to different phenomena and care should be taken on the appropriate use of each. Furthermore, CT-specific dose metrics have been developed to address CT's complex nature, in which the acquisition geometry is time-varying, and there are many acquisition parameters that affect the resulting dose. However, changes in CT technology, such as the introduction of CBCT, have resulted in some of these CT-specific dose concepts to be challenging, if not impossible, to apply. Therefore, CT dosimetry is a field of continuing study and development, with new metrics and measurement procedures still being proposed and developed.

5.2.1 EXPOSURE AND AIR KERMA

Exposure, normally denoted with the letter X, is in fact not a measure of energy deposition or dose but of ionization in air caused by x-rays. Its formal traditional definition is the number of electron–ion pairs formed per unit *volume* of dry air by incident x-rays. Its traditional unit is the roentgen (R), in honor of the discoverer of x-rays, the German physicist Wilhelm Conrad Röntgen.

Under the Système International definition, exposure measures the amount of charge, in coulombs, per unit *mass* of air created by the incident x-rays, so the SI unit for exposure is coulomb per kilogram (C/kg). Although the roentgen has been used for many years when expressing exposure, today the SI unit should be used in all communications. The exact conversion between the two measures of exposure is $1 \text{ R} \equiv 2.58 \times 10^{-4}$ C/kg.

Exposure has been widely used due to its ease of measurement with the use of an ionization chamber and an electrometer (Figure 5.1) as a measure of x-ray tube output. However, as is clear from its definition, exposure is not a measure of dose, and it is not defined for any material other than air. In addition to the change in the units to be used, exposure has been phased out as a metric and has been replaced by the *air kerma*.

Air kerma, an acronym for *k*inetic *e*nergy *r*eleased in *ma*tter, is defined as the amount of kinetic energy gained by charged particles (in this case, electrons) from the incident radiation per unit mass of matter (in this case, air). The SI unit for air kerma is the joule per kilogram (J/kg), and has been assigned the special name gray (Gy), in honor of Louis Harold Gray, a British physicist who made important contributions to radiation biology. Air kerma, like exposure, also can be measured directly using an ionization chamber or electrometer and is now the recommended metric to express the intensity of an x-ray beam. To convert exposure to air kerma, the following conversion factors are used: 1 Gy air kerma = 114.2 R = 2.944×10^{-2} C/kg.

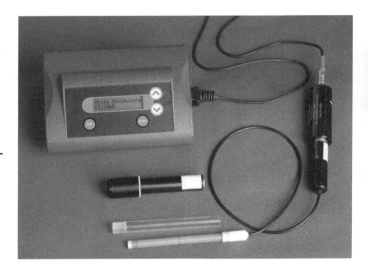

Figure 5.1 Ionization chambers, digitizer, and electrometer used to measure exposure and air kerma. The chamber on the bottom of the photograph is the CT 100-mm pencil-shaped ionization chamber, shown with the plastic sheath used when inserting the pencil chamber in the CT dose phantom depicted in Figure 5.4. The black cylinder above it is a general-purpose ionization chamber. The digitizer used to connect the chambers to the electrometer is at right.

5.2.2 ABSORBED DOSE

Absorbed dose[*] is the amount of energy deposited in matter per unit mass; therefore, it also is expressed in gray (1 Gy = 1 J/kg). Absorbed dose relates directly to the potential for biological damage as long as the values compared are for the same type of radiation (e.g., x-rays) and in the same organ. This is because the biological damage caused by the same amount of absorbed dose from different types of particles (e.g., neutrons vs. x-rays) can vary substantially,[†] whereas the radiosensitivity of organs to a certain amount of absorbed dose also varies. This latter variation is addressed by the *effective dose* (see Section 5.2.3). One of the complexities of absorbed dose is that it cannot be measured directly, but rather it is always estimated. The methods used to estimate absorbed dose are discussed later.

5.2.3 EFFECTIVE DOSE

Effective dose provides a way to estimate the increased risk of stochastic (cancer and genetic damage) effects due to exposure of various organs to radiation. To obtain this estimate, effective dose is computed using the following equation:

$$E = \sum_T w_T w_R D_T \qquad (5.1)$$

where E is the effective dose, w_T is the tissue weighting factor for organ T, w_R is the particle weighting factor which for x-rays equals unity, and D_T is the absorbed dose for organ T. The tissue weighting factors reflect the difference in relative radiosensitivity

---

[*] This metric is the metric commonly referred to when the term dose is used with no qualifier.

[†] This variation in *relative biological effectiveness* (RBE) among different types of ionizing radiation is taken into account by the *radiation weighting factor* (W_R) in the metric denoted *equivalent dose*. However, because CT involves the use of x-rays only, equivalent dose is not discussed further.

of the different organs in the human body; these weighting factors are established by the International Commission on Radiological Protection (ICRP), based on epidemiological studies, empirical data, and expert judgments. The original tissue weighting factors were first published in 1977 (ICRP 1977), revised in 1990 (ICRP 1991), and then revised again in 2007 (ICRP 2007). In its current form, the ICRP recommends the use of specific tissue weighting factors for 14 specific organs, with the remaining organs grouped into one additional tissue weighting factor. The sum of the 15 factors is unity, so that a uniform absorbed dose distribution throughout the body results in the same value for effective dose. Because the tissue weighting factors are dimensionless, effective dose also has units of joule per kilogram, but for effective dose this unit has been assigned the special name sievert (Sv), in honor of Rolf Maximilian Sievert, a Swedish medical physicist who studied radiation biology.

Effective dose has both limiting and advantageous features. Effective dose is not a metric based on physical principles, but an estimated value whose definition was arrived at by convention. As such, because the recommended tissue weighting factors have changed twice since first introduced, the resulting effective dose for a given set of absorbed dose distribution also has changed with time. In addition, effective dose characterizes the increased risk only for stochastic effects, not providing direct information on the possibility of deterministic effects on organs, such as erythema. Finally, the tissue weighting factors are sex- and age-averaged, so effective dose is not intended to be used as a personalized metric, but rather as a measure for a standardized person.

However, because effective dose allows for the consolidation of absorbed dose values for various organs into a single value, it provides a convenient method to compare differences in dosimetric consequences due to different protocols, acquisition geometries, imaging modalities, or a combination. For example, if a CT-based diagnostic test had similar clinical performance to a single-photon emission computed tomography (SPECT)–based test, and we had complete information on the absorbed dose resulting from both tests for each organ of the body, how could we decide which test carries the lower risk and therefore should be adopted for clinical use? It is very possible that the CT-based test would result in higher absorbed dose values for a few organs that are in or close to the primary x-ray beam during acquisition, and low absorbed dose values for organs far away from the beam. The SPECT-based test, because it involves the systemic delivery of a radiopharmaceutical by intravenous (IV) injection, will probably have a more homogeneous distribution of organ absorbed dose. If the somewhat homogeneous distribution of SPECT dose is somewhere in between the high and low organ doses for the CT test, which test should be adopted on the basis of dose alone? The effective dose, or the dose that would yield a single number that represents overall increase in stochastic risk for each test, in this case, may provide the information necessary to decide.

In addition, because effective dose provides a single value that in some way reflects the dose involved during an imaging study of any type (e.g., CT, SPECT, radiography) on any part of the body, it could be a useful metric to record in each patient's medical history along the corresponding imaging study. Of course, care must be taken on what significance and usefulness these records actually have. Because effective dose should not be used for personalized risk assessment, these records could be useful to perform retrospective studies on protocol improvements, changes in use trends of tests involving ionizing radiation, and possibly identifying single cases in which incorrect protocols were possibly used. However, other uses, such as suggesting which type of imaging test should be ordered [e.g., CT vs. magnetic resonance imaging (MRI)] due to a patient's previously received cumulative effective dose, would be beyond the scope of use of this information.

5.2.4 ESTIMATING ABSORBED AND EFFECTIVE DOSE

5.2.4.1 Monte Carlo simulations

One method to estimate absorbed dose is the use of Monte Carlo computer simulations in which the acquisition conditions, imaged object (the relevant portion of or the whole patient), and relevant physics are represented in a computer, and the total energy deposition resulting from the interaction of simulated x-rays with the tissue in question is calculated. These computer simulations typically provide conversion factors between the air kerma measured at a reference point or area and the resulting absorbed dose in the organs or tissues of interest.

Monte Carlo simulations for estimating absorbed dose to different organs or absorbed and/or effective dose to the entire body are very powerful tools for various reasons. These simulations can provide dosimetry results with a very high level of detail, because local dose distributions within an organ or throughout the entire body can be estimated. Furthermore, these computer-based simulations make comprehensive characterization of dosimetry studies possible, allowing for the understanding of how absorbed dose to an organ or a set of organs varies with a large number of different acquisition parameters. These comprehensive studies are important to better understand the dosimetric consequences of imaging technologies, and to allow for the optimization of parameters to minimize dose. Finally, Monte Carlo simulations allow for the estimation of absorbed and effective dose for a proposed imaging technology before the actual system is built, helping determine whether such new technology is feasible, and optimizing the system design, at least from the dosimetric point of view.

However, Monte Carlo simulations do have limitations. The most important of them is not technically a limitation but a requirement: for the results of Monte Carlo simulations to be representative of the real world, the simulation specifications also have to be representative of the real world. In other words, the simulation results are only as good as the input simulation details. Therefore, the description of the imaging system components, acquisition geometry, patient anatomy and composition, and relevant physics must be as exact as possible for the results to be reliable. In many cases, simplifications and assumptions can (and usually must) be made in some of these aspects to make the simulations feasible, but the consequences of these should be well understood when interpreting the results. In general, the use of the correct relevant physics is not problematic, because most Monte Carlo simulations for diagnostic imaging are performed using as a basis one of a few, well-validated software packages such as

Geant4, MCNP, EGSnrc, or Penelope. These packages have been used and tested by various investigators, so their physics-related behavior has been verified repeatedly. However, the specifics of each simulation, namely, the description of the imaging system and of the imaged object are up to each investigator; therefore, care must be taken when implementing these details. Furthermore, to ensure the correct behavior of a new Monte Carlo simulation, validation of some of its predictions against empirical measurements or at least previously published simulation results should be performed. One other previously noted limitation of Monte Carlo simulations is that some simplifications and assumptions need to be made for almost all simulations. For example, it is usually impossible to describe an imaging system to its last detail, and it is always impossible to represent the human anatomy to the smallest detail, even with the use of very detailed patient models, such as those described by Xu and Eckerman (2010), so simplifications of geometry are always made. As mentioned, however, the consequences of these simplifications and assumptions on the results should be well understood.

5.2.4.2 Empirical estimations

Another method to estimate absorbed and effective dose is using point dosimeters, such as thermoluminescent dosimeters (TLDs) and radiophotoluminescence glass dosimeters (RPLs), among others. TLDs are small detectors whose main component is a crystal (e.g., lithium fluoride or calcium fluoride) in which the absorbed energy from its interaction with an incident x-ray excites an electron in the crystal to a higher energy state. The electron retains this excited state until it is heated, whereupon it returns to its ground energy state by releasing a light photon with the energy equivalent to that originally absorbed. Therefore, the total light output by the TLD crystal when heated is directly proportional to the energy absorbed by the crystal. To obtain an actual measure of absorbed dose from the measurement of the light output, TLDs response to varying dose levels is measured to obtain calibration curves that relate light output to absorbed dose and also should account for the energy dependent response of the TLDs, especially at diagnostic energies. RPLs work in a similar way, with the main difference being that ultraviolet light is used to release the energy absorbed by the crystal, as opposed to heating.

To obtain an estimate of the absorbed dose in one or several organs due to an image acquisition (planar or tomographic) with these types of dosimeters, several dosimeters are placed in cavities located throughout the organ or organs of interest in an anthropomorphic phantom. These phantoms mimic the shape and attenuation properties of humans and might vary in size so as to represent males or females of different age groups. With the dosimeters in place, the phantom is irradiated using the acquisition technique being tested and the dose absorbed by the dosimeters during acquisition is measured. In general, the dose absorbed by each specific organ is estimated as the average of the readouts of the dosimeters that were located inside it. Because the absorbed dose will vary spatially throughout the body and within each organ, depending on the size of the organ, it is important to sample each organ at various points. The estimated organ doses can be combined to obtain the effective dose.

5.3 CT-SPECIFIC DOSE UNITS

There are two main differences between CT and planar radiography that result in the need for additional CT-specific dose metrics. The most apparent of these differences is the time-varying acquisition geometry. Specifically, the x-ray source during a CT acquisition is continuously rotating around the imaged body; therefore, the x-ray entrance surface varies continuously. This results in an absorbed dose distribution throughout the axial plane of the body that is very different from that in planar radiography (Figure 5.2). Clearly, the dose in radiography decreases considerably from the x-ray entrance side of the body to the x-ray exit side. In CT, however, because of the rotating gantry, this averages out resulting in a more uniform, and, assuming a homogeneous body composition, radially symmetric dose distribution. The dose difference between points close to the surface of the body and its center depends on the size of the imaged body, its composition, and the x-ray spectrum used.

Another difference between CT and radiography dosimetry is the relative contribution of x-ray scatter to the overall absorbed dose. In radiography, the primary x-ray field is rectangular and in most cases its sides are a few centimeters (20–40) long, so the extent of tissue outside the primary x-ray field that is irradiated by scattered x-rays is relatively small compared with that irradiated directly by the primary beam. Therefore, when estimating dose in radiographic applications the absorbed dose to tissues outside the primary x-ray field is usually ignored. In CT, however, the primary x-ray field is normally several millimeters thick (10–40), at most (excluding CBCT and some multislice systems with very high number of slices). This results in a large volume of tissue being exposed only to scattered x-rays. As seen in Figure 5.3, a typical CT absorbed dose profile includes long tails beyond the directly irradiated tissue that extend well beyond the primary beam width. Therefore, in CT, the contribution to the total absorbed dose from scattered x-rays must be taken into account. In other words, when estimating the dose due to a CT acquisition, it is not enough to take into account the area under the narrow-peaked curve depicted in Figure 5.3 from dose due to primary x-rays, the area under the long tails from dose from scattered x-rays must also be included in the estimation.

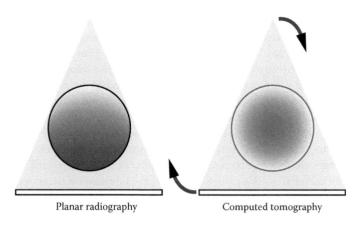

Figure 5.2 Comparison of dose distribution inside the body between planar radiography and CT. Due to the rotation of the gantry, the dose distribution in CT is more homogeneous and for a homogeneous body it is rotationally symmetric.

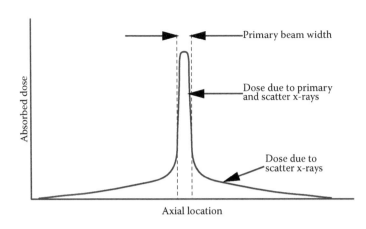

Figure 5.3 Typical absorbed dose profile for one axial slice in CT, including the dose due to primary and scattered x-rays. As shown, the scattered x-rays result in deposition of dose a substantial distance from the acquired slice.

Figure 5.4 Photos of the three phantoms used to measure $CTDI_{100}$, representing a pediatric head (10 cm in diameter), both a pediatric body and an adult head (16 cm), and an adult body (32 cm). This phantom consists of the smallest cylinder plus two rings to form the bigger cylinders when nested together. The photo on the right shows the phantoms partially nested together. The photos also show the 100-mm pencil ionization chamber and the thin cylinders used to fill the holes not occupied by the chamber.

5.3.1 CT DOSE INDEX

To address these issues, the CT dose index (CTDI) has been developed and adopted to reflect the *amount of radiation* involved in the acquisition of a CT scan. As discussed here, CTDI refers to the absorbed dose (in air) in portions of a specific phantom, not to the imaged patient. This is the reason why it is called an *index*; CTDI should not be viewed as an estimate of patient absorbed dose, but as a relative measure of the CT system's x-ray output and the dose measured in a phantom of fixed size (McCollough et al. 2011). This metric can therefore be used only as indicative of patient dose and specifically to compare dose in relative terms when evaluating different imaging technique factors, protocols and/or CT systems.

Although various definitions have been used for CTDI in its most basic form, currently, the most commonly used CTDI basic metric is the $CTDI_{100}$. The $CTDI_{100}$ is obtained using a pencil-shaped ionization chamber 100 mm in length (Figure 5.1) to measure the total exposure in a phantom from the acquisition of one axial CT scan. Using a chamber that is 100 mm in length, $CTDI_{100}$ attempts to include in its measurement the dose to the phantom beyond the extent of the primary x-ray field due to scattered x-rays (Figure 5.3). Therefore, $CTDI_{100}$ is defined as follows:

$$CTDI_{100} = \frac{1}{nT} \int_{-50\,mm}^{+50\,mm} D(z)\,dz \qquad (5.2)$$

where n is the number of data channels acquired per axial scan, T is the width of one data channel (always measured at the centerline of the field of view), and $D(z)$ is the dose in air at axial location z. Note that although the ionization chamber is placed inside a phantom to measure $CTDI_{100}$, the metric is defined as the dose absorbed in the volume of air occupied by the ionization chamber, so the exposure or air kerma measured by the electrometer must be converted to absorbed dose in air.

The phantoms used to determine $CTDI_{100}$ are cylindrical phantoms composed of polymethyl methacrylate (PMMA) of 10-, 16-, or 32-cm diameter. These sizes of phantoms represent, respectively, a pediatric head (10 cm), both a pediatric body and

an adult head (16 cm), and an adult body (32 cm). Sometimes these phantoms are built as a set, with only the 10-cm-diameter cylinder included and two additional rings to form the two larger phantoms. As shown in Figure 5.4, each cylinder has 1.41-cm-diameter holes along its entire length at the center and in four points 1 cm from the edge for the CT pencil chamber, in which the 100-mm pencil chamber is placed to measure $CTDI_{100}$.

With the pencil chamber inserted in the hole in the phantom's center, the index is called the $CTDI_{100,center}$, whereas with the pencil chamber inserted in one of the perimeter holes the index is called $CTDI_{100,perimeter}$. As mentioned previously, depending on the size of the phantom, these two indices will not be the same, while the $CTDI_{100,perimeter}$ values for all four peripheral hole positions should be approximately equal. To obtain an averaged CTDI value for the whole axial plane, the weighted CTDI has been defined as follows:

$$CTDI_W = \frac{1}{3} CTDI_{100,center} + \frac{2}{3} CTDI_{100,perimeter} \qquad (5.3)$$

As mentioned, $CTDI_{100}$ and $CTDI_W$ reflect the absorbed dose from the acquisition of one axial image. However, to obtain a dose index for a complete CT acquisition that involves a number of axial images, the separation between each axial image must be taken into account. For example, an axial CT protocol that involves the acquisition of contiguous images results in a higher dose than one in which there is a gap between acquired images. To reflect this variation, the volume CTDI has been defined as follows:

$$CTDI_{VOL} = \frac{nT}{I} CTDI_W \qquad (5.4)$$

where I is the table's travel distance between x-ray tube rotations. Given the definition of pitch for helical CT (I/nT), $CTDI_{VOL}$ can be expressed for helical CT as follows:

$$CTDI_{VOL} = \frac{CTDI_W}{pitch} \qquad (5.5)$$

Exam Description: CT Head w/o Contrast					
		Dose Report			
Series	Type	Scan Range (mm)	CTDIvol (mGy)	DLP (mGy–cm)	Phantom cm
1	Scout	–	–	–	–
2	Axial	I209.750–1151.450	78.06	474.88	Head 16
2	Axial	I150.000–153.680	55.76	565.33	Head 16
2	Axial	I192.500–I157.013	55.76	226.13	Head 16
			Total Exam DLP:	1266.34	

1/1

Figure 5.5 Screen capture of a typical dose report showing the $CTDI_{VOL}$ and the DLP resulting from a head CT scan. Dose reports similar to this one are now recorded with the patient's images by current CT systems.

$CTDI_{VOL}$ is the most commonly used and communicated CT dose index since it provides a protocol-specific (but not patient-specific) dose metric. It is now displayed by CT systems after image acquisition and recorded with the patient's images (Figure 5.5). However, $CTDI_{VOL}$ reflects only the dose to a phantom due to one CT gantry rotation, not a complete multiple-image patient scan. This allows for the comparison of techniques and protocols, but the $CTDI_{VOL}$ value will be the same no matter how many axial images were acquired for a single patient scan.

5.3.2 DOSE LENGTH PRODUCT

To obtain a measure that *relates* to actual patient absorbed dose for a complete CT scan, the total length of the CT scan must be considered in addition to the protocol used to acquire each image.[*] Clearly, if the same protocol is used for the acquisition of a 10-cm-long scan or a 20-cm-long scan, the latter results in double the total energy deposition than the former. However, the $CTDI_{VOL}$ will be the same for both cases. To take into account the difference in total volume scanned, the dose length product (DLP) has been defined as follows:

$$DLP = CTDI_{VOL} \times \text{total scan length} \qquad (5.6)$$

As expected, the DLP has units of mGy·cm. Given its useful information, the DLP also is displayed and recorded by CT systems after image acquisition (Figure 5.5).

5.3.3 COMPUTING EFFECTIVE DOSE IN CT

Recall that the metric that relates amount of irradiation to possible increase in stochastic risk is the effective dose. For CT acquisitions, there are two methods to estimate the effective dose resulting from a scan. The more involved method, similar to the method to estimate effective dose from conventional radiography, requires the estimation of each individual organ absorbed dose and the conversion of these dose values to a total effective dose using the ICRP-recommended tissue weighting factors. As noted

---

[*] As has been mentioned, it is important to remember that $CTDI_{100}$ is based on the measurement of absorbed dose to a cylindrical PMMA phantom; therefore, all the metrics derived from it, including the DLP, provide information on dose to this phantom and can be said to provide only *relative* information of actual patient-absorbed dose.

previously, the estimation of individual organ absorbed dose can be performed using Monte Carlo–based computer simulations or experimentally using anthropomorphic phantoms and point dosimeters (Fujii et al. 2009). For CT imaging, three European groups have compiled Monte Carlo–based results into easy-to-use computer programs to obtain estimates of effective dose for various CT systems and a large number of acquisition conditions (Kalender et al. 1999; Stamm and Nagel 2002; ImPACT Group 2009).

However, for CT-related effective dose estimations, it has been found that direct estimation of effective dose from DLP data can be performed using the following equation:

$$E = k \times DLP \qquad (5.7)$$

where k is a conversion factor, in units of mSv·mGy^{-1}·cm^{-1}, and varies only with the body region scanned (Huda et al. 2008; McCollough et al. 2008). Huda et al. (2008) have shown that the conversion factors k do not vary substantially for CT systems of different generations and manufacturers, making it a convenient method to estimate effective dose. Of course, the conversion factors k have been computed using simulated anthropomorphic phantoms representing average-sized patients of different ages, so the effective dose values estimated with these factors will underestimate risk for small patients and overestimate risk for large patients.

5.3.4 DOSIMETRY IN CBCT

The concept behind $CTDI_{100}$ and its derivative metrics ($CTDI_W$, $CTDI_{VOL}$, and DLP) is that the entire dose profile due to a single axial scan is included in the measurement. In other words, the dose function $D(z) = 0$ beyond $|z| = 50$ mm. As can be expected, with the introduction of multidetector CT, and especially in CBCT, this condition is not met; therefore, $CTDI_{100}$ and its derivatives result in a substantial underestimation of dose. As an example, Mori et al. (2005) have shown that only 76% and 60% of the dose profile would be captured when integrating a total of 100 mm of the profile if the x-ray beam's total width was 20 and 138 mm, respectively. Of course this issue could be avoided using a longer cylindrical phantom and a longer ionization chamber, but it is unrealistic to define a CTDI metric that would require people to use ionization chambers that are 40 cm in length.

To address this shortcoming and other limitations, an American Association of Physicists in Medicine (AAPM) Task Group recently proposed in its Report No. 111 a new method to estimate CT dose that is applicable to axial, helical, fan-beam, multislice, and cone beam technologies (Dixon et al. 2010). For axial scanning involving bed translation and for helical scanning, the basic concept behind this proposal is that instead of attempting to measure the integral of the entire dose profile simultaneously with a long ionization chamber without axial movement by the CT gantry, the dose profile should be measured using a point measurement translated axially until an *equilibrium length* is reached. The equilibrium length is defined as the scanning length beyond which the measured integral of the dose profile does not increase significantly. To perform this measurement, a Farmer-type ionization chamber (0.6-cm^3 active volume) is placed in the centerline of a long (at least 45 cm in length) cylindrical phantom and repeated measurements of the cumulative air kerma for scans with increasing axial displacement recorded. Given the shape of a dose profile (Figure 5.3), for a long enough axial displacement, the measured cumulative air kerma will not increase further. The length of this axial displacement is the equilibrium length, and the absorbed dose converted from the measured maximum air kerma is denoted the equilibrium dose. Report No. 111 specifies additional derivative metrics that can be obtained from the equilibrium dose to take into account variations in other imaging parameters.

For CBCT, in which no bed translation is involved during acquisition, the AAPM Task Group Report No. 111 proposes that only the value for the dose profile at the axial center of the x-ray field, denoted f(0), needs to be measured, again using a Farmer-type ionization chamber placed in the centerline of a cylindrical phantom. It has been shown that f(0) in CBCT with no bed translation is the equivalent to CTDI for narrow-beam axial CT with bed translation or helical CT (Dixon and Boone 2010).

As with CTDI, the measurement of f(0) provides enough information to compare the dosimetric consequences of different imaging protocols, and to identify system variations with time during quality assurance testing. However, as with CTDI and its related indices, both equilibrium dose (for axial scanning with bed translations and helical scanning) and f(0) (for CBCT with no bed translation) should not be viewed as a patient-specific dose metric, because, as mentioned, they are all based on phantom measurements that do not take into account differences in anatomy.

5.4 IMAGE ACQUISITION PARAMETERS AFFECTING DOSE

CT imaging involves a high number of acquisition parameters that affect not only image quality but also patient dose. Therefore, the optimization of the acquisition protocol involves a delicate balance between achieving the image quality required for the diagnostic task in question and minimizing the dose to the patient. The following are the acquisition parameters that affect patient dose and techniques to optimize them.

5.4.1 TUBE VOLTAGE

The x-ray tube voltage, denoted by the kilovoltage, peak (kVp), is the voltage that accelerates the electrons in the x-ray tube between the cathode and the anode. Typical tube voltage values in CT range from 80 to 140 kVp, mostly depending on the body region being scanned, with 120 kVp being the most common setting. Higher attenuating regions such as head and pelvis are normally acquired with the highest kVp settings, whereas the lowest kVp values are normally used for small and pediatric patients.

The x-ray spectrum output by the x-ray tube is directly affected by the tube voltage, with a higher tube voltage resulting in a higher mean and maximum x-ray energy output (Figure 5.6). As also can be seen in Figure 5.6, a higher tube voltage results in a higher overall output of x-rays at all energies, with the overall increase in tube output increasing by approximately the square of the increase in kVp. In addition, due to the overall higher penetrability of the higher-energy output x-ray spectrum, the signal at the detector, after attenuation of the x-ray beam by the

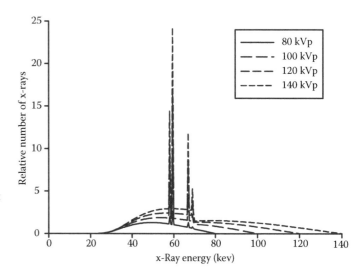

Figure 5.6 Typical x-ray spectra used in CT imaging. It can be seen how the overall energy of the spectra and the total number of x-rays increase with increasing tube voltage.

patient, increases even faster with increasing kVp. Therefore, care must be taken when considering the effect of varying tube voltage on dose and image quality.

Specifically, if all parameters, including tube current, are kept constant, then dose increases with increasing tube voltage, as can be seen in Table 5.1. The table shows the results of imaging a CT phantom (QRM-Cardio-CT-Phantom, QRM GmbH, Moehrendorf, Germany) with a 64 slice clinical CT system (LightSpeed VCT, GE Healthcare, Waukesha, WI). However, the increase in tube voltage also results in an increase in the x-ray flux emitted by the x-ray tube and arriving at the detector, so the image noise is greatly reduced with increasing tube voltage (Table 5.1 and Figure 5.7).

If tube current is adjusted downward when tube voltage is increased so as to maintain a constant image noise, then the dose to the patient will decrease with increasing tube voltage, as shown in Table 5.2. However, the attenuation difference (i.e., contrast) between different tissues decreases with higher energy, so higher tube voltage also results in lower contrast images (Figure 5.8).

Therefore, it should not be assumed that all images should be acquired with the highest tube voltage possible, because the trade-off between image noise, contrast, and patient dose, and consequently the decisions on tube voltage and current, must be analyzed carefully considering the diagnostic task at hand. For example, it has been determined that lower tube voltages (as low as 80 kVp), with their corresponding increased noise, result in lower doses for pediatric patients while still yielding diagnostically adequate images (Kim and Newman 2010). In addition, in large patients or when scanning high attenuation body regions, the use of higher tube voltages is necessary, because obtaining an adequate signal-to-noise ratio in this high attenuation conditions would result in impossibly high tube currents or unacceptably long scan times if low tube voltages are used. Furthermore, for contrast enhanced CT studies the use of lower tube voltages is beneficial, because this will result in an x-ray spectrum with an overall x-ray energy closer to the contrast agent's K absorption edge (33.2 keV for I), resulting

Table 5.1 Relationship between tube voltage, dose, image noise, and contrast with a constant tube current when acquiring the images in Figure 5.7

TUBE VOLTAGE (kVp)	TUBE CURRENT (mA)	RELATIVE DOSE	RELATIVE IMAGE NOISE	RELATIVE SIGNAL DIFFERENCE
80	200	31.0	178.3	136.0
120	200	100.0	100.0	100.0
140	200	140.0	82.0	89.4

Note: For all three metrics, the values for the 120-kVp image are used as the reference values. As expected, an increase in tube voltage while maintaining the same tube current results in an increase in dose and a decrease in image noise (measured as the standard deviation in a homogeneous region of interest in the image).

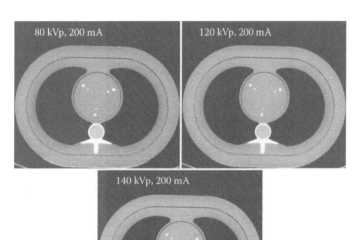

Figure 5.7 CT images of an anthropomorphic phantom acquired with different tube voltages while all other settings were kept constant. A decrease in image noise and contrast with increase in tube voltage can be seen. The values in Table 5.1 also show an important increase in dose with increasing tube voltage.

in a considerable higher image contrast. This higher contrast compensates for the increase in noise due to the lower x-ray tube output, yielding higher image quality for the required diagnostic task while resulting in lower patient dose (Schindera et al. 2008).

5.4.2 TUBE CURRENT AND TUBE CURRENT-EXPOSURE TIME PRODUCT

The tube current, measured in milliamperes (mA), determines how many electrons travel from the x-ray tube's cathode to the anode per unit time and therefore is related linearly to the x-ray flux. Therefore, if all other parameters, including the exposure time, are kept constant, the tube current is linearly and directly related to the absorbed dose. The tube current-exposure time product, expressed in milliamperes-seconds (mAs), is the product of the tube current and the length of time that the tube current is on and therefore the time the x-ray tube is emitting x-rays, so it is related linearly to the x-ray fluence. Thus, the tube current-exposure time product is also linearly and directly related to the absorbed dose.

In addition to its relation to patient dose, the tube current and the tube current-exposure time product are very important parameters in CT acquisition because they inversely affect to image

noise. Therefore, for the same exposure time, a decrease in tube current, while lowering the patient dose, results in a noisier image. Specifically, a change in the tube current results in an inverse change in the image noise by a factor of $1/\sqrt{mA}$. Due to this ever-present trade-off between patient dose and image noise, the choice of tube current and exposure time is very important, and should be given careful consideration when planning a CT acquisition protocol. In general, because rotation time, and therefore exposure time, is minimized for all protocols, the actual parameter that can be adjusted per protocol is the tube current.

The method for specifying the tube current for a CT protocol depends on how advanced the CT system is and whether the most advanced features of the most modern CT systems are used. Specifically, in protocols in which the tube current is specified explicitly, the overall patient size should be considered, especially in the case of pediatric patients. Due to their lower attenuation, body CT scanning of very small pediatric patients may be performed with a tube current level that is as much as four to five times *lower* than that used for a normal adult patient, while obese adult patients may be scanned with up to double the tube current used for a normal adult (McCollough et al. 2009). For head CT scanning, pediatric patients may be scanned with a tube current between 2 to 2.5 times lower than that for an adult, whereas no adjustment is necessary for obese adults (McCollough et al. 2009).

In modern CT systems, an advanced algorithm, called *tube current modulation*, is available to optimize the trade-off between image noise and patient dose. Recall that with all other conditions and parameters constant, image noise is inversely affected by tube current. However, during a CT scan of a patient, the object in the path of the x-ray field does not remain constant; the patient's anatomy and consequently its x-ray attenuation vary in a complex manner. For example, the general elliptical shape of the human body results in considerably lower attenuation when the x-ray field travels through the body in the anterior/posterior direction compared to the lateral direction. Therefore, if a constant tube current is used during the acquisition, the signal arriving at the detector varies considerably with projection angle, being much higher when the x-ray tube is positioned in the anterior/posterior locations and decreasing considerably when positioned in the lateral directions. This variability led to the development of the concept of *tube current modulation*, in which the tube current is varied almost continuously in a somewhat sinusoidal manner as the x-ray tube rotates around the patient (Figure 5.9). The use of this angular tube current modulation results in a substantial reduction in the total patient dose with little or no reduction in image quality.

Table 5.2 Relationship between tube voltage, dose, and tube current with a constant image noise when acquiring the images shown in Figure 5.8

TUBE VOLTAGE (kVp)	TUBE CURRENT (mA)	RELATIVE DOSE	RELATIVE IMAGE NOISE	RELATIVE SIGNAL DIFFERENCE
80	675	104.6	98.0	135.7
120	200	100.0	100.0	100.0
140	125	87.5	98.7	90.4

Note: For all three metrics, the values for the 120-kVp image are used as the reference values. The large decrease in tube current necessary to maintain image noise with increasing tube voltage results in a decrease in the dose, but the higher energy spectrum also results in a decreased image contrast.

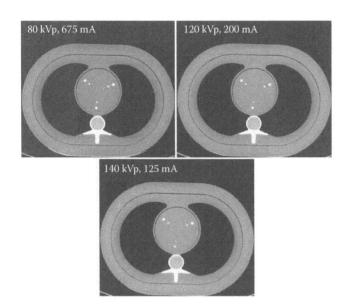

Figure 5.8 CT images of an anthropomorphic phantom acquired with different tube voltages while all other settings were kept constant, except for the tube current, which was decreased with increasing tube voltage so as to result in the same image noise in all three images. As can be seen, this required an important variation in tube current, which resulted in a decrease in dose with increasing tube voltage (see Table 5.2).

In addition to the variability in attenuation between the anterior-posterior and the lateral directions of the human body, the overall attenuation of the organs and tissues in the human body also vary in the longitudinal direction. For example, due to the presence of more bone tissue at the level of the shoulders the overall attenuation can be expected to be higher than at the level of the lungs. Therefore, a second, independent form of tube current modulation can be used that results in a variation of the tube current with x-ray tube position in the longitudinal direction (Figure 5.9).

Modern CT scanners include both of these tube current modulation features to optimize patient dose and image quality. The use of both angular and longitudinal tube current modulation has been found to result in important dose reductions, with studies reporting decreases in dose of 17% to 35% in adults (Tack et al. 2003; Graser et al. 2006) and more than 50% for pediatric applications (Herzog et al. 2008).

To define the desired level of image quality, different CT manufacturers require the user to specify different target metrics, including specifying the desired image noise level or the desired equivalent tube current-exposure time product that would be used for a standard patient or homogeneous phantom. Again depending

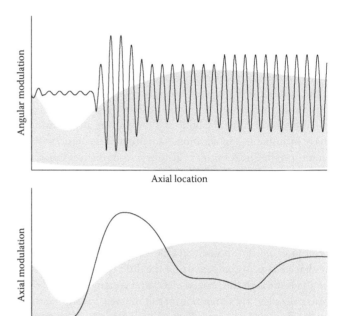

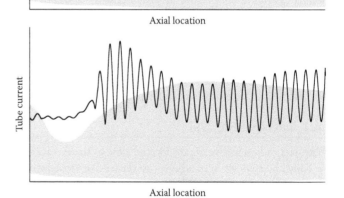

Figure 5.9 Typical tube current modulation. (Top) Variation in tube current due to angular modulation. (Middle) Variation in tube current due to longitudinal modulation. (Bottom) Actual tube current used for acquisition due to the combination of the angular and longitudinal modulation. The shaded shape shows a profile of the patient's anatomy.

on the system manufacturer, how the current is modulated is determined by analyzing a planar scout image acquired before the CT scan; in addition, some systems use pseudo-real-time analysis in which the detected signal from a previous projection, offset either 180° or 360° from the current signal, is used to determine the tube current for the current projection.

Fundamental principles and techniques

In CBCT systems, in which there is no movement of the gantry in the longitudinal direction during acquisition, the second form of tube modulation is not applicable. However, in applications in which there is considerable asymmetry in the anatomy being imaged, angular current modulation is applicable and should be considered (Daly et al. 2006). However, for many CBCT applications the implementation of angular current modulation will be challenging because, given the large amount of anatomy included in the field of view in each projection (e.g., along the z direction), it could be difficult to determine how to modulate the tube current when there is a large variability in the attenuation of objects included within each projection.

5.4.3 X-RAY FILTERS

The use of filters to absorb the lower energy x-rays and therefore *harden* the beam substantially reduces patient dose, especially to the skin (most important for deterministic effects), without substantially affecting image quality, since most of these x-rays would be absorbed by the body and not provide any additional image information. In addition, hardening of the beam with a filter installed next to the x-ray tube output reduces beam hardening in the body, which can introduce cupping (lower brightness in the center of the imaged object) and beam hardening artifacts (streaks due to highly attenuating features).

In addition to the use of flat filters to modify the shape of the x-ray spectrum of the entire x-ray beam, the use of bow-tie filters is common in CT, especially in CBCT. These filters (Figure 5.10) are shaped so as to compensate for the different path length of the x-rays through the imaged object and therefore reduce both patient dose and the dynamic range of the signal at the detector. In essence, bow-tie filters introduce an intraprojection x-ray intensity modulation that is not possible to introduce with tube current modulation. For example, in dedicated breast cone beam CT, it has been found that the use of a bow-tie filter can reduce the dose to the imaged breast by as much as 40% (Boone et al. 2004). To be effective, however, the shapes and the relative sizes of the bow-tie filter and the object being imaged must be matched, so in general there are various bow-tie filters available in each CT system.

5.4.4 ORGAN SHIELDS

To protect radiosensitive organs located close to the surface of the body and that are in or adjacent to the field of view during image acquisition, it is possible to use special, highly attenuating shields. Typical examples of organs that can be protected with organ shields are the breasts, thyroid, eyes, and testes. The shields, normally composed of bismuth, attenuate the direct x-ray fluence incident on these organs when the x-ray tube is on the same side of the body as these organs. Therefore, they could be most useful for organs in the field of view. For organs close but outside the field of view, their entire organ dose is due to scattered x-rays, and most of these originate inside the body, so shields placed on the surface of the body have limited utility.

The use of organ shields is the subject of intense research to better understand their impact on organ dose, image noise, and presence of artifacts. According to studies that have used anthropomorphic phantoms and dose estimating tools such as TLDs, organ dose reductions on the order of 20% to 36% can be

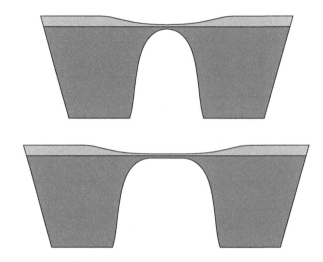

Figure 5.10 Diagram of two bow-tie filters used in CT systems. The two different shades of gray represent the possible use of two different materials in the filter, for example, graphite and Al.

achieved, with an accompanying increase in image noise of 10% to 50% (Hohl et al. 2006; Coursey et al. 2008; Hurwitz et al. 2009) and the introduction of some diagnostically important artifacts. However, another study that performed the dosimetry estimations using computer-based Monte Carlo methods, and therefore resulted in more comprehensive data, found that the reduction in the total dose imparted to the whole body due to organ shields is minor, and, more importantly, that the equivalent levels of organ dose reductions can be achieved with overall tube current reductions that results in a lower and more uniform increase in image noise than the presence of the organ shields, resulting in overall higher image quality (Geleijns et al. 2006).

The use of an overall reduction in tube current rather than organ shields has two more advantages. First, the x-ray signal attenuated by an organ shield when the tube is on the opposite side of the body will still be attenuated by the shield *after* the organs have been exposed, therefore providing no benefit in organ dose reduction but reducing image quality (Geleijns et al. 2006). In addition, using an organ shield with tube current modulation can be challenging. If the CT system determines the modulation to be used based on a scout image, then the shield should be placed after acquisition of the scout image to maximize the reduction in dose (Coursey et al. 2008; Hurwitz et al. 2009).

5.4.5 HALF-REVOLUTION ACQUISITIONS

In cone beam CT, where the acquisitions normally do not involve any axial movement of the CT gantry, it is possible to perform a complete CT acquisition with projection acquisitions that encompass only 180° (plus fan angle) around the imaged object. This capability introduces the possibility of achieving dose reductions to especially radiosensitive organs that are located toward one side of the body by selecting which side of the body should be the entrance side (Daly et al. 2006). For example, during a head CT acquisition, the projections acquired should be those in which the x-ray tube is positioned posterior to the head, therefore avoiding the direct irradiation of the lens of the eyes by the incident x-ray beam. Another potential example of this organ dose reduction technique would be in chest CT wherein direct irradiation of the breasts can be minimized by acquiring projections only when the x-ray tube is on the posterior side of the body.

5.4.6 RECONSTRUCTED IMAGE THICKNESS

The minimum nominal thickness of the axial images that will be reconstructed and viewed from the CT projection acquisitions determines the x-ray beam width in single-slice CT or the thickness of each data acquisition system channel used in multislice CT. In the case of the latter, although each detector element has a specific fixed size, the data acquisition system channel can be varied so as to read out each single detector element individually or by binning several detector elements together. For example, 0.75-mm detector elements can be read out individually to yield 0.75-mm-thick axial images or can be binned in pairs to result in a data acquisition system channel 1.5 mm thick to yield 1.5-mm-thick axial images. Because the projection information was binned before readout, thinner axial images from binned scans cannot be reconstructed. However, the acquisition of CT images with thicker slice or data acquisition system channels results in lower image noise, allowing for a decrease in patient dose or an improvement in image quality for a constant patient dose.

5.4.7 X-RAY BEAM WIDTH, GEOMETRIC EFFICIENCY, AND PITCH

In multidetector CT, to aid in the reconstruction, the x-ray beam's penumbra, although contributing to the patient's dose, is not included in the recorded signal. Therefore, when narrow x-ray beams are used, in which the relative size of the penumbra is large compared with the main portion of the beam, the contribution to dose from the unused penumbra is high compared with the total dose. This results in a larger x-ray beam width being associated with savings in dose due to the concept of *geometric efficiency*. In CBCT, however, by definition the x-ray beam width is very large, so geometric efficiency is very high and the relative contribution of the penumbra to the overall dose is low.

In helical CT, pitch describes the ratio between the bed's axial displacement per x-ray tube rotation and the x-ray beam width. Therefore, a pitch lower than unity results in overlap in acquired slices and therefore higher dose (assuming all other factors are held constant) but increased sampling in the longitudinal direction, whereas a pitch higher than unity decreases dose (again, assuming all other factors are held constant). It should be noted that some systems use the effective milliamperes-seconds or milliamperes-seconds/slice concept (where effective mAs = [(tube current × rotation time)/pitch] and that as pitch is increased, the tube current is increased by the same amount to keep a constant effective milliamperes-seconds. In CBCT, however, acquisition is performed with one x-ray tube revolution with no axial displacement, so pitch is not a factor in dose and image quality.

5.5 SPECIFIC APPLICATIONS OF CBCT

The dosimetry estimation methods described previously are not application-specific and therefore provide dose estimation tools that can be used, as appropriate, in almost any clinical application of axial, helical, or CBCT. However, some applications, due to either the specific organ being imaged or the imaging system used, require or can take advantage of special metrics and

estimation methodologies that are not applicable to CT imaging in general. These applications, specifically dedicated breast CT imaging and position verification for radiation therapy are discussed here.

5.5.1 CONE BEAM DEDICATED BREAST CT

As opposed to most other cone beam CT applications, dedicated breast CT typically includes a single organ in the field of view, the breast. This makes estimating the dosimetric characteristics of dedicated breast CT somewhat simpler than in body CT. For breast dosimetry studies, including those involving mammography, breast tomosynthesis and breast CT, the breast has been assumed to consist of three tissue types: a mixture of adipose tissue and glandular tissue, surrounded by a layer of skin tissue. Because cancer risk is minimal in adipose tissue, breast dosimetry is mainly concerned with the absorbed dose in the glandular tissue portions of the breast; therefore, the studied metric of absorbed dose in breast imaging is called the *glandular dose*. However, although this three-tissue model of the organ is quite simple, one challenge that is encountered in breast dosimetry is that the mixture of adipose and glandular tissue in a human breast is heterogeneous and pseudorandom (Figure 5.11), making it impossible to develop a model to represent all or most breasts encountered clinically. To overcome this problem, in breast dosimetry the breast is modeled as a volume consisting of a homogeneous mixture of adipose and glandular tissue, surrounded by skin. To study breasts with different proportions of adipose and glandular tissues, the homogeneous material's chemical composition is varied from the equivalent of 100% adipose tissue to the equivalent of 100% glandular tissue. The chemical composition of each of the three tissues typically used for breast dosimetry studies is that reported by Hammerstein et al. (1979).

As with most breast dosimetry studies, dosimetry in cone beam dedicated breast CT has been studied using computer-based Monte Carlo simulations. This method is preferred over empirical

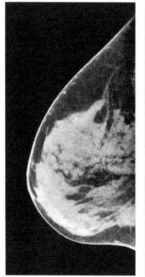
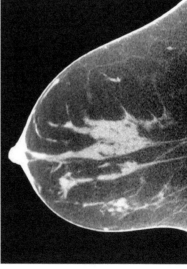

Figure 5.11 Dedicated breast CT slices of two different patients' breasts. The substantial possible variation among patients in amount of glandular tissue (bright gray) compared with adipose tissue (dark gray) and its pseudorandom distribution is apparent.

methods because it allows for the most accurate estimation of the glandular dose from the estimated total absorbed dose and it allows for easily studying the variation in glandular dose with different imaging conditions (such as source-to-object distance or x-ray energy) and for breasts of different characteristics (such as size, glandular fraction, and shape). The values reported in these studies are normalized mean glandular dose in CT (D_gN_{CT}), and provide the relationship between absorbed dose to the glandular portion of the homogeneous mixture of breast tissue, averaged throughout the whole breast, per unit air kerma, measured at a reference point for the same imaging technique (Boone et al. 2004; Thacker and Glick 2004; Sechopoulos et al. 2010). This reference air kerma is defined as the air kerma measured, without the breast present, with a ionization chamber placed at the center of the field of view (in the source-to-detector direction), with the end of the chamber positioned at the central ray (in the anterior-posterior direction) (Figure 5.12). Therefore, the mean absorbed dose to the glandular portion of the homogeneous breast can be estimated by multiplying the D_gN_{CT}, found in published graphs and tables (Boone et al. 2004; Thacker and Glick 2004; Sechopoulos et al. 2010), for the breast size and x-ray spectrum of interest by the measured air kerma resulting for the imaging technique of interest.

To obtain estimates of effective dose from cone beam dedicated breast CT, it is not enough to estimate the absorbed dose to the imaged breast, it is also necessary to take into account the absorbed dose to the rest of the organs of the body because these organs receive some dose from x-rays that scattered from the breast tissue. Due to the position of the breasts, this is especially important given their proximity to especially radiosensitive organs such as the lungs. Sechopoulos et al. (2008) used Monte Carlo simulations to estimate the dose to the other organs of the body during breast CT acquisition and provided values for a metric that relates glandular dose to effective dose, in units of millisieverts

per milligray. In this study, it also was established that the use of a lead shield between the x-ray source and the body, with an opening allowing for the imaged breast to pass through, did not decrease substantially the effective dose or the absorbed dose to the organs. This showed that most of the scattered x-rays that deposit energy in the rest of the body enter the trunk through the breast, making them impossible to shield against.

Recall that breast dosimetry provides estimates for a breast presumed to be comprised of a homogenous mixture of adipose and glandular tissue. Of course, this does not reflect any clinically encountered human breast. Therefore, the concept of normalized mean glandular dose is not to attempt to provide patient-specific dosimetry but as a relative measure of the amount of radiation involved in the imaging of a breast that can be used to perform relative comparisons between different techniques, technologies, and even among different breast imaging modalities.

5.5.2 CBCT FOR RADIOTHERAPY POSITIONING

The introduction of image-guided radiation therapy has greatly improved tumor targeting during the delivery of the planned therapy dose, improving tumor control and reducing dose delivery to normal tissues. With the introduction of on-board flat-panel imagers, the acquisition of a pair of orthogonal images for verification of patient positioning and identification of possible anatomical variations has not only become more convenient, but has also given way for the acquisition of tomographic images for this task. Two methods are used to acquire these tomographic images for image-guided radiation therapy: megavoltage CBCT (MV CBCT) that uses an on-board flat-panel imager opposite the system's existing radiotherapy linear accelerator to acquire the projections needed to reconstruct the imaged volume, or kilovoltage cone beam computed tomography (kV CBCT) that involves the addition of a traditional x-ray tube and an on-board flat-panel imager orthogonal to the linear accelerator to perform

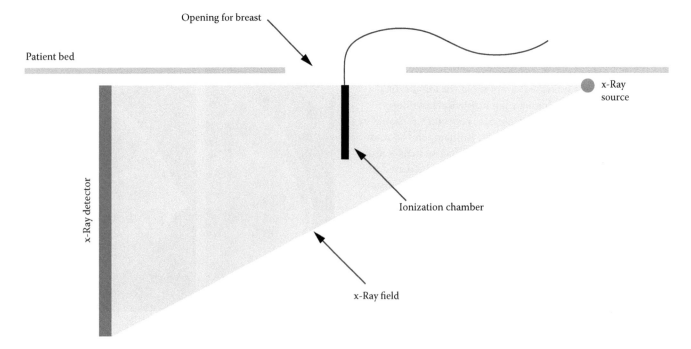

Figure 5.12 Schematic showing the correct positioning of the CT pencil ionization chamber for the measurement of the reference air kerma used for estimation of the mean glandular dose of a cone beam dedicated breast CT acquisition.

the projection acquisitions. As expected, these two imaging modes have different dosimetric consequences.

5.5.2.1 kV CBCT

Given its similarity to other CBCT applications, the same metrics and estimation methods for patient dose as those in diagnostic CT imaging are used in kV CBCT. However, in radiation therapy it is common to use the *maximum dose* as a dosimetry metric. The maximum dose refers to the highest local absorbed dose value estimated, normally using Monte Carlo simulations, to a small volume included either inside an organ or anywhere inside the irradiated tissue volume. The maximum dose therefore differs from the concept of organ absorbed dose, because the latter is the mean of the absorbed dose of the entire organ, and the former refers to a small volume in a particular location.

Various studies have been published in which the absorbed dose due to kV CBCT in patients undergoing image-guided radiation therapy is estimated. To obtain estimates of patient dose, in general these studies use either anthropomorphic phantoms with TLDs or perform computer-based Monte Carlo simulations. Of course, the absorbed dose estimates vary considerably depending on the therapy, and therefore imaging, site. For example, the organ dose to the eye lens during imaging for head and neck radiotherapy has been estimated at 8 cGy (80 mGy), while the dose to the prostate during prostate therapy has been estimated at 4 cGy (40 mGy) (Ding et al. 2008). In this same study, a table of various organ absorbed doses for various therapy sites was provided. In terms of effective dose, these have been estimated to range from 3.2 mSv for kV CBCT imaging related to head and neck therapy, up to 19.4 mSv for breast therapy (Sawyer et al. 2009). The considerable higher effective dose when imaging for breast therapy is expected due to the inclusion of many more highly radiosensitive organs in the field of view and the inclusion of larger anatomy in the projections, requiring a higher x-ray signal during acquisition.

When comparing the patient dose from kV CBCT to that from the acquisition of two orthogonal planar images acquired with the MV linear accelerator, it has been shown that the former technology results in considerably lower dose (Islam et al. 2006), in addition to the fact that the dose distribution from kV CBCT is more uniform to that from MV planar imaging. However, the dose due to MV imaging could be taken into account and compensated for by reducing the planned therapeutic dose, while the dose from kV CBCT, being from a very different x-ray energy range would be a lot more challenging to compensate for (Islam et al. 2006). This point might be minor though, since it has been reported that the dose due to kV CBCT imaging is approximately only 2%–10% of the typical planned therapy dose (Ding et al. 2008), so its impact on the patient, while not negligible, is secondary compared with the ability to better target the tumor site with the therapeutic dose thanks to the acquisition of the kV CBCT images.

5.5.2.2 MV CBCT

The use of the radiation therapy system's linear accelerator as the source of x-rays for the acquisition of the cone beam CT image results in a simplification of the system due to the avoidance of adding a second x-ray source. Therefore, the use of MV CBCT and its dosimetric characteristics has been a topic of considerable research.

One major difference between characterizing dosimetry for MV CBCT and for standard diagnostic imaging is the normalization metric. In diagnostic imaging, it is common to use the air kerma, in units of mGy, measured with a predefined procedure and location, as the normalization metric when reporting normalized dose values. However, in MV CBCT, normalized dose values are reported using the *monitor unit* (MU) as the normalization factor. One MU corresponds to the x-ray intensity that results in a maximum absorbed dose of 1 cGy (10 mGy) at the center of a 10×10 cm^2 beam in water using a source-to-surface distance of 100 cm.

Using this methodology, the x-ray intensity used when acquiring MV CBCT projections has been reported to be between 0.01 MU/projection and 0.05 MU/projection (Pouliot et al. 2005; Morin et al. 2006, 2007). Furthermore, a typical MV CBCT image involves the acquisition of 90–200 projections, each spanning between 1° and 2°. Therefore, a total emission of approximately 2–10 MU is used for a complete MV CBCT acquisition.

Because the linear accelerator is used to acquire the projections, it is possible to use standard treatment planning software to determine the patient dose from the MV CBCT acquisition. With such software, dose levels have been reported to various organs due to imaging for various target tumor types. For example, an absorbed dose to the head of 0.9 cGy/MU during head and neck cancer treatment and to the prostate of 0.75 cGy/MU during prostate cancer treatment has been reported (Morin et al. 2006). Considering the MU used for a complete MV CBCT acquisition, absorbed dose values of approximately 5 cGy (50 mGy) and 7 cGy (70 mGy) to the head and prostate (for head-and-neck and prostate cancer treatment), respectively, are common. For comparison purposes, a typical diagnostic head CT with a CTDI$_{VOL}$ of approximately 40–50 mGy results in an organ absorbed dose to the brain of approximately 2.3 cGy (23 mGy) to 3.8 cGy (38 mGy) (Jaffe et al. 2010), approximately half of the absorbed dose of the MV CBCT acquisition. In contrast, these dose levels for MV CBCT are typically less than 4% of the planned therapy dose; furthermore, it is possible to take into account this additional imaging-related dose when performing the treatment planning for the therapeutic dose delivery (Morin et al. 2007).

ACKNOWLEDGMENTS

I thank Dr. Xiangyang Tang for assistance in acquiring the CT images, Mr. Eric Jablonowski for the photographs, and Drs. Srini Tridandapani and Michael McNitt-Gray for reviewing the manuscript and for scientific discussions.

REFERENCES

Ambrose, J. 1973. Computerized transverse axial scanning (tomography): part 2. clinical application. *Br J Radiol* 46: 1023–47.

Boone, J.M., Shah, N. and Nelson, T.R. 2004. A comprehensive analysis of DgN(CT) coefficients for pendant-geometry cone-beam breast computed tomography. *Med Phys* 31: 226–35.

Coursey, C., Frush, D.P., Yoshizumi, T., et al. 2008. Pediatric chest MDCT using tube current modulation: effect on radiation dose with breast shielding. *Am J Roentgenol* 190: W54–61.

Daly, M.J., Siewerdsen, J.H., Moseley, D.J., et al. 2006. Intraoperative cone-beam CT for guidance of head and neck surgery: assessment of dose and image quality using a C-arm prototype. *Med Phys* 33: 3767–80.

Ding, G.X., Duggan, D.M. and Coffey, C.W. 2008. Accurate patient dosimetry of kilovoltage cone-beam CT in radiation therapy. *Med Phys* 35: 1135–44.

Dixon, R.L., Anderson, J.A., Bakalyar, D.M., et al. 2010. *Comprehensive Methodology for the Evaluation of Radiation Dose in X-Ray Computed Tomography*. AAPM Report 111. College Park, MD: American Association of Physicists in Medicine.

Dixon, R.L. and Boone, J.M. 2010. Cone beam CT dosimetry: a unified and self-consistent approach including all scan modalities—with or without phantom motion. *Med Phys* 37: 2703–18.

Fujii, K., Aoyama, T. Yamauchi-Kawaura, C., et al. 2009. Radiation dose evaluation in 64-slice CT examinations with adult and paediatric anthropomorphic phantoms. *Br J Radiol* 82: 1010–18.

Geleijns, J., Salvadó Artells, M., Veldkamp, W., et al. 2006. Quantitative assessment of selective in-plane shielding of tissues in computed tomography through evaluation of absorbed dose and image quality. *Eur Radiol* 16: 2334–40.

Graser, A., Wintersperger, B.J., Suess, C., et al. 2006. Dose reduction and image quality in MDCT colonography using tube current modulation. *Am J Roentgenol* 187: 695–701.

Hammerstein, G.R., Miller, D.W., White, D.R., et al. 1979. Absorbed radiation dose in mammography. *Radiology* 130: 485–91.

Heggie, J., Kay, J. and Lee, W. 2006. Importance in optimization of multi-slice computed tomography scan protocols. *Australas Radiol* 50: 278–85.

Herzog, C., Mulvihill, D.M., Nguyen, S. A., et al. 2008. Pediatric cardiovascular CT angiography: radiation dose reduction using automatic anatomic tube current modulation. *Am J Roentgenol* 190: 1232–40.

Hohl, C., Wildberger, J.E., Süß, C., et al. 2006. Radiation dose reduction to breast and thyroid during MDCT: effectiveness of an in-plane bismuth shield. *Acta Radiol* 47: 562–7.

Hounsfield, G.N. 1973. Computerized transverse axial scanning (tomography): part 1. Description of system. *Br J Radiol* 46: 1016–22.

Huda, W., Ogden, K.M. and Khorasani, M.R. 2008. Converting dose-length product to effective dose at CT. *Radiology* 248: 995–1003.

Hurwitz, L.M., Yoshizumi, T.T., Goodman, P.C., et al. 2009. Radiation dose savings for adult pulmonary embolus 64-MDCT using bismuth breast shields, lower peak kilovoltage, and automatic tube current modulation. *Am J Roentgenol* 192: 244–53.

ICRP (International Commission on Radiological Protection). 1977. *ICRP Publication 26: Recommendations of the International Commission on Radiological Protection*. Oxford, NY: ICRP by Pergamon Press.

ICRP (International Commission on Radiological Protection). 1991. *ICRP Publication 60: 1990 recommendations of the International Commission on Radiological Protection*. Oxford, NY: ICRP by Pergamon Press.

ICRP (International Commission on Radiological Protection). 2007. *ICRP Publication 103: The 2007 Recommendations of the International Commission on Radiological Protection*. Oxford, NY: ICRP Protection by Pergamon Press.

ImPACT Group. 2009. *ImPACT's ct Dosimetry Tool*. 2010. Available from: http://www.impactscan.org/ctdosimetry.htm (Accessed August 1, 2010).

Islam, M.K., Purdie, T.G., Norrlinger, B.D., et al. 2006. Patient dose from kilovoltage cone beam computed tomography imaging in radiation therapy. *Med Phys* 33: 1573–82.

Jaffe, T.A., Hoang, J.K., Yoshizumi, T.T., et al. 2010. Radiation dose for routine clinical adult brain CT: variability on different scanners at one institution. *Am J Roentgenol* 195: 433–8.

Kalender, W.A., Schmidt, B., Zankl, M., et al. 1999. A PC program for estimating organ dose and effective dose values in computed tomography. *Eur Radiol* 9: 555–62.

Kim, J.-E. and Newman, B. 2010. Evaluation of a radiation dose reduction strategy for pediatric chest CT. *Am J Roentgenol* 194: 1188–93.

McCollough, C., Cody, D., Edyvean, S., et al. 2008. *The Measurement, Reporting, and Management of Radiation Dose in CT*. AAPM Report 96.

McCollough, C.H., Leng, S., Yu, L., et al. 2011. CT dose index and patient dose: they are not the same thing. *Radiology* 259: 311–16.

McCollough, C.H., Primak, A.N., Braun, N., et al. 2009. Strategies for reducing radiation dose in CT. *Radiol Clin North Am* 47: 27–40.

Mori, S., Endo, M., Nishizawa, K., et al. 2005. Enlarged longitudinal dose profiles in cone-beam CT and the need for modified dosimetry. *Med Phys* 32: 1061–9.

Morin, O., Gillis, A., Chen, J., et al. 2006. Megavoltage cone-beam CT: system description and clinical applications. *Med Dosim* 31: 51–61.

Morin, O., Gillis, A., Descovich, M., et al. 2007. Patient dose considerations for routine megavoltage cone-beam CT imaging. *Med Phys* 34: 1819–27.

NCRP (National Council on Radiation Protection and Measurements). 2009. *Report No. 160 – Ionizing Radiation Exposure of the Population of the United States*. Bethesda, MD: NCRP.

Nekolla, E.A., Griebel, J. and Brix, G. 2009. *Population Exposure from Medical X-Rays in Germany: Time Trends – 1996 to 2005*. World Congress on Medical Physics and Biomedical Engineering, September 7–12, 2009, Munich, Germany, Springer Berlin Heidelberg. 25/3: 301–4.

Pouliot, J., Bani-Hashemi, A., Josephine, C., et al. 2005. Low-dose megavoltage cone-beam CT for radiation therapy. *Int J Radiat Oncol Biol Phys* 61: 552–60.

Sawyer, L.J., Whittle, S.A., Matthews, E.S., et al. 2009. Estimation of organ and effective doses resulting from cone beam CT imaging for radiotherapy treatment planning. *Br J Radiol* 82: 577–84.

Schindera, S.T., Nelson, R.C. Mukundan, S., et al. 2008. Hypervascular liver tumors: low tube voltage, high tube current multi–detector row CT for enhanced detection—phantom study. *Radiology* 246: 125–32.

Sechopoulos, I., Feng, S.S.J. and D'Orsi, C.J. 2010. Dosimetric characterization of a dedicated breast computed tomography clinical prototype. *Med Phys* 37: 4110–20.

Sechopoulos, I., Vedantham, S., Suryanarayanan, S., et al. 2008. Monte Carlo and phantom study of the radiation dose to the body from dedicated CT of the breast. *Radiology* 247: 98–105.

Stamm, G. and Nagel, H. 2002. CT-expo—a novel program for dose evaluation in CT. *RöFo: Fortschritte auf dem Gebiete der Röntgenstrahlen und der Nuklearmedizin* 174: 1570–6. [German]

Tack, D., De Maertelaer, V. and Gevenois, P.A. 2003. Dose reduction in multidetector CT using attenuation-based online tube current modulation. *Am J Roentgenol* 181: 331–4.

Thacker, S.C. and Glick, S.J. 2004. Normalized glandular dose (DgN) coefficients for flat-panel CT breast imaging. *Phys Med Biol* 49: 5433–44.

Xu, X.G. and Eckerman, K.F. 2010. *Handbook of Anatomical Models for Radiation Dosimetry*. Boca Raton, FL: CRC Press.

Fundamental principles and techniques

Part II

Advanced techniques

6

Image simulation

Iacovos S. Kyprianou

Contents

6.1 INTRODUCTION

Accurate models of computed tomography (CT) image projection have several important applications. The traditional utility of simulation was to generate data for developing and evaluating reconstruction algorithms. With the introduction of model-based iterative reconstruction algorithms, accurate projection models have become an important part of the reconstruction process. Realistic models of the projection geometry and physics result in reconstructions that more effectively extract information from projection data and improve image quality over reconstructions using traditional methods. Modeling the imaging chain of a CT concept system can aid in its design; the effects that the focal spot unsharpness, scatter, polyenergetic sources, detector blur, correlated noise, and realistic backgrounds have on the reconstructed image quality can be evaluated without the cost of hardware modifications. Improved detector models, geometries, scatter rejection techniques (Sisniega et al. 2013), and dose-reduction strategies (Rupcich et al.

2013) can be tested and evaluated before being implemented in hardware (Rupcich et al. 2013). Accurate modeling of all the physical properties involved in the generation of a projection image when combined with detailed anthropomorphic phantoms that include realistic pathological models (S. Gu and IS Kyprianou 2011) can allow clinical trials to be run computationally. Such capabilities would allow computational validation of new technology to great extent, thereby reducing the costs of clinical trials.

6.2 PROJECTION MODELS

A projection model is the core of CT simulation, and it can be generally divided into two categories: deterministic ray-tracing and probabilistic Monte Carlo–based models. In ray-tracing, a model of the source shape and energy spectrum is used to cast rays that are transported through the object to be imaged. The objects consist of databases of attenuation coefficients and geometrical boundaries. Scatter and detector blur, when modeled, are usually approximated and convolved as a kernel on the resulting projection image. Because of its nature, ray-tracing generates noise-free images. When quantum noise must be simulated, it is usually added once the projection images are generated (Rupcich et al. 2013).

Monte Carlo radiation transport approaches the image formation process as a set of probabilities. The x-ray source, including focal spot shape, energy spectrum, and any added spatially varying filtration such a bow-tie filter, is defined as a set of probability density functions. An x-ray, for example, is generated with a probability of having a specific energy according to the specified energy spectrum, is initiated from a location specified by a probability density function describing the spatial distribution of the focal spot, and has a direction according to an angular emission probability. Given those initial conditions, the location of the next interaction is determined based on the interaction cross sections of the current material. Once the location of the next interaction is determined, the particle is transported to that location using the geometric model used by the Monte Carlo code. Monte Carlo packages also test for material boundary intersections and reevaluate the probability of interaction. Once an interaction has occurred, the type of interaction, such as coherent, incoherent, photoelectric, or pair production, is determined; secondaries are generated and added to the particle stack. A particle is usually followed until its initial energy drops below a certain threshold. Detector models also make use of spatial and statistical probability distributions. For each interaction, Monte Carlo codes use both analytical and empirical probability distributions and scatter cross-section models. The materials involved in the imaging chain are usually defined using databases of published properties. Each particle arriving at the detector will have a log with information about the interactions it has experienced, the materials it has crossed, and its current state. The particle log is used to develop numerous detector models, such as ideal energy-integrating, detectors with simple spread and depth of interaction models or complex indirect optical and direct scintillator models. Because of the statistical nature of each interaction, any resulting image will have inherent noise and will include scatter from the object to be imaged as well as scatter in the detector. Depending on the accuracy of the physical models used, the resulting image, quantum noise, and scatter estimates can be very accurate and realistic. Because of the electron tracking, Monte Carlo can report the energy deposited in each of the materials or phantom voxels in the simulation; therefore, organ and patient dose can be estimated at the same time as the image generation (Rupcich et al. 2012).

6.2.1 MONTE CARLO PARTICLE TRANSPORT PACKAGES

In this chapter, we focus on using Monte Carlo to generate projection images; however, many of the ideas and concepts discussed are applicable to deterministic ray-tracing simulations. The Monte Carlo packages of this section can all generate separate images of the primary and the different types of scatter involved in radiation transport. Such images can help in system design because they can identify the effects of each scatter type on image quality (Sisniega et al. 2013). Furthermore, all packages referenced in this section have the capability to generate deterministic (noise-free) ray-traced projection images of the simulation geometries. A ray-traced-simulated projection can be

obtained quickly (typical projections can be obtained within a minute) with the same, noise-free transmission probability as the expected Monte Carlo–based scatter-free projection.

The Monte Carlo codes referenced in this chapter are based on the PENELOPE (version 2006; Salvat et al. 2003),* which is a general-purpose code that performs Monte Carlo simulation of coupled electron–positron–photon transport in arbitrary materials and in the energy range from 50 eV to 1 GeV. The standard geometry model used by PENELOPE (PENGEOM) is based on defining objects by the volume limited by a set of quadric surfaces.

6.2.1.1 penEasy imaging

penEasy_Imaging† is an extension of the general-purpose Monte Carlo particle transport simulation toolkit penEasy‡ that provides new capabilities that allow the simulation of medical imaging systems. Currently, penEasy_Imaging is based on the penEasy version 2008-06-15 and uses the atomic physics modeling subroutines from PENELOPE 2006.

6.2.1.2 penMesh

penMesh (Kyprianou et al. 2007; Badal et al. 2009)§ is a software package that combines a state-of-the-art Monte Carlo simulation algorithm and a modern computer-aided design (CAD) geometry model with the purpose of simulating radiation transport across complex, free-form objects. penMesh uses the particle transport physics from the PENELOPE 2006 package and a geometry model in which objects are defined by closed triangle meshes. The main program, reporting tally and source routines are based on the penEasy 2006 package, extended with the tools from the penEasy_Imaging package to enable generation of radiographic images. Currently, the main application of penMesh is the simulation of clinical x-ray imaging systems using detailed anatomical phantoms. However, the code is general enough to simulate any other situation involving photon, electron or positron transport, as long as the simulation stays within the applicability range of PENELOPE.

Figure 6.1 shows the algorithmic flowchart of penMesh; this flowchart uses a hybrid geometry model that combines the simplicity of objects defined by quadric surfaces and the geometric detail of objects represented by triangle meshes. penMesh uses two ray-casting methods: ray-tracing through quadrics derived from penEasy 2006 and a novel ray-tracing method through triangular meshes. When a ray is cast from the current position to the projected position of the particle, a determination must be made as to which quadric surface or triangular mesh (or which triangles) the ray intersects. To avoid determining the intersections between the ray and the millions of triangles in the simulation universe, penMesh implements a spatial data structure called octree. To generate an octree, penMesh iteratively subdivides three-dimensional (3D) space into boxes that contain triangles until a certain condition is met, that is, only one triangle is left in each box, or a certain number of subdivisions have been reached (Figure 6.2). Ray-tracing is then

---

* The PENELOPE subroutines are copyrighted by the Universitat de Barcelona and can be obtained for free at http://www.nea.fr/abs/html/nea-1525.html or at http://www-rsicc.ornl.gov/

† The code can be downloaded from http://code.google.com/p/peneasy-imaging/

‡ The latest version of penEasy can be obtained at the website http://inte.upc.edu/downloads

§ The code can be downloaded from http://code.google.com/p/penmesh/

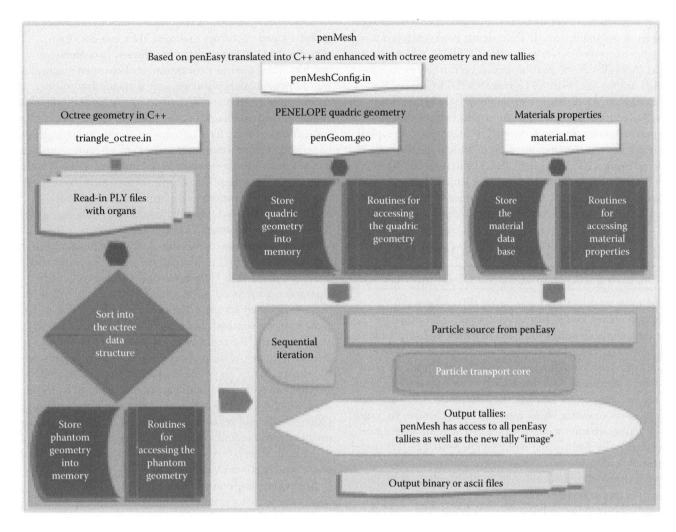

Figure 6.1 Flowchart of the penMesh Monte Carlo code.

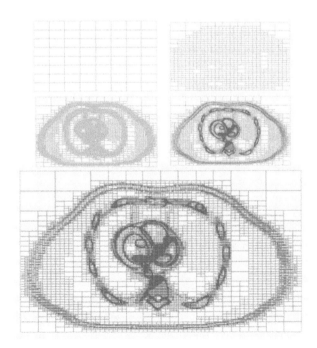

Figure 6.2 Octree node structure.

reduced to simple tracing only through the occupied voxels and the triangles that are in the voxels crossing the path of the ray. This method significantly accelerates the algorithm and makes simulation through complex, mesh-based objects possible within realistic timeframes.

One benefit of using penMesh for generating projections is the possibility of simulating with realistic, complex anthropomorphic phantoms with nonuniform resolution, incorporating geometries of imaging systems derived from their CAD drawings. Another benefit of the capabilities of penMesh is the ability to record images of the object of interest using different detector models and to separately record scatter and primaries.

6.2.1.3 MC-GPU

MC-GPU (Badal and Badano 2009; Badal et al. 2010) is a graphics processing unit (GPU)–accelerated x-ray transport simulation code that can generate clinically realistic radiographic projection images and CT scans of the human anatomy.

MC-GPU implements a massively multithreaded Monte Carlo simulation algorithm for the transport of x-rays in voxelized geometry and uses the x-ray interaction models and cross sections from PENELOPE 2006. The code can handle large, realistic voxel phantoms representing human anatomy. Electron transport, however, is not implemented. The code has been developed using the compute unified device architecture (CUDA) programming model (NVIDIA 2009) and the simulation can be executed in parallel in state-of-the-art GPUs from NVIDIA Corporation (Santa Clara, CA). A message parsing interface (MPI) library is used to address multiple GPUs in parallel (within a single computer or across networks) during CT simulations. In typical diagnostic imaging simulations, a 15- to 30-fold speed-up is achieved using a GPU compared with a central processing unit (CPU) execution. As Sisniega et al. 2013 demonstrate, the benefit of using GPU accelerated Monte Carlo transport is the simulation of CT configurations under realistic conditions within minutes.

6.3 IMAGING SYSTEM MODELING

6.3.1 GEOMETRICAL MODELING OF CT SIMULATION

To set up an image simulation sequence that will generate projection images for reconstruction, the user must define the location and orientation of the detector, object, and source for each projection in the image acquisition scheme. When developing a geometrical modeling framework for CT simulations, we must allow for a number of configurations that can represent different acquisition schemes: for example, the detector–source system can be fixed, with a rotating object placed between them, or the object can be stationary and rotate the source and detector. In the more general case, the detector and source can be rotated and translated independent of each other. The schematic of Figure 6.3 lays out a typical setup for cone beam CT (CBCT). The detector plane is centered on the **z**-axis of the laboratory reference frame, and is perpendicular to the central x-ray coming from the x-ray tube. The object lies at the origin between the detector and the source and can rotate about its rotation axis **q**, which usually coincides with the **y** axis of the laboratory reference system. For all image projection simulation codes we mentioned earlier in this chapter, one must define the source, detector, and object locations and orientations for each point in their trajectories about the patient. In this section, we describe a mathematical framework that can simplify the definitions of arbitrary CT geometries.

We use homogeneous coordinates to describe our imaging system. Homogeneous coordinates have a natural application

to x-ray imaging modeling; they form a basis for the projective geometry used extensively to project 3D scenes onto two-dimensional (2D) image planes (the display) in computer graphics. They also unify the treatment of common graphical transformations and operations, such as translation, scaling, rotation, and projection of geometric objects. With this in mind, computer-chip manufacturers for CPUs and GPUs have hard-coded instruction sets to perform calculations directly in homogeneous coordinates by performing calculations of four-dimensional vectors and matrices in single-step processes. Furthermore, 3D graphics languages such as openGL take advantage of these instruction sets and provide functions to perform the aforementioned transformations in real time. The graphical use of homogeneous coordinates can be historically attributed to Roberts (1965).

The homogeneous coordinates $[x\ y\ w]$ provide a scale invariant representation for points $(x',\ y')$ in the Euclidean plane, with $x'{\sim}x/w$, $y'{\sim}y/w$, $w \neq 0$. Homogeneous points with $w = 0$ are regarded as corresponding to points in the ordinary plane because they are infinitely far away. The need for such points at infinity arose from the sixteenth and seventeenth-century work of Kepler and Desargues. The ordinary plane augmented with points at infinity is known as the projective plane. The use of homogeneous coordinates in CT for generating projections, experimental system calibration and reconstruction was reported by Müller-Merbach (1996), Karolczak et al. (2001), and Kyprianou et al. (2006). As we will see later, homogeneous coordinates also are used to define nonuniform rational Bézier splines (NURBSs), mathematical tools used in the description of anthropomorphic phantoms (Segars et al. 2003).

We begin our geometry description by first defining the laboratory reference frame

$$\mathbf{R} = [x\ y\ z\ o] = \mathbf{I},\qquad(6.1)$$

where **x**, **y**, and **z** are orthogonal vectors and **o** is the system origin which for our system has been set to $o = [0\ 0\ 0\ 1]^T$. Here, we define the laboratory reference frame **R** to be **I**, the unitary matrix; however, **R** can be more general.

Let **D** be the detector reference system defined as the dot product:

$$\mathbf{D} = [\mathbf{a}\ \mathbf{b}\ \mathbf{c}\ \mathbf{d}] = \mathbf{RD},\qquad(6.2)$$

where **a**, **b**, and **c** are orthogonal vectors. **c** is a vector perpendicular to the detector, and **d** is the center of the detector reference system. Here, we assume that the detector is planar; however, our treatment can be generalized for curved detectors. We set the detector center at $\mathbf{d} = [0\ 0\ z_D\ 1]^T$ with **c** parallel to **z**. We set the detector at the negative **z**-axis; therefore, z_D takes negative values. The extent of the detector is defined by the scalars l_a and l_b.

The rotation $\mathbf{Rot}(\alpha, \beta, \gamma)$ with the Euler angles α, β, and γ about the origin **o** can be written as follows:

$\mathbf{Rot}(\alpha, \beta, \gamma)$

$$= \begin{bmatrix} 1 & 0 & 0 & 0 \\ 0 & \cos\alpha & -\sin\alpha & 0 \\ 0 & \sin\alpha & \cos\alpha & 0 \\ 0 & 0 & 0 & 1 \end{bmatrix} \begin{bmatrix} \cos\beta & 0 & \sin\beta & 0 \\ 0 & 1 & 0 & 0 \\ -\sin\beta & 0 & \cos\beta & 0 \\ 0 & 0 & 0 & 1 \end{bmatrix} \begin{bmatrix} \cos\gamma & -\sin\gamma & 0 & 0 \\ \sin\gamma & \cos\gamma & 0 & 0 \\ 0 & 0 & 1 & 0 \\ 0 & 0 & 0 & 1 \end{bmatrix}.$$

$$(6.3)$$

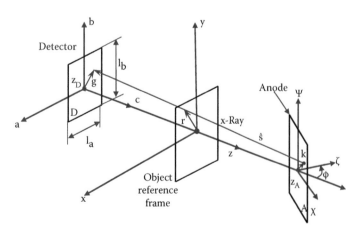

Figure 6.3 Scanner geometry for a typical CBCT acquisition.

The rotational transformation modifies vectors and coordinate systems. Notice that we follow the right-hand rule.

The translation transformation can be written as follows:

$$\mathbf{Trl}(x_t, y_t, z_t) = \begin{bmatrix} 1 & 0 & 0 & x_t \\ 0 & 1 & 0 & y_t \\ 0 & 0 & 1 & z_t \\ 0 & 0 & 0 & 1 \end{bmatrix}. \qquad (6.4)$$

Most x-ray tube anodes are at an angle with respect to the **z**-axis. They are designed in this way so that the projection images of large focal spots on the anode surface appear much smaller from the detector's point of view. The side effect, however, is that the size and shape of the focal spot appears different at different locations on the detector. We can define the focal spot reference frame **S** as a rotation about the **y**-axis (the angle φ is related to the anode angle as $\phi = \pi/2 - $ anode angle)* and a translation of the reference system by z_S:

$$\mathbf{S} = \mathbf{RTrl}(0, 0, z_s)\mathbf{Rot}(0, \phi, 0). \qquad (6.5)$$

For consistency with the detector negative location, we use positive values for z_S. Additional transformations can be added for tilted x-ray tubes such as in the case of breast CT where the tube radiation cone is rotated about the **z**-axis and tilted away from the patients' body.

6.3.1.1 Typical CT acquisition schemes

To construct a general CT trajectory, the detector **D** and source **S** reference frames can be rotated about their axes by the Euler angles $(\alpha_i^D, \beta_i^D, \gamma_i^D)$, and $(\alpha_i^s, \beta_i^s, \gamma_i^s)$, as well as translated by, (x_i^d, y_i^d, z_i^d) and (x_i^s, y_i^s, z_i^s). The index i denotes the ith angle of the discrete trajectory of the source–detector system.

For a system where the detector and source are equidistant from the rotational axis $\mathbf{q} = [x_q\, y_q\, z_q\, 0]^T$, we set $-z_D = z_s = R$, where R is the radius of the source–detector trajectory. For simplicity, we set x_q and z_q equal to zero for a rotation about the **y**-axis. Note that when the fourth component of the vector is set to zero, the vector points to infinity, thereby becoming a directional point vector. The location and orientation of the detector and source for the ith projection can then be obtained with the following operators:

$$\mathbf{D}_i = \mathbf{Rot}(0, \theta_i, 0)\mathbf{D} \qquad (6.6)$$

and

$$\mathbf{S}_i = \mathbf{Rot}(0, \theta_i, 0)\mathbf{S}, \text{ where } \theta_i = i\frac{360°}{N}. \qquad (6.7)$$

These matrix operators can be applied to the detector $\mathbf{d} = [0\ 0\ -R\ 1]^T$ and source center $\mathbf{s} = [0\ 0\ R\ 1]^T$ as well as their directional vectors (i.e., normal to the their planes) $\hat{\mathbf{d}} = [0\ 1\ 0\ 0]^T$ and $\hat{\mathbf{s}} = [0\ -1\ 0\ 0]^T$. The results can be given as inputs to the Monte Carlo codes referenced in this chapter. To simulate spiral CT as seen in Figure 6.4, a z-translation operation must be performed for each projection angle.

---

* The angle ϕ is related to the anode angle as $\phi = \pi/2 - $ anode angle.

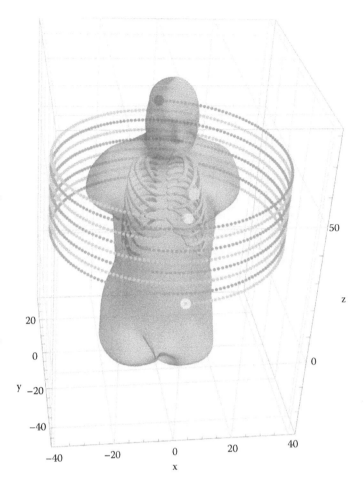

Figure 6.4 Spiral CT trajectories of source and detector for cardiac CBCT.

In the case of a bench-top CT system, where the source–detector system is fixed, then the object **f** is rotated about the rotational axis **q**. The object at the ith projection is obtained by the following formula:

$$\mathbf{f}_i = \mathbf{Rot}(0, \theta_i, 0)\mathbf{f}. \qquad (6.8)$$

6.3.2 MODELING A REALISTIC SOURCE

When modeling the x-ray source for simulating CBCT projections, there are two considerations that affect projection and CT image quality: the spatial distribution of the source radiance, commonly known as the *focal spot shape* and *size*, and the spectral distribution and intensity along the **x**- and **y**-axes resulting from bow-tie filtration and the Heel effect. The focal spot spatial extend degrades image quality by creating a penumbra around objects. The penumbra shape depends on the object location in the field of view. Bow-tie filters and the Heel effect, in contrast, affect both the intensity and the spectrum of the source radiance depending on the location in the field of view.

With Monte Carlo, we can simulate an x-ray tube, including the focal spot filament, electric field and anode, thereby generating very realistic x-ray tube spectra and x-ray radiance spatial distributions. However, such a detailed simulation for the typical numbers of x-ray photons involved in CBCT imaging would be very time-consuming, and adding a bow-tie filter in the simulation path would increase simulation time further. The approach we outline here creates a mathematical model for the spatial and energy probabilistic distributions of the source radiance that models the focal spot shape, Heel effect, and added bow-tie-shaped filtration. This approach alters the probability of photons generated at the source to have a given direction and energy in a similar way as would have resulted from an x-ray tube and added shaped filtration.

6.3.2.1 Realistic focal spot

Kyprianou et al. (2006) developed a Gaussian mixture model of a focal spot radiance spatial distribution based on the recorded shape of the focal spot at the detector plane as well as from models developed by Wagner et al. (1974) The measured image of the focal spot radiance at the detector plane was modeled by a linear combination of Gaussians:

$$L(\mathbf{r}_D) = \sum_{i=1}^{3} v_i e^{-(q_{a,i}\mathbf{r}_D - q_{b,i}\mathbf{r}_{0,i})^2} \tag{6.9}$$

By fitting the measured focal spot image, we can obtain the constants $\mathbf{r}_{0,i}$, $q_{a,i}$, $q_{b,i}$, and v_i. This model with three Gaussians describes the focal spot image, with a broad Gaussian centered at the origin modeling the general shape of the focal spot, and two narrow Gaussians modeling the effects of the focusing cup on the tube, which produces two sharp peaks at the top and bottom of the focal spot image.

To develop a focal spot source model for simulating projections, Equation 6.9 must be projected onto the focal spot plane. Here, for simplicity, we assume that in the vicinity of the focal spot image on the anode, the surface of the anode is relatively flat and can be considered planar. A ray $\mathbf{r} = [x\, y\, z\, 1]^T$ starting from a point $\mathbf{\kappa} = [\chi\, \psi\, 0\, 1]^T$ on the focal spot plane $\mathbf{S}$ intersecting the detector plane at the point $\mathbf{g} = [a\, b\, 0\, 1]^T$, for $|a| \leq l_a/2$, $|b| \leq l_b/2$ is defined as a function of the parameter t:

$$\mathbf{r} - \mathbf{RS\kappa} = (\mathbf{RDg} - \mathbf{RS\kappa})t. \tag{6.10}$$

The transformation $\mathbf{RDg}$ transforms the coordinates from the detector to the laboratory reference frame, whereas $\mathbf{RS\kappa}$ transforms the coordinates from the focal spot plane to the laboratory frame.

After solving Equation 6.10 for $\mathbf{g}$ and eliminating t, it can be shown that

$$\begin{pmatrix} a \\ b \end{pmatrix} \rightarrow \begin{pmatrix} x \\ y \end{pmatrix} + \begin{pmatrix} \dfrac{(z - z_D)(x - \chi)}{z_s - z + \chi \tan\phi} \\ \dfrac{(z - z_D)(y - \psi)}{z_s - z + \chi \tan\phi} \end{pmatrix}, \tag{6.11}$$

representing the mapping from a point on the focal spot plane via a point in the object space (laboratory reference frame) to the detector plane.

Using the mapping defined by Equation 6.11, we can map the focal spot radiance model (Equation 9.9) on the anode plane and define the spatial source model of our projection simulation code.

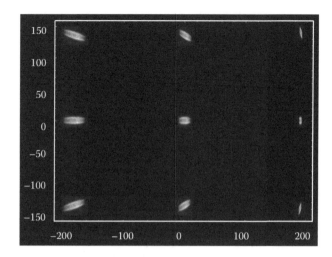

Figure 6.5 Demonstration of the significant differences of the source luminance projection (focal spot image) at different locations on the detector for the focal spot model of Equation 6.9.

Figure 6.5 shows example projections of the focal spot image through a pinhole positioned at various locations near the focal spot plane.

6.3.2.2 Bow-tie filter and Heel effect

Various studies have investigated how the Heel effect (Dixon 2005) and bow-tie-shaped filters (Turner et al. 2009; Abboud et al. 2013) affect the spatial intensity and spectral shape of CT x-ray source radiance. Using the method developed by Abboud et al. (2013), the angle-dependent spectra based on a model of the bow-tie filter installed in a clinical Toshiba Aquilion 64 CT scanner are shown in Figure 6.6 for 120 kVp.

With free-in-air measurements of air kerma, the method makes use of an ionization chamber and added Al filtration. With two measurements at each angle of interest (with and without added Al filtration), and with published spectral data, equivalent filter thicknesses of polymethyl methacrylate (PMMA) and Al were calculated to yield angle-dependent spectra. The Monte Carlo codes penEasy_Imaging and penMesh can use the angle-dependent spectra to define a source model based on the probability that each photon is emitted at a given angle and a given energy.

Using the models that Dixon (2005) developed for the Heel effect in CT, the intensity and spectral shape variation (beam hardening) of the x-ray source radiance due to the anode angle (Heel effect) can be accounted for a more complete source model.

6.3.3 MODELING THE DETECTOR

To generate projection images for a CT reconstruction, one must first define how the image is generated in the detector. The following detector models have been implemented within Tally Image, part of both penMesh and penEasy_Imaging.

6.3.3.1 Simple energy-integrating detector

The simplest model implemented in Tally Image used by both penEasy_Imaging and penMesh is an energy-integrating detector. Assuming a planar detector, images are generated by scoring the 2D energy deposition inside the detector object with a

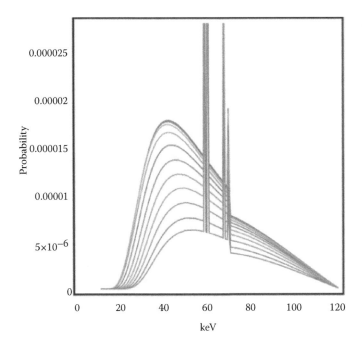

Figure 6.6 Angle-dependent spectra for the small bow-tie filter at 120 kVp of the Toshiba Aquilion 64 CT commonly used for head CT. The spectra are given here from 0 to 11° about the ψ axis starting from the highest to the lowest curve. For a symmetric filter, the curves would be mirrored to obtain the negative angles. The spectra were clipped for better display.

user-defined uniform grid of pixels. The pixel values have absolute units of eV/mm^2 per simulated history. If the detector material is set to be completely absorbent, the image values will be equivalent to the energy fluence passing through each pixel. With this tally, the user can choose one of the following three options: generate a single output image including all radiation; tally only primaries (the simulation is sped up, not tracking scatter); or tally separately the primary, the secondary (electrons, fluorescence) and the scattered particles, differentiating photons that suffered Rayleigh, Compton, and multiple scattering events in the track. As can be seen in Figure 6.7, the quantum noise of the images generated by this model is highly uncorrelated. The images therefore appear grainy compared with real images from scintillator-based energy-integrating detectors.

6.3.3.2 Simple optical model for indirect energy-integrating detector

With this model, images are generated modeling the transport of optical photons created after each x-ray interaction inside the scintillator. The optical photons are sampled using a triple Gaussian spatial distribution model fitted to an experimentally measured point spread function described by Kyprianou et al. (2008a) The model works by first identifying the location on the surface of the detector material where an interaction with energy absorption occurs. The number of optical photons is determined by the total energy deposited in the detector material by the particle and its secondaries. The optical photons are then spatially distributed according to the Gaussian-mixture point-spread-function and sampled in the detector pixels. This model generates realistic images that incorporate the detector

spread and have noise correlations similar to those found in physical detectors; however, the model is limited in that it does not account for depth of interaction.

6.3.3.3 Depth-dependent optical model for indirect energy-integrating detector

The optical photons in this model (Kyprianou et al. 2008b; Figure 6.8) are generated using a depth-of-interaction-dependent point spread function developed with the Monte Carlo radiation transport package called MANTIS (Badano and Sempau 2006).[*] The optical photon, electron/positron Monte Carlo package MANTIS was used to generate optical photon response φ and collection efficiency η as a function of the x-ray/electron interaction depth for a realistic scintillator geometry. The detector geometry used for the simulations was reported in the Badano et al. (2006a, 2006b, 2007) and is based on a 500-μm-thick columnar cesium iodide (CsI) scintillator. The resulting depth-dependent optical photon responses φ were fit to a parametrized Gaussian mixture model. The model parameters are the depth-dependent radial shift of the response peak, the depth dependent widths of the Gaussians, and the depth-dependent magnitude of the Gaussians in the mixture. The depth-dependent optical spread has a maximum spatial shift of 53 μm. The optical collection efficiency η at the photodiode layer follows a power law varying from 90% for interactions at the scintillator exit surface to 20% for interactions at the detector entrance. The number of optical photons γ generated at each energy deposition point is determined by the energy deposited in the detector material by the electrons and is governed by a Poisson distribution. The depth of interaction model can generate images that more accurately model detector spread and noise correlations.

6.3.3.4 Analytical ray-tracing detector model

This model disables Monte Carlo sampling and uses the penEasy transport routines as an analytical ray-tracing engine to generate scatter- and noise-free images. It has a major benefit over the previous detector models in that it can generate a ray-traced projection image of the geometry in a very short time however scatter is not accounted for. In this model, rays are cast from the focal spot to different positions inside each pixel to estimate the probability of x-rays arriving at the pixel without interacting in the track. This process is repeated by each energy bin given in the input energy spectrum. The final image is scaled by the probability of emitting the x-rays within the solid angle covering each pixel, assuming that the source is collimated to exactly cover the detector surface. The reported pixel values have units of eV/mm^2.

Comparing the three images of Figure 6.9, we can see that the image generated with the depth of interaction optical detector model appears the most realistic. Because it is highly uncorrelated, the noise in the first image appears grainy even though the two images were obtained with the same number of particles. The third image is scatter- and noise-free, and it appears less realistic. To give an idea of the timing involved in the simulations, the ray-traced image was obtained within 1 min using a single CPU, the first image within a few hours, and the second in about 10 hr with a typical 2008 personal computer.

---

[*] Monte Carlo x-rAy electroN opTical Imaging Simulation: http://code.google.com/p/mantismc/

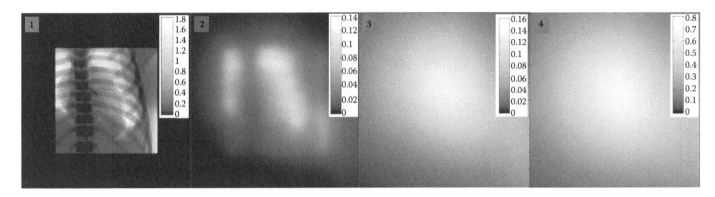

Figure 6.7 Images showing the primary-only (nonscattered) photons (1), the secondary (electrons, fluorescence) and the scattered particles, differentiating photons that suffered Rayleigh (2), Compton (3), and multiple scattering events (3) as obtained from the simple energy-integrating detector model.

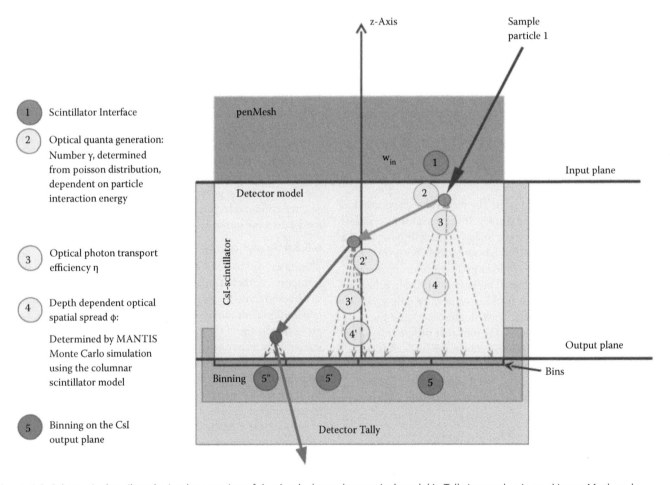

Figure 6.8 Schematic describes the implementation of the depth-dependent optical model in Tally Image that is used in penMesh and penEasy_Imaging.

6.4 OBJECT/PATIENT MODELS

Depending on the needs of a research project, there are phantoms available of varying degrees of complexity, simplicity, and realism. Simple stylized geometric objects made from cylinders, spheres, and other quadrics are simple to generate and serve many purposes. One of the first examples of such phantoms is the Shepp–Logan head phantom (Shepp and Logan 1974). Typical uses of stylized phantoms range from educational purposes, radiation dosimetry, and evaluation of CT reconstruction algorithms.

Phantom detail and complexity follow the trends of computational capacity, which continuously grows. Highly detailed phantoms can achieve higher degrees of realism. The closer anthropomorphic phantoms approximate human anatomy and physiology the more realistic research studies

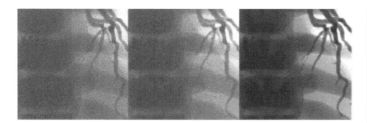

Figure 6.9 From left to right are simulated coronary artery angiograms using the simple energy-integrating, depth-dependent optical, and the analytical ray-tracing detector models, respectively.

can be performed, reducing the need for *in vivo* trials. Current trends in phantom development include very realistic anatomical phantoms developed by segmenting organs from CT or magnetic resonance (MR) images. There are phantoms with varying degrees of resolution that exhibit motion such as breathing and beating hearts. Phantoms of different ages, sizes, and sex are also available. Such phantoms allow for very realistic simulations, permitting studies that can evaluate imaging performance of novel CT imaging systems for specific imaging protocols and also for specific subpopulations, leading toward personalized medicine. For a reference on the current state-of-the-art on digital phantoms for simulations, please see Xu and Eckerman (2010).

In this chapter, we focus on two phantoms that we frequently use for image quality and dosimetry studies for CBCT, the NURBS-based cardiac torso (NCAT) phantom (Segars et al. 2003)* and the triangular mesh–based virtual family (Christ et al. 2010) or Virtual Population[†] set of phantoms.

6.4.1 NCAT PHANTOM

The NCAT phantom is based on the visible male project[‡] and was developed for nuclear heart imaging. Kyprianou et al. (2007) adapted the NCAT phantom, as shown in Figure 6.10, for penMesh by converting the individual NURBS surfaces of each organ into closed manifold and non-self-intersecting tessellated surfaces. NURBS is a mathematical model commonly used in computer graphics for generating and representing curves and surfaces and is defined using homogeneous coordinates (Piegl and Tiller 1995–1997). Additional surfaces were included for skin, bone marrow, intestinal contents, and inner lung matter and a high-resolution model of a heart with coronaries. The resulting 330 surfaces of the phantom organs and tissues are now comprised of about 5×10^6 triangles whose size depends on the individual organ surface normals.

A database of the elemental composition of each organ was generated, and material properties such as density and scattering cross sections were defined for PENELOPE. The resulting surface models of each organ were converted to the industry standard PLY format[§] and were passed to penMesh

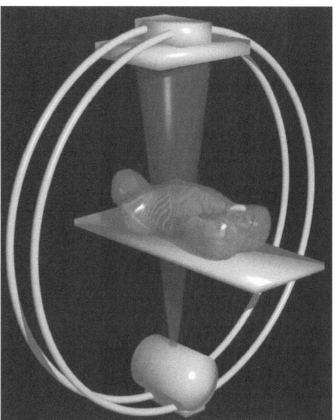

Figure 6.10 Depiction of the geometry for coronary artery CBCT angiography using a tessellated version of the NCAT phantom that includes a high-resolution heart phantom we developed.

simulation package as input. The benefit of mesh-based over voxel-based phantoms is that organs with different resolutions and detail can be intermixed to conserver memory and computational power. One example application is the simulation of high-resolution blood vessels within a lower resolution torso phantom.

6.4.2 VIRTUAL FAMILY PHANTOMS

The Virtual Family phantoms includes anatomically accurate whole-body human models of an adult male (34 years old), an adult female (26 years old), and two children (an 11-year-old girl and a 6-year-old boy). They were originally designed for the evaluation of electromagnetic exposure. The phantoms were developed by segmenting high-resolution MR images of healthy volunteers and include 80 different tissue types or organs. The software included with the phantoms can export voxelized versions of the phantoms with each voxel index representing one of the 80 organs. In a similar manner as with the NCAT phantom, PENELOPE was used to generate databases of the material compositions of each organ. The voxel-based versions of the Virtual Family phantoms can be used with both penEasy_Imaging and MC-GPU. Triangular mesh surfaces were fitted to each organ and converted into the PLY format to be used with penMesh (Figure 6.11). Additional surfaces were added for the skin as well as high-resolution heart phantoms (Rupcich 2013, Sisniega 2013, Rupcich 2013, Gu 2011).

---

* XCAT, MOBY and ROBY Phantom Series: http://deckard.mc.duke.edu/xcatmobyrobyphantom.html

† The Virtual population project: www.itis.ethz.ch/itis-for-health/virtual-population/

‡ The Visible Human Project: http://www.nlm.nih.gov/research/visible/visible_human.html

§ The Stanford 3D Scanning Repository: http://www-graphics.stanford.edu/data/3Dscanrep/

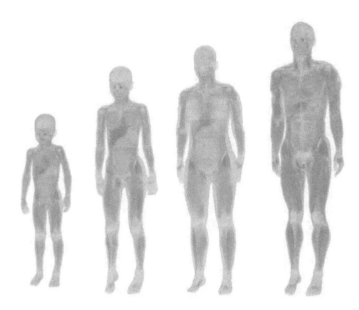

Figure 6.11 Volume rendering of the Virtual Family showing the level of detail and realism the phantoms have.

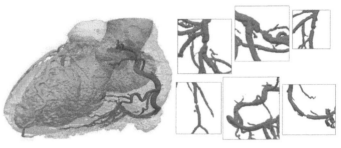

Figure 6.12 High-resolution heart phantom and coronary stenosis modeling.

6.4.3 HIGH-RESOLUTION HEART PHANTOMS

The heart phantoms included in the Virtual Family and the NCAT phantoms are low resolution and are inadequate for cardiac CBCT. Gu et al. (2011) developed an open source heart phantom development software platform called HADES,* with a graphical user interface (GUI) that can be used for developing high-resolution heart phantoms for image simulation purposes. They also have developed seven high-resolution cardiac/coronary artery phantoms for imaging and dosimetry, two of which are freely available.* To extract a phantom from a coronary CTA, the relationship between the intensity distribution of the myocardium, the ventricles, and the coronary arteries was identified via histogram analysis of the CTA images. By further refining the segmentation using anatomy-specific criteria such as vesselness, connectivity criteria required by the coronary tree, and image operations such as active contours, they were able to capture excellent detail within their heart phantoms. For example, in one of the female heart phantoms, as many as 100 coronary artery branches were identified. The left image on Figure 6.12 shows the triangular meshes that were fitted to the segmented high-resolution CTA data of the female heart phantom.

HADES includes a visualization and manipulation tool for adding stenotic lesions to the coronaries (coronary artery disease) as can be seen in the left image of Figure 6.12. Images of the stenotic lesions can be used in detectability studies. The resulting male and female heart phantoms have been registered within the mesh-based Virtual Family as well as the NCAT phantoms for high-resolution cardiac imaging simulations.

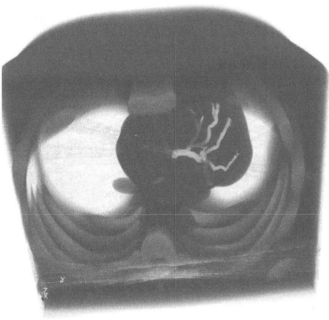

Figure 6.13 3D rendering of the coronary artery CT angiogram whose acquisition geometry is shown in Figure 6.10.

6.5 IMAGE GENERATION AND RECONSTRUCTION

We have described some of the tools that can be used to set up a CT simulation. Figure 6.10 shows the geometry we used for this example simulation for cardiac CBCT. The phantom was prepared by generating a material for the coronaries that included blood and 10% per volume iodine contrast. The simulation parameters were as follows: 120 kVp, 10° W anode spectra with 3-mm Al filtration without bow-tie filter, and 10^8 particles per degree for a circular trajectory 360° about the patient centered on the heart. The source to image distance was 140 cm, and the flat panel detector was 500 × 256 pixels or 110 × 11 cm using 600-μm-thick CsI scintillator. The negative logarithm of the normalized projection images were reconstructed using filtered back-projection. Figure 6.13 shows the volume rendering of the reconstruction.

---

* Heart phantom development tool and two male/female high-resolution heart phantoms: http://code.google.com/p/hades/

6.6 CONCLUSIONS

In this chapter, we have outlined concepts and several readily available tools that can be used either for Monte Carlo or ray-tracing-based projection simulations. The geometric description of the imaging system using homogeneous coordinates simplifies the definition of CBCT acquisition schemes. We have described readily available realistic anthropomorphic phantoms and the modifications needed to make them suitable for high-resolution CBCT imaging. A simple cardiac CBCT-simulated projection image acquisition and reconstruction was presented to demonstrate the simulation packages capabilities.

REFERENCES

Abboud, S., Jennings, R.J., Stern, S.H., et al. 2013. A method to characterize the shaped filtration associated with angle-dependent spectra of clinical computed tomography (CT) systems. *Med Phys*.

Badal, A. and Badano, A. 2009. Accelerating Monte Carlo simulations of photon transport in a voxelized geometry using a massively parallel graphics processing unit. *Med Phys* 36: 4878.

Badal, A., Kyprianou, I., Sharma, D. and Badano, A. 2010. Fast cardiac CT simulation using a graphics processing unit-accelerated Monte Carlo code. *Proc SPIE* 7622. In: Samei, E. and Pelc, N.J. (eds.). *SPIE Medical Imaging 2010: Physics of Medical Imaging*, 762231.

Badal, A., Kyprianou, I.S., Badano, A., et al. 2009. penMesh—Monte Carlo radiation transport simulation in a triangle mesh geometry. *IEEE Trans Med Imaging* 28(12): 1894–901.

Badano, A., Kyprianou, I.S., Jennings, R., et al. 2007. Anisotropic imaging performance in breast tomosynthesis. *Med Phys* 34(11): 4076–91.

Badano, A., Kyprianou, I.S. and Sempau, J. 2006a. Three-dimensional columnar CsI model for x-ray imaging system simulations using mantis: validating for noise, blur, and light output. *Proc SPIE* 6142. In: Flynn, M.J. and Hsieh, J. (eds.). *Medical Imaging 2006: Physics of Medical Imaging*. 61420W.

Badano, A., Kyprianou, I.S. and Sempau, J. 2006b. Anisotropic imaging performance in indirect x-ray imaging detectors. *Med Phys* 33(8): 2698–713.

Badano, A. and Sempau, J. 2006. MANTIS: combined x-ray, electron and optical Monte Carlo simulations of indirect radiation imaging systems. *Phys Med Biol* 51: 1545–61.

Christ, A., Kainz, W., Hahn, E.G., et al. 2010. The Virtual Family—development of surface based anatomical models of two adults and two children for dosimetric simulations. *Phys Med Biol* 55(2): N23–38.

Dixon, R.L. 2005. An improved analytical model for CT dose simulation with a new look at the theory of CT dose. *Med Phys* 32: 3712.

Gu, S., Gupta, R. and Kyprianou, I.S. 2011. Computational high-resolution heart phantoms for medical imaging and dosimetry simulations. *Phys Med Biol* 56: 5845.

Karolczak, M., Schaller, S., Engelke, K., et al. 2001. Implementation of a cone-beam reconstruction algorithm for the single-circle source orbit with embedded misalignment correction using homogeneous coordinates. *Med Phys* 28(10): 2050–69.

Kyprianou, I.S., Badal, A., Badano, A., et al. 2007. Monte Carlo simulated coronary angiograms of realistic anatomy and pathology models. *Proc SPIE* 6509. In: Cleary, K.R. and Miga, M.I. (eds.). *Medical Imaging 2007: Visualization and Image Guided Procedures*, 65090O.

Kyprianou, I.S., Badano, A., Gallas, B.D., et al. 2006. A practical method for measuring the H matrix of digital x-ray and cone beam CT imaging systems. *Proc SPIE* 6142. In: Flynn, M.J. and Hsieh, J. (eds.). *Medical Imaging 2006: Physics of Medical Imaging*. 61421U.

Kyprianou, I.S., Badano, A., Gallas, B.D., et al. 2008a. Singular value description of a digital radiographic detector: theory and measurements. *Med Phys* 35(10): 4744–56.

Kyprianou, I.S., Brackman, G., Myers, K.J., et al. 2008b. An efficient depth- and energy-dependent Monte Carlo model for columnar CsI detectors. *Proc SPIE* 6913. In: Hsieh, J. and Samei, E. (eds.). *SPIE Medical Imaging 2008: Physics of Medical Imaging*. 69130.

Kyprianou, I.S., Paquerault, S., Gallas, B.D., et al. 2006. Framework for determination of geometric parameters of a cone beam CT scanner for measuring the system response function and improved object reconstruction. *IEEE Transactions on Information Technology in Biomedicine*, Crystal City, VA, 1248–51.

Müller-Merbach, J. 1996. *Simulation of X-Ray Projections for Experimental 3D Tomography*. Report No. LiTH-ISY-R-1866, ISSN 1400-3902. Sweden: Linköping University.

NVIDIA, "NVIDIA CUDATM Programming Guide (Version 2.2)," tech. rep., NVIDIA Corporation (2009). Available from: http://www.nvidia.com/cuda.

Piegl, L. and Tiller, W. 1995–1997. *The NURBS Book*, 2nd edn. Heidelberg, Germany: Springer-Verlag.

Roberts, L.G. 1965. *Homogeneous Matrix Representation and Manipulation of N-Dimensional Constructs*. Technical Report MS-1405, Cambridge, MA: Lincoln Laboratory, MIT.

Rupcich, F., Badal, A., Kyprianou, I. S., Gilat Schmidt, T. A. 2012. Database for estimating organ dose for coronary angiography and brain perfusion CT scans for arbitrary spectra and angular tube current modulation. *Med phys* 39, 5336.

Rupcich, F., Badal, A., Popescu, L.M., Kyprianou, I. and Schmidt, T.G. 2013. Reducing radiation dose to the female breast during CT coronary angiography: a simulation study comparing breast shielding, angular tube current modulation, reduced kV, and partial angle protocols using an unknown-location signal-detectability metric. *Med Phy* 40: 081921.

Salvat, F., Fernandez-Varea, J.M. and Sempau, J. *PENELOPE, A Code System for Monte Carlo Simulation of Electron and Photon Transport*. Workshop Proceeding Issy-les-Moulineaux, OECD/NEA 7–10 July 2003. ISBN: 92-64-02145-0. Available from: http://www.oecd-nea.org/tools/abstract/detail/nea-1525 (accessed September 2013).

Segars, W.P., Tsui, B.M.W., Frey, E.C., et al. 2003. Extension of the 4D NCAT phantom to dynamic X-ray CT simulation. *IEEE Nucl Sci Symp Conf* 5: 3195–9.

Shepp, L.A. and Logan, B.F. 1974. The Fourier reconstruction of a head section. *IEEE Trans Nucl Sci* 21(3): 21–43.

Sisniega, A., Zbijewski, W., Badal, A., Kyprianou, I.S., Stayman, J.W., Vaquero, J.J. and Siewerdsen, J.H. 2013. Monte Carlo study of the effects of system geometry and antiscatter grids on cone-beam CT scatter distributions. *Med Phy* 40: 051915.

Turner, A.C., Zhang, D., Kim, H.J., et al. 2009. A method to generate equivalent energy spectra and filtration models based on measurement for multidetector CT Monte Carlo dosimetry simulations. *Med Phys* 36: 2154–64.

Wagner, R.F., Weaver, K.E., Denny, E.W., et al. 1974. Toward a unified view of radiological imaging systems. *Med Phys* 1: 11–24.

Xu, X.G. and Eckerman, K.F. 2010. *Handbook of Anatomical Models for Radiation Dosimetry*. Boca Raton, FL: Taylor & Francis.

7 3D image processing, analysis, and visualization

Kenneth R. Hoffmann, Peter B. Noël, and Martin Fiebich

Contents

7.1 INTRODUCTION

Computed tomography (CT) data is obtained from numerous sources for several applications. The sources determine the quality of the CT data, and the diagnostic questions drive the analysis and visualization tools developed for presentation of the information contained in the CT data. In this chapter, we present an overview of some of the image processing, analysis, and visualization approaches used for three-dimensional (3D) data, CT data in particular. We begin with an introduction of some of the aspects of the CT data.

7.1.1 SOURCES OF CT DATA

A primary, if not the primary, source of clinical CT data is helical CT. In helical CT, also called spiral CT, the CT gantry (i.e., the x-ray source and detectors) rotates about the patient as the patient is moved forward along the axis parallel to the axis of rotation of the CT gantry (z-axis). As a result, the projection data are obtained along a helix of a cylinder. The data on the helix are then interpolated to generate CT slice data with which standard CT reconstructions can be performed (Calhoun et al. 1999; Schaller et al. 2000). Over the past decade, multiple detector rows have been added to allow faster scanning of the patients. Faster scans have opened the door for many applications, including imaging of moving structures, such as the heart or the lungs. In addition, helical CT changed CT presentation and analysis substantially and significantly. Before helical CT, individual slices were obtained as the gantry rotated about the patient, the patient was moved, and the data for the next slice were obtained. Thus, the CT data resembled a stack of pennies, with each slice being "independent" of the other slices, often disrupting the continuity of the various structures. With helical CT, the data in the slices are coupled due to the interpolation of the acquired data (see Chapters 2 and 3). Indeed, CT slices can be reconstructed for arbitrarily small sampling distances between slices. However, improved sampling does not in general improve resolution that is determined by the detectors and the pitch (distance moved along z during one rotation). Because of this coupling, the structures in helical CT data sets are continuous (when there are no artifacts; see the following section and Chapter 4 and 11). In addition, the voxel size along the z-axis can be chosen relatively arbitrarily, so that isotropic voxel sizes (the same along all three axes) can be chosen. The coupling of the CT data across slices and the "isotropy" of the voxel dimensions give rise to a continuity of the structures that was not possible in standard CT.

A second major CT breakthrough came with the publication of the paper by Feldkamp, Davis, and Kress (1984). Before the publication of this paper, CT data were restricted to single or a small number of slices because the data from the out-of-plane detectors could result in artifacts in the reconstruction. Feldkamp, Davis, and Kress (1984) proposed a method by which these out-of-plane data could be used to generate high-quality 3D data sets. Their approach opened the door for multidetector CT, that when coupled with helical CT, increased the scan rate and the quality of the resulting reconstructions (see Chapter 3 and 11). Their paper also led to the development of cone beam CT. In cone beam CT, a "wide-angle" or large-area detector array is used as the x-ray detector, for example, an image intensifier or flat-panel detector. As a result, an entire volume (e.g., the head) can be imaged and then reconstructed. Like helical CT, the reconstructed CT data from cone beam is continuous through the volume, and the voxel sizes tend to be isotropic. In general, the acquisition time of cone beam CT systems is on the order of seconds to minutes, thereby influencing the types of applications for which it is used.

Helical and cone beam CT form the basis of all current CT acquisitions. Each has strengths and weaknesses in terms of speed, resolution, dynamic range, and noise (see Chapters 3, 4 and 11). Both can be used for volume-of-interest CT (VOI-CT; see Chapter 8) and in methods using fewer views reducing patient dose, or both (see Chapter 5; McNitt-Gray 2002; Angel et al. 2009; Nievelstein et al. 2010). But, it is the speed of acquisition and the quality of the reconstructed data (quality that reflects closely the anatomy itself) that have led to an explosion of CT applications over the last decade and that have affected dramatically the analyses that can be and have been attempted.

7.1.2 CT APPLICATIONS

The speed of acquisition and the continuity of the reconstructed structures are the primary reasons for the ubiquity of CT in patient work up in just about every aspect of patient care (Table 7.1). Research and patient evaluation using CT colonoscopy (Fletcher et al. 1999; Macari et al. 1999) and bronchoscopy (Vining et al. 1996; Fleiter et al. 1997; Lee et al. 1997) are feasible because of helical CT. Evaluations of pulmonary emboli, lung nodules, and lung motion (in radiation

Table 7.1 Number of research articles on various CT-related topics sited by PubMed

TOPIC	NO. OF ARTICLES
CT analysis	47,046
CT computer analysis	4967
CT computer-aided diagnosis	6672
CT computer analysis bone	1045
CT computer analysis head	407
CT computer analysis cranial	455
CT computer analysis colon	67
CT computer analysis polyp	76
CT computer analysis lesions	432
CT computer analysis lung	648
CT computer analysis nodules	177
CT computer analysis vessels	382
CT computer analysis arteries	333
CT computer analysis heart	285
CT computer analysis liver	272
CT computer analysis bronchi	31
CT computer analysis parenchyma	65
CT image processing	7554
CT visualization	3251
CT rendering	963

oncology; see Chapter 15) are enhanced. Cone beam CT of the breast (see Chapter 16) is being investigated as a means to improve breast cancer detection. CT is used for a wide variety of vascular evaluations, diagnostic as well as interventional, in several organs, including the heart (see Chapter 13), in the assessment and evaluation of aneurysms and stenoses. The quality of the CT data have spawned several computer-assisted diagnosis (CAD) techniques in the lung, liver, colon, bronchi, coronary vessels, cranial vessels, and the breast. In addition, cone beam CT is appearing in dental schools across the United States, with many dentists, primarily oral surgeons and endodontists, purchasing cone beam CT system for their practices.

7.1.3 CT IMAGE BASICS

So what makes CT data so useful? In Chapter 4, the various aspects of image quality are discussed in detail. In Chapter 10, methods for correcting the projection and CT data are presented. Here, we briefly review those aspects that are relevant to processing, analysis, and visualization.

CT data sizes are relatively large. Each image in the CT data set usually contains 512 × 512 pixels (but 1 K × 1 K are appearing, e.g., for inner ear studies; Mayer et al. 1997; Caldemeyer et al. 1999; Van Spaendonck et al. 2000), and the number of images can range from a few tens of images to thousands. Note, cone beam CT studies are acquired using N × M arrays, where N and M are usually approximately 1 K, so that N³-sized CT 3D data sets are standard. CT data for each pixel are generally stored in 2 bytes, so CT data sets can range from 100 megabytes (MB) to over 4 gigabytes (GB). The size of the data set may preclude loading the entire data set and results of analysis into memory at once. Thus, partial CT volumes or lower resolution volumes may be used in the analyses. The voxel sizes tend to be fractions of millimeters (0.1–0.6 mm) depending on the acquisition and applications, with micro-CT voxel sizes ranging from 1 to 50 Mm. The resolution of the data depends on the acquisition parameters, but the full-width-at-half-maximum of the resolution function ranges from 0.3 to 0.5 mm for clinical systems to 2 to 50 Mm for micro-CT systems. The system resolution and the sampling used in the reconstructions impact the quality of the CT, producing artifacts such as partial volume effects (Turnier et al. 1979; Glover and Pelc 1980; Schneiders and Bushong 1980) and aliasing (Stockham 1979; Crawford 1991; Hathcock and Stickle 1993; Yen et al. 1999; Boone 2001; Zou et al. 2004). In addition, inconsistencies in the projection data as a result of patient motion during the acquisition, from the imaged object being larger than the field of view, and/or from variations in the detector sensitivities, result in artifacts in the reconstruction (see Chapter 3) that must be corrected before analysis and presentation. One last consideration is that of noise in the images. In CT, because of the reconstruction process, the noise in the reconstructed data is highly correlated spatially; and because of the reconstruction process (at least if standard techniques are used), the noise in the reconstructed data is linearly related to the noise in the projection images. Thus, increasing the noise in the projection data (e.g., from lowering the x-ray dose) results in increased noise in the reconstructed data. Each of these aspects of the image data needs to be taken into account in some way in the development of the analysis and

subsequent visualization. For the rest of this chapter, we assume that artifacts have been dealt with (see Chapters 3 and 4) and that we have a pretty good idea of what the resolution and noise in the reconstructed data are. Note that CT data, like all other radiographic data, are stored and transferred using the Digital Imaging and Communications in Medicine (DICOM) standard, in which the acquisition parameters as well as the image data are part of the data set.

7.2 3D IMAGE PROCESSING

Volumetric data can be processed with a huge variety of intensity transformations, resampling procedures, and spatial filtering. Combinations of these methods are frequently applied for subsequent segmentation and registration tasks. In this section, image processing techniques are presented. To enhance readability and depiction in most cases, the methods are presented in two dimensions (2D), but they can easily be extended to 3D. When necessary, the 3D case will be used. In CT imaging, a lot of image processing tasks can be implemented in the image space or in Fourier space.

7.2.1 INTENSITY TRANSFORMATIONS

Intensity transformations are the most commonly used image processing techniques and belong to the simpler methods. The value of a voxel is usually given in Hounsfield units (HU), sometimes called CT number. The HU scale is defined such that –1000 corresponds to air and 0 corresponds to water. The conversion from attenuation coefficient to HU is given by

$$HU = K\,(\mu_{material} - \mu_{water})/\mu_{water}, \quad (7.1)$$

where K is a constant (usually 1000), $\mu_{material}$ is the linear attenuation coefficient of the material, and μ_{water} is the linear attenuation coefficient of water. Note that the attenuation coefficients are energy dependent, so what is measured is an effective attenuation coefficient that depends on the x-ray beam quality. CT machines are usually calibrated on a regular basis so that this relationship is known and the HU values in the 3D data are stable. The eye and the display for CT images are not able to handle 4096 (2^{12}) gray values; therefore, the image data are adjusted for appropriate visual appreciation.

The HU scale has to be mapped to a displayed intensity value that extends from 0 to 255, for example, air is shown as black (0) and bone as white (255). This conversion process is in most cases done with a piecewise linear function. For example, a function to convert the voxel values lying between a lower limit (A) and an upper limit (B) to a scale of 256 gray values is

$$
f(HU) = \begin{cases}
= 0 & \text{for } HU \le A \\
= (HU - A) \times 255/(B - A) & \text{for } A \le HU \le B, \\
= 255 & \text{for } HU > B
\end{cases} \quad (7.2)
$$

This relationship can be expressed in terms of the window width [W = (B − A)] and the window level (L = (A + B)/2) (Note: another term for window level is window center).

$$= 0 \quad \text{for HU} \leq L - W/2$$

$$f(HU) = (HU - (L - W/2)) \times 255/(W)$$
$$\text{for } L - W/2 \leq HU \leq L + W/2 \quad (7.3)$$

$$= 255 \quad \text{for HU} > L + W/2.$$

As the window width decreases, the contrast in the displayed image increases; as the window level moves up (down), the image becomes darker (lighter). This operation is called windowing or leveling, and it adjusts brightness and contrast for the visualization. It is important to note that the original CT data are not changed; only the presentation of the data is changed by the technique. In practice, the lower and upper HU limits are defined by the setting of window width (W) and window level (L), for example, W/L 200/0 stands for a window width of 200 HU and a window center of 0. This example would yield a lower HU limit of −100 and an upper HU limit of 100. In this example, the structures with HU values between −100 HU and 100 HU will be displayed with a good contrast. However, all values lower than −100 HU will appear black and all values larger than 100 HU will appear white. The effect of various window and level settings on a CT data set is shown in Figures 7.1 and 7.2. The transformation of HU to display intensities usually proceeds via a look-up table (LUT) that allows computer hardware to handle the transformation in real time. In CT scanners, standard settings for window width and level are stored for the various examinations types. These standard settings are included in the DICOM files to ensure a standardized presentation and visualization of these CT images.

7.2.2 HISTOGRAMS

The distribution of the frequency of occurrence of the HU in an image or in a volume is called a histogram. A histogram allows an appreciation of the distribution of the HU in the image or volume. It is often obtained and analyzed as one of the first steps in image processing or analysis. In Figure 7.3, a slice from a chest CT is presented along with its histogram.

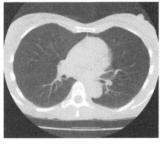 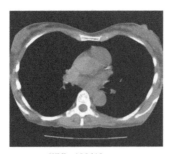

W/L: 1500/−600   W/L: 400/40

Figure 7.1 Examples of the effect of window width and window level on the intensities of the presented CT data. (Left) Window and level are set to 1500 and −600, respectively, so that black corresponds to HU −1350 and below and white corresponds to HU 150 and above. As a result, the lung tissue is gray and visible along with the pulmonary vessels. (Right) Window and level are set to 400 and 40, respectively, so that black corresponds to HU −360 and below and white corresponds to HU 240 and above. As a result, the lungs are black, and the components in the heart can be visually differentiated.

The histogram can be used to guide the window and level process. The histograms of two images can be used to equalize the contrast in the two data sets, called histogram equalization (Gonzales 2008; Sonka et al. 2007). To equalize the histograms of two images, the following steps are taken: (1) the respective histograms, $h_1(i)$ and $h_2(i)$, are obtained, where i is the HU value; and (2) the cumulative histogram is then obtained by integrating (summing) each histogram:

$$Ch_1(i) = \Sigma_j \, h_1(j), \quad (7.4)$$

$$Ch_2(i) = \Sigma_j \, h_2(j), \quad (7.5)$$

The percentage cumulative histogram is obtained as

$$Pch_1(i) = Ch_1(i)/Ch_1(max) \times 100, \quad (7.6)$$

$$Pch_2(i) = Ch_2(i)/Ch_2(max) \times 100. \quad (7.7)$$

The values of Pch_1 and Pch_2 are used as the y and x coordinates to generate a curve which is then used as an LUT. An HU value of image 2 (say, 81) is input, the corresponding point on the curve is determined, and the y value of that point is taken as the new HU (say, 220) for the previous HU value of 81. Histogram equalization is often used to convert data to similar intensity and contrast distributions to allow application of the same image processing or analysis to a group of data sets. Note that histogram equalization can be performed on subregions in a data set (Pizer et al. 1984; Lehr and Capek 1985; Zimmerman et al. 1988; Fayad et al. 2002).

7.2.3 IMAGE/VOLUME RESAMPLING

CT volumes are usually generated with voxel dimensions that are anisotropic, for example, with the x and y dimensions being equal but the z dimension being different (usually larger). For a number of applications, isotropic voxel dimensions simplify the computations and analysis, for example, geometric analyses such as curvature. In addition, comparison of various CT data sets is facilitated if the data sets have common voxel dimensions. Thus, CT data must be generated on new voxel meshes.

The simplest method for generating data on a new voxel mesh is trilinear interpolation. The mathematics associated with trilinear interpolation is probably best presented and understood by considering the 2D case of a series of discrete points on a curve, $g(x_i)$, where i is the index of the point on the curve at location x_i, and f(i) is the value at that point (Figure 7.4). To generate the data, g(x), at an arbitrary location x, we find the x_i points just before and after the point location x and perform a linear interpolation:

$$g(x) = (1.0 - f_x) \times g(x_i) + f_x \times g(x_{i+1}),$$
$$\text{where } f_x = (x - x_i)/(x_{i+1} - x_i). \quad (7.8)$$

This linear interpolation approach can be extended to 2D. The mesh that you have is a 2D set of points (x_i, y_i) and their values, $g(x_i, y_i)$, and you desire to calculate the value at an arbitrary location (x, y). Consider the points and data at (x, y_i) and (x, y_{i+1}), where y_i and y_{i+1} straddle y. The data at these points can be calculated from the linear case as

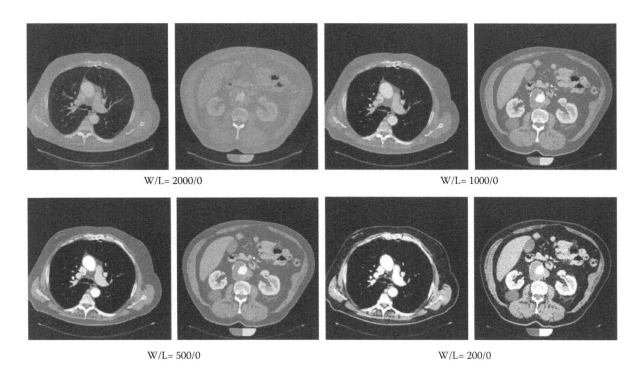

W/L= 2000/0   W/L= 1000/0

W/L= 500/0   W/L= 200/0

Figure 7.2 Effect of window width on contrast in CT images. CT slices from thoracic and abdominal studies, presented with various window settings; level settings were set to zero for all images. As the window width decreases, the contrast in the images increases; however, some structures seen in the low-contrast images disappear as the window width decreases.

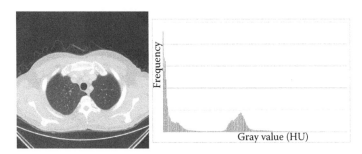

Figure 7.3 Slice from a CT examination of the chest and its histogram. The peak in the lower portion of the histogram corresponds to the air and lung region that has lower HU and comprises a significant portion of the image. The other peak corresponds to the tissue. The tail above the tissue peak corresponds to the bone region.

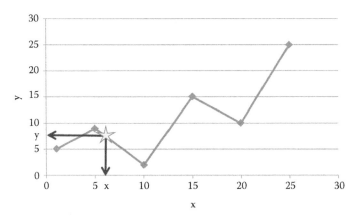

Figure 7.4 Illustration of linear interpolation. Data between the acquired data points (diamonds) are generated by linear interpolation, for example, the y value of the point (star) located at x.

$$g(x, y_i) = (1.0 - f_x) \times g(x_i, y_i) + f_x \times g(x_{i+1}, y_i),$$
$$\text{where } f_x = (x - x_i)/(x_{i+1} - x_i) \tag{7.9}$$

and

$$g(x, y_{i+1}) = (1.0 - f_x) \times g(x_i, y_{i+1}) + f_x \times g(x_{i+1}, y_{i+1}),$$
$$\text{where } f_x = (x - x_i)/(x_{i+1} - x_i). \tag{7.10}$$

Linear interpolation will then yield $g(x, y)$ from the $g(x, y_i)$ and $g(x, y_{i+1})$ as

$$g(x, y) = (1.0 - f_y) \times g(x, y_i) + f_y \times g(x, y_{i+1}),$$
$$\text{where } f_y = (y - y_i)/(y_{i+1} - y_i). \tag{7.11}$$

The extension to 3D is

$$g(x, y_i, z_i) = (1.0 - f_x) \times g(x_i, y_i, z_i) + f_x \times g(x_{i+1}, y_i, z_i),$$
$$\text{where } f_x = (x - x_i)/(x_{i+1} - x_i) \tag{7.12}$$

and

$$g(x, y_{i+1}, z_i) = (1.0 - f_x) \times g(x_i, y_{i+1}, z_i) + f_x \times g(x_{i+1}, y_{i+1}, z_i),$$
$$\text{where } f_x = (x - x_i)/(x_{i+1} - x_i) \tag{7.13}$$

and

$$g(x, y, z_i) = (1.0 - f_y) \times g(x, y_i, z_i) + f_y \times g(x, y_{i+1}, z_i),$$
$$\text{where } f_y = (y - y_i)/(y_{i+1} - y_i). \tag{7.14}$$

Similar sets of equations are used to obtain $g(x, y, z_{i+1})$ and

$$g(x, y, z) = (1.0 - f_z) \times g(x, y, z_i) + f_z \times g(x, y, z_{i+1}),$$
$$\text{where } f_z = (z - z_i)/(z_{i+1} - z_i). \tag{7.15}$$

Advanced techniques

Note that this approach generalizes easily to n dimensions. Note also that it could be generalized for one to n-dimensional polynomial fits, with the number of coefficients increasing rapidly with the number of dimensions. It should be remembered that the data contains noise, so that interpolations that go through the data, such as splines, can introduce artifacts.

With the ability to interpolate onto arbitrary meshes, the 3D data can be zoomed or shrunk in a semicontinuous manner. If interpolation is not preferred for zooming, voxel replication can be performed; similarly, when shrinking, voxel averaging or sampling can be performed. Figure 7.5 shows the result of zooming an image using bilinear interpolation and pixel replication. Although the results of bilinear interpolation appear smoother and are perhaps more appealing, it should be remembered that the presented interpolated data have been calculated and do not actually exist.

7.2.4 3D IMAGE PROCESSING: SOME PHILOSOPHY

CT data sets may not be optimal in the sense that they are artifact-free, noiseless, and perfect resolution. Patient motion, fluctuations in the detectors or the x-ray beam, and the acquisition technique can give rise to artifacts (see Chapters 3 and 4). The "as low as reasonably achievable" (ALARA) principles demand keeping exposures low; thus, CT data can be noisy. The detectors have finite resolution. However, the CT data and the associated images must have sufficient quality for answering diagnostic questions. In general, the human brain can compensate for many if not most of these issues; however, image processing can be used to further enhance our human capabilities. Usually, noise and resolution issues can be improved using appropriate filtering (discussed in the following). In addition, image processing can be performed to enhance the detectability of the objects or structures of interest, but image processing to enhance one type of structure can frequently reduce the quality of other structures in the data set. Thus, images are sometimes processed using multiple techniques to enhance detectability of the various structures. Let us begin by discussing the more "philosophical" aspects of image processing.

Image processing is most frequently used to increase the visibility of structures of interest, to enhance their contrast, or to separate them from other structures before subsequent image analysis. To select the image processing approach, the structure of the object of interest must be known and preferably understood. The characteristic that is most apparent is difference in HU value from the surrounding objects. Objects such as bones, contrast-filled vessels, or the airspaces in the lung are readily recognizable because they stand out from the surrounding background structures, and their boundaries are relatively easy to discern. Objects with significant contrast differences from their neighboring structures can be enhanced or identified using simple thresholding techniques, for example, selecting all those voxels above or below a chosen HU value. A second characteristic that is common to most objects is that of contiguity or connectivity, that is, the parts of the object are connected or in contact with one another. Although thresholding selects all objects satisfying the threshold criterion, contiguity can be added to the threshold criterion and enforced by region growing. The next characteristic that allows us to differentiate structures is shape; nodules tend

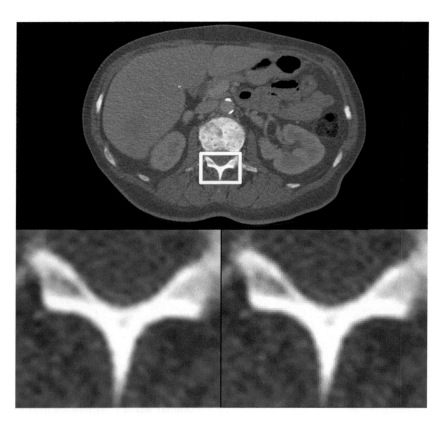

Figure 7.5 Examples of an original image (left) and a section of the image after zooming using bilinear interpolation (middle) and pixel replication (right). In the pixel replication image, blocky artifacts can be seen especially near regions of rapid changes in HU (high gradient regions); however, these artifacts are seen only when relatively high zoom is used.

to be spherical and blood vessels tend to cylindrical. Thus, if the first two characteristics do not provide sufficient enhancement or identification of the desired structure, the shape criteria can be used. One of the most common image processing tools is that of the matched filter. A matched filter has characteristics (e.g., shape) similar (in the spatial and/or Fourier domain) to that of the object to be enhanced. Thus, one of the first steps in identifying lung nodules is frequently to convolve (see the following section) the CT data set with a spherical filter that is about the size and shape of the lung nodules being sought. If the boundary of an object is desired, an edge-enhancing filter is used to convert the change in the HU values at the boundaries to a high-contrast surface that appears as line-like structures in the CT slices. Once the contrast is enhanced using these "edge-matched" filters, thresholding or region growing can be used to extract the objects.

Use of these characteristics, contrast, continuity, and shape, will usually do 95% of the job producing satisfying results as a first shot. However, CT data suffer from two realities that have been studied extensively in 2D images, resolution and noise (see Chapters 3 and 4). Resolution blurs the boundaries of structures, in that if two neighboring (contiguous) structures have different HU values, say H1 and H2, the values in the CT data across the boundary will change gradually (Figure 7.6 shows the effect of resolution on an edge). But, if the resolution function or image processing filter is spatially symmetric, the boundaries determined by thresholding will be fairly accurate. Noise confounds the problem further in that the HU values fluctuate from voxel to voxel throughout the data set, making differentiation even more difficult. Thus, resolution and noise affect the results of the thresholding and region growing. Moreover, matched filters of extended objects applied to the image data degrade the resolution blurring the edges, and edge-enhancing images enhance the noise in the image. If the difference in HU values at the boundary between the two structures is comparable to the noise level, edge enhancement can increase the noise variance, resulting in obscuring the boundary. If noise is the issue, more specifically if the signal-to-noise ratio (SNR) is the issue, a good first step is to reduce the noise, perhaps by using an averaging or a median filter (with a small cost of degrading the resolution). If the noise is random and uncorrelated spatially, the variance of the noise can be reduced by about a factor of N, where N is the number of

elements in the averaging filter, improving the SNR by a factor of sqrt(N). However, simple filtering can blur edges and reduce the intensity of the edges generated when edge enhancing filters are applied.

The next level of image processing is to use more completely the local features of the object of interest. Take advantage of what you know about the object of interest. What is its shape, elongated like vessel or bronchi or compact like a nodule? To achieve more optimal separation of an object from the surrounding background structures, use filters that take advantage of these shape differences. Each application is different and requires different combinations of filters and intermediate analyses, some of which we discuss in the following section. But in developing image processing approaches, you should analyze how you yourself extract the objects from the scene. The human eye–brain system is probably the best image processing system in existence. Thus, if you can figure out and perhaps verbalize and then convert to computer code how *you* (or your clinical colleagues) extract a given object from the scene, your algorithms could perform as well or even better than you. Ask yourself, What am I myself using to distinguish it from all the other things around it? Its contrast, the boundaries, its continuity in 3D, the slight difference in noise variance or characteristics? What do I know about the object that I am using, its expected shape or contrast?

7.2.5 SPATIAL FILTERING IN THE IMAGE DOMAIN

Before moving into filters and filtering, we should first point out that there are two types of filters: linear filters and nonlinear filters. Linear filters perform linear operations on the data, so they also can be performed in Fourier space (Gasquet et al. 1998; Russ 2007; Sonka et al. 2007; Chu 2008; Gonzales and Woods 2008). The data in Fourier space provide a representation of the spatial frequency distribution of the original 3D CT data in image space. It is sometimes computationally advantageous to perform image processing operations in Fourier space. In the discussions that follow, we "stay" for the most part in image space. However, when filtering (applying spatial filters to image data), it is useful, if not important, to be aware of the spatial frequency content of the structures you are processing or analyzing as well as the spatial frequency content of the filters you are using. This is true

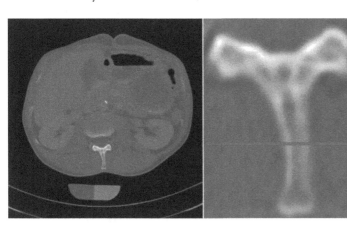
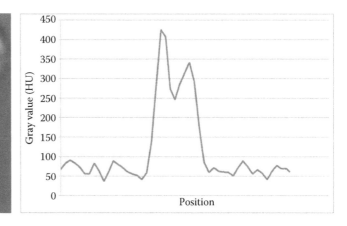

Figure 7.6 Effects of resolution on the edge of structures in the CT image. An original image (left), a zoomed-in view of the vertebra (middle), and the profile (right) taken from the data along the red line indicated in the middle image. We see that the edges in the image appear blurred and that the values in the profile do not drop sharply but decrease over to the tissue value over several pixels.

even when using nonlinear filters, because each nonlinear filter has correspondences with certain linear filters.

Let us begin with the linear filters. The common feature of all filtering techniques is that the neighboring pixels or voxels are used in the calculation of the new pixel value. As a start, let us consider a one-dimensional (1D) data set and a 1D smoothing filter in which the new value, $P_{new}(x_j)$, is the weighted sum of the element, $P(x_j)$ and its neighbors, $P(x_i)$.

$$P_{new}(x_j) = \Sigma_i \, w_i \, P(x_{i+j}), \qquad (7.16)$$

where the elements i are distributed about and include the point j, and have corresponding weights w_i. This operation is repeated for all points j. Note: the result should always be stored in a new array, not the same array as that being filtered. The simplest filter is the running-mean filter in which the values of w_i are uniform and equal to 1/N, where N is the number of elements being averaged. The elements i are usually distributed symmetrically about j, so than N is odd. The values of the weights are also usually distributed symmetrically about j, but need not be. The weights should sum to 1 (for smoothing filters), so that the average value over all the data does not change upon filtering. Obviously, the weights can be varied to generate various types of filters, such as Gaussians

$$w_i = \text{Norm exp } (-(j - i)^2/2 \times \sigma^2),$$
$$\text{where Norm} = 1/(\Sigma_i \exp \, (-(j - i)^2/2 \times \sigma^2)), \qquad (7.17)$$

where σ is the standard deviation of the Gaussian. Equation 7.16 can be easily generalized to 3D as

$$P_{new}(x_j, y_k, z_l) = \Sigma_{imn} \, w_{imn} \, P(x_{j+i}, y_{k+m}, z_{l+n}), \qquad (7.18)$$

with each direction having potentially different σ's.

Smoothing filters are generally used to increase the SNR of structures that are larger than the size of the smoothing filter. But they blur the image data, and in Fourier space they reduce the high-frequency components in the data. As a result, the sharpness of structures that vary rapidly from pixel to pixel, such as edges or noise, is reduced. The extent or size of the filter (i.e., N) is called the kernel or mask. The larger the size of the mask is, the larger the smoothing and the noise reduction will be.

Another frequently used filter type is the derivative filter. For derivative filters, the sum of the weights is often 0 (i.e., some of the weights are negative), although it need not be. A simple 1D derivative filter has weights, w(−1) = 1, w(0) = 0, w(1) = −1. As with the smoothing filters, various derivative filters can be designed by varying the weights in the filter. Other common derivative filters are Laplacians or the difference of Gaussians (DoG) [obtained by taking the differences of two Gaussians (Equation 7.17) with different standard deviations]. Derivative filter will increase the contrast of structures that vary rapidly spatially, such as edges and noise. Examples of the effect of smoothing are shown in Figure 7.7, and the effects of derivative filters are shown in Figures 7.8 and 7.9.

One commonly used filtering process is that of unsharp masking (Figure 7.10). In this approach, the data are first smoothed using a smoothing filter, reducing the strength of the high-frequency structures in the image. This result [e.g., $P_{new}(x_j,$ $y_k, z_l)$ from Equation 7.18], a low-frequency version of the original data, is then subtracted from the original data [$P(x_j, y_k, z_l)$] for all voxels to generate a high-frequency image, that is,

$$P_{high}(x_j, y_k, z_l) = P(x_j, y_k, z_l) - P_{new}(x_j, y_k, z_l) \quad \text{for all j, k, and l.} \qquad (7.19)$$

$P_{high}(x_j, y_k, z_l)$ is then added to the original image to generate an edge-enhanced image, but with the structures of the original image still visible.

$$P_{usm}(x_j, y_k, z_l) = P(x_j, y_k, z_l) + \beta \times P_{high}(x_j, y_k, z_l)$$
$$\text{for all j, k, and l,} \qquad (7.20)$$

where β is the weight given to the high frequency image. β equal to 1 corresponds to a standard derivative approach, with strongly enhanced edges. As β goes from 0 to 1, the strength of the high-frequency structures is increased. Thus, the degree of enhancement of the edges can be selected.

Note, there are a number of "standard" derivative filters, such as the Sobel filter and the Canny filter, each filter has its strengths and weaknesses, too many and too varied to go over here. The best way to learn is to try them on different data, see what happens, and note at least mentally what occurred as you tried the different filters.

In addition to linear filters, there are nonlinear filters (Figure 7.11) that cannot be represented as a series of linear operations. Examples of nonlinear filters are the median filter and the various morphological filters, both binary and grayscale. For the median filter, the values of the pixels in the neighborhood (i ± n) are sorted, and the median value is placed in the new array at location i, that is,

Median filter: $P_{new}(x_j) = \text{Median } (P(x_{i+j})) \qquad \text{for i} = -n, n.$ (7.21)

A median filter's primary strength is that it is good at eliminating shot noise (single pixels or pixel clumps with large deviations from the average); however, it also is used to reduce the noise in the data while preserving (better than smoothing linear filters) the edges in the image. Two examples of morphological filters are the maximum and minimum filters (also called the dilation and erosion operators). Maximum (minimum) filters expand the extent of high (low)-intensity structures in the image while reducing the extent of the low (high)-intensity structures:

Maximum filter: $P_{new}(x_j) = \text{MAX } (P(x_{i+j})) \qquad \text{for j} = -n, n$ (7.22)

Minimum filter: $P_{new}(x_j) = \text{MIN } (P(x_{i+j})) \qquad \text{for j} = -n, n.$ (7.23)

Other morphological filters are the dilation and erosion filters. As with other filters, one defines a mask (say 3 × 3 or 5 × 5). The mask is moved over the entire image, one pixel at a time. At a given pixel, the mask is centered on that pixel and the values of the pixels which overlap the mask are analyzed. For the

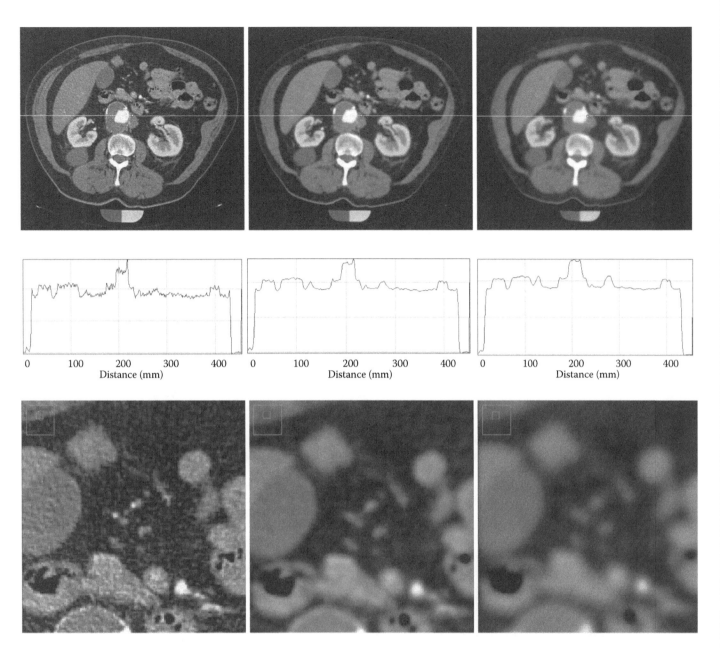

Figure 7.7 Effects of smoothing. (Left to right) Original image, a 5 × 5-average smoothed image, and a 9 × 9-average smoothed image. (Top) Original images. (Middle) HU of voxels along the line in the images. (Bottom) Zoomed regions of the images. With increased smoothing, the fluctuations in the data decrease, voxel values are more uniform within structures, but the edges are blurred.

dilation (erosion) operation, the maximum (minimum) pixel value overlapping the mask is placed in the pixel on which the mask is centered. Remember, the results of the operations are not placed in the original array but in a "results" array. A nice feature of morphological operators is that sequential application of the operators produce useful results, such as elimination of structures. For example, consider low-intensity structures surrounded by high intensity pixels. If the mask is chosen to be larger than some of these structures and a dilation is applied, the smaller structures will disappear. If an erosion operation is then applied, the structures that "survived" the dilation will grow back to their original size. This operation, dilation followed by an erosion, is called a "close" operation, named because it removes or closes up small holes in the data. Similarly, if we erode first then dilate

(called an "open" operation), high-intensity regions smaller than the mask will disappear, cleaning up or opening low-intensity regions. The results of dilation, erosion, closing, and opening are shown in Figure 7.12. Because these operations change the extent of the structures, these filters can be used to enhance the edges in the data. As shown in Figure 7.13, subtraction of the results of an erosion from the results of a dilation yields boundaries of the various structures.

Note that like the linear filters, extension to 2D and 3D (or higher dimensions) is straightforward. A size and shape of the region for the filter is specified in each dimension and the operation is performed using convolution. For both smoothing and derivative filters (linear or nonlinear), the simpler the filter is, the better the results will usually be, with the caveat that

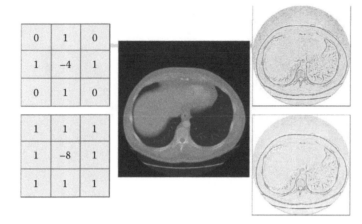

Figure 7.8 Effects of Laplace filtering. (Left) Laplace filter (four neighbors, top; eight neighbors, bottom). (Middle) CT slice of the lower thorax/upper abdomen. (Right) Results of application of the filters.

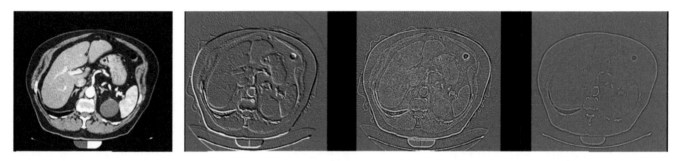

Figure 7.9 Effect of derivative filters. (Top to bottom) Original image, and the images obtained using the derivative along the diagonal, using a Laplacian filter and a DoG filter. Clearly, the filter used influences the resulting image, specifically the type of edges (major and/or minor) that one obtains.

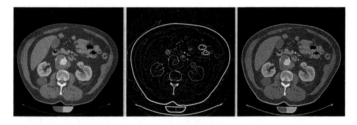

Figure 7.10 Effects of unsharp masking. (Left) Original image. (Middle) Difference image of original image and smoothed image, edges are significantly enhanced. (Right) Sum of the original and edge-enhanced image; edges are enhanced as is the noise.

the filters should be designed to match the structures you are trying to enhance.

7.2.6 COMPUTATIONAL SPEED UPS AND INTUITION

Note that the computational time increases as the size of the filter region (mask) increases. For 1D, the computational time increases linearly, for 2D quadratically, for 3D as the cube, that is, 3D filtering can be compute intense. Thus, filter size should be chosen carefully. There are some tricks that can be used to reduce the computational time. If the filter is separable, that is, it is the combination of a series of linear filters in each dimension (like the running mean filter or the Gaussian filter), you can perform 1D

operations for each dimension, using the output of one operation as the input for the subsequent operation, that is,

$$P_{new1}(x_j, y_k, z_l) = \Sigma_i\, w_i\, P(x_{j+i}, y_k, z_l), \tag{7.24}$$

$$P_{new2}(x_j, y_k, z_l) = \Sigma_m\, w_m\, P_{new1}(x_j, y_{k+m}, z_l), \tag{7.25}$$

$$P_{new3}(x_j, y_k, z_l) = \Sigma_n\, w_n\, P_{new2}(x_j, y_k, z_{l+n}), \tag{7.26}$$

effectively making an N^3 problem a $3 \times N$ problem. For a 3D $5 \times 5 \times 5$ filter, this reduces the relative number of operations from 125 to 15, over an eightfold time savings. A second trick is to subsample the data. Suppose you wanted to use a large mask, say a $21 \times 21 \times 21$ filter. The number of operations scales as 21^3, or 9261. Now the data do not vary spatially very fast, so you could use every second or third voxel for a time savings of 8 or 27, respectively. Another trick is to apply a series of $3 \times 3 \times 3$ filters to get the effect from filters with larger mask sizes, for example, 5^3–9^3. Other trick is to spend time on the areas of interest such as identification of local regions to which the filtering should be applied (e.g., boundaries). Be aware that filtering can change the location of boundaries and affect volume calculations.

Before using these types of techniques, you should try them on a sample data set to see what happens, applying the standard

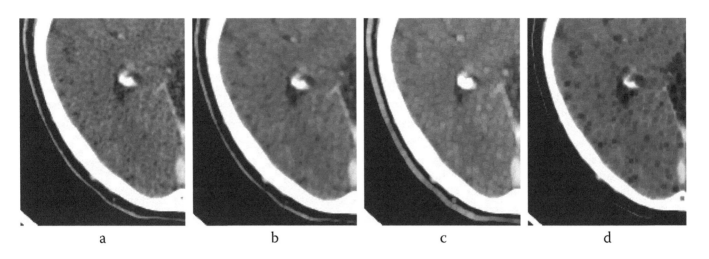

Figure 7.11 Effects of nonlinear filters on the CT data. (a) Original image. (b) Median-filtered image. (c) Maximum-filtered image. (d) Minimum-filtered image. A 3 × 3 pixel mask was used for each. After median filtering, the noise seen in the original image is reduced, but the data in the tissue region appear to be blotchier. Note, however, that the edges between the objects of different contrast still appear relatively sharp. With the maximum filter, the high (low)-intensity structures appear larger (smaller), whereas with the minimum filter, the low (high)-intensity structures appear larger (smaller). Subtraction of the original from the maximum- or minimum-filtered images would reveal edges of the various structures.

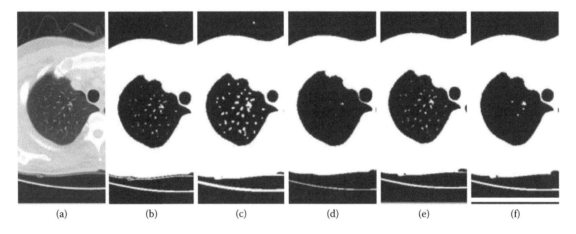

Figure 7.12 Effects of morphological filtering. (a) Original image. (b) Binarized version (using thresholding). (c) After dilation of image in b, the high-intensity regions are expanded. (d) After erosion of image in b, the high-intensity regions are smaller. (e) Dilation then erosion (closing), similar to b in this case. (f) Erosion then dilation (opening); only those high-intensity structures that survived the erosion are still in the image. The mask size here was 3 × 3.

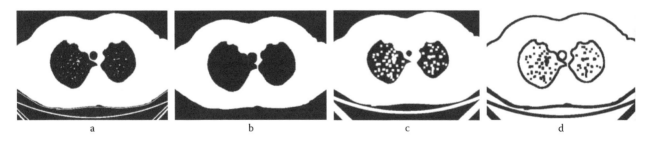

Figure 7.13 Edge enhancement. (Top to bottom) Original image, erosion, dilation, subtraction of the results of dilation from results of erosion. In erosion, the boundaries of the high-intensity regions move in toward the center of the high-intensity regions, with small high-intensity regions disappearing. In dilation, the high-intensity regions expand. The difference image shows the edges of the large high-intensity regions and the dilated small regions. The mask size here was 5 × 5. Because the mask size here was larger than that used for the images in Figure 7.12, more regions in the lung are eliminated upon erosion, the results of dilation are larger. The size of the edge also is related to the size of the mask.

approach and the faster approach. Subtract the results from the two approaches and see the differences. See what happens for various filters and then change the parameters slightly and compare the results of the new approach with that of the other approach. Remember, each of the various filters and combination of filters have strengths and weaknesses, get to know how they affect your data and your results, if you learn of a new filter, play with it on some known data sets, and build your understanding and intuition as to how it affects your data and your quantitative results.

7.2.7 SEGMENTATION

Segmentation is one technique used to separate structures or objects from each other in the data set. In CT, the primary parameter or feature that differentiates the various objects is the HU. Thus, the simplest method of segmentation is thresholding which involves determining a minimum (or maximum) HU and designating all voxels with HU above (below) this HU to belong to the object. Frequently, a histogram of the volume is first obtained, and this HU threshold value is determined from the histogram, either manually or automatically. In addition to thresholding, contiguity can be used, that is, the voxels corresponding to the structure of interest are connected or touch each other. It should be noted that structures that are connected in 3D may not be connected in a given 2D slice. If the thresholding is combined with contiguity constraints, the object of interest is "grown," that is, a seed is placed in the object (either manually or automatically), the voxels contiguous with the seed voxel are inspected [usually the six nearest neighbors (along the axes) or the 26 nearest neighbors in a 3 × 3 × 3 volume centered on the seed voxel]. Those that satisfy the threshold criterion are accepted as being in the object, and their locations are placed in a list of "future seeds" whose nearest neighbors will be inspected. This process (of inspecting voxels that are nearest neighbors to the "seeds," identifying them, and entering their locations into the "future seed" list) continues until the future seed list is exhausted. Examples of the results of thresholding and region growing are shown in Figure 7.14. Other techniques similar to thresholding or region growing, such as watershed (Higgins and Ojard 1993; Sijbers et al. 1997; Grau et al. 2004; Beare 2006; Burger and Burge 2008) and level sets (Yang et al. 2004; Li and Wang 2005; Li et al. 2006; Way et al. 2006; Li and Yezzi 2007; Sundaramoorthi et al. 2008; Krishnan et al. 2010), can also be used. In addition to HU as a threshold, other parameters, such as texture values, can be used. New segmentation techniques appear in the literature each day, and we cannot do justice to them all here. In general, these newer techniques tend to take advantage of *a priori* knowledge about the structures being segmented, for example, smoothness, continuity, and similarity of one or more features. Each technique has strengths and weaknesses that vary based on the underlying assumptions, the noise and resolution in the data sets, and the features of the structures themselves.

7.2.8 DATA FITTING

So far you have data, you have identified structures in the CT data set, say a boundary or a vessel, but it is noisy. The body usually does not have sharp structures, that is, structures that change abruptly spatially. Structures in the body tend to be smooth, slowly changing spatially. Thus, it makes sense to fit the structures that you detect with slowly varying functions. Fitting can improve the look of the structure, and it can improve the SNR of the fitted structure as you can use the fit to improve the local analysis, perhaps by taking the boundary into account when filtering. There are several techniques for fitting; here, we discuss briefly linear, polynomial, spline, and snakes. Note that one can fit entire curves or surfaces using these techniques, but because the number of data points can be large and because curves and surfaces can be complicated, high-order fits are often required and can lead to overfitting and anomalous results. A simpler approach is to fit the curves or surfaces in a local region, moving the fitting region along the curve or surface similar to the running-mean approach in filtering.

The simplest fit is a linear or straight line fit of the data points. Generation of a smoothed curve or surface using simple running mean on the coordinates of data points is equivalent to a linear fit of the data points. For each point in the curve or surface, select the N points surrounding the point of interest, calculate the new vector as the average of the selected points, and store that average vector in a new array, for example,

$$\mathbf{x}'(i) = 1/N \sum_i \mathbf{x}(i). \tag{7.27}$$

Note that the linear fit is conceptually simple and generally produces the results that are congruent with the original data. This approach can be modified so that the median of the selected vectors is used, by using the median of the coordinates along each direction of the selected vectors, that is,

$$\mathbf{x}'(i) = \text{med}_i [\mathbf{x}(i)]. \tag{7.28}$$

Median fitting, like median filtering, is very useful in removing spikes or unusually deviant data points from the data. Remember, median fitting is a nonlinear operation.

The next level of fitting uses polynomials to fit the curves or surfaces. For these fits, the selection of the independent variable is the first step. The independent variable should be monotonic, that is, not double valued, and usually the one that has the greatest range. When you solve the equation for 2D curves, you are solving for the coefficients, a_j, of the equation (assuming x is the independent variable) using matrix inversion techniques (Numerical Recipes).

$$y(i) = \sum_j a_j x^j(i). \tag{7.29}$$

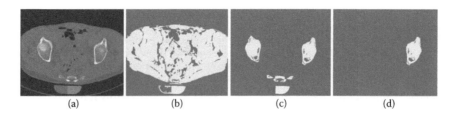

Figure 7.14 Results of thresholding and region growing. (a) Original image. (b) Thresholding for those HU values between 0 and 1000. (c) Thresholding for those HU values between 100 and 1000 (only the bony regions appear). (d) Region growing using this last threshold with a seed in the bony region on the right; only those voxels contiguous with the seed are grown.

For 3D curves, you could select one of the coordinates as the independent variable and fit the other two coordinates; however, this approach is not easily generalizable. An approach that works well and is relatively simple is to fit each of the coordinates independently as a function of point index or path length, t, along the curve, for example,

$$x(i) = \Sigma_j \, a_j \, t^j(i), \quad y(i) = \Sigma_j \, b_j \, t^j(i), \quad z(i) = \Sigma_j \, c_j \, t^j(i). \quad (7.30)$$

This approach gets around choosing the independent variable from the coordinates and thus can be used independent of the complexity of the curve and nonmonotonicity of the coordinates. The problem with running-polynomial fits is that they can introduce discontinuities in the data and the derivatives. A fitting technique that enforces continuity of the data and the first and second derivatives are spline-based techniques [see De Boor (2001) for a more complete explanation of splines, their properties, and how to use splines to fit data]. Because of these properties, splines are frequently used in data fitting.

So far, we have been considering curve fitting without bringing in what we specifically know about the structures themselves, for example, smoothness and continuity. One set of techniques, which incorporates the knowledge or assumptions about the curves or surfaces in the fitting itself, are called snakes or active contours (Lobregt and Viergever 1995; Germain and Réfrégier 1996; Finet et al. 1998; Boscolo et al. 2002; Young and Levy 2005; Way et al. 2006). A snake-based fitting approach makes use of two concepts, internal forces or constraints and external forces or constraints. The internal forces are usually the data themselves, the external forces are those constraints imposed by the algorithm developer, for example, smoothness, continuity of the derivative, or bounds of excursion. The fits mentioned earlier are basically snakes without the external forces [see Blake and Isard (1998) for more information about snakes and their uses].

However, it must be remembered that the data are really all we have (along with our understanding of the basic structure of the curve or surface we attempt to determine). These data have resulted from our previous analyses, for example, thresholding, region growing, and boundary detection, and can be affected by resolution and noise in the original 3D data. Thus, the data will (more than likely) *not* lie on the "true" curve or surface but only approximate it, and specifically, you will not know the truth for any given experimentally determined data set. And so, you will not be able to determine whether the selected fitting approach results in a better estimation of the underlying truth, although it may be more visually pleasing. Thus, you probably should try a couple of these techniques on any given data set (or data sets) and see what happens. Each of the techniques has its own strengths and weaknesses, become familiar with these by correlating these with the fitted data you generate. As you become familiar with the results, you will gain intuition about what to use and the results to trust when fitting your data.

7.2.9 APPLICATIONS

The techniques described earlier have been used extensively in analysis of CT data sets. Here, we describe briefly application of these techniques to extraction of clinically relevant data from the CT data sets.

7.2.9.1 Colon

Colonoscopy is used to inspect endoscopically the colon and to detect colon cancer or colon polyps that are often precursors of colon cancer. CT provides a means of inspecting the colon without endoscopy. In CT colonoscopy, the colon is distended by inflating the colon with a gas (air or nitrogen) as in colonoscopy; the air effectively functions to provide contrast and allow visualization of the colon wall. The CT scan is acquired and reconstructed and then inspected by the radiologist. In the early and mid-1990s, techniques were developed to extract the colon surface from the CT data sets. The low HU values of the air allowed for selection of threshold values for region growing. Because of the large difference in HU between the tissue and air, the segmentation is relatively insensitive to threshold values; however, partial volume effects can alter the local shapes of structures of interest (Figure 7.15). The grown surface was then rendered for the radiologist for inspection. As in colonoscopy, the inspection took the form of an endoscopic view (see 7.18 and 7.19) and was called virtual colonoscopy. Navigation through the virtual colon was difficult; to facilitate inspection, techniques were developed to determine the centerline of the segmented colon (Ge et al. 1999; Samara et al. 1999; Iordanescu and Summers 2003; Frimmel et al. 2005; Summers et al. 2009).

7.2.9.2 Bronchi

As with the colon, the air in the bronchi also functions as contrast. Thresholds are set, and region growing is used to segment the bronchi (Mori et al. 2000, 2009; Mayer et al. 2004; Schmidt et al. 2004; Socha et al. 2004; Venkatraman et al. 2006; du Plessis et al. 2009; Fabijańska 2009). For the larger bronchi, the segmentation is relatively insensitive to the set thresholds. However, as the bronchi narrow, the sensitivity of the segmentation results on the threshold used increases due to partial volume effects and noise in the data. In an attempt to achieve accurate segmentation farther down the bronchial tree, iterative region growing can be used. In iterative region growing, an initial threshold is selected such that the main trunk of the structure is segmented. A series of more careful region growings

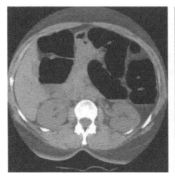 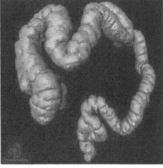

Figure 7.15 Colon segmentation. (Left) Axial slice through the abdomen: the colon is dark, the vertebrae and ribs are right, and the tissue is gray. The two kidneys can be seen on either side of the vertebrae, and the liver is visible on the left. (Right) Colon data after segmentation and rendering presented for visualization.

are used with a slightly different thresholds (higher for bronchi), and the results are visually inspected. To select the thresholds more automatically, the histogram of the data just outside the grown surface (either locally or globally) can be obtained and analyzed. The SNR of the data just outside the surface can be improved by directed smoothing of the CT data parallel to the local surface (e.g., elongated asymmetric filters) or along the local centerline of the grown bronchi. To estimate the local centerline, fit the grown voxels local to the end of the grown region with a straight line. To do this well, local branching relationships should be taken into account.

7.2.9.3 Liver

Liver disease can be local such as a tumor or global such as in hepatitis or cirrhosis. Tumors tend to have slightly different attenuation coefficients, and thereby HU, than normal liver. Thus, segmentation techniques can be applied to extract them from the liver data (Lang et al. 2005). Again, the results can be sensitive to thresholds used due to partial volume effects and noise, especially if the HU of the tumor is close to those of the normal liver. To the degree that the segmentation is noise sensitive, the SNR of the tumor can be improved using averaging filters. However, these can blur the boundaries. So, if boundaries are important, boundary-sensitive smoothing (Bae et al. 1993; Selle et al. 2002; Boskamp et al. 2004; Lang et al. 2005; Friman et al. 2010) or iterative approaches are recommended. For identification of vessels in the liver, the approaches described for the bronchi can be used (Selle et al. 2002; Lang et al. 2005). Iterative approaches, especially for the smaller vessels in the hepatic tree, are recommended.

7.3 3D DATA ANALYSIS

CT data, because it is multidimensional (three coordinates, HU, and time), can provide a wealth of information about the patient and the various organs and structures in the body. Generally the information acquired or used to date is visual and subjective, with some quantitation being performed, for example, measurement of sizes, diameters, areas, and angle, but most are performed manually. However, once the data associated with a structure are identified and extracted, more complete analysis of the data can begin, for example, quantitation of parameters to describe the structure. Although the clinician is often interested in sizes, diameters, volumes, and relative locations, several other parameters can be calculated often as part of CAD algorithms (Doi 2007), such as the center of mass, volume, sphericity, opacity, texture, relative locations, motion, and surface curvature. In this section, we discuss several of these analyses.

The first step in analysis is to determine what you are looking for or more specifically what are the relevant characteristics of the object of interest. As with decisions regarding image processing, the best path is to understand how you yourself analyze the image and identify the object and its extent. The first question is usually, what differentiates this object from the objects around it? Higher HU, lower HU? Are the edges visible? Are they sharp? Are there texture differences that can be taken advantage of? Would it be easier to identify and remove the objects that are not the object of interest? The next question is often, what is the shape of the object? Amorphous? Tubular? Circular? Spherical? How does it change as we move through the volume? Is the surface important (as in CT colonoscopy) or are the structures in the volume important (as in the brain)? Another question you probably should answer is how automated do you want the analysis to be? Finding a needle in a haystack is easier and faster if you know where to look, that is, indication of an object or region of interest can often significantly speed up the analysis. With your questions laid out and answered, the analysis can begin.

7.3.1 BRANCHING RELATIONSHIPS

The bronchial system in the lungs and the vascular system in the various organs of the body consist of many connected tree-like luminal structures, branching from a truck (e.g., aorta) and subsequently proceeding through a number of branching levels. Boskamp et al. (2004) proposed a method for determining the centerlines of vessels and the branching relationships of the liver vasculature during the region-growing process. At each iteration of the region-growing process, the grown regions are inspected for self-contiguity. Remember that region growing propagates outward from already identified (grown) voxels and that the region-growing process involves several steps or iterations. In each of these iterations, a new region R(i) (a set of "future seeds") is grown from the voxels identified in the previous iteration R(i − 1) (a previous set of "future seeds"), where i is the iteration number. Note that at each iteration, the center of mass of the grown voxels, R(i), can be calculated and taken as the centerline of the structure (Samara et al. 1999; Selle et al. 2002). Now for a compact structure, the region grown at a given iteration of the region growing process is self-contiguous, that is, starting at any point in R(i) you can reach all other points in R(i) using only voxels in R(i). This is true for convex structures and most compact structures, except near the end of the region-growing process. However, for branching structures, self-contiguity is disrupted at each bifurcation. Imagine a tree planted in the ground and water flooding the ground. As the water level rises, the bifurcation, that is, the location where the branching occurs, disappears under the water and all you see above the water is two separate branches which are no longer visibly connected. Similarly, as region growing proceeds up a major trunk or branch, the R(i) are self-contiguous and continue so into bifurcations. But after the bifurcation, there comes a point where there are two distinct regions both belonging to R(i), but spatially separated and noncontiguous (Figure 7.16). By performing a second region growing starting at a seed point in R(i), the self-contiguity of R(i) can be checked. Once, R(i) fails the test of self-contiguity, the second region (not grown in the second region growing) is regrown from a seed point in its volume, that is, by a third region growing. Subsequently, the two separate regions (grown in region growing two and three) are identified as branches from the one branch, and the vessel hierarchy is updated to include this bifurcation and the arising branches. The grown branches and their associated hierarchies can then be used to facilitate virtual endoscopic viewing, or analysis of the branching structures, for example, quantifying the vessel sizes or volume as a function of the hierarchical level or for fractal analysis (Kassab 2006; Tawhai et al. 2006; Cassot et al. 2009).

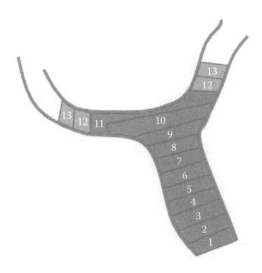

Figure 7.16 Region growing in a vessel and identification of a bifurcation. As the region growing proceeds, the set of voxels found in one iteration is used as seeds in the subsequent iterations. The voxels grown at each iteration are labeled with the number of the iteration. The boundaries of these iterations of the region growing are shown as lines inside the vessel. The darker gray regions are those regions that are all mutually continuous, that is, those voxels grown during one iteration can be grown from any voxel in that region. At the bifurcation, voxels grown at iteration 12 are not mutually contiguous, that is, there are now two separate regions, these voxels are indicated by the lighter gray and are labeled as new branches in the vessel hierarchy.

7.3.2 DETERMINATION OF VOLUMES OF STRUCTURES

Once an object is segmented, its volume (and/or shape) can be calculated relatively easily say by summing the volumes of the grown voxels (or via selected shape measures). However, the question and challenge are determination of the accuracy and reliability of this calculated volume. In clinical cases, accuracy is always a problem as truth is usually not known. However, reliability of the calculated volume (or other parameters) can be at least investigated by determining the variation in the volume and/or shape of the structure as a function of the threshold (or other criteria) used in region growing. Stability of the calculated parameters is a pretty good indication of the reliability (and perhaps accuracy) of the calculated parameters.

7.3.3 EXTRACTION OF VASCULAR INFORMATION

Once a vessel or vessel tree has been identified in the CT data set, say by region growing, the most frequent parameters extracted from CT vessel data are the vessel diameter and cross-sectional area, either along the length of the vessel or at specified points along its length. Both parameters should be measured perpendicular to the vessel axis or centerline. Thus, the centerline must be determined first. Vessel centerlines can be determined based on manual indications that can then be refined or automatically using a variety of techniques, such as skeletonization (Blum and Nagel 1978; Zhou and Toga 1999; Sorantin et al. 2002; Boskamp et al. 2004; Bouix et al. 2005) and distance transform (Ragnemalm 1993; Boskamp et al. 2004; Rakshe et al. 2008), or during the region-growing process (Samara et al. 1999). Once the centerline is calculated, the plane

normal to the centerline and vectors in that plane are generated. If working with region-grown data, the data are binary, and vessel size and area are determined by identifying the boundaries of the vessel, by using either the binary region-grown data or an established threshold. Note the size and area measured depend on the threshold used and partial volume effects, and remember that the errors at the boundary of the vessel (or any region) propagate strongly into the calculated area, that is, area error is proportional to r × Δr, where Δr is the error in radious r.

Other vessel parameters that have potential clinical implications are vessel curvature (Hoffmann et al. 2005) or tortuosity (Bullitt et al. 2003, 2005, 2007). Vessel curvature can be determined from the vessel centerline data and can be presented to the clinician for visualization (Hoffmann et al. 2005) or used as one of several methods to evaluate tortuosity (Bullitt et al. 2003, 2005, 2007). Vessel tortuosity has been found to decrease during radiation treatment in vessels associated with the cancer being treated (Bullitt et al. 2007) (Figure 7.17).

7.3.4 SURFACE CHARACTERIZATION

The morphology of a grown and identified anatomic structure often contains valuable information about the structure. One of the most evident examples is the case of the mucosal surface analysis of the colon in CT colonography, also known as virtual colonoscopy (Pickhardt and Kim 2009). Morphologically, the inner surface of the colonic lumen has a relatively simple structure: it consists of folds and colonic wall as normal anatomic structures and polyps—a precursor of colorectal cancer—as abnormal structures (Pickhardt and Kim 2009). On CT colonography images, polyps tend to appear as bulbous, cap-like structures adhering to the colonic wall, whereas folds appear as elongated, ridge-like structures, and the colonic wall appears as a large, nearly flat, cup-like structure (Figure 7.18a). Therefore,

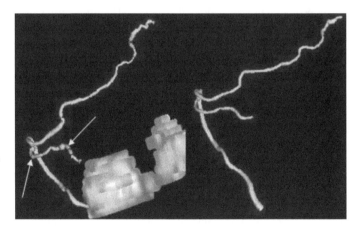

Figure 7.17 (See color insert.) Normalization of vessel shape during radiation treatment. (Left) 3D rendering of the segmented tumors (gray) and of the basilar, left superior cerebellar, and left posterior cerebral arteries at baseline as shown from a lateral point of view. Note the abnormal tortuosity of the left superior cerebellar artery (arrows). The posterior cerebral artery is also abnormal, although to a lesser degree. (Right) Rendering of the same vessels from the same point of view at month 2. Note the arterial "straightening." (From figure 2 of Bullitt, E., et al., *Radiology*, 245, 824–30, 2007; Burger, W. and Burge, M.J., *Digital Image Processing: An Algorithmic Introduction Using Java*, Springer, Berlin, 2008. With permission; Courtesy of Elizabeth Bullitt, University of North Carolina.)

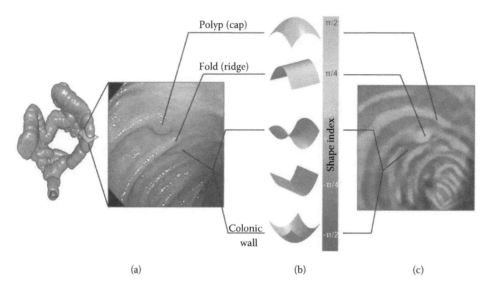

(a)   (b)   (c)

Figure 7.18 (See color insert.) Schematic illustration of the morphologic characterization of the structures in the colonic lumen. (a) Colonoscopy image of a 6-mm polyp from endoscopy. (b) Morphologic characterization of the structures in the colonic lumen based on the shape index. (c) Color coding of the colonic lumen in a computer-generated endoscopic view based on the shape index value, in which polyps, folds, and the colonic wall are shown in green, pink, and brown, respectively. (From Perumpillichira, J.J., et al., *Canc Imag*, 5, 11–16, 2005. With permission; Courtesy of Hiro Yoshida, Harvard University.)

morphological analysis that differentiates among these distinct types of shapes and scales is effective in the identification of pathologies such as polyps. To this end, various methods have been developed, including the use of a volumetric shape index (Yoshida and Nappi 2001; Yoshida et al. 2002; Perumpillichira et al. 2005), second-order curvature flow (van Wijk et al. 2010), and Hessian-based structure analysis (Cai et al. 2008), each of which has been shown to be effective in identifying colonic structures. For example, the volumetric shape index (Yoshida and Nappi 2001; Yoshida et al. 2002) analyzes the vicinity of a voxel and determines to which of the following five topologic classes a voxel belongs: cup, rut, saddle, ridge, or cap (Figure 7.18b). Color coding of the anatomic structures in the colonic lumen based on the shape index can thus differentiate polyps, folds, and colonic walls effectively (Figure 7.18c). A polyp can be identified as a region representing the shape index values corresponding to a cap shape (Figure 7.19).

7.3.5 VOLUME TRACKING AND 4D-CT

Identification and characterization of temporal changes of volumes are becoming more important as cancer treatments improve. One application is measuring the size and volume of tumors of cancers being treated. The simplest and clinically the most frequently used tracking approach compares the segmented volumes over time. Care must be taken that the data sets are properly normalized (say, by using histograms; see the previous section), so that the grown regions accurately correspond to the tumor tissue in all the data sets. The data sets can be registered (see the next section) and, after proper histogram normalization, subtracted for visual appreciation and quantification of any changes that have occurred.

Another application is tracking the locations of lung tumors. In radiation therapy, motion of the target volume due to breathing (Sarrut et al. 2006; Li et al. 2008), and to a less extent cardiac motion, affects the quality of the images used

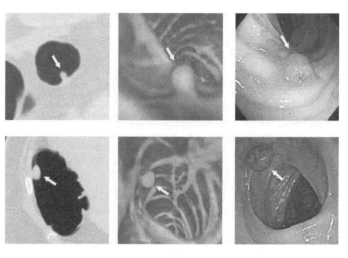

Figure 7.19 (See color insert.) Example of identification of polyps in CT colonographic data. In each row, the left image shows an axial CT image of a polyp (arrow), the middle image shows a computer-generated 3D endoluminal view that is color-coded based on the shape index values in Figure 7.18, and the right image shows a colonoscopy image of the polyp. Top row: 7-mm polyp in the transverse colon. Bottom row: 11-mm sessile polyp in the hepatic flexure. The polyp (green) is clearly differentiated from the folds (pink) and the colonic wall (brown). (Courtesy of Hiro Yoshida, Harvard University.)

for treatment plans and subsequent dose delivery. This effect has lead to the development of *respiratory-correlated*, or more commonly termed *four-dimensional* (4D), CT imaging (Ford et al. 2003; Vedam et al. 2003; Pan et al. 2004) in which the respiratory signal and oversampled CT images are synchronously acquired. During postprocessing, each CT image is placed into a discrete respiratory bin (typically 8–15 bins are used) depending on the phase (and/or displacement) of the respiratory signal when the image was acquired. Although 4D-CT reduces artifacts compared with standard free-breathing CT, artifacts

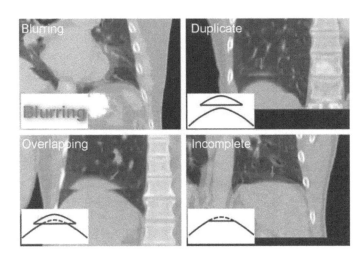

Figure 7.20 Artifacts in 4D-CT images. Four types of artifacts are shown: blurring, duplicate structure, overlapping structure, and incomplete structure. Corresponding artifacts are indicated by arrows in respective images. (From figure 2 of Yamamoto, T., et al., *Int J Radiat Oncol Biol Phys*, 72, 1250–8, 2008. With permission; Courtesy of Paul Keall, Stanford University.)

(Figure 7.20) remain due to irregular breathing. The significant relationship between the respiratory irregularity and artifacts suggests that breathing training (George et al. 2006; Neicu et al. 2006; Venkat et al. 2008) could reduce the artifacts in 4D-CT images. Improved sorting methods (Fitzpatrick et al. 2006; Lu et al. 2006; Rietzel and Chen 2006; Abdelnour et al. 2007; Ehrhardt et al. 2007; Mutaf et al. 2007; Langner and Keall 2010), acquiring CT data only when the breathing is regular, ensuring data sufficiency (Langner and Keall 2010), and more advanced postscan image processing invoking deformable registration (Schreibmann et al. 2006; Ehrhardt et al. 2007) can be used. Many modern linear accelerators have additional gantry-mounted kilovoltage imagers that allow both 3D (Jaffray et al. 2002) and more recently 4D (Sonke et al. 2005) cone beam CT images of the patient anatomy near to the time of treatment. These volumetric images have substantially increased the accuracy with which soft tissue can be visualized and targeted during a course of radiation therapy. They also facilitate adaptive radiation therapy, where the temporal changes in anatomy and anatomic motion during a course of therapy can be visualized and accounted for by updating the treatment plan to reflect the most recent anatomical measurement. There are several applications of 4D-CT images in the clinical radiation oncology workflow: motion inclusive, respiratory gated and tumor tracking treatments (Keall et al. 2006) with the majority of current treatments using motion inclusive methods and a small but growing number using gating and tumor tracking.

7.3.6 REGISTRATION

When CT data are acquired at different times or if we want to take advantage of the complementary information of different modalities, for example, CT, magnetic resonance (MR), positron emission tomography (PET), single-photon emission computed tomography (SPECT), and ultrasound, the data sets should be registered or aligned to each other. Registration involves determining the transformation relating the "internal"

or "intrinsic" coordinate systems associated with each study. The transformation is given by

$$\mathbf{x}_2(i) = S\,R\,(\mathbf{x}_1(i) - \mathbf{t}), \tag{7.31}$$

where $\mathbf{x}_1(i)$ and $\mathbf{x}_2(i)$ are the position vectors of the corresponding points in system 1 and 2, respectively; S is the scale factor relating the two systems (say reflecting voxel size differences); and R and **t** are, respectively, the rotation and translation vectors relating the two systems. If corresponding points (a minimum of four are required) are known in the two data sets, several approaches have been proposed (Maurer et al. 1997; West et al. 1997; Fitzpatrick et al. 1998; Fitzpatrick and West 2001), many of which have been tested extensively. Some important points have been raised by Fitzpatrick and his group (Fitzpatrick et al. 1998; Fitzpatrick and West 2001) regarding the location of the points used for registration and the objects of interest (e.g., target volumes), indicating that the object of interest should lie within or near the volume defined by the registered points. One registration approach is based on the Procrustes algorithm (Schoeneman 1969; Schoeneman and Carroll 1970) that minimizes the root-mean-squared (RMS) distance between corresponding points. The registration proceeds as follows. The centers of mass (COMs) of the corresponding points are determined in both studies, and one study is translated by the difference of the COMs so that both COMs are aligned. The scale, S, is calculated as the ratio of the RMS distance of the points in each system from the respective COM. The rotation matrix is calculated by generating a 3 × 3 matrix as the outer product of the coordinates of the COM-adjusted points and passing it through a singular-value-decomposition algorithm (GSL Team 2007). The Procrustes algorithm basically solves a series of linear equations and so is very fast (<<1 s). Note that the errors in the locations of your corresponding points propagate directly into the errors in your results, more or less averaging these errors, thus the more points input the more accurate your result (if your errors are random and not systematic).

If corresponding points are not known, but regions or surfaces are known (say, by segmentation), the surfaces can be registered by minimizing the distance between surfaces using a technique introduced by Pelizzari et al. (1989) (Figure 7.21), similarly to the iterative closest point technique (Besl 1992; Cao et al. 2004; ITK; VTK). The distance between the surfaces at a given point is determined by calculating the surface normal at a point on one of the surfaces, finding the point of intersection of that normal with the second surface and then calculating the distance between these two corresponding points. Note that this calculation involves casting a large number of rays between surfaces and requires a nonlinear search of the space of transformation parameters, so it is more compute-intensive than the Procrustes algorithm that requires only a 3 × 3 matrix inversion. This may be viewed as the "cost" of not having sufficient point correspondences. The technique is fastest and most robust when the two surfaces are initially somewhat to fairly well aligned.

If segmentation is not an option or not desired, one can take advantage of the information that is similar or shared in the two data sets, say, by using mutual information algorithms

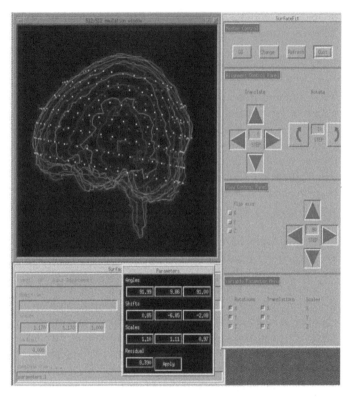

Figure 7.21 Example user interface from a surface-registration program. Translation, rotation, and scale parameters can be varied manually or via a nonlinear search, to minimize the mean-squared distance between the MRI brain surface (wireframe of sagittal contours) and the PET brain surface (set of points from transaxial contours). (Courtesy of Charles Pelizzari, University of Chicago.)

(Esteban and Morales 1995; Pluim et al. 2003). For data from the same modality (e.g., CT/CT), there is a large amount of shared or mutual information. As the overlap of information in the modalities decreases, the quality and speed of the registration can degrade. Remember that CT data are basically a mapping of the electron density in the body, so CT presents primarily anatomic information. Registration of CT with functional images (such as PET or SPECT whose information may not spatially correspond well with the anatomic information of CT) may require modification of the input data or weighting of corresponding structures in the algorithm. Mutual information is also compute-intense, so an initial approximate alignment of the structures is a good idea. Because mutual information approaches more completely use the 3D data, registrations can be very accurate. Note, Insight Segmentation and Registration Toolkit (ITK) provides several tested registration techniques as open source.

Thus far, we have been talking about linear transformations. However, if there are anatomic changes in the patient (e.g., due to breathing or response to radiation treatment) or if members of a population are to be mapped into a single patient system, the data sets must not only be aligned but also mapped (or warped) into the single patient system. Techniques for these mapping have been described for some organs (Brock et al. 2003; Li et al. 2003; Christensen et al. 2006; Chakravarty et al. 2008; Li et al. 2008; Joshi et al. 2010). One approach to this problem is linear registration as a first step, followed by manual or automated

determination of local corresponding points or regions on the surface or the volume, determination of a parameterized transformation relating those regions, and then generation of a mapping function that takes into account the parameterized transformations.

7.3.7 FEATURE EXTRACTION AND MEASUREMENT

In feature extraction and measurements, the goal is characterization of the structures you have identified usually to facilitate subsequent algorithm development (to improve segmentation) or the measure changes in the structures over time by comparison with subsequently acquired data sets (e.g., for assessment of treatment; Buckler et al. 2010; Hadjiiski et al. 2010; Yee et al. 2011). Some simple features of structures include widths along specified directions, average radius, binary COM, intensity-based center of mass, volume, average HU, and RMS variation about the average HU, all of which are relatively simple to calculate. More complex features extracted are circularity and deviations from circularity (Giger et al. 1988; Yin et al. 1994) or sphericity (Dehmeshki et al. 2008), roughness of the surface, higher order moments, HU texture (Chabat et al. 2003; Miles et al. 2009), various shape-specific metrics (Yoshida and Nappi 2001; Perumpillichira 2005), and clustering that are usually associated with a specific characterization problem. As the features calcuated become more complex, the methods used to determine them also tend to become more complex, for example, curve-fitting or surface-fitting techniques will be used. For each of these features, it is important that you understand the reliability of the calculated quantities as a function of the parameters used in the extraction of the structure. Also, it is important to perform these calculations in the physical parameter space (e.g., mm, degrees) and not in pixels or voxels.

With these measurements performed, the change of these quantities across data sets [e.g., patient populations (here registration most likely needs to be performed) or temporally for a given patient] can be quantified. Here, in particular, knowing the reliability of your measurements is important. With this knowledge, techniques can be developed to differentiate and classify structures automatically, say, for CAD (Doi 2007) or to provide the clinician with more information than previously available.

7.4 VISUALIZATION IN CT

So, you have all this great information, now you have to present it for appreciation and interpretation. The human being is a great image processor and a great scene interpreter. Thus, perhaps the most important aspect of the data analysis is presentation for visualization, appreciation, and interpretation of the data and the results of the analysis. Many different visualization options are available for diagnostics and treatment planning. In the following section, we discuss the main 3D strategies and their uses during routine clinical practice.

7.4.1 CT SLICES

The most frequently used mode of presentation of CT data is the "tomographic view," that is, presentation of the CT slices in sequence (Figure 7.22). Although presentation of axial

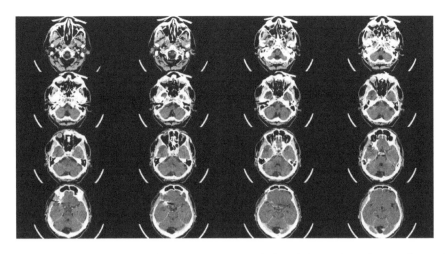

Figure 7.22 Series of axial slices through the head. The slice sequence proceeds from left to right and top to down. The bone is high intensity; the tissue is gray; and the ventricles, fluid, and air spaces are dark.

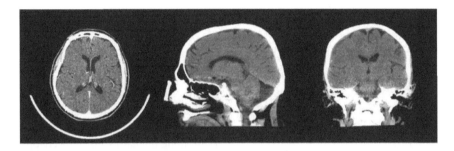

Figure 7.23 Examples of data displayed for various cut planes. (Left) Axial slice. (Middle) Sagittal slice. (Right) Coronal slice.

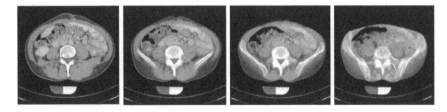

Figure 7.24 Slabbed CT data. (Left to right) One, 5, 10, and 20 CT slices are combined. As the number of CT slices increases, more out-of-plane information is contained in the image, becoming more like a projection image.

slices (perpendicular to the patient long axis) is perhaps the most frequently used, coronal, sagittal, and multiplanar slices (Figure 7.23) are also used. The term used for these planes that effectively cut the data and allow viewing of the data in those planes is "cut-planes." These data are generated by defining planes in the volume, determining the pixel location in the volume, determining the HU value (voxel value) corresponding to that pixel location (usually using interpolation schemes, such as trilinear interpolation as described in the previous section). Note that a single plane is commonly used, but one can include data from nearby planes (a "slab" of data) by summing along rays perpendicular to the plane. This process involves generating a line perpendicular to the plane centered on each pixel, determining the HU values at points along that line by interpolation, and storing the summation or average of these values at that pixel location. Creating this slab of data for presentation decreases the resolution in the direction of the slabbing but allows appreciation

of nearby structures in the data and can reduce noise somewhat (Figure 7.24).

7.4.2 RAY CASTING

Similar to the method of creating slabs of data for presentation, there are many techniques for "integrating" data along selected directions through the entire volume. These techniques fall under the general label of projection techniques as they project the entire volume (or selected portions of the volume) onto single planes. Imagine a 3D volume, a source of x-rays, and a detector plane (Figure 7.25). Consider one pixel in that plane. The data in that pixel correspond (after proper conversions) to a combination (e.g., summation or averaging) of the attenuation values along the path from the source to the pixel. So, you generate a line; calculate the HU values in the data volume corresponding to each point along that line; and, for a projection image, sum (or average) these values. This approach is generally known as ray

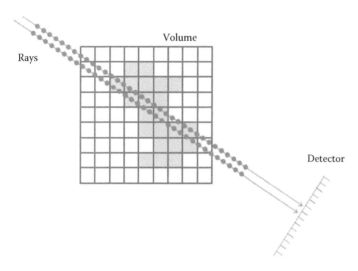

Figure 7.25 Schematic of ray casting. The 3D volume is represented as voxels (squares), and their HU values are represented by intensities (gray or white). Rays are projected through the volume into the pixels in the detector. The data in the volume are sampled along the rays at specified (usually uniform) intervals (gray dots). The results of the sampled data along the rays are stored in the corresponding pixel locations.

casting (Roth 1982; Levoy 1990; Watt and Watt 1992), as one is casting rays through the volume. Although the entire volume is now displayed, the 3D information has been lost; however, by presenting several different projection views in sequence, some appreciation of the 3D relationships can be gained. Note that the source could be a single point, so that a cone beam projection is generated or an "area" source so that a parallel beam projection is generated. For the parallel beam projection, all rays are perpendicular to the detector plane. Note also that you need to sample the volume with a sufficient frequency to obtain an accurate representation of the projection of the data in the volume, but not too frequently because the computation time increases with the sampling frequency. A good rule (related to the Nyquist criterion) is about or at least two samples per voxel, that is, the step size along each line is about half the voxel size. Because of the number of points inspected (number of pixels in the projection image × the number of steps along the projection lines), projection techniques tend to be computationally intense. But this issue is substantially ameliorated when graphic processing units (GPUs) are used (Smelyanskiy et al. 2009; Noël et al. 2010; see the following section). Also, as with the CT images, cut-planes can be used to select a subvolume of the object for rendering, such as a slab with a chosen location and thickness along the ray can be used to reduce the effects of overlapping structures.

7.4.3 SIMPLE PROJECTION TECHNIQUES

In the simplest projection technique, a source type and position and detector orientation are chosen, the rays corresponding to each pixel are generated, and the average voxel value along each given ray is calculated and placed in the corresponding pixel in the image. The projections allow visualization of all structures in the volume simultaneously, allowing appreciation of their continuity, just like in standard projection imaging. By varying the angles of the projection, the spatial relationships can be

better appreciated, and projection views that allow an "optimal" appreciation of the structures of interest can be identified.

7.4.4 MAXIMUM INTENSITY PROJECTION AND MINIMUM INTENSITY PROJECTION

Simple integration effectively generates images that appear similar to standard projection images; however, structures that may not be of interest are included and can reduce appreciation of structures of interest. A technique frequently used to increase the contrast of the objects of interest, for example, bones or contrast-filled vessels, is the "maximum intensity projection" (MIP) technique. For calculation of the MIP, the source and detector geometry are chosen, and the maximum value along each ray is placed in the corresponding pixel in the image. In the simple 2D illustration here (see the following), the MIP values are [5, 3, 9] using projections along the horizontal line.

$$
\begin{bmatrix} 2 & 5 & 4 \\ 3 & 2 & 1 \\ 8 & 1 & 9 \end{bmatrix} \rightarrow \begin{bmatrix} 5 \\ 3 \\ 9 \end{bmatrix}
\tag{7.32}
$$

The MIP approach effectively selects out the high-intensity structures, improving dramatically their SNR, even for small thin structures. It must be remembered that the highest voxel value along the path wins, so overlapping structures compete for visualization (Figure 7.26 illustrates this issue). In this case, a thorax CT was obtained during injection of a contrast agent, and the ribs obscure the vasculature in the MIP image. One way to overcome this "bone problem" is to generate MIP images for slabs in the object. As presented in Figure 7.27, these so-called slab-MIPs allow visualization of the vascular structures of interest; in particular, the vascular structure of the lung becomes more visible and can be better appreciated with increasing slab thickness. With respect to such projection techniques, many variations exist;

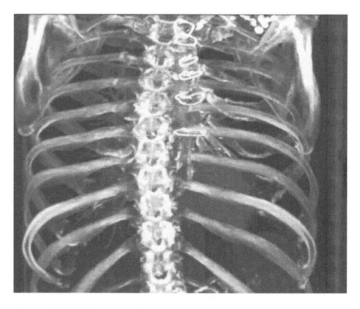

Figure 7.26 MIP image of a thorax study.

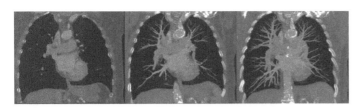

Figure 7.27 Slabbed MIP projections with varying slab thicknesses: 0 mm (left), 25 mm (middle), and 50 mm (right). The extent and complexity of the vasculature is more completely visible as the thickness of the slab increases.

one example is the Minimum Intensity Projection (min-IP). The min-IP is computationally similar to the MIP, but instead of the maximum value, the minimum value along the ray is carried into the 2D projection. This technique is often used to visualize regions with low attenuation, such as airways in the lung. Here, care must be taken that the min-IP is acquired along rays inside the body because air outside the body will corrupt the min-IP.

7.4.5 VOLUME RENDERING

Volume rendering is another projection technique but with a twist. After the viewing angle (projection direction) is selected, the voxel values along the rays are converted to opacity values and color values via corresponding (user-selected) LUTs, the summation of opacity continues until either the volume is exited or a threshold is reached. The opacity LUT allows the user to select the opacity (high value) or translucency (low value) of various structures, for example, for muscle or bone. High-opacity structures will take precedence over low-opacity/high-translucency structures. The color LUT is selected so as to allow meaningful interpretation of the scene; for example, bone (high HU) can be presented in white or ivory in the final image. Next, every pixel is shaded, according to the surface orientation relative to the source(s) of light. Figures 7.28 and 7.29 illustrate volume renderings obtained using various color and opacity LUTs.

7.4.6 SURFACE RENDERING

In contrast to the ray-casting projection techniques, surface rendering is not a true projection technique, but rather it makes use of reflections from surfaces and depth cues similar to our standard way of viewing the world. Surface-rendering techniques present surfaces of presegmented volumes making use of graphic card capabilities by which triangles can be rapidly rendered. The first step is to obtain a segmented volume or volumes, say, by using thresholding or region growing. The next step is determination of the surface of the volume(s), using an algorithm such as marching cubes (Lorensen and Cline 1987). Marching cubes determine the voxels on the surface and converts these surface voxels into vertices of triangles, their surface normals, and their connectivity relationships. As with volume rendering, the user can set the viewing angle and lighting position. The light reflection from the surface is enhanced with shading so that the surface-rendered 2D image facilitates the impression of depth (Holway and Boring 1941; Kilpartrick and Ittleson 1953; Gibson 1954; Berbaum et al. 1983; Udupa et al. 1991; Grevera et al. 2007; Svakhine et al. 2009). These data are then used by the graphics cards to present the 3D surfaces for visualization.

Figure 7.28 (See color insert.) Volume rendering of the thorax studies shown in Figure 7.26. (Left) Opacity was chosen so that tissue was opaque, allowing visualization of the skin. (Middle) high voxel values are opaque, allowing visualization of the bone [note the difference between the volume-rendered and MIP-rendered image (see Figure 7.26)]. (Right) LUT of the middle image was used but a cut-plane was placed so as to allow visualization of the inside ribcage.

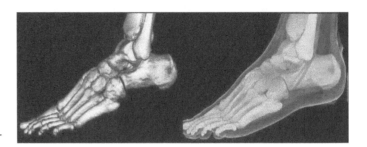

Figure 7.29 Volume renderings of CT data of the foot. (Left) Only high-attenuation data are rendered. (Right) Soft tissue was made partially opaque, so that the relationships between the bony structures and tissue can be better appreciated.

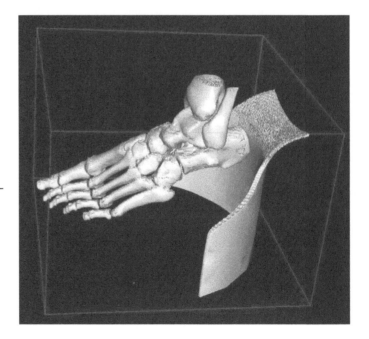

Figure 7.30 Surface rendering of the bones of a foot.

In Figure 7.30, we present a surface rendering of the bones of the foot obtained by selecting those voxels with HU values above 225.

7.4.7 RENDERING COMMENTS

Before going to the next topic, we want to discuss the clinical differences between presented techniques thus far. Each has its own unique clinical advantages. In comparison to the other two techniques, volume rendering not only can illustrate the high contrast anatomy (such as bones and contrast-filled vascular structures) but can also allow a display of soft tissue, muscle, and other regions. This extra information may provide the clinician with more comprehensive details of the pathology, and by doing so may also help in making diagnostic and treatment decisions. In contrast, the use of MIP in CT angiography is very advantageous because of the fast display of a vascular map and easy visualization of the small intraparenchymal branches (Kang et al. 2002; Raman et al. 2003; Sparacia et al. 2007). Furthermore, MIPs are rapidly available, and this is one of the reasons why the MIP technique is so often used clinically. The reason for the rapid availability lies in the fact that no display parameters, such as opacity, are necessary for MIPs. The only parameter needed for MIPs is the slab thickness, and slab thickness can be preset for each CT protocol used. Because of the necessary user interaction, the quality of the volume-rendered structures can vary across observers (Brindle et al. 2006; Honda et al. 2009; Hein et al. 2010; Perandini et al. 2010) and depends on user's mastery in the optimization of rendering parameters. Another important concern is the use of computational resources such as running time and memory use. From this point of view, surface rendering is superior to the other techniques, because data are reduced to a simple set of vertices and triangles. For the illustration of skeletal pathology, surface rendering is clinically preferred, because in many cases these renderings appear more 3D than when used with other techniques. However, this technique is limited by poor image fidelity and the inability to show subcortical details (Kuszyk et al. 1996).

7.4.8 TEMPORAL VISUALIZATION

In CT, we are often challenged with diagnostic questions that concern temporally changing structures. Recently, 4D data has become available via the introduction of flat-panel CT and wide detector multislice CT systems. One area of 4D research is cardiac imaging that takes advantage of the latest development in CT systems (320 and more slices and rotation time below 300 ms). With these systems, it is possible to reconstruct with a high temporal resolution and upon rendering visualize the full cardiac anatomy in motion. Figure 7.31 illustrates one example of the volume rendering of the human heart. In the 4D case, such a volume rendering is calculated for each time point, and then data are joined to a multimedia clip.

A further example for temporal acquisitions is CT perfusion. Perfusion (the distribution of blood in an organ) is used, for example, to determine the existence and magnitude of a stroke; in this case, the CT data provide information about tissue perfusion, blood flow patterns, and tissue status. In cardiac imaging, cardiac motion can be appreciated. In perfusion, the temporal quantity is, for example, uptake of the contrast agent, either in terms of the speed of uptake (related to time to peak or cerebral blood flow) or the quantity of uptake (cerebral blood volume) (Harrigan et al. 2005; de Lucas et al. 2008; Wintermark et al. 2008; Konstas 2009a,b). Time to peak is the interval after injection that the HU in a given pixel is maximum. Cerebral blood flow is calculated using the change of contrast over time in the various parenchymal regions. Cerebral blood volume is calculated based on the sum of the HU values (after background subtraction) over a specified

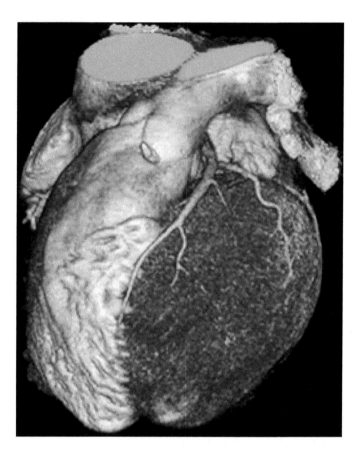

Figure 7.31 One frame out of a series of volume rendering from a cardiac CT acquisition.

time interval (e.g., start of injection to 30 s after injection). This "fourth dimension" can be color coded on a pixel by pixel basis for presentation (Figure 7.32). The color-coded data (e.g., yellow corresponding to high perfusion and blue to low perfusion) are presented on 3D data sets to provide a direct anatomic overview of the full functional pathology, so that say brain region that is affected by a stroke could be quickly diagnosed. Using the visualization approaches described here, clinicians can more completely understand the complex spatial/anatomic and temporal/functional aspect of pathologies.

7.4.9 VISUALIZATION TOOLS

In any stage of developing reconstruction algorithms or pre- or postprocessing tools, one comes to the point where a 3D visualization is necessary for evaluations. One of the most common tools used for evaluation is the Visualization Toolkit (VTK). VTK is an open-source, freely available software and cross-platform system for 3D computer graphics, image processing, and visualization. VTK is a C++ class library that is operated in a pipeline scheme. This library is a developer-level system and thus does not often directly allow a "quick and dirty" visualization. Another open source system called OsiriX (Rosset et al. 2004) incorporates all the convenient functions included in ITK and VTK and many more in a graphical user interface. OsiriX functions simultaneously as a DICOM PACS workstation for image display and image-processing software for medical research, functional imaging, 3D imaging, and molecular imaging. One who developed his or her own medical imaging

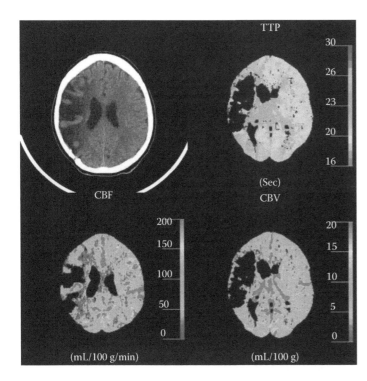

TTP

30
26
23
20
16
(Sec)

CBF

CBV

200
150
100
50
0
(mL/100 g/min)

20
15
10
5
0
(mL/100 g)

Figure 7.32 (See color insert.) Evaluation of cerebral perfusion. In the color scale for the images, red indicates good perfusion and blue indicates poor perfusion. (Top left) CT slice. (Top right) Time to peak (TTP). (Bottom left) Cerebral blood flow. (Bottom right) Cerebral blood volume. The regions of deficit correlate fairly well with each parameter providing additional information to the clinician regarding the perfusion of the brain.

algorithm could easily combine his or her work with VTK and ITK by incorporating it into OsiriX. Another practical tool for analysis and visualization of medical images is ImageJ. ImageJ is a public domain, Java-based image-processing program developed at the National Institutes of Health (Collins 2007). ImageJ was designed with an open architecture that provides extensibility via Java plug-ins and recordable macros (Girish and Vijayalakshmi 2004). The advantages of ImageJ is its cross-platform availability; however, the integration of ImageJ with VTK and ITK is a little involved because of its native Java language.

7.4.10 GPU MEETS MEDICAL IMAGING

The medical imaging community was one of the earliest to take advantage of GPU computing to achieve computational acceleration. The start was made not in postprocessing but in reconstruction of magnetic resonance imaging (MRI) and CT data. In particular, the introduction of flat-panel CT and wide-area detector multislice CT systems and the resulting increase in raw data inspired the use of GPU-based reconstruction algorithms (Sharp et al. 2007; Xu and Mueller 2007; Zeng et al. 2007; Kyriakou et al. 2008; Noël et al. 2010). Nevertheless, other areas of the medical imaging community followed swiftly to accelerate their work with GPUs. The secret behind using GPUs is to use a CPU and GPU together in a heterogeneous computing model. This means that sequential parts of a method run on a CPU and the computationally intensive parts run on GPUs. As in the Nvidia/CUDA or Open CL architecture, the computationally intensive

part is massively parallelized. From the user's point of view, the GPU uses several hundred cores instead of a one or four core CPU. In any given area of medical imaging, GPUs have been introduced for image vision, reconstruction, simulations, and rendering. Thus, the use of GPUs in this field has matured to the point that there are now several medical modalities delivered with GPUs.

7.5 CONCLUSIONS

CT is one of the main-stay modalities for assessment of patient status and direction. Although visual inspection is the primary mode used for this assessment, the number and variety of processing and analysis techniques are increasing along with their use in clinical practice. Because of the plethora of techniques, we have selected and described only a small sample of the available techniques, to give a flavor of the possibilities. For both the visual inspection and development of the analysis techniques, understanding the object of your search and efforts is key. For development and optimization of the analyses, understanding the quality of your data (in terms of noise and resolution) and understanding the effect of your data-processing and data analysis techniques on your data and your calculated properties and metrics is essential. Decisions regarding the presentation of the data, specifically the type of visualization approaches and tools you use, should be developed with your end user in mind, and preferably, involved. As with any number of advances, the fun has just begun and will continue for quite a while. As a result, patients' lives will be positively affected and treatments will be improved.

ACKNOWLEDGMENTS

We acknowledge the assistance of Elizabeth Bullitt, PhD (University of North Carolina–Chapel Hill), Paul Keall, PhD (Stanford University), Charles Pelizzari, PhD (University of Chicago), and Hiro Yoshida, PhD (Harvard University) in the preparation of the manuscript, and for providing images and text.

REFERENCES

Abdelnour, A.F., Nehmeh, S.A., Pan, T., Humm, J.L., Vernon, P., Schoder, H., Rosenzweig, K., Mageras, G., Yorke, E., Larson, S.M. and Erdi, Y. 2007. Phase and amplitude binning for 4D-CT imaging. *Phys Med Biol* 52: 3515–29.

Angel, E., Yaghmai, N., Jude, C.M., DeMarco, J.J., Cagnon, C.H., Goldin, J.G., McCollough, C.H., Primak, A.N., Cody, D.D, Stevens, D.M. and McNitt-Gray, M.F. 2009. Dose to radiosensitive organs during routine chest CT: effects of tube current modulation. *Am J Roentgenol* 193: 1340–5.

Bae, K.T., Giger, M.L., Chen, C.T. and Kahn, C.E., Jr. 1993. Automatic segmentation of liver structure in CT images. *Med Phys* 20: 71–8.

Beare, R. 2006. A locally constrained watershed transform. *IEEE Trans Pattern Anal Mach Intell* 28: 1063–74.

Berbaum, K., Tharp, D. and Mroczek, K. 1983. Depth perception of surfaces in pictures: looking for conventions of depiction in Pandora's box. *Perception* 12: 5–20.

Besl, P. and McKay, N. 1992. A method for registration of 3-D shapes. *IEEE Trans Pattern Anal Mach Intell (PAMI)* 14: 239–56.

Blake, A. and Isard, M. 1998. *Active Contours: The Application of Techniques from Graphics, Vision, Control Theory and Statistics to Visual Tracking of Shapes in Motion.* Berlin: Springer Verlag.

Blum, H. and Nagel, R.N. 1978. Shape description using weighted symmetric axis features. *Pattern Recogn* 10: 167–80.

Boone, J.M. 2001. Determination of the presampled MTF in computed tomography. *Med Phys* 28: 356–60.

Boscolo, R., Brown, M.S. and McNitt-Gray, M.F. 2002. Medical image segmentation with knowledge-guided robust active contours. *Radiographics* 22: 437–48.

Boskamp, T., Rinck, D., Link, F., Kümmerlen, B., Stamm, G. and Mildenberger, P. 2004. New vessel analysis tool for morphometric quantification and visualization of vessels in CT and MR imaging data sets. *Radiographics* 24: 287–97.

Bouix, S., Siddiqi, K. and Tannenbaum, A. 2005. Flux driven automatic centerline extraction. *Med Image Anal* 9: 209–21.

Brindle, J.M., Trindade, A.A., Pichardo, J.C., Myers, S.L., Shah, A.P. and Bolch, W.E. 2006. CT volumetry of the skeletal tissues. *Med Phys* 33: 3796–803.

Brock, K.M., Balter, J.M., Dawson, L.A., Kessler, M.L. and Meyer, C.R. 2003. Automated generation of a four-dimensional model of the liver using warping and mutual information. *Med Phys* 30: 1128–33.

Buckler, A.J., Mulshine, J.L., Gottlieb, R., Zhao, B., Mozley, P.D. and Schwartz, L. 2010. The use of volumetric CT as an imaging biomarker in lung cancer. *Acad Radiol* 17: 100–6.

Bullitt, E., Gerig, G., Pizer, S. and Aylward, S.R. 2003. Measuring tortuosity of the intracerebral vasculature from MRA images. *IEEE Trans Med Imaging* 22: 1163–71.

Bullitt, E., Lin, N.U., Smith, J.K., Zeng, D., Winer, E.P., Carey, L.A., Lin, W. and Ewend, M.G. 2007. Blood vessel morphological changes as visualized by MRA during treatment of brain metastases. *Radiology* 245: 824–30.

Bullitt, E., Zeng, D., Gerig, G., Aylward, S., Joshi, S., Smith, J.K., Lin, W. and Ewend, M.G. 2005. Vessel tortuosity and brain tumor malignancy: a blinded study. *Acad Radiol* 12: 1232–40.

Burger, W. and Burge, M.J. 2008. *Digital Image Processing: An Algorithmic Introduction Using Java.* Berlin: Springer.

Cai, W., Zalis, M.E., Näppi, J., Harris, G.J. and Yoshida, H. 2008. Structure-analysis method for electronic cleansing in cathartic and noncathartic CT colonography. *Med Phys* 35: 3259–77.

Caldemeyer, K.S., Sandrasegaran, K., Shinaver, C.N., Mathews, V.P., Smith, R.R. and Kopecky, K.K. 1999. Temporal bone: comparison of isotropic helical CT and conventional direct axial and coronal CT. *Am J Roentgenol* 172: 1675–82.

Calhoun, P.S., Kuszyk, B.S., Heath, D.G., Carley, J.C. and Fishman, E.K. 1999. Three-dimensional volume rendering of spiral CT data: theory and method. *Radiographics* 19: 745–64.

Cao, Z., Pan, S., Li, R., Balachandran, R., Fitzpatrick, J.M., Chapman, W.C. and Dawant, B.M. 2004. Registration of medical images using an interpolated closest point transform: method and validation. *Med Image Anal* 8: 421–7.

Cassot, F., Lauwers, F., Lorthois, S., Puwanarajah, P. and Duvernoy, H. 2009. Scaling laws for branching vessels of human cerebral cortex. *Microcirculation* 16: 331–44.

Chabat, F., Yang, G.Z. and Hansell, D.M. 2003. Obstructive lung diseases: texture classification for differentiation at CT. *Radiology* 228: 871–7.

Chakravarty, M.M., Sadikot, A.F., Germann, J., Bertrand, G., Collins, D.L. 2008. Towards a validation of atlas warping techniques. *Med Image Anal* 12: 713–26.

Christensen, G.E., Johnson, H.J. and Vannier, M.W. 2006. Synthesizing average 3D anatomical shapes. *Neuroimage* 32: 146–58.

Chu, E.C. 2008. *Discrete and Continuous Fourier Transforms: Analysis, Applications and Fast Algorithms.* Boca Raton, FL: CRC Press.

Collins, T.J. 2007. ImageJ for microscopy. *BioTechniques* 43(Suppl 1): 25–30.

Crawford, C.R. 1991. CT filtration aliasing artifacts. *IEEE Trans Med Imaging* 10: 99–102.

De Boor, C. 2001. *Practical Guide to Splines* (Applied Mathematical Sciences 27). Berlin: Springer.

Dehmeshki, J., Amin, H., Valdivieso, M. and Ye, X. 2008. Segmentation of pulmonary nodules in thoracic CT scans: a region growing approach. *IEEE Trans Med Imaging* 27: 467–80.

de Lucas, E.M., Sánchez, E., Gutiérrez, A., Mandly, A.G., Ruiz, E., Flórez, A.F., Izquierdo, J., Arnáiz, J., Piedra, T., Valle, N., Bañales, I. and Quintana, F. 2008. CT protocol for acute stroke: tips and tricks for general radiologists. *Radiographics* 28: 1673–87.

DICOM. Available from: http://medical.nema.org

Doi, K. 2007. Computer-aided diagnosis in medical imaging: historical review, current status and future potential. *Comput Med Imag Graph* 31: 198–211.

du Plessis, J., Goussard, P., Andronikou, S., Gie, R. and George, R. 2009. Comparing three-dimensional volume-rendered CT images with fibreoptic tracheobronchoscopy in the evaluation of airway compression caused by tuberculous lymphadenopathy in children. *Pediatr Radiol* 39: 694–702.

Ehrhardt, J., Werner, R., Saring, D., Frenzel, T., Lu, W., Low, D. and Handels, H. 2007. An optical flow based method for improved reconstruction of 4D CT data sets acquired during free breathing. *Med Phys* 34: 711–21.

Esteban, M.D. and Morales, D. 1995. A Summary of entropy statistics. *Kybernetika* 31: 337–46.

Fabijańska, A. 2009. Two-pass region growing algorithm for segmenting airway tree from MDCT chest scans. *Comput Med Imag Graph* 33: 537–46.

Fayad, L.M., Jin, Y., Laine, A.F., Berkmen, Y.M., Pearson, G.D., Freedman, B. and Van Heertum, R. 2002. Chest CT window settings with multiscale adaptive histogram equalization: pilot study. *Radiology* 223: 845–52.

Feldkamp, L.A., Davis, L.C. and Kress, J.W. 1984. Practical cone-beam algorithm. *J Opt Soc Am A* 1: 612–19.

Finet, G., Maurincomme, E., Reiber, J.H., Savalle, L., Magnin, I. and Beaune. J. 1998. Evaluation of an automatic intraluminal edge detection technique for intravascular ultrasound images. *Jpn Circ J* 62: 115–21.

Fitzpatrick, J.M. and West, J.B. 2001. The distribution of target registration error in rigid-body point-based registration. *IEEE Trans Med Imaging* 20: 917–27.

Fitzpatrick, J.M., West, J.B. and Maurer, C.R., Jr. 1998. Predicting error in rigid-body point-based registration. *IEEE Trans Med Imaging* 17: 694–702.

Fitzpatrick, M.J., Starkschall, G., Antolak, J.A., Fu, J., Shukla, H., Keall, P.J., Klahr, P. and Mohan, R. 2006. Displacement-based binning of time-dependent computed tomography image data sets. *Med Phys* 33: 235–46.

Fleiter, T., Merkle, E.M., Aschoff, A.J., Lang, G., Stein, M., Görich, J., Liewald, F., Rilinger, N. and Sokiranski, R. 1997. Comparison of real-time virtual and fiberoptic bronchoscopy in patients with bronchial carcinoma: opportunities and limitations. *Am J Roentgenol* 169: 1591–5.

Fletcher, J.G., Johnson, C.D., MacCarty, R.L., Welch, T.J., Reed, J.E. and Hara, A.K. 1999. CT colonography: potential pitfalls and problem-solving techniques. *Am J Roentgenol* 172: 1271–8.

Ford, E.C., Mageras, G.S., Yorke, E. and Ling, C.C. 2003. Respiration-correlated spiral CT: a method of measuring respiratory-induced anatomic motion for radiation treatment planning. *Med Phys* 30: 88–97.

Friman, O., Hindennach, M., Kühnel, C. and Peitgen, H.O. 2010. Multiple hypothesis template tracking of small 3D vessel structures. *Med Image Anal* 14: 160–71.

Frimmel, H, Näppi, J. and Yoshida, H. 2005. Centerline-based colon segmentation for CT colonography. *Med Phys* 32: 2665–72.

Advanced techniques

Gasquet, C., Ryan, R.D. and Witomski, P. 1998. *Fourier Analysis and Applications: Filtering, Numerical Computation, Wavelets. Texts in Applied Mathematics.* New York: Springer.

Ge, Y., Stelts, D.R., Wang, J. and Vining, D.J. 1999. Computing the centerline of a colon: a robust and efficient method based on 3D skeletons. *J Comput Assist Tomogr* 23: 786–94.

George, R., Chung, T.D., Vedam, S.S., Ramakrishnan, V., Mohan, R., Weiss, E. and Keall, P.J. 2006. Audio-visual biofeedback for respiratory-gated radiotherapy: impact of audio instruction and audio-visual biofeedback on respiratory-gated radiotherapy. *Int J Radiol Oncol Biol Phys* 65: 924–33.

Germain, O. and Réfrégier, P. 1996. Optimal snake-based segmentation of a random luminance target on a spatially disjoint background. *Opt Lett* 21: 1845–7.

Gibson, J.J. 1954. A theory of pictorial perception. *Av Commun Rev* 1: 3–23.

Giger, M.L., Doi, K. and MacMahon, H. 1988. Image feature analysis and computer-aided diagnosis in digital radiography. 3. Automated detection of nodules in peripheral lung fields. *Med Phys* 15: 158–66.

Girish, V. and Vijayalakshmi, A. 2004. Affordable image analysis using NIH Image/ImageJ. *Indian J Cancer* 41: 47.

Glover, G.H. and Pelc, N.J. 1980. Nonlinear partial volume artifacts in x-ray computed tomography. *Med Phys* 7: 238–48.

Gonzales, R.C. and Woods, R.E. 2008. *Digital Image Processing.* Upper Saddle River, NJ: Prentice Hall.

Grau, V., Mewes, A.U., Alcañiz, M., Kikinis, R. and Warfield, S.K. 2004. Improved watershed transform for medical image segmentation using prior information. *IEEE Trans Med Imaging* 23: 447–58.

Grevera, G., Udupa, J., Odhner, D., Zhuge, Y., Souza, A., Iwanaga, T. and Mishra S. 2007. CAVASS: a computer-assisted visualization and analysis software system. *J Digit Imaging* 20(Suppl 1): 101–18.

GSL Team. 2007. *§13.4 Singular Value Decomposition, GNU Scientific Library.* Reference Manual. Available from: http://www.gnu.org/software/gsl/manual/html_node/Singular-Value-Decomposition.html, (Accessed 2007).

Hadjiiski, L., Mukherji, S.K., Ibrahim, M., Sahiner, B., Gujar, S.K., Moyer, J. and Chan, H.P. 2010. Head and neck cancers on CT: preliminary study of treatment response assessment based on computerized volume analysis. *AJR Am J Roentgenol* 194: 1083–9.

Harrigan, M.R., Leonardo, J., Gibbons, K.J., Guterman, L.R. and Hopkins, L.N. 2005. CT perfusion cerebral blood flow imaging in neurological critical care. *Neurocrit Care* 2: 352–66.

Hathcock, J.T. and Stickle, R.L. 1993. Principles and concepts of computed tomography. *Vet Clin North Am Small Anim Pract* 23: 399–415.

Hein, P.A., Romano, V.C., Rogalla, P., Klessen, C., Lembcke, A., Bornemann, L., Dicken, V., Hamm, B. and Bauknecht, H.C. 2010. Variability of semiautomated lung nodule volumetry on ultralow-dose CT: comparison with nodule volumetry on standard-dose CT. *J Digit Imaging* 23: 8–17.

Higgins, W.E. and Ojard, E.J. 1993. Interactive morphological watershed analysis for 3D medical images. *Comput Med Imag Graph* 17: 387–95.

Hoffmann, K.R., Rudin, S., Meng, H., Hopkins, L.N., Guterman, L. and Levy, E. 2005. Three-dimensional analysis of the cerebral vasculature: concepts and applications, In: Armato, S.G. and Brown, M.S. (eds.) *Multidimensional Image Processing, Analysis, and Display.* Oak Brook, IL: Radiological Society of North America, 173–84.

Holway, A.H. and Boring, E.G. 1941. Determinants of apparent visual size with distance variant. *Am J Psychol* 54: 21–37.

Honda, O., Kawai, M., Gyobu, T., Kawata, Y., Johkoh, T., Sekiguchi, J., Tomiyama, N., Yoshida, S., Sumikawa, H., Inoue, A., Yanagawa, M., Daimon, T. and Nakamura, H. 2009. Reproducibility of temporal volume change in CT of lung cancer: comparison of computer software and manual assessment. *Br J Radiol* 82: 742–7.

Iordanescu, G. and Summers, R.M. 2003. Automated centerline for computed tomography colonography. *Acad Radiol* 10: 1291–301.

ITK (Insight Toolkit). Available from: http://www.itk.org/

Jaffray, D.A., Siewerdsen, J.H., Wong, J.W. and Martinez, A.A. 2002. Flat-panel cone-beam computed tomography for image-guided radiation therapy. *Int J Radiat Oncol Biol Phys* 53: 1337–49.

Joshi, A.A., Chaudhari, A.J., Li, C., Dutta, J., Cherry, S.R., Shattuck, D.W., Toga, A.W. and Leahy, R.M. 2010. DigiWarp: a method for deformable mouse atlas warping to surface topographic data. *Phys Med Biol* 55: 6197–214.

Kang, H.K., Jeong, Y.Y., Choi, J.H., Choi, S., Chung, T.W., Seo, J.J., Kim, J.K., Yoon, W. and Park, J.G. 2002. Three-dimensional multi-detector row CT portal venography in the evaluation of portosystemic collateral vessels in liver cirrhosis. *Radiographics* 22: 1053–61.

Kassab, G.S. 2006. Scaling laws of vascular trees: of form and function. *Am J Physiol Heart Circ Physiol* 290: H894–903.

Keall, P.J., Mageras, G.S., Balter, J.M., Emery, R.S., Forster, K.M., Jiang, S.B., Kapatoes, J.M., Low, D.A., Murphy, M.J., Murray, B.R., Ramsey, C.R., Van Herk, M.B., Vedam, S.S., Wong, J.W. and Yorke, E. 2006. The management of respiratory motion in radiation oncology report of AAPM Task Group 76. *Med Phys* 33: 3874–900.

Kilpatrick, F.P. and Ittleson, W.H. 1953. The size-distance invariance hypothesis. *Psychol Rev* 60: 223–31.

Konstas, A.A., Goldmakher, G.V., Lee, T.Y. and Lev, M.H. 2009a. Theoretic basis and technical implementations of CT perfusion in acute ischemic stroke, part 1: theoretic basis. *Am J Neuroradiol* 30: 662–8.

Konstas, A.A., Goldmakher, G.V., Lee, T.Y. and Lev, M.H. 2009b. Theoretic basis and technical implementations of CT perfusion in acute ischemic stroke, part 2: technical implementations. *Am J Neuroradiol* 30: 885–92.

Krishnan, K., Ibanez, L., Turner, W.D., Jomier, J. and Avila, R.S. 2010. An open-source toolkit for the volumetric measurement of CT lung lesions. *Opt Express* 18: 15256–66.

Kuszyk, B.S., Heath, D.G., Bliss, D.F. and Fishman, E.K. 1996. Skeletal 3-D CT: advantages of volume rendering over surface rendering. *Skeletal Radiol* 25: 207–14.

Kyriakou, Y., Lapp, R.M., Hillebrand, L., Ertel, D. and Kalender, W.A. 2008. Simultaneous misalignment correction for approximate circular cone-beam computed tomography. *Phys Med Biol* 53: 6267–89.

Lang, H., Radtke, A., Hindennach, M., Schroeder, T., Frühauf, N.R., Malagó, M., Bourquain, H., Peitgen, H.O., Oldhafer, K.J., Broelsch, C.E. 2005. Impact of virtual tumor resection and computer-assisted risk analysis on operation planning and intraoperative strategy in major hepatic resection. *Arch Surg* 140: 629–38.

Langner, U.W. and Keall, P.J. 2010. Quantification of artifact reduction with real-time cine four-dimensional computed tomography acquisition methods. *Int J Radiat Oncol Biol Phys* 76: 1242–50.

Lee, K.S., Yoon, J.H., Kim, T.K., Kim, J.S., Chung, M.P. and Kwon, O.J. 1997. Evaluation of tracheobronchial disease with helical CT with multiplanar and three-dimensional reconstruction: correlation with bronchoscopy. *Radiographics* 17: 555–67.

Lehr, J.L. and Capek, P. 1985. Histogram equalization of CT images. *Radiology* 154: 163–9.

Levoy, M. 1990. A hybrid ray tracer for rendering polygon and volume data. *IEEE Comput Graph Appl* 10: 33–40.

Li, B., Christensen, G.E., Hoffman, E.A., McLennan, G. and Reinhardt, J.M. 2003. Establishing a normative atlas of the human lung: intersubject warping and registration of volumetric CT images. *Acad Radiol* 10: 255–65.

Li, B., Christensen, G.E., Hoffman, E.A., McLennan, G. and Reinhardt, J.M. 2008. Pulmonary CT image registration and warping for tracking tissue deformation during the respiratory cycle through 3D consistent image registration. *Med Phys* 35: 5575–83.

Li, H. and Yezzi, A. 2007. Local or global minima: flexible dual-front active contours. *IEEE Trans Pattern Anal Mach Intell* 29: 1–14.

Li, H. and Wang, Z. 2005. A seepage flow model for vertebra CT image segmentation. *Conf Proc IEEE Eng Med Biol Soc* 6: 6364–7.

Li, S., Fevens, T., Krzyzak, A., Jin, C. and Li, S. 2006. Fast and robust clinical triple-region image segmentation using one level set function. *Med Image Comput Comput Assist Interv* 9(Pt 2): 766–73.

Lobregt, S. and Viergever, M.A. 1995. A discrete dynamic contour model. *IEEE Trans Med Imaging* 14: 12–24.

Lorensen, W.E. and Cline, H.E. 1987. Marching cubes: a high resolution 3D surface construction algorithm. *Computer Graph* 21: 163–9.

Lu, W., Parikh, P.J., Hubenschmidt, J.P., Bradley, J.D. and Low, D.A. 2006. A comparison between amplitude sorting and phase-angle sorting using external respiratory measurement for 4D CT. *Med Phys* 33: 2964–74.

Macari, M., Berman, P., Dicker, M., Milano, A., Megibow, A.J. 1999. Usefulness of CT colonography in patients with incomplete colonoscopy. *Am J Roentgenol* 173: 561–4.

Maurer, C.R., Jr, Fitzpatrick, J.M., Wang, M.Y., Galloway, R.L., Jr, Maciunas, R.J. and Allen, G.S. 1997. Registration of head volume images using implantable fiducial markers. *IEEE Trans Med Imaging* 16: 447–62.

Mayer, D., Bartz, D., Fischer, J., Ley, S., del Río, A., Thust, S., Kauczor, H.U. and Heussel, C.P. 2004. Hybrid segmentation and virtual bronchoscopy based on CT images. *Acad Radiol* 11: 551–65.

Mayer, T.E., Brueckmann, H., Siegert, R., Witt, A. and Weerda, H. 1997. High-resolution CT of the temporal bone in dysplasia of the auricle and external auditory canal. *Am J Neuroradiol* 18: 53–65.

McNitt-Gray, M.F. 2002. AAPM/RSNA physics tutorial for residents: topics in CT. Radiation dose in CT. *Radiographics* 22: 1541–53.

Miles, K.A., Ganeshan, B., Griffiths, M.R., Young, R.C. and Chatwin, C.R. 2009. Colorectal cancer: texture analysis of portal phase hepatic CT images as a potential marker of survival. *Radiology* 250: 444–52.

Mori, K., Hasegawa, J., Suenaga, Y. and Toriwaki, J. 2000. Automated anatomical labeling of the bronchial branch and its application to the virtual bronchoscopy system. *IEEE Trans Med Imaging* 19: 103–14.

Mori, K., Ota, S., Deguchi, D., Kitasaka, T., Suenaga, Y., Iwano, S., Hasegawa, Y., Takabatake, H., Mori, M. and Natori, H. 2009. Automated anatomical labeling of bronchial branches extracted from CT datasets based on machine learning and combination optimization and its application to bronchoscope guidance. *Med Image Comput Comput Assist Interv* 12(Pt 2): 707–14.

Mutaf, Y.D., Antolak, J.A. and Brinkmann, D.H. 2007. The impact of temporal inaccuracies on 4DCT image quality. *Med Phys* 34: 1615–22.

Neicu, T., Berbeco, R., Wolfgang, J. and Jiang, S.B. 2006. Synchronized moving aperture radiation therapy (SMART): improvement of breathing pattern reproducibility using respiratory coaching. *Phys Med Biol* 51: 617–36.

Nievelstein, R.A., van Dam, I.M. and van der Molen, A.J. 2010. Multidetector CT in children: current concepts and dose reduction strategies. *Pediatr Radiol* 40: 1324–44.

Noël, P.B., Xu, J., Corso, J.J., Walczak, A.M., Hoffmann, K.R. and Schafer S. 2010. GPU-based cone beam computed tomography. *Comput Methods Programs Biomed* 98: 271–7.

Numerical Recipes. Available from: http://www.nr.com

OsiriX. Available from: http://www.osirix-viewer.com/

Pan, T., Lee, T.Y., Rietzel, E. and Chen, G.T. 2004. 4D-CT imaging of a volume influenced by respiratory motion on multi-slice CT. *Med Phys* 31: 333–40.

Pelizzari, C.A., Chen, G.T., Spelbring, D.R., Weichselbaum, R.R. and Chen, C.T. 1989. Accurate three-dimensional registration of CT, PET, and/or MR images of the brain. *J Comput Assist Tomogr* 13: 20–6.

Perandini, S., Faccioli, N., Inama, M. and Pozzi, M.R. 2010. Freehand liver volumetry by using an electromagnetic pen tablet: accuracy, precision, and rapidity. *J Digit Imag* in 24: 360–5.

Perumpillichira, J.J., Yoshida, H. and Sahani, D.V. 2005. Computer-aided detection for virtual colonoscopy. *Canc Imag* 5: 11–16.

Pickhardt, P.J. and Kim, D.H. 2009. *CT Colonography: Principles and Practice of Virtual Colonoscopy*. Philadelphia, PA: Elsevier/Saunders.

Pizer, S.M., Zimmerman, J.B. and Staab, E.V. 1984. Adaptive grey level assignment in CT scan display. *J Comput Assist Tomogr* 8: 300–5.

Pluim, J.P.W., Maintz, J.B.A. and Viergever, M.A. 2003. Mutual information based registration of medical images: a survey. *IEEE Trans Med Imaging* 22: 986–1004.

Ragnemalm, I. 1993. The euclidean distance transform in arbitrary dimensions. *Pattern Recognition Letters* 14: 883–8.

Rakshe, T., Fleischmann, D., Rosenberg, J., Roos, J.E., Straka, M. and Napel, S. 2008. An improved algorithm for femoropopliteal artery centerline restoration using prior knowledge of shapes and image space data. *Med Phys* 35: 3372–82.

Raman, R., Napel, S. and Rubin, G.D. 2003. Curved-slab maximum intensity projection: method and evaluation. *Radiology* 229: 255–60.

Rietzel, E. and Chen, G.T. 2006. Improving retrospective sorting of 4D computed tomography data. *Med Phys* 33: 377–9.

Rosset, A., Spadola, L. and Ratib, O. 2004. OsiriX: an open-source software for navigating in multidimensional DICOM images. *J Digit Imag* 17: 205–16.

Roth, S.D. 1982. Ray casting for modeling solids. *Comput Graph Image Process* 18: 109–44.

Russ, J.C. 2007. *The Image Processing Handbook*. Boca Raton, FL: CRC Press.

Samara, Y., Fiebich, M., Dachman, A.H., Kuniyoshi, J.K., Doi, K. and Hoffmann, K. 1999. Automated calculation of the centerline of the human colon in CT images. *Acad Radiol* 6: 352–60.

Sarrut, D., Boldea, V., Miguet, S. and Ginestet, C. 2006. Simulation of four-dimensional CT images from deformable registration between inhale and exhale breath-hold CT scans. *Med Phys* 33: 605–17.

Schaller, S., Flohr, T., Klingenbeck, K., Krause, J., Fuchs, T. and Kalender, W.A. 2000. Spiral interpolation algorithm for multislice spiral CT – part I: theory. *IEEE Trans Med Imaging* 19: 822–34.

Schmidt, A., Zidowitz, S., Kriete, A., Denhard, T., Krass, S. and Peitgen, H.O. 2004. A digital reference model of the human bronchial tree. *Comput Med Imag Graph* 28: 203–11.

Schneiders, N.J. and Bushong, S.C. 1980. A precise CT phantom alignment procedure. *Med Phys* 7: 549–50.

Schoeneman, P.H. 1969. A generalized solution of the orthogonal procrustes problem. *Psychometrika* 31: 1–10.

Schoeneman, P.H. and Carroll, R.M. 1970. Fitting one matrix to another under choice of a central dilation and a rigid motion. *Psychometrika* 35: 245–54.

Schreibmann, E., Chen, G.T. and Xing, L. 2006. Image interpolation in 4D CT using a BSpline deformable registration model. *Int J Radiat Oncol Biol Phys* 64: 1537–50.

Selle, D., Preim, B., Schenk, A. and Peitgen, H.O. 2002. Analysis of vasculature for liver surgical planning. *IEEE Trans Med Imaging* 21: 1344–57.

Sharp, G.C., Kandasamy, N., Singh, H. and Folkert, M. 2007. GPU-based streaming architectures for fast cone-beam CT image reconstruction and demons deformable registration. *Phys Med Biol* 52: 5771–83.

Sijbers, J., Scheunders, P., Verhoye, M., van der Linden, A., van Dyck, D. and Raman, E. 1997. Watershed-based segmentation of 3D MR data for volume quantization. *Magn Reson Imag* 15: 679–88.

Smelyanskiy, M., Holmes, D., Chhugani, J., Larson, A., Carmean, D.M., Hanson, D., Dubey, P., Augustine, K., Kim, D., Kyker, A., Lee, V.W., Nguyen, A.D., Seiler, L. and Robb, R. 2009. Mapping high-fidelity volume rendering for medical imaging to CPU, GPU and many-core architectures. *IEEE Trans Vis Comput Graph* 15: 1563–70.

Socha, M., Duplaga, M. and Turcza, P. 2004. Methods of bronchial tree reconstruction and camera distortion corrections for virtual endoscopic environments. *Stud Health Technol Inform* 105: 285–95.

Sonka, M., Hlavac, V. and Boyle, R. 2007. *Image Processing, Analysis, and Machine Vision*. Toronto: Thomson Learning.

Sonke, J., Zijp, L., Remeijer, P. and Van Herk, M. 2005. Respiratory correlated cone beam CT. *Med Phys* 32: 1176–86.

Sorantin, E., Halmai, C., Erdöhelyi, B., Palágyi, K., Nyúl, L.G., Ollé, K., Geiger, B., Lindbichler, F., Friedrich, G. and Kiesler, K. 2002. Spiral-CT-based assessment of tracheal stenoses using 3-D-skeletonization. *IEEE Trans Med Imaging* 21: 263–73.

Sparacia, G., Bencivinni, F., Banco, A., Sarno, C., Bartolotta, T.V. and Lagalla, R. 2007. Imaging processing for CT angiography of the cervicocranial arteries: evaluation of reformatting technique. *Radiol Med* 112: 224–38.

Stockham, C.D. 1979. A simulation study of aliasing in computed tomography. *Radiology* 132: 721–6.

Summers, R.M., Swift, J.A., Dwyer, A.J., Choi, J.R. and Pickhardt, P.J. 2009. Normalized distance along the colon centerline: a method for correlating polyp location on CT colonography and optical colonoscopy. *Am J Roentgenol* 193: 1296–304.

Sundaramoorthi, G., Yezzi, A. and Mennucci, A. 2008. Coarse-to-fine segmentation and tracking using Sobolev active contours. *IEEE Trans Pattern Anal Mach Intell* 30: 851–64.

Svakhine, N.A., Ebert, D.S. and Andrews, W.M. 2009. Illustration-inspired depth enhanced volumetric medical visualization. *IEEE Trans Vis Comput Graph* 15: 77–86.

Tawhai, M.H., Burrowes, K.S. and Hoffman, E.A. 2006. Computational models of structure-function relationships in the pulmonary circulation and their validation. *Exp Physiol* 91: 285–93.

Turnier, H., Houdek, P.V. and Trefler, M. 1979. Measurements of the partial volume phenomenon. *Comput Tomogr* 3: 213–9.

Udupa, J.K., Hung, H.M. and Chuang, K.S. 1991. Surface and volume rendering in three-dimensional imaging: a comparison. *J Digit Imag* 4: 159–68.

Van Spaendonck, M.P., Cryns, K., Van De Heyning, P.H., Scheuermann, D.W., Van Camp, G. and Timmermans, J.P. 2000. High resolution imaging of the mouse inner ear by microtomography: a new tool in inner ear research. *Anat Rec* 259: 229–36.

van Wijk, C., van Ravesteijn, V.F., Vos, F.M. and van Vliet, L.J. 2010. Detection and segmentation of colonic polyps on implicit isosurfaces by second principal curvature flow. *IEEE Trans Med Imaging* 29: 688–98.

Vedam, S.S., Keall, P.J., Kini, V.R., Mostafavi, H., Shukla, H.P. and Mohan, R. 2003. Acquiring a four-dimensional computed tomography dataset using an external respiratory signal. *Phys Med Biol* 48: 45–62.

Venkat, R.B., Sawant, A., Suh, Y., George, R. and Keall, J. 2008. Development and preliminary evaluation of a prototype audiovisual biofeedback device incorporating a patient-specific guiding waveform. *Phys Med Biol* 53: N197–208.

Venkatraman, R., Raman, R., Raman, B., Moss, R.B., Rubin, G.D., Mathers, L.H. and Robinson, T.E. 2006. Fully automated system for three-dimensional bronchial morphology analysis using volumetric multidetector computed tomography of the chest. *J Digit Imag* 19: 132–9.

Vining, D.J., Liu, K., Choplin, R.H. and Haponik, E.F. 1996. Virtual bronchoscopy. Relationships of virtual reality endobronchial simulations to actual bronchoscopic findings. *Chest* 109: 549–53.

VTK (Visualization Toolkit). Available from: http://www.vtk.org/

Watt, A. and Watt, M. 1992. *Advanced Animation and Rendering Techniques: Theory and Practice*. Reading: Addison-Wesley.

Way, T.W., Hadjiiski, L.M., Sahiner, B., Chan, H.P., Cascade, P.N., Kazerooni, E.A., Bogot, N. and Zhou, C. 2006. Computer-aided diagnosis of pulmonary nodules on CT scans: segmentation and classification using 3D active contours. *Med Phys* 33: 2323–37.

West, J., Fitzpatrick, J.M., Wang, M.Y., Dawant, B.M., Maurer, C.R. Jr, Kessler, R.M., Maciunas, R.J., Barillot, C., Lemoine, D., Collignon, A., Maes, F., Suetens, P., Vandermeulen, D., van den Elsen, P.A., Napel, S., Sumanaweera, T.S., Harkness, B., Hemler, P.F., Hill, D.L., Hawkes, D.J., Studholme, C., Maintz, J.B., Viergever, M.A., Malandain, G., Woods, R.P., et al. 1997. Comparison and evaluation of retrospective intermodality brain image registration techniques. *J Comput Assist Tomogr* 21: 554–66.

Wintermark, M., Sincic, R., Sridhar, D. and Chien, J.D. 2008. Cerebral perfusion CT: technique and clinical applications. *J Neuroradiol* 35: 253–60.

Xu, F. and Mueller, K. 2007. Real-time 3D computed tomographic reconstruction using commodity graphics hardware. *Phys Med Biol* 52: 3405–19.

Yamamoto, T., Langner, U., Loo, B.W., Jr, Shen, J. and Keall, P.J. 2008. Retrospective analysis of artifacts in four-dimensional CT images of 50 abdominal and thoracic radiotherapy patients. *Int J Radiat Oncol Biol Phys* 72: 1250–8.

Yang, J., Staib, L.H. and Duncan, J.S. 2004. Neighbor-constrained segmentation with level set based 3-D deformable models. *IEEE Trans Med Imaging* 23: 940–8.

Yee, D., Rathee, S., Robinson, D. and Murray, B. 2011. Temporal lung tumor volume changes in small-cell lung cancer patients undergoing chemoradiotherapy. *Int J Radiat Oncol Biol Phys* 80: 142–7.

Yen, S.Y., Yan, C.H., Rubin, G.D. and Napel, S. 1999. Longitudinal sampling and aliasing in spiral CT. *IEEE Trans Med Imaging* 18: 43–58.

Yin, F.F., Giger, M.L., Doi, K., Vyborny, C.J. and Schmidt, R.A. 1994. Computerized detection of masses in digital mammograms: investigation of feature-analysis techniques. *J Digit Imag* 7: 18–26.

Yoshida, H., Masutani, Y., MacEneaney, P., Rubin, D.T. and Dachman, A.H. 2002. Computerized detection of colonic polyps at CT colonography on the basis of volumetric features: pilot study. *Radiology* 222: 327–36.

Yoshida, H. and Nappi, J. 2001. Three-dimensional computer-aided diagnosis scheme for detection of colonic polyps. *IEEE Trans Med Imaging* 20: 1261–74.

Young, Y.N. and Levy, D. 2005. Registration-based morphing of active contours for segmentation of CT scans. *Math Biosci Eng* 2: 79–96.

Zeng, K., Bai, E. and Wang, G. 2007. A fast CT reconstruction scheme for a general multi-core PC. *Int J Biomed Imag* 2007: Article ID 29160.

Zhou, Y. and Toga, A.W. 1999. Efficient skeletonization of volumetric objects. *IEEE Trans Vis Comput Graph* 5: 196–209.

Zimmerman, J.B., Pizer, S.M., Staab, E.V., Perry, J.R., McCartney, W. and Brenton, B.C. 1988. An evaluation of the effectiveness of adaptive histogram equalization for contrast enhancement. *IEEE Trans Med Imaging* 7: 304–12.

Zou, Y., Sidky, E.Y. and Pan, X. 2004. Partial volume and aliasing artefacts in helical cone-beam CT. *Phys Med Biol* 49: 2365–75.

Volume-of-interest imaging

Chris C. Shaw

Contents

8.1 INTRODUCTION

Since ionizing radiation was used in medical x-ray imaging, radiation dose to the patient has been a major concern in designing the imaging systems and configuring the imaging techniques for clinical applications (Cho and Han 2012, Dietrich et al. 2013, Paul et al. 2013, Yu and Butson 2013), especially for procedures requiring higher exposure levels, longer exposure times, and/or large number of exposures (Boone 1999, Boone et al. 2000, Kim et al. 2012, Paul et al. 2013). Cone beam computed tomography (CBCT) requires a large number of exposures, ranging from a couple of hundred to over a thousand. Although it is often performed at an exposure level leading to an overall dose level lower than those of regular computed tomography (CT) procedures, the dose level is still on the higher side, requiring careful consideration of the techniques used and motivating research and developments to contain radiation dose to the patient (Boone et al. 2001, 2004). Accompanied with the consideration of radiation dose is the need to maintain the image quality at an acceptable level (Paul et al. 2013). The selection of the exposure techniques is often the result of compromise between the radiation dose and image quality for a specific imaging task. It should be noted that techniques capable of reducing the radiation dose while achieving the same image quality can often be used to achieve better image quality while keeping the patient dose at the same level.

Radiation dose may be reduced by increasing the detector efficiency of the imaging system (see Chapter 2) and often involves more efficient methods to detect and convert x-rays into image signals. This type of improvement is fundamental but does not come easily. Notable examples include the development of modern detectors used in CT and the flat panel detectors used for projection imaging as well as CBCT (see Chapter 1). Actually, the efficiency of modern flat panel detectors probably motivated and led to the development of many CBCT systems and techniques for clinical applications many years after the use of less efficient CBCT systems for industrial and small-animal imaging. A second strategy is to match the spatial resolution of the imaging system to the imaging tasks intended for. High-resolution capability often requires the use of higher exposures. Conversely, lower resolution, if acceptable for the imaging tasks intended for, allows lower exposures to be used, thus saving the radiation dose to the patient. Examples include the use of binning in CBCT for radiation treatment verification or other applications that do not require high spatial resolution (see Chapter 15).

Another strategy involves selective or variable distribution of exposures in the spatial or temporal domain. Examples for using variable exposures in the temporal domain include the use of variable exposure during the scan to acquire projection images with similar overall detector exposures or signal-to-noise ratios (SNRs). This approach would optimize the image quality and reduce the exposures and radiation dose required. The use of spatially varying exposure in image acquisition was widely practiced in projection imaging. It is customary to insert an Al or Cu filter into the lung area to equalize the exposures to the detector (image intensifier or screen film combination) in fluoroscopy or angiography so that higher exposures can be used to improve the image quality without saturating signals in the lung area. As a side benefit, the radiation dose also is reduced. This idea was pushed to a higher level with the region-of-interest (ROI) fluoroscopy or angiography with which a circular collimator and a small field high-resolution detector are moved into the x-ray field to image small details (e.g., metallic stems) in a small ROI with extremely high resolution in neuroradiological procedures (Rudin and Bednarek 1980). This requires substantially elevated exposure level for the ROI. However, the area outside the ROI is spared from direct exposure, thereby keeping the overall dose to the patient at an acceptable level. Volume-of-interest (VOI) CBCT may be viewed as a three-dimensional (3D) version of this technique. Like the ROI fluoroscopy or angiography, the VOI CBCT technique is developed to image a small VOI with better quality but without increasing the overall (averaged over the entire object) radiation dose. Thus, VOI CBCT is useful only for situations when a VOI can be identified outside which the image quality is of little or no importance. This may be a small volume where a lesion has been identified but needs to be examined in more detail for malignancy. Thus, it may be less useful for screening but more useful for later stages of diagnosis or treatment. However, unlike ROI fluoroscopy or angiography, the VOI CBCT technique does not prevent the regions outside the VOI from being directly exposed. Therefore, dose reduction in VOI CBCT is more complicated to estimate or quantify. In this chapter, I describe and discuss the principle and implementation of the VOI CBCT technique. I also discuss the potential improvement of image quality and the dose reduction associated with the use of the VOI CBCT technique.

8.2 PRINCIPLE OF VOI CBCT

The principle of VOI CBCT is to acquire projection data through the VOI with higher exposures and acquire projection data of regions outside the VOI with lower exposures (Chityala et al. 2004, Chen et al. 2008). CBCT images reconstructed from such projection image data result in better image quality inside the VOI and poorer image quality outside the VOI. This concept was first proposed and demonstrated for CBCT used in radiation therapy and later for CBCT used for neuroradiological and breast imaging (Chityala et al. 2004, Chen et al. 2008). Notice that

the 3D images of the VOI are reconstructed from projection data corresponding to x-rays passing through voxels inside the VOI only. These data are all acquired with high exposures and therefore have higher SNRs that help reduce noise levels in the reconstructed images. Projection data corresponding to x-rays passing through voxels outside the VOI are acquired with high or low exposures depending on the view angles but with overall lower exposures, thus resulting in higher noise levels in the reconstructed images in regions outside the VOI. However, the purpose in using the VOI CBCT technique is to improve the image quality inside the VOI. Therefore, poorer image quality outside the VOI is expected and acceptable. Actually, there is no need to reconstruct 3D images for voxels outside the VOI unless they are needed to provide anatomical reference. The high-exposure projection image data for the VOI may be acquired with a small field high-resolution detector to increase the spatial resolution. The higher level exposures used provide the greater fluence required in achieving the same level of SNRs in using such detectors with their smaller pixel sizes.

8.3 IMPLEMENTATION OF VOI CBCT

8.3.1 X-RAY EXPOSURES

The nonuniform exposures required for the VOI CBCT technique may be delivered with a filtering or a collimating mask. With the filtering approach, a partially transmitting filter with an opening at the center is used to allow full intensity x-rays to pass through and expose the VOI while using filtered x-rays to expose regions outside the VOI. The filter material and thickness may be selected to achieve a preselected reduced exposure level for the x-ray spectrum used. This approach is illustrated by Figure 8.1. In some ways, this approach is very similar to the use of a bow-tie filter in CT or CBCT in its goal to allow part of the object to be exposed with higher exposures. By using the bow-tie filter, the exposure is gradually reduced toward the peripheries to make the detector exposure more uniform, thus helping avoid signal saturation

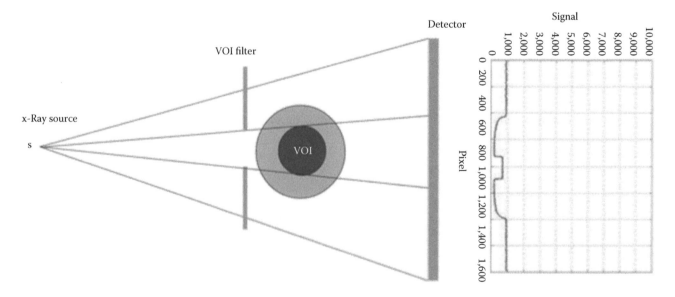

Figure 8.1 VOI CBCT with the filtering approach. A filtering mask is used to reduce exposures to regions outside the VOI and keep the overall radiation dose at an acceptable level. (From Chen et al., *Med Phys* 35, 3482–90, 2008. With permission.)

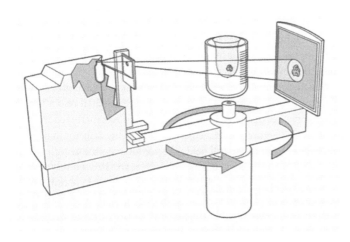

Figure 8.2 A breast CBCT system fitted with a moving VOI mask for implementing the VOI CBCT technique. (From Chen et al., *Med Phys* 35, 3482-90, 2008. With permission.)

toward and outside the peripheries and reducing the overall patient dose. The filtering approach was first proposed and investigated by Chityala et al. (2004) for neuroangiography applications.

A second approach is to use a collimator to follow and direct the x-ray beam to directly expose the VOI only. This is referred to as the VOI scan. The VOI scan by itself results in incomplete projection image data that if directly used for reconstruction would result in truncation artifacts. Therefore, a low-exposure full field scan, with the collimator mask removed, is performed to obtain projection data outside the VOI, which may be combined with the high-exposure projection data obtained from the VOI scan to form a complete projection image set. This approach has been proposed and investigated by Chen et al. (2008) for breast imaging application. A schematic drawing for a VOI CBCT system for breast imaging is shown in Figure 8.2. The filtering approach has the advantage that only a single scan is required. The collimation approach would require a separate scan. However, it may be used to perform a closer examination of a specific VOI with higher image quality after a full field scan is performed. In this situation, the full field projection data are already available and no additional scan is necessary.

8.3.2 SCANNING GEOMETRY

CBCT is typically performed with circular scans with which the x-ray source and detector gantry revolve around the so-called isocenter, usually located at or near the center of the object. It is possible to use the same geometry for VOI CBCT (Figure 8.3a) (Chen et al. 2009). This approach is referred to as the object-centered scan. If the VOI selected happens to be aligned with or near the isocenter, it would stay stationary at or move slightly around the isocenter in the projection images. However, more likely the VOI would be off center, and it would move around in the projection images as the angle of view changes. Thus, the VOI mask needs to be moved around to track and align the collimated or filtered x-ray beam with the VOI. Another issue is that the VOI would look bigger or smaller depending on the angular view and the relative position of the VOI between the x-ray source and the detector. Thus, the opening of the VOI mask needs to be varied for different angular views. How the position and opening

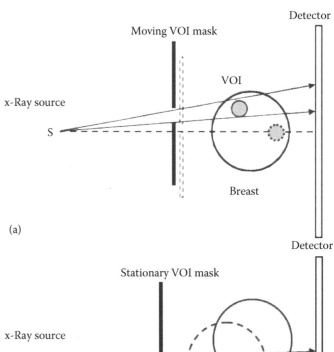

(a)

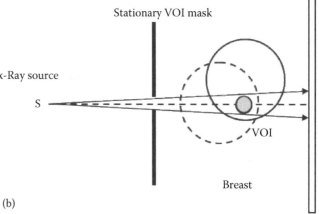

(b)

Figure 8.3 VOI CBCT implemented with an object-centered (a) and VOI-centered (b) scanning technique. (From Chen, L. et al., *Med Phys*, 36, 4007–14, 2009. With permission.)

size of the VOI mask vary with the angular view, the VOI size, and location during the scan must be computed and programmed before the scan starts. It is a complicated but doable endeavor.

Obviously, it is not a trivial task to keep the object centered during the VOI scan. An alternative approach is to keep the VOI centered with the isocenter, thus avoiding the need to move the VOI mask and vary the size of its opening (Figure 8.3b) (Chen et al. 2009). This approach is referred to as the VOI-centered scan. Because the object is no longer centered with the isocenter, the position of the object relative to the x-ray source and detector would vary with the angle of view. This could mean that part of the object could be excessively close to the x-ray source in some views, resulting in excessively larger field of view or higher radiation dose for those views. Thus, the size of the detector must be sufficiently large to cover the larger field of view. Greater focal spot blurring resulting from closer distance between the object and the x-ray source could be a concern, too. With the collimator approach, the full field projection images also need to be acquired with the VOI-centered scanning geometry before being combined with the VOI scan data. This makes it more difficult to use projection data previously acquired with the object-centered scanning geometry because the data must be reconstructed first and then reprojected for the VOI-centered scanning geometry.

8.3.3 NORMALIZING AND COMBINING PROJECTION IMAGE SETS

Before image reconstruction, the low-exposure image data for regions outside the VOI must be normalized to compensate for the exposure differences and then combined with the high-exposure data of the VOI into a complete set of projection image data. The normalization process is complicated by the presence of scatter component in different proportions in the two sets of projection images. With the filter mask approach, the scatter-to-primary ratios (SPRs) in the low-exposure image data for regions outside the VOI are higher than those in the regular full field projection images because the VOI receives higher exposures that generate more scatter to regions outside the VOI, relative to the primary x-rays there. However, inside the VOI, the SPRs are lower because the lower exposures outside the VOI generate less scatter to the VOI, relative to the primary x-rays there. This mismatch of SPRs complicates the process of normalizing the low-exposure image signals outside the VOI to the high-exposure signals inside the VOI. With the collimator mask approach, the scenario is slightly different. Because the full field and VOI scans are performed separately, scatter signals generated from one scan do not have to be included in image signals acquired with the other scan, resulting in slightly lower SPRs in both sets of projection images. However, the issue with the mismatch of the SPRs between signals in the VOI and those outside the VOI is still there because the SPRs in the VOI scan images are considerably lower due to the use of collimated x-rays.

8.3.4 ACQUISITION WITH TWO DETECTORS

Use of higher exposures for the VOI scan results in projection images with higher SNRs that may allow smaller objects to become visible, but it does not improve the spatial resolution of the system. With the collimator approach, it is possible to use a small field, high-resolution detector to acquire projection images of higher spatial resolution, which may be used to obtain high-resolution reconstructed images of the VOI (Figure 8.4) (Chen et al. 2009). This detector may be inserted in front of the full field detector for acquisition with smaller geometric magnification during the VOI scan. This would result in a set of projection images that have higher pitch and are unlikely to be aligned with the full field projection images. Thus, in addition to the processing described in the previous section, the low-resolution, low-exposure full field exposure images must be interpolated to the same pitch as and aligned with the high-resolution, high-exposure projection images of the VOI. It is also possible to use a large area high-resolution detector for image acquisition in both the VOI and full field scans. However, the latter approach may be performed in the binning mode to reduce the spatial resolution and to maintain the pixel SNRs at a reasonable level with the low exposures used. This approach would greatly simplify the pitch matching and alignment of the two sets of projection images.

8.3.5 IMAGE RECONSTRUCTION

Image reconstruction for VOI CBCT is straightforward. Once the low-exposure and high-exposure image data are combined, the

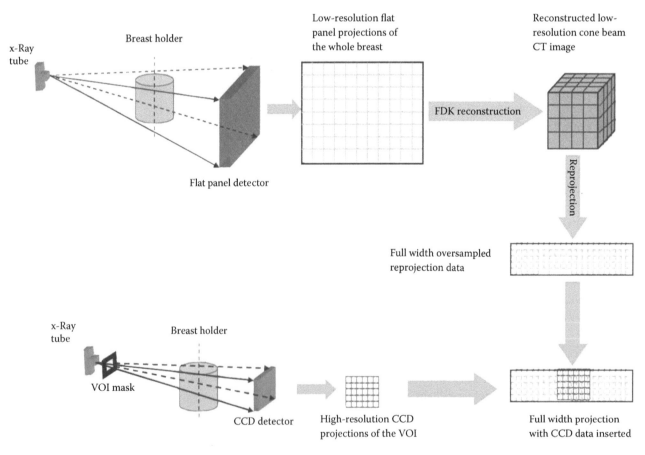

Figure 8.4 Schematic drawing showing VOI CBCT using an additional small field high-resolution detector for the VOI scan and the normalization, alignment, and pitch matching process in synthesizing the projection images for reconstruction. (From Chen, L. et al., *Med Phys*, 36, 4007–14, 2009. With permission.)

Feldkamp (FDK)-filtered backprojection reconstruction algorithm may be directly used to reconstruct 3D images of the entire object, although it may be unnecessary to reconstruct images for volumes outside the VOI (Feldkamp et al. (1984) (see Chapter 3).

8.4 IMPROVEMENT OF IMAGE QUALITY

The goal in using the VOI CBCT technique is to obtain high-quality images of a preselected VOI without increasing the overall radiation dose to the object. An example is to select a VOI surrounding a suspected tumor in the breast and to obtain high-quality images of the tumor to allow for more accurate characterization, classification, or delineation. A major improvement with the VOI CBCT images (reconstructed images of the VOI) is the lower noise level that allows more detail to become visible in the images. A major contributing factor to this improvement is obviously the use of high exposures for the VOI scan. However, another major, but not as significant, contributing factor is the lower SPRs resulted from the relatively lower exposures used for regions outside the VOI during image acquisition (Lai et al. 2009). Figure 8.5 shows an example of scatter reduction by plotting and comparing the SPRs in the projection images of a 15-cm-diameter Lucite cylinder for VOI CBCT (blue), high-exposure full field CBCT (red), and low-exposure full field CBCT (green). Although this demonstration was based on VOI CBCT using the collimation method, similar but probably slightly smaller scatter reduction effect should be obtained with the filter method. The degree of scatter reduction effect obviously varies with the location of the VOI and the VOI size relative to the object. The scatter reduction effect is usually greater if the VOI is located near the center of the object and smaller toward the periphery. The effect also decreases as the VOI

becomes larger as the required use of a larger opening in the VOI masks would result in less scatter rejection due to the use of a wider collimated x-ray beam. Scatter reduction would not only result in lower noise level in the reconstructed VOI images but also more accurate CT numbers because the projection image signals are less biased by the relatively smaller scatter components. An example of image quality improvement resulting from the VOI CBCT technique is shown in Figure 8.6 with the image quality measured by the figure-of-merit, defined as contrast-to-noise ratio squared divided by the average dose.

Lower noise level helps make low contrast object (e.g., a small tumor) more visible in the reconstructed VOI images. However, it does not change the inherent spatial resolution of the images. There are imaging tasks that require higher spatial resolution to allow small objects to be "resolved." Higher spatial resolution helps preserve the size and shape when imaging small objects. It also helps image closely spaced small objects as separate objects instead of having them merged into a bigger object due to image blurring. A detector with higher pitch is required to increase the spatial resolution. For scintillator-based detectors, it also helps to use a thinner scintillator. However, a thinner scintillator results in lower quantum detection efficiency and may not be desirable for improving the overall image quality. Even without using a thinner scintillator, the use of higher pitch detector is still helpful because it results in a better aperture function and a higher Nyquist frequency (Giger and Doi, 1984). The former helps improve the presampling modulation transfer function, and the latter helps capture image signals at higher frequencies and reduce aliasing artifacts. However, the use of higher pitch detector also requires higher photon fluence to keep the pixel SNRs at an acceptable level. This is necessary for detection and visualization of smaller because it helps suppress the effects of detector noises that may begin to dominate if the number of photons detected

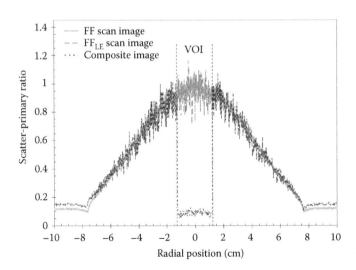

Figure 8.5 Reduction of SPRs in VOI CBCT. The SPRs in the projection image combined from a high-exposure scan of the VOI and a low-exposure scan of the entire object appear as a solid line (light grey). The SPRs in the corresponding image from a regular scan of the entire object appear in the dotted line (darker grey). Although the SPRs are similar in regions outside the VOI, the SPRs in the former are substantially reduced in the VOI. The phantom is a 15-cm-diameter Lucite cylinder. (From Lai et al., *Phys Med Biol*, 54, 6691–709, 2009. With permission)

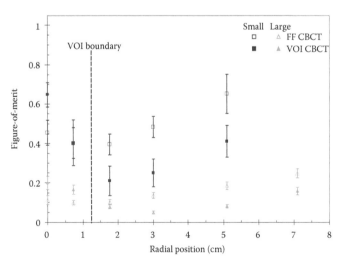

Figure 8.6 Improve of figures-of-merit (FOMs), defined as contrast-to-noise ratios squared divided by the average doses. Higher FOMs for the VOI (solid line) were obtained by populating the high-exposure VOI scan data for the VOI and low-exposure full field scan for regions outside the VOI. Lower FOMs for the VOI (dotted line) were obtained by using data from a regular scan for the entire images. Lowest FOMs for the VOI (dashed line) were obtained when the low-exposure full field scan data are added to the VOI scan data inside the VOI. (From Lai et al., *Phys Med Biol*, 54, 6691–709, 2009. With permission.)

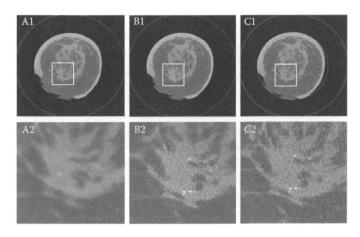

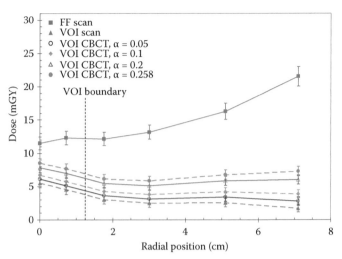

Figure 8.7 Images of a mastectomy breast specimen reconstructed from low-resolution full field scan (A1), high-resolution VOI scan combined with low-resolution full field scan (B1), and high-resolution full field scan (C1). The bottom images (A2, B2, and C2) are the magnified views of the highlighted area in the top row of images. The scans were obtained at 80 kVp with 0.6-mm focal spot size and an estimated mean dose of 15 mGy. The low-resolution projection images were simulated by binning the high-resolution projection images (0.194-mm pitch) with a 4 × 4 kernel. Notice that the low-resolution projection image data may be acquired with reduced exposures to keep the average dose for VOI CBCT at an acceptable level. The full field high-resolution images require the entire object to be scanned with high exposures, thus resulting in the higher radiation dose (15 mGy).

Figure 8.8 Dose profiles for a full field scan (square) and a VOI scan (triangle) performed at an increased exposure level for improved image quality. Also shown are dose profiles for combinations of the high-exposure VOI scan with various reduced exposure full field scans. (From Lai et al., *Phys Med Biol*, 54, 6691–709, 2009. With permission.)

by each image pixel and the resulting signal level become too low. The VOI CBCT technique provides a viable solution when the spatial resolution needs to be improved while the overall radiation dose to the object must be maintained at the same level. This is achieved by using a high-resolution detector to acquire high-resolution projection images of the VOI with high exposure and combine them with low-exposure, low-resolution full field projection images for image reconstruction. Notice that the inclusion of the latter helps eliminate the truncation artifacts without affecting the high-resolution quality of the reconstructed VOI images, as demonstrated by the reconstructed images of a mastectomy breast specimen in Figure 8.7.

8.5 REDUCTION OF RADIATION DOSE

The VOI CBCT technique seeks to improve the image quality of the reconstructed VOI images without increasing the overall dose to the object, as compared to full field CBCT. This is achieved by using higher exposures to acquire projection images of the VOI and lower exposures to acquire projection images for regions outside the VOI. The aim is to maintain the overall dose level at the same or even reduced level, even though the dose to the VOI is increased. Because the x-ray source and detector gantry, along with the filtering or collimating mask, rotate around the object, it is impossible to completely block exposures to the volume outside the VOI, which will be directly exposed to the x-rays in some of the projection views. The closer a region is to the VOI, the more views in which it is directly exposed.

One important question for VOI CBCT is how high the exposures can be used for the VOI scan while maintaining the overall dose at an acceptable level (Lai et al. 2009). This obviously

depends on how low the exposures for scanning the entire object (for the collimation approach) or regions outside the VOI (for the filtering approach) are used. The lower they are, the higher the exposures for the VOI scan can be used. Figure 8.8 shows measured dose profiles for a full field scan and a VOI scan performed at an increased exposure level to allow the quality of the reconstructed images to be improved over the entire object. Although the full field scan results in unacceptable doses (11.5–21.6 mGy from center to periphery), the VOI scan results in much lower doses (5.5 to 1.7 mGy from center to periphery) due to the collimation of x-rays to cover the VOI only during the scan. Notice that even when exposed at the same level, the VOI is subject to lower radiation dose due to decrease of scatter exposures. By combining a lower (5%–25.8%) exposure full field scan with a full exposure VOI scan, the dose levels can be maintained at an acceptable level while providing a complete set of projection images to avoid truncation effect. As with the scatter reduction effect, the answer also depends on the location of the VOI and the VOI size relative to the object. For larger VOIs, the allowable exposure increase for the VOI scan is more limited for the same exposure reduction used for scanning regions outside the VOI. The allowable exposure increase for the VOI is also more limited if the VOI is located closer to the periphery.

REFERENCES

Boone, J.M., Glandular breast dose for monoenergetic and high-energy mammographic x-Ray beams: Monte Carlo assessment. *Radiology*, 1999. 213: p. 23–37.

Boone, J.M., M. buonocore, and V. Cooper III, Monte Carlo validation in diagnostic radiological imaging. *Med Phys*, 2000. 27(6): p. 1294–1304.

Boone, J.M., T.R. Nelson, and K.K. Lindfors, Dedicated Breast CT: Radiation Dose and Image Quality Evaluation. *Radiology*, 2001. 221(3): p. 657–667.

Boone, J.M., N. Shan, and T.R. Nelson, A comprehensive analysis of DgNCT Coefficients for pendant-geometry cone-beam breast computed tomography. *Med. Phys.*, 2004. 31(2): p. 226–235.

Chen, L. et al. 2008. Feasibility of volume-of-interest (VOI) scanning technique in cone beam breast CT- a preliminary study. *Med Phys*, 2008. 35(8): p. 3482–90.

Chen, L., et al., 2009. Dual resolution cone beam breast CT: a feasibility study. *Med Phys* 36(9): 4007–14.

Cho, J.Y. and W.J. Han, The reduction methods of operator's radiation dose for portable dental X-Ray machines. *Restor Dent Endod*, 2012. 37(3): p. 160–4.

Chityala, R., et al., Region of Interest(ROI) Computed Tomography. *SPIE*, 2004. 5368: p. 534–541.

Dietrich, T.J., et al., Comparison of radiation dose, workflow, patient comfort and financial break-even of standard digital radiography and a novel biplanar low-dose X-Ray system for upright full-length lower limb and whole spine radiography. *Skeletal Radiol*, 2013. 42(7): p. 959–67.

Feldkamp, L.A., L.C. Davis, and J.W. Kress, Practical cone-beam algorithm. *J. Opt. Soc. Am.*, 1984. 1: p. 612–619.

Giger, M.L., Doi, K. 1984. Investigation of basic imaging properties in digital radiography. I. Modulation transfer function. *Med Phys* 11(3): 287–95.

Kim, S., et al., Radiation dose from 3D rotational X-Ray imaging: organ and effective dose with conversion factors. *Radiat Prot Dosimetry*, 2012. 150(1): p. 50–4.

Lai, C.J. et al., 2009. Reduction in x-ray scatter and radiation dose for volume-of-interest (VOI) cone-beam breast CT—a phantom study. *Phys Med Biol* 54: 6691–709.

Paul, J., et al., Radiation dose and image quality of X-Ray volume imaging systems: cone-beam computed tomography, digital subtraction angiography and digital fluoroscopy. *Eur Radiol*, 2013. 23(6): p. 1582–93.

Paul, J., E.C. Mbalisike, and T.J. Vogl, Radiation dose to procedural personnel and patients from an X-Ray volume imaging system. *Eur Radiol*, 2013.

Rudin, S. and D. Bednarek, Improving contrast in special procedures using a rotating aperture wheel (RAW) device. *Radiology*, 1980. 137(2): p. 505–510.

Yu, P.K. and M.J. Butson, Measurement of effects of nasal and facial shields on delivered radiation dose for superficial x-Ray treatments. *Phys Med Biol*, 2013. 58(5): p. N95–N102.

Four-dimensional cone beam computed tomography

Tinsu Pan

Contents

9.1 INTRODUCTION

Respiratory motion poses a significant challenge to the imaging and treatment of tumors in the lungs or abdomen in radiation therapy. Motion can distort the shape of an object and cause a suboptimum planning for treatment if not accounted for. Inaccuracy in estimating the target volume due to the respiratory motion distorts the treatment volume and prevents dose escalation for the treatment of a target tumor (Keall et al. 2006). Four-dimensional (4D) computed tomography (CT) is a standard technique to image a tumor in respiratory motion during radiation treatment planning, and it is available on multislice computed tomography (MSCT) scanners with either the cine CT scan (Low et al. 2003; Pan et al. 2004) in which each location is scanned over one breath cycle plus 1 s or with the helical CT at a very low pitch of about 0.1 or less (Keall et al. 2004). Technical differences between the cine and helical 4D-CTs have been reported previously (Pan 2005). Incorporation of iodine-contrast in 4D-CT imaging of the liver tumor also has been explored previously (Beddar et al. 2008). However, the role of 4D-CT in verification of treatment delivery is not clear because a repeat patient setup for 4D-CT on a MSCT scanner will be difficult in the therapy session unless in-room multislice CT is used. In contrast, integration of cone beam computed tomography (CBCT) in a linear accelerator provides a platform to allow verification of patient setup and assessment of tumor motion before or after radiation treatment for image-guided radiation therapy (IGRT). With the integration of CBCT with linear accelerator, it is feasible to achieve a similar 4D-CT imaging capability on the CBCT. The advantage of 4D-CBCT over 4D-CT is its capability of imaging the tumor motion without repositioning the patient. Alternatively, the advantages of 4D-CT over 4D-CBCT are image quality with less photon scattering and higher detector quantum efficiency.

9.2 ACQUISITION STRATEGIES

In an earlier implementation of 4D-CBCT, the x-ray CBCT projection data were acquired, and the diaphragm position in each projection data was identified to assign the projection data into multiple respiratory phases for reconstruction of the 4D-CBCT images over a respiratory cycle (Sonke et al. 2005). The 4D-CBCT required the patient's diaphragm in the imaging field of view, eliminating the need of a respiratory monitoring device during data acquisition. Maintaining the diaphragm in each angle of projection data may not be always feasible if the tumor is positioned in the central plane of the CBCT for alignment and treatment. In addition, fixed data acquisition time and fixed sampling rate in 4D-CBCT could result in different image quality for the patients with different durations of breath cycle. Image quality tends to get worse for the lung tumor with a longer breath cycle because the projection data of the same phase are more far apart in projection angle for a longer breath cycle and can cause more streak artifacts in the reconstructed images for 4D-CBCT (Figure 9.1).

Another implementation of 4D-CBCT was based on either a single CBCT (Dietrich et al. 2006) or multiple CBCT scans (Li et al. 2006) at a rotation cycle of about 1 min per revolution. The projection data were correlated with either a strain-gage sensor or with the diaphragm position. There are about 10–20 respiratory cycles per min for the respiratory cycles of 3–6 s. The projection data of the same respiratory phase between two consecutive cycles are spaced between 18° and 36° with the gantry rotation of 1 min per revolution (Dietrich et al. 2006; Li et al. 2006). The CBCT scan could be repeated multiple times to increase data sampling of the projection data to remedy insufficient sampling of

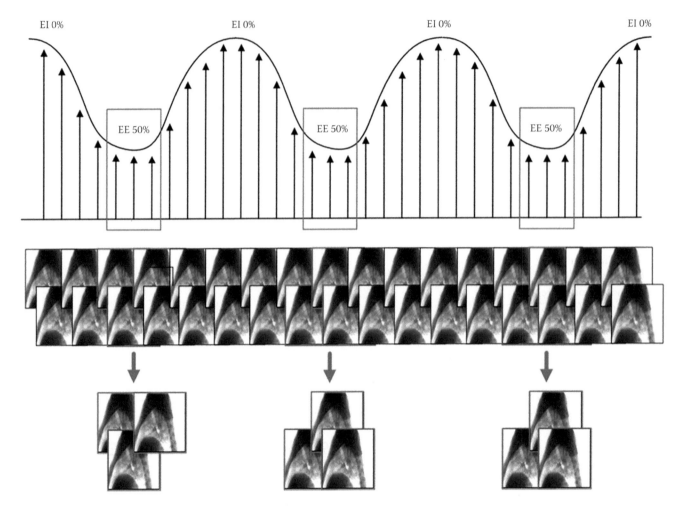

Figure 9.1 Illustration of 4D-CBCT data acquisition, typically performed over several minutes to increase data sampling and reduce the gap of projection data between two adjacent respiratory cycles. The projection data at end-expiration (50%) are collected for the image reconstruction of 4D-CBCT at 50% phase. End-inspiration (0%) or other phases of data can be derived similarly. The densely sampled projection data from a single averaged CBCT becomes the sparsely sampled projection data for the multiple phases of 4D-CBCT. Streaking artifacts typically occur in the reconstruction of the sparsely sampled 4D-CBCT data.

the projection data from a single rotation of CBCT (Li et al. 2006). One drawback of this approach was the uncertainty of data sampling at each projection angle due to the stop and restart of the CBCT scan between two successive acquisitions.

Varying the gantry rotation time to maintain a similar image quality for different patient respiratory cycles may be necessary and has been implemented on a slow-rotation CBCT with the Varian Real-Time Position Management (RPM) system (Lu et al. 2007). This approach used a slower scan to collect the projection data of a longer breath cycle, which causes the angular spacing to be smaller between the projection data of the same phase between two consecutive breath cycles. Radiation exposure per projection can be reduced for a longer acquisition time to maintain the similar radiation dose to the patient.

9.3 IMAGE QUALITY

Image quality is strongly dependent on the spacing of the projection data in the image reconstruction. There are about 600 projections for a 360° rotation of 1-min gantry rotation,

standard in CBCT for IGRT. Because the projection data of the same respiratory phase are used in the image reconstruction of a single phase of the 4D-CBCT data, it is important to understand the sampling requirement for the projection data of the same respiratory phase in two consecutive respiratory cycles.

When an insufficient number of projection data is acquired, the azimuthally spatial resolution becomes poor. The International Electrotechnical Commission (IEC) body phantom (Figure 9.2) was scanned with the CBCT technique of 115-kV, 50-mA, 10 frames per second (fps), and 6-ms pulsed x-ray per projection. The total number of projections for a 360° rotation was 579. The number of projection data was artificially reduced from 579 to 73 in multiple steps, corresponding to an increase of the angular interval between two consecutive projections of 0.6°, 1.2°, 2.5°, 3.1°, 3.7° to 5.0°. Figure 9.3 shows the images of the phantom in the axial view and demonstrates the dependence of image quality as a function of the number of projections. Image quality deteriorates as the angular spacing increases. Image quality is stabilized once angular spacing is less than 3°.

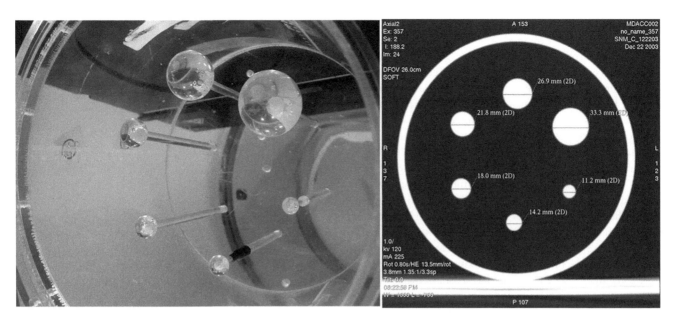

Figure 9.2 IEC phantom on the left and the CT image at the largest cross section of the spheres. The six spheres are 1.1, 1.4, 1.8, 2.2, 2.7, and 3.3 cm in diameter.

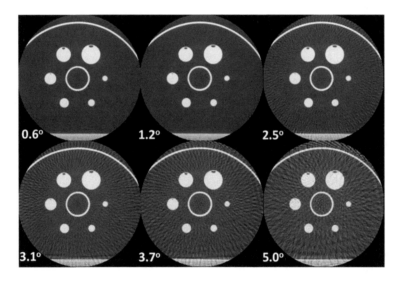

Figure 9.3 Dependence of image quality on the spacing between projections. The numbers of projections over 360° from upper left to lower right are 579, 290, 145, 116, 97, and 73, corresponding to the angles between two consecutive projections of 0.6°, 1.2°, 2.5°, 3.1°, 3.7°, and 5.0°, respectively. Image quality deteriorates as the number of projections decreases or the angular spacing increases. (Reproduced from Figure 5 of Lu, J. et al., *Med Phys*, 34, 3520–9, 2007. With permission.)

9.4 RECONSTRUCTION STRATEGIES

Various reconstruction algorithms have been proposed to address sparse-view sampling in 4D-CBCT. One approach was to incorporate all data, rather than just the binned subset, into the reconstruction. This approach was taken in the autoadaptive phase-correlated (AAPC) (Bergner et al. 2009) and Mc Kinnon–Bates (MB) (Mc Kinnon and Bates 1981; Leng et al. 2008a) reconstruction algorithms. Both algorithms are used to reduce the undersampling artifacts, but each has certain disadvantages. The AAPC reconstruction requires an algorithm for segmenting the detector pixels into two groups: pixels affected and unaffected by motion. Some parameters in this segmentation require optimization. The MB algorithm can eliminate streak artifacts (Leng et al. 2008a) at the expense of

noise and artifacts. The temporal resolution from the use of the MB algorithm may not be as good as that in the standard 4D-CBCT algorithm (Bergner et al. 2010).

Reconstruction of the CBCT data from a very few samples of projection data is a sparse-view or an underdetermined problem. There are many solutions to the underdetermined problem. Some constraints have to be introduced to reduce the solution space to a unique solution such as the use of total-variation minimization (Sidky et al. 2006; Sidky and Pan 2008). However, the image quality of this approach tends to be blocky and cartoonish. Prior image constrained compressed sensing (PICCS) is a promising technique and it uses a prior image constraint to improve the reconstruction accuracy and convergence speed (Chen et al. 2008; Leng et al. 2008b).

Evaluations of the image quality, robustness, convergence, and computational aspects of these reconstruction algorithms using sparse data were provided by Bergner et al. (2010) and Bian et al. (2010). The reconstruction algorithms referenced earlier are also promising for sparse-view reconstructions, but the prohibitive factor may be the computational cost. In the application of IGRT, the images have to be available immediately after the data acquisition. Computational cost may make iterative algorithms less attractive in the clinical environment (Pan et al. 2009). However, computational cost will not become an issue in the future as advanced graphics processing units are incorporated in the computation. One region-of-interest reconstruction approach to use all the projection data for the reconstruction of an average CBCT data for alignment of the patient's bony structure, and to extract the sparse-sampled data of each respiratory phase for the reconstruction of the planning target volume (PTV) has been proposed (Ahmad et al. 2011). This approach has the advantages of providing the images of no streaks outside the tumor area and the motion images from the spare-sampled data in the target tumor region with reduced streaking artifacts.

9.5 EXAMPLE OF 4D-CBCT

We use the CBCT on the Varian Trilogy system to demonstrate an implementation of 4D-CBCT (Li et al. 2006). This implementation is not vendor specific and can be applied to other IGRT systems with CBCT. The x-ray beam axis of the x-ray flat panel detector or on-board imager (OBI) is orthogonal to the treatment megavoltage (MV) x-ray beam axis. The OBI has 2048 × 1536 square pixels of 194 mm. Both the isocenters of the OBI and the MV x-ray are calibrated to the same position. In the CBCT application with OBI, a 2 × 2 binning mode is used to down-sample the data to 1024 × 768 pixels, with a square pixel size of 388 mm. A dual-gain technique is applied to the OBI readout to extend the dynamic range from 14-bit digital to analog conversion to their 16-bit data equivalent. Maximum data sampling rate is 15 fps, typical gantry rotational time is 1 min per revolution, and x-ray focal spot size is 0.8 mm in diameter. The OBI gantry orbits a circular trajectory. The distance between the x-ray source and the OBI is 150 cm, including 100-cm source-to-isocenter distance and 50-cm isocenter-to-OBI distance (Figure 9.4). The reconstruction field of view (RFOV) is 25 cm in diameter with a 17-cm cranial–caudal coverage in the full-fan mode, and the RFOV is 45 cm with a 14-cm coverage in the half-fan mode.

The Varian RPM system is used to record the respiratory signal of the patient. The respiratory signal is measured with an infrared reflector box placed on the patient's abdomen. The reflected signals are detected by an optical camera and analyzed to determine the end-inhalation phases based on the motion amplitude in the respiratory signal. The respiratory phases of the respiratory waveform are determined by a linear relationship between two adjacent end-inspiration phases of 0% and 100%.

The IEC body phantom was placed on a moving platform to simulate the respiratory motion of different tumor sizes. The phantom has six hollow acrylic spheres of 1.1–3.3 cm in diameter. The spheres were filled with water to simulate

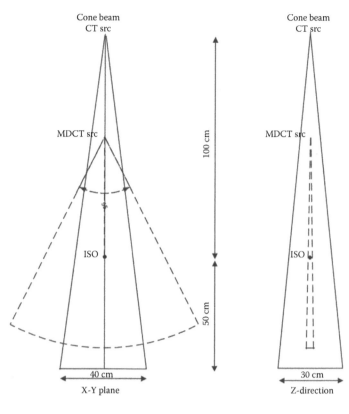

Figure 9.4 Illustration of the CBCT geometry in the full-fan mode and in comparison with a General Electric (GE) 64-slice multislice CT. The CBCT x-ray source is 150 cm to the detector, and the RFOV of 25 cm in CBCT is smaller than the 50 cm in the 64-slice CT. The comparison of the X-Y plane geometries is on the left, and the comparison of the Z-direction geometries is on the right. The 64-slice CT has a total of 4 cm coverage at isocenter (ISO) in the Z-direction.

the soft-tissue density. The moving platform was programmed to simulate a regular breathing motion, with a superior-to-inferior motion amplitude of 2 cm and a breathing cycle of 4 s. The centers of the spheres were on the same axial or imaging plane, perpendicular to the long axis of the phantom or parallel to the rods holding the spheres.

9.6 4D-CBCT DATA

The 4D-CBCT images of the IEC phantom are shown in Figure 9.5. The RFOV was 25 cm in diameter, and the slice thickness was 2.5 mm. There are motion artifacts in both the axial and sagittal views in the CBCT image, and the artifacts are reduced appreciably in the 4D-CBCT images of the 0%, 30%, 50%, 70%, and 90% phases. The degradation in the CBCT images was due to the reconstruction of inconsistent projection data from all phases of the projection data of the motion phantom. The motion artifacts caused a smearing of the spheres. In comparison, the 4D-CBCT images were almost free of motion artifacts. This is a demonstration that 4D-CBCT can suppress most of the gross motion artifacts.

Figure 9.6 shows the axial, coronal, and sagittal views of a patient data by 4D-CBCT. The scan duration was 4.5 min. The average breath cycle for the patient was 3 s. The number of projection data was 2000 over 200° of data collection.

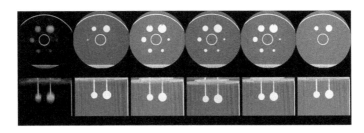

Figure 9.5 Axial (top) and sagittal (bottom) views of the ungated (averaged) CBCT and 4D-CBCT images of the IEC phantom in a 4-s motion cycle. The columns from left to right correspond to the ungated (averaged) CBCT, 0%, 30%, 50%, 70%, and 90% 4D-CBCT images. The motion artifacts in the ungated CBCT were appreciably reduced in the 4D-CBCT images.

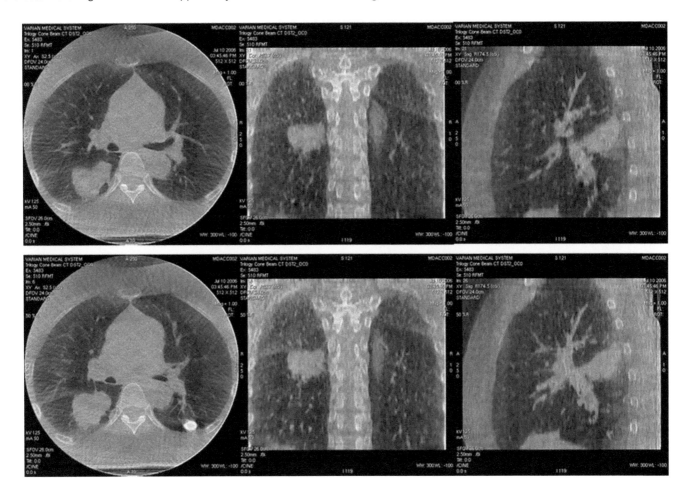

Figure 9.6 End-inspiration (0%) and end-expiration (50%) images of a 4D-CBCT patient study are shown in the top and bottom, respectively. The images from left to right are axial, coronal, and sagittal. (Reproduced from Figure 5 of Lu, J. et al., *Med Phys*, 34, 3520–9, 2007. With permission.)

9.7 SUMMARY

4D-CBCT can image the tumor motion before or after radiation treatment while the patient is still in the treatment position. However, incorporating 4D-CBCT in a treatment session is challenging because each treatment session is about 20 min, including patient setup and treatment. Regular CBCT for IGRT has been accepted in the clinic for its relatively short 1-min scan time and short reconstruction time. Without any spare-view iterative reconstruction algorithms such as AAPC, MB, and PICCS to reduce the scan duration, a 4D-CBCT would take up to 4 min for data acquisition and another 1–3 min for reconstruction and alignment of the tumor and other anatomical structures. Spending up to 7 min for 4D-CBCT is not acceptable in a clinical environment when most of the patients are not able to keep still for more than 20 min. Reducing the scan time to 1–2 min makes 4D-CBCT attractive but at the expense of streaking artifacts in the images and degradation of image quality in IGRT. In addition, advanced sparse view reconstruction algorithms have to be used to minimize the streaking artifacts with increased reconstruction time. A solution to reduce the scan time for IGRT may be that we need only 4D-CBCT to reconstruct the tumor region for the assessment of tumor motion. The same 1-min CBCT data can also be used for alignment of the patient anatomy.

REFERENCES

Ahmad, M., Balter, P. and Pan, T. 2011. Four-dimensional volume-of-interest reconstruction for cone-beam computed tomography-guided radiation therapy. *Med Phys* 38(10): 5646–56.

Beddar, A.S., et al. 2008. 4D-CT imaging with synchronized intravenous contrast injection to improve delineation of liver tumors for treatment planning. *Radiother Oncol* 87(3): 445–8.

Bergner, F., et al. 2009. Autoadaptive phase-correlated (AAPC) reconstruction for 4D CBCT. *Med Phys* 36(12): 5695–706.

Bergner, F., et al. 2010. An investigation of 4D cone-beam CT algorithms for slowly rotating scanners. *Med Phys* 37(9): 5044–53.

Bian, J., et al. 2010. Evaluation of sparse-view reconstruction from flat-panel-detector cone-beam CT. *Phys Med Biol* 55(22): 6575–99.

Chen, G.H., Tang, J. and Leng, S. 2008. Prior Image Constrained Compressed Sensing (PICCS). *Proc Soc Photo Opt Instrum Eng* 6856: 685618.

Dietrich, L., et al. 2006. Linac-integrated 4D cone beam CT: first experimental results. *Phys Med Biol* 51(11): 2939–52.

Keall, P.J., et al. 2004. Acquiring 4D thoracic CT scans using a multislice helical method. *Phys Med Biol* 49(10): 2053–67.

Keall, P.J., et al. 2006. The management of respiratory motion in radiation oncology report of AAPM Task Group 76. *Med Phys* 33(10): 3874–900.

Leng, S., et al. 2008a. High temporal resolution and streak-free four-dimensional cone-beam computed tomography. *Phys Med Biol* 53(20): 5653–73.

Leng, S., et al. 2008b. Streaking artifacts reduction in four-dimensional cone-beam computed tomography. *Med Phys* 35(10): 4649–59.

Li, T., et al. 2006. Four-dimensional cone-beam computed tomography using an on-board imager. *Med Phys* 33(10): 3825–33.

Low, D.A., et al. 2003. A method for the reconstruction of four-dimensional synchronized CT scans acquired during free breathing. *Med Phys* 30(6): 1254–63.

Lu, J., et al. 2007. Four-dimensional cone beam CT with adaptive gantry rotation and adaptive data sampling. *Med Phys* 34(9): 3520–9.

Mc Kinnon, G.C. and Bates, R.H.T. 1981. Towards imaging the beating heart usefully with a conventional CT scanner. *IEEE Trans Biomed Eng* 28(2): 123–7.

Pan, T. 2005. Comparison of helical and cine acquisitions for 4D-CT imaging with multislice CT. *Med Phys* 32(2): 627–34.

Pan, T., et al. 2004. 4D-CT imaging of a volume influenced by respiratory motion on multi-slice CT. *Med Phys* 31(2): 333–40.

Pan, X.,Sidky, E.Y. and Vannier, M. 2009. Why do commercial CT scanners still employ traditional, filtered back-projection for image reconstruction? *Inverse Probl* 25(12): 1230009.

Sidky, E.Y., Kao, C.M. and Pan, X. 2006. Accurate image reconstruction from few-views and limited-angle data in divergent-beam CT. *J X-Ray Sci Tech* 14(2): 119–39.

Sidky, E.Y. and Pan, X. 2008. Image reconstruction in circular cone-beam computed tomography by constrained, total-variation minimization. *Phys Med Biol* 53(17): 4777–807.

Sonke, J.J., et al. 2005. Respiratory correlated cone beam CT. *Med Phys* 32(4): 1176–86.

10 Image corrections for scattered radiation

Cem Altunbas

Contents

10.1 INTRODUCTION

Introduction of amorphous silicon (a-Si) x-ray flat-panel detectors (FPDs) in the late 1990s has led to a big leap in research and development of novel cone beam computed tomography (CBCT) imaging systems. FPD's large field of view (FOV) along the gantry rotation axis allows adequate volumetric coverage to image clinically relevant anatomy using gantry rotation of 360° or less, thereby reducing or eliminating the need for slip ring–based, continuously rotating gantry and moving patient couch. Simplicity of volumetric imaging with an FPD enabled the development of novel CBCT scanners with a small footprint for many clinical applications, such as image-guided radiation therapy (Jaffray et al. 2002), intraoperative three-dimensional (3D) volumetric imaging (Siewerdsen et al. 2005; Petrov et al. 2011), and dedicated CBCT scanners for maxillofacial imaging and breast cancer screening (Boone et al. 2001; Sukovic 2003). Besides being electromechanically simple, FPD-based CBCT scanners bring their own trade-offs

to computed tomography (CT) image quality. Compared with conventional multidetector computed tomography (MDCT), CBCT images have inferior low-contrast resolution, decreased accuracy in CT numbers, and increased image artifacts. There are two major reasons for this difference in quality. (1) A CBCT image acquisition geometry that uses single, circular source trajectory cannot sufficiently sample the imaged object outside the source trajectory plane (Tuy 1983), and the reconstructed image is said to be "approximate." The magnitude of incomplete sampling and the degradation in image quality increase for image voxels further away from the source trajectory plane. Further discussion of sampling issues in CBCT can be found in other publications (Smith 1985; Buzug 2008). (2) Commonly used filtered backprojection algorithms, such as Feldkamp–Davis–Kress (FDK), assume that the line integral of attenuation coefficients along the x-ray path and the log-attenuation value of transmitted x-ray fluence through the object have a linear relationship (Feldkamp et al. 1984; Kak and Slaney 1988). Unless the projection images are acquired under narrow beam geometry using monoenergetic x-rays, this assumption breaks down severely in FPD-based CBCT imaging applications.

Image quality degradation associated with reason (1) is an intrinsic problem of CT image reconstruction due to cone beam geometry and circular source trajectory; this topic is outside the scope of this chapter. Reason (2) is largely related to inaccurate and inconsistent x-ray attenuation measurements via projection images.

Scattered radiation is one of the causes for inaccurate attenuation measurements. Its contribution to total image signal and its effect on image quality are less pronounced in MDCT scanners. However, in many CBCT applications, the magnitude of the scatter fluence may well exceed the primary x-ray fluence on the detector plane, making the detected scatter fluence one of the leading causes for image quality degradation in CBCT images. Compared with MDCT, CBCT images have more shading cupping artifacts, less low-contrast sensitivity, and reduced accuracy in CT numbers.

Much of the CBCT image quality improvement efforts are therefore focused on suppressing the effects of scattered radiation (Ruhrnschopf and Klingenbeck 2011a,b). In addition, beam hardening, image lag due to inherent properties of a-Si, and patient and organ motion further increase the inconsistencies among projection images. Depending on the imaged anatomy, organ or patient motion also can be a significant factor causing inconsistent x-ray attenuation measurement and image artifacts due to the slow rotation speed of the CBCT gantries (CBCT scan acquisition time is generally between 10 and 120 sec). In Figure 10.1, MDCT and CBCT image slices of the same patient are displayed. The loss of low-contrast resolution and shading artifacts in the liver and around vertebral bodies is particularly noticeable. Breathing motion artifact manifests itself as streaks in the CBCT image. The dark circular band in the peripheral region pointed with an arrow is due to the image lag associated with a-Si FPD. Finally, the skin line in the CBCT image is not clearly visible, partly due to detector glare and saturation originating from unattenuated x-ray beam incident on the detector near the tissue–air boundary.

In this chapter, I briefly explain the major image corrections in CBCT, namely, beam hardening, image lag, and scatter corrections. Then, I further elaborate on the methods to reject scatter from projections and how to correct the effects of scatter after its inclusion into the image signal. In a clinical CBCT scanner, additional corrections may be performed, such as truncation corrections and geometric correction to account for gantry flex. A comprehensive discussion of CT image corrections can be found in Hsieh (2009).

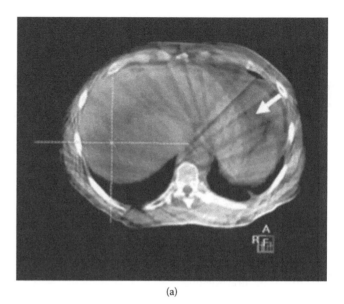

(a)

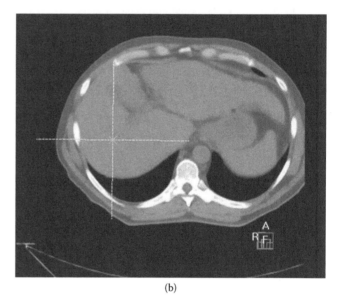

(b)

Figure 10.1 (a) CBCT image acquired for radiation treatment localization. (b) Helical CT image of the same anatomical region is shown as reference. The CBCT isocenter is marked with a crosshair. Both image sets were acquired using similar radiation doses and tube voltages. The loss of low-contrast sensitivity, shading artifacts, and undersampling streaks due to organ motion are apparent in the CBCT image. Darker, circular band (i.e., radar artifact) due to FPD's image lag is marked with an arrow. Also note that skin line in the CBCT image is not clearly visible, due to detector glare and pixel saturation.

10.2 CORRECTION OF ATTENUATION MEASUREMENT ERRORS

10.2.1 BEAM HARDENING

The problem of beam hardening and correction methods have been heavily investigated since early days of CT imaging (Brooks and DiChiro 1976; Herman 1979). Beam hardening artifacts arise from the use of polyenergetic x-ray beams, leading to a nonlinear relationship between the logarithm of projection image signal and the line integral of attenuation coefficients along the x-ray path. Beam hardening introduces spatially varying bias in CT numbers that appears as shading and streak artifacts in CBCT images. Beam filters are the first line of defense to reduce the low-energy x-rays in the spectrum and to reduce beam hardening artifacts. A balancing act is needed because thicker filters reduce beam hardening artifacts while reducing x-ray output and low-contrast sensitivity (Benitez and Ning 2010). Quasi-monoenergetic beams also can be generated by exploiting K-edge filtering methods (Crotty et al. 2006), but such methods also suffer from reduced x-ray output due to heavy beam filtration.

Beam filtration by itself has limited efficacy in reducing beam hardening artifacts, because the polyenergetic nature of the x-ray beam is not effectively suppressed. To further reduce beam hardening, measured log-attenuation values in projections (P_p) can be linearized by mapping P_p to ideal, monoenergetic log-attenuation values (P_m), where log-attenuation value is the log ratio of incident fluence, I_0, to detected fluence, I:

$$P = \ln\left(\frac{I_0}{I}\right) \quad (10.1)$$

The mapping function can be expressed as Nth order polynomial series (Hsieh 2009)

$$P_m = \sum_{i=1}^{N} a_i P_p^i \quad (10.2)$$

To determine a_i, P_p is measured in water-equivalent, uniform phantoms of varying thicknesses. The choice for P_m can be arbitrary as long as the correlation between the values of P_m and corresponding phantom thicknesses is linear. The method described in the preceding is the so-called first-order beam hardening correction, and its various interpretations were used in CBCT imaging (Spies et al. 2001; Matsinos 2005; Grimmer and Kachelriess 2011). Higher order corrections to account for high-density structures also have been developed for MDCT and micro-CBCT scanners (Kijewski and Bjarngard 1978; Hsieh et al. 2000b; Kyriakou et al. 2010a). However, CBCT systems for humans are more challenging for implementation of high-order beam hardening corrections as inconsistencies in log-attenuation values are dominated by high fraction of scatter.

FPDs may exhibit pixel-to-pixel sensitivity variations due to beam hardening that lead to ring artifacts. Conventional flood field gain calibration has limited efficacy to correct such pixel-to-pixel sensitivity variations. Ring artifacts can be reduced by applying the first-order beam hardening correction using pixel-specific beam hardening calibration data (Altunbas et al. 2006), or they are corrected by using image processing methods (JanSijbers and Andrei 2004; Zellerhoff et al. 2005; Star-Lacket al 2006; Prell et al. 2009; Anas et al. 2011).

10.2.2 IMAGE LAG AND GHOSTING

Image lag is the residual signal retained in the detector pixels after readout that is released in the subsequent image frames. As a result, the signal generated in each pixel is biased by the propagation of residual image signal originating from prior image frames. Image lag is a function of x-ray exposure and frame rate, and it is typically quantified as the percentage residual signal amplitude transferred from one frame to the subsequent frames. First frame image lag is about 3%–4% for indirect FPDs and drops down to less than 1% at the 10th frame (Siewerdsen and Jaffray 1999b; Granfors et al. 2003). The other effect closely related to image lag is FPD ghosting and refers to more long-term effects of x-ray exposure. Ghosting appears as a shadow of the imaged object even several hours after the x-ray exposure. Residual image signal due to ghosting has very small, but still detectable, amplitude.

Origins of image lag in a-Si FPDs differ from those in MDCT detectors. In MDCT detectors, image lag is largely due to the slow release of trapped charge carriers in the scintillator, known as afterglow. In FPDs, image lag is attributed to charge trapping and its subsequent release in a-Si photodiodes (Siewerdsen and Jaffray 1999b). Other effects, such as afterglow of scintillator in indirect FPDs and incomplete transfer of charge from pixels to readout electronics, also contribute to image lag, but their role is minor.

The magnitude of image lag may appear to be small at first glance, but artifacts associated with it are noticeable in CBCT images. This issue comes from the fact that FPD pixels are exposed to a wide range of x-ray intensities during CBCT image acquisition; pixels that are exposed to unattenuated x-ray fluence in some projections may register the beam fluence in heavily attenuated regions in subsequent projections. Given that attenuated primary intensity can be less than 1% of unattenuated field intensity in the diagnostic energy range, magnitude of image lag signal originating from prior projections is nonnegligible with respect to image signals in heavily attenuated regions. As a result, circular dark bands, also known as radar artifacts, tend to appear in the reconstructed image (Figure 10.1). Also, blurring and spatial distortion of high-contrast objects may be observed due to image lag (Siewerdsen and Jaffray 1999a).

To reduce image lag, several unexposed image frames can be acquired in between exposed frames (Siewerdsen and Jaffray 1999b). A more practical approach is based on removal of image lag from projections using recursive filtering (Hsieh et al. 2000a). In this approach, each projection image is corrected by subtracting a weighted sum of images acquired before the corrected image. To obtain weighting factors, propagation of lag from one image frame to the subsequent ones is modeled empirically using temporally stationary impulse response function (IRF). Mail et al. (2008) and Starman et al. (2011) have both implemented this approach to model image lag and reduce radar artifacts. One of the limitations of recursive filtering methods is the modeling of image lag using stationary IRFs. Characteristics of image lag depend on the x-ray exposure; hence, exposure variations due to imaged object size and shape lead to temporally nonstationary behavior of image lag. As a result, a stationary IRF may lead to under- or overcorrection of image lag in CBCT projections.

Because the image lag is attributed to charge trapping in a-Si, hardware-based methods were proposed to reduce defect states during image acquisition. Starman et al. (2012) developed a technique to forward bias photodiodes to fill defect states in between two x-ray exposures. Also, Mail et al. (2008) demonstrated that bow-tie filters reduce image lag and associated radar artifacts by reducing the flood field x-ray fluence in the periphery of the object.

10.2.3 SCATTERED RADIATION AND DETECTOR GLARE

In CBCT imaging, scattered radiation is far more dominant in causing errors for attenuation measurement, compared with other factors such as beam hardening or image lag. Scattered radiation may contribute to more than 80% of total image signal in high-scatter imaging situations. The largest source of scattered radiation is the imaged object itself, but scatter also may be generated from system components, such as patient couch, beam filters, and FPD cover. In indirect FPDs, detector glare associated with light dispersion in the x-ray scintillator and x-ray backscattering also generates biases and contributes to errors in x-ray attenuation measurement. To reduce the adverse effects of scatter, two scatter reduction approaches have been used: (1) scatter rejection that refers to removal of scatter before it is included into the image signal and (2) scatter correction that refers to recovery of "scatter-free" image signal after inclusion of scattered radiation into the image signal. In the rest of this chapter, I first provide an overview of the effect of scattered radiation on CBCT image quality and then review scatter rejection and correction methods in FPD-based CBCT imaging.

10.3 EFFECT OF SCATTERED RADIATION AND SCATTER CORRECTIONS ON CBCT IMAGE QUALITY

In the diagnostic imaging x-ray energy range (up to 140 keV), both coherent and incoherent scattering are major interaction mechanisms in soft tissue. Although coherent scattering has smaller cross section than incoherent scattering, its contribution to total scatter intensity is not negligible, especially at lower energies. Also, coherent single scatter is largely responsible for spatial variation in scatter distribution due to its forward-peaked differential cross section. Multiple scatter dominates scatter signals for larger objects and higher energies, and when compared with single scatter, it has a much smoother spatial distribution (Glover 1982; Johns and Yaffe 1983; Kyriakou and Kalender 2007b). Fraction of scatter in total image signal primarily depends on object size (thickness and exposed area) and composition, whereas it has weaker dependence on beam energy in the diagnostic range (Kalender 1981; Niklason et al. 1981).

General characteristics of scattered radiation distribution are explained in Figure 10.2. The anterior–posterior (AP) projection image of an anthropomorphic chest phantom in Figure 10.2a is decomposed into primary and scatter components using a multiple-slit scatter measurement method. As shown in Figure 10.2b, the scatter component is a spatially slowly varying function, and its intensity pattern approximately follows the primary signal intensity. As I discuss in the next section, the scatter-to-primary ratio (SPR) is a more important quantity than the absolute scatter intensity itself; the SPR determines the degree of image quality deterioration. SPR (Figure 10.2d) is a fast-varying function driven by the primary intensity variation (Figure 10.2c), and it is inversely proportional to the primary signal. As evident from the SPR profiles (Figure 10.2e and 10.2f), the SPRs may reach very high values (>5), not because of the increase in scatter but because of attenuation of primary x-rays by high-density structures, such as vertebral bodies.

10.3.1 DEGRADATION IN CT NUMBERS AND CONTRAST-TO-NOISE RATIO (CNR)

To explain the effects of scatter on CT image quality, I first evaluate how contrast and noise are transferred from the projections to the reconstructed image domain. To understand

the correlation between scatter intensity and basic image quality metrics, I use a simplified example where CT image is reconstructed from a single projection. Also, I assume that monoenergetic x-rays with fluence I_0 are incident on a uniform cylindrical phantom with diameter d, and transmitted fluence is detected by a photon counting image detector. Under narrow beam geometry, projection of the attenuation coefficients along a ray passing through the central axis of the cylinder is

$$\mu_T d = \ln \frac{I_0}{P} \tag{10.3}$$

where μ_T is the true linear attenuation coefficient of the phantom's material, and P is the transmitted primary fluence. Under the broad-beam geometry, scattered radiation would be detected by the image receptor, and it would appear as spatially uniform and additive fluence S. Therefore, measured projection value in the presence of scattered radiation is

$$\mu_M d = \ln \frac{I_0}{P + S} \tag{10.4}$$

where μ_M is the measured attenuation coefficient. By combining Equations 10.3 and 10.4

$$\mu_M = \mu_T + \frac{1}{d} \ln \left(\frac{1}{1 + S/P} \right) \tag{10.5}$$

As evident from Equation 10.5, in the presence of scatter, measured attenuation coefficient (or the measured CT number in a CT image) is less than the true attenuation coefficient, and the magnitude of contrast degradation is a function of SPR. In addition, Equation 10.5 indicates that the spatial variations of SPR values, not the absolute scatter intensity, in the projections determine the magnitude of CT number nonuniformities and shading artifacts. Using the same approach, the measured contrast, C_M, between an insert with attenuation coefficient μ_{M2} and diameter d_2, and the surrounding uniform cylinder with attenuation coefficient μ_{M1} and diameter d_1 would be (Siewerdsen and Jaffray 2001)

$$C_M = \mu_{M1} - \mu_{M2} = C_T + \frac{1}{\alpha d} \ln \left(\frac{1 + \left(\frac{S}{P} \right) e^{-\alpha C_T d}}{1 + \frac{S}{P}} \right) \tag{10.6}$$

where C_T is the true contrast, $d = d_1 + d_2$, and $d_2 = \alpha d$. If C_T is small, Equation 10.6 can be approximated as

$$C_M = C_T \left(\frac{1}{1 + S/P} \right) \tag{10.7}$$

Thus, contrast between two different materials is reduced due to scatter. Image noise is defined as standard deviation of measured attenuation coefficients, $\alpha(\mu_M)$, and it can be derived from Equation 10.5

$$\sigma(\mu_M) = \frac{\partial \left(\ln \frac{I_0}{(P + S)} \right)}{\partial (P + S)} \sigma(P + S) \tag{10.8}$$

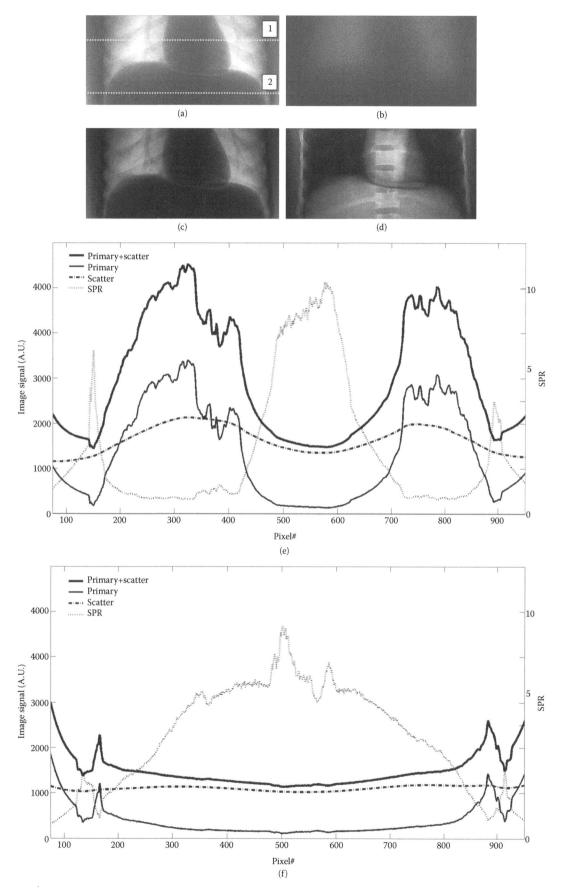

Figure 10.2 (a) CBCT projection of an anthropomorphic chest phantom. The projection image is decomposed into scatter-only (b) and primary-only (c) images using a multiple-slit scatter measurement technique. (d) Scatter-to-primary ratio (SPR) map as a gray scale image. Signal profiles along two dotted lines (marked as 1 and 2) in (a) are shown in panels (e) and (f), respectively.

where $\sigma(P + S)$ is the standard deviation of projection image signal. (For clarity, d in Equation 10.8 was dropped.) We assume that the noise in the projection image has no additive components such as electronic readout noise, and that it follows Poisson statistics; thus, $\sigma(P + S) = \sqrt{P + S}$. Equation 10.8 becomes

$$\sigma\left(\mu_M\right) = \frac{1}{P^{1/2}\left(1 + \dfrac{S}{P}\right)^{1/2}} \qquad (10.9)$$

The low-contrast object detectability in images is generally assessed using the metric contrast-to-noise ratio (CNR). By taking the ratio of Equations 10.7 and 10.9, measured CNR is

$$\mathrm{CNR}_M = C_T \frac{P^{1/2}}{\left(1 + \dfrac{S}{P}\right)^{1/2}} \qquad (10.10)$$

As seen in Equation 10.9, scattered x-rays increase the total number of x-ray quanta in the projection image; hence, image noise is reduced in the reconstructed image; at a given gray window/level, a CT image with higher SPR appears to be "smoother" than a CT image with lower SPR. However, the loss of contrast due to increased SPR (Equation 10.7) is larger than the noise reduction, and it leads to a drop in CNR in a CT image (Equation 10.10).

As mentioned previously in this section, the magnitude of SPR can vary greatly depending on the anatomy being imaged. For example, in pelvis CBCT scans for image-guided radiation therapy, values of maximum SPR can easily exceed 5 in projections, whereas in dedicated breast CBCT systems, maximum SPR is typically in the order of 0.2 to 1 (Liu et al. 2005a; Chen et al. 2009). The degradation of CNR due to increase in SPR was experimentally demonstrated by Siewerdsen and Jaffray (2001) for low-contrast objects in water equivalent phantoms. They reported that CNR could drop as much as factor of 2 when SPR was varied from 0 to 1.

10.3.2 EFFECT OF SCATTER CORRECTIONS ON IMAGE QUALITY

A scatter correction method suppresses the effects of scatter after the scatter x-ray fluence is acquired by the detector as part of the image signal. It recovers the magnitude of scatter-free image signal, but the noise component due to scattered x-rays cannot be decoupled from image noise, and its stochastic effects remain a factor to degrade the image quality. To better understand the effect of scatter correction on the CNR, we assume that the scatter signal intensity is already estimated in a projection image. After subtraction of scatter intensity, SPR would be zero, and Equation 10.7 becomes

$$C_{CM} = C_T \qquad (10.11)$$

where C_{CM} is the measured and scatter-corrected contrast. After correction for scatter signal intensity, the image noise due to scatter would still be present in the corrected projection image. Equation 10.8 becomes

$$\sigma\left(\mu_{CM}\right) = \frac{\partial\left(\ln\dfrac{I_0}{(P)}\right)}{\partial(P)}\sigma(P + S) \qquad (10.12)$$

$$\sigma\left(\mu_{CM}\right) = \frac{\left(1 + S/P\right)^{1/2}}{P^{1/2}} \qquad (10.13)$$

Measured CNR after scatter correction is obtained by taking the ratio of Equations 10.11 and 10.13

$$\mathrm{CNR}_{CM} = C_T \frac{P^{1/2}}{\left(1 + S/P\right)^{1/2}} \qquad (10.14)$$

which is the same as measured CNR without scatter correction in Equation 10.10. Thus, subtractive scatter corrections help to restore true CT image values and reduce image artifacts in the expense of increased CT image noise (Equation 10.13), and CNR remains unaffected.

10.4 SCATTER REJECTION METHODS

The scatter rejection methods are inherently hardware based. With them, scattered radiation is physically prevented from reaching the image receptor or erased before final data recording as in slot scan imaging (Liu et al. 2006a). The advantage of scatter rejection over scatter correction is that it prevents injection of stochastic noise due to scatter into the projection image signal, therefore CNR can be potentially improved. The majority of scatter rejection methods in CBCT imaging are inherited from projection radiography, and their characteristics were extensively investigated in the past. Here, I review their characteristics in the context of CBCT.

10.4.1 AIR GAP

The air gap method, where the gap refers to the space between the imaged object and the FPD, is one of the simplest scatter rejection methods. This method exploits the fact that scattered radiation emanating from the imaged object appears as a secondary x-ray source. As the air gap between the object and the image receptor increases, the intensity of x-rays from the secondary x-ray source falls faster than the intensity of the primary beam reaching the detector due to inverse square law. Thus, reduction in SPR is primarily a function of air gap, and to a lesser extent, source to isocenter distance (SID). For an SID of 75–100 cm and an air gap of 20–30 cm, SPR can be reduced by a factor of 2 to 6 with respect to no air gap (Sorenson and Floch 1985). Beyond 50 cm of air gap, the efficacy of air gap method goes down as the differential effect of inverse square law between the x-ray source and scattered x-ray source diminishes. The main advantage of air gaps is the ability to reduce SPR and to increase CNR at the same time. Neitzel (1992) has shown that CNR of low-contrast objects in projections could be increased as much as factor of 2 using the air gap method under a wide range of SPR conditions. In CBCT, the air gap method is used inherently due to physical clearance needed between the patient and image receptor during image acquisition. The selection of the air gap or, instead, the image acquisition geometry requires a balance between different system design requirements imposed by the imaging task (Siewerdsen and Jaffray 2000). For example, excessively large air gap leads to increased object magnification on the detector plane, thereby reducing the reconstructed image volume. In applications that require high spatial resolution, increased air gap also leads to image unsharpness due to magnification of the x-ray focal spot size.

10.4.2 ANTISCATTER GRIDS

First developed by Bucky (1913) for projection radiography, antiscatter grids (ASGs) are commonly used scatter rejection devices along with air gap methods. Standard fluoroscopy and radiography ASGs are made of unidirectional radiopaque strips interleaved by fiberglass, aluminum, or paper. Their effect on image quality in radiography has been well studied (Chan and Doi 1982; Kalender 1982; Boone et al. 2002), and ASGs used in radiography are also used in CBCT systems. Scatter rejection performance of ASGs depend on angular distribution and energy of scattered x-rays, a function of CBCT system parameters and

the imaged object. For example, for large objects, the multiple scatter component and the angle of incidence of scattered x-rays on the ASG become more oblique. As a result, ASGs, especially the ASGs with higher grid ratios, can more efficiently block scatter. The air gap distance affects the performance of the antiscatter grid; at larger air gaps, the efficacy of ASGs goes down due to smaller fraction of multiple scatter in total scatter fluence. Also, ASGs are less efficient at higher energies due to increased transmission of scatter through the radiopaque strips. A thorough discussion of ASGs' scatter rejection characteristics can be found elsewhere (Johns and Yaffe 1983; Neitzel 1992; Wiegert et al. 2004).

With the use of ASGs, SPR in CBCT projections is reduced by a factor of 2 to 5 depending on the imaging conditions (Wiegert et al. 2004; Kyriakou and Kalender 2007a). As an example, Monte Carlo (MC) simulation of SPR in a CBCT projection profile is shown in Figure 10.3a (Lazos and Williamson 2010). In this study, CBCT system geometry mimicked a clinical linac-mounted CBCT scanner. As seen in

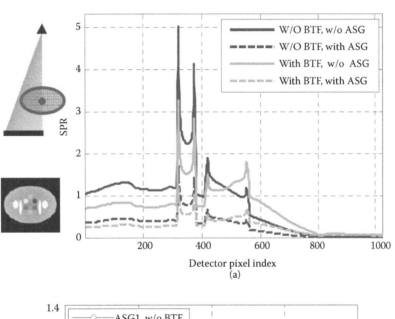

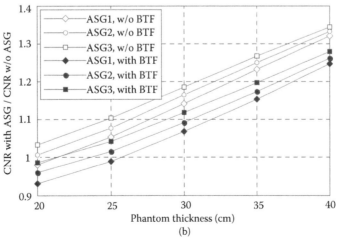

Figure 10.3 MC simulation of (a) SPR profiles and (b) CNR improvement ratios with and without ASGs, and bow-tie filters (BTFs) in half-fan geometry. System parameters mimic a linac-mounted CBCT scanner (OBI, Varian Medical Systems, Palo Alto, CA). The inset in panel (a) shows the depiction of SPR simulation geometry and the cross section of the pelvis phantom. CNR in panel (b) is displayed as a function of circular water equivalent phantom diameter. The transmission ratios (primary, scatter) for each grid design are ASG1 (0.7, 0.28), ASG2 (0.75, 0.31), and ASG3 (0.78, 0.4). (Reprinted from Lazos, D. and Williamson, J.F., *Med Phys*, 37, 5456–70, 2010. With permission.)

the figure, maximum SPR for a pelvis phantom scanned in half-fan geometry was reduced by about a factor of 3 throughout the projection profile with the ASG. Similarly, Kwan et al. (2005) measured up to factor of 2 reduction in SPR in dedicated breast CBCT geometry. The use of ASG translates into reduction in shading and cupping artifacts (Siewerdsen et al. 2004; Kyriakou and Kalender 2007a). Kyriakou et al. reported that the percentage of CT number deviations from ground truth CT numbers without the ASG were in the order of 30% and 12% for body and head phantoms, respectively. With the ASG, nonuniformities dropped down below 15% and 5%, respectively.

The main deficiency of ASGs is that the improvement of CNRs is limited. In fact, the CNRs are reduced by the ASG in low to moderate SPR conditions (e.g., SPR < 1), where air gaps may be more favorable for improving CNR (Neitzel 1992). Modest CNR improvement was achievable only in high SPR scenarios, such as imaging of the pelvis. In MC simulations by Lazos and Williamson (2010) (Figure 10.3b), ASGs helped to improve CNR, when the phantom thickness was more than 20 cm in half-fan scanning geometry. Wiegert et al. (2004) simulated a C-arm-based CBCT system and reported that signal to noise ratio in projections degraded with the use of ASGs for head and thorax phantoms, and a small improvement was observed for the pelvis phantom. The lack of improvement in CNRs is due to the fact that ASGs also attenuate the primary beam intensity by 30%–40%; the recovery of image contrast by scatter rejection is offset by increased image noise due to primary beam attenuation. If imaging dose is not a concern, image noise can be reduced by increasing the incident x-ray exposure. Schafer et al. (2012) showed that the loss of soft tissue CNRs due to ASG in a C-arm-based CBCT requires an increase in exposure level by a factor of 1.6 to 2.5 depending on the ASG's grid ratio.

10.4.3 BOW-TIE FILTERS

A bow-tie filter compensates for reduced attenuation of x-rays in the periphery of a patient to reduce skin dose, and to reduce the range of radiation exposure incident on the detector in CBCT imaging. It is not strictly a scatter rejection device by itself, because SPR is suppressed via modulation of incident x-ray fluence distribution in the lateral direction. Depending on the imaging conditions, bow-tie filters reduce the SPR by 30% to 50% and increase the CNR (Kwan et al. 2005; Graham et al. 2007; Mail et al. 2009; Lazos and Williamson 2010; Menser et al. 2010). Moreover, bow-tie filter makes the spatial variation of the SPR more uniform and helps to reduce the shading artifacts. As shown in Figure 10.3a, the bow-tie filter reduces the SPR mainly in the central section of the pelvis phantom, while increasing the SPR in the periphery. Bow-tie filter's scatter reduction efficacy is less than ASG, but if it is used in conjunction with an ASG, the SPR may be further reduced. Also, as shown in Figure 10.3b, combined use of bow-tie filter and ASG may result in small losses of CNRs due to the reduced scatter rejection efficiency of ASG (Lazos and Williamson 2010). Due to the fixed geometry of bow-tie filter and symmetric exposure modulation pattern, bow-tie filter's SPR reduction efficiency depends on the object shape and positioning. For example, even if an object is placed asymmetrically in relation to the gantry rotation axis, the exposure may be over- or undercompensated by the bow-tie filter.

This issue would lead to less uniform SPR distribution, and it may even cause an increase in the SPRs in some regions.

10.4.4 VOLUME OF INTEREST SCANNING

Because the SPR strongly depends on the size of FOV, limiting the FOV to a clinically relevant volume of interest (VOI) can significantly reduce the SPRs and increase the low-contrast sensitivity. Smaller FOV can be simply achieved by collimating the beam along the CBCT gantry rotation axis. Scatter rejection is further improved by limiting the FOV in the axial plane, known as the VOI scanning technique (Chityala et al. 2004; Chen et al. 2008; Kolditz et al. 2010). This technique typically relies on two CBCT scans. The first scan is a full field scan that serves as a guide to locate the clinically relevant VOI. After the location of VOI is determined, either the patient is moved to align the VOI with the isocenter or the CBCT gantry is moved to align its isocenter with the VOI. The second scan covers the VOI or the clinically relevant region only. To eliminate truncation artifacts in the reconstructed image, projections from both CBCT scans are combined, and composite projections are used for reconstruction (Figure 10.4).

The VOI techniques have two advantages in achieving increased contrast sensitivity. First, the SPR can be significantly reduced, directly translating into an increase in the contrast sensitivity. Second, with the full view scan performed with lower exposures and the VOI scan performed with higher exposures, the image noise level may be reduced and the CNRs increased within the VOI. As a result, dose to the patient is better managed in VOI imaging. Lai et al. (2009) investigated the application of the VOI scanning technique for dedicated breast CBCT, and they demonstrated that collimated scan of VOIs with several centimeters in size can decrease the SPRs by an order of magnitude in the projection images of a breast phantom. In Figure 10.4, the contrast sensitivity improvement for low-contrast structures is shown in a VOI with a diameter of 2.5 cm within a polycarbonate phantom with a diameter of 15 cm (Lai et al. 2009).

10.5 SCATTER CORRECTION METHODS

The performance of scatter rejection methods is still insufficient to achieve high levels of CT number accuracy and improved low-contrast sensitivity. For further improvements in image

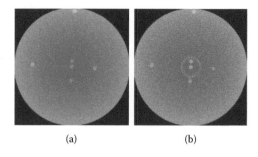

(a)   (b)

Figure 10.4 Full field (a) VOI and (b) CBCT scans of a polycarbonate phantom with a diameter of 15 cm with I contrast inserts. The VOI has a diameter of 2.5 cm, and the radiation dose is approximately 30% with respect to the dose at the center of the full field scan. (Reprinted from Lai, C.J. et al., *Phys Med Biol*, 54, 6691–709, 2009. With permission.)

quality, scatter correction methods can be used in combination with scatter rejection methods. Because scatter correction is performed after detection of scattered radiation, the noise due to scatter cannot be eliminated, and the CNRs cannot be improved without suppressing the image noise.

During scatter correction, scatter distribution is estimated first, and the image is corrected in the second step. As the scatter signal is modeled as an additive term, the projection image signal at pixel (u,v), $I(u,v)$, is expressed as

$$I(u,v) = P(u,v) + S(u,v) \qquad (10.15)$$

where $P(u,v)$ and $S(u,v)$ are the primary and scatter signals, respectively. If $S(u,v)$ is known, the primary signal is simply estimated as

$$P(u,v) = I(u,v) - S(u,v) \qquad (10.16)$$

In iterative scatter correction algorithms, the multiplicative correction scheme can be used instead to calculate primary during iteration steps (Zellerhoff et al. 2005)

$$P^{(m+1)}(u,v) = I(u,v) \cdot \left[\frac{P^{(m)}(u,v)}{P^{(m)}(u,v) + S^{(m)}(u,v)} \right] \qquad (10.17)$$

where m is the iteration step number. Multiplicative correction scheme avoids negative values for $P(u,v)$ due to overestimation of scatter signal.

For clinical applications, an ideal scatter correction method should be accurate and general enough to be used in a variety of imaging conditions. Also, it is preferable to perform scatter corrections close to real time for clinical applications, that is, within seconds after CBCT image acquisition. It is not trivial to achieve all the requirements in one scatter correction method, and trade-offs are often made particularly in the scatter estimation step.

Scatter estimation may be performed in the projection domain or in the CT image domain. The latter method uses the reconstructed images to derive a physical object model, and estimates scatter in forward projections of the object model. The choice between the two methods depends on the trade-offs between scatter correction accuracy and speed. Projection domain-based methods are faster, but only limited information about the object shape and composition can be extracted as an input to scatter estimation algorithms. CT image domain-based methods are inherently more accurate, as the knowledge of the object's physical shape and composition in 3D improves scatter estimation at the expense of computation speed.

A simplified flow diagram for image signal corrections is shown in Figure 10.5. After acquisition of raw projections, FPD dead pixel, gain, and offset corrections are applied to equalize pixel-to-pixel variations in sensitivity and flood field fluence (Seibert et al. 1998; Schmidgunst et al. 2007; Abella et al. 2012). Image lag can be corrected after these stages. Because the scatter distribution has a slowly varying spatial pattern, projections are downsampled to increase computational speed during scatter estimation. To further

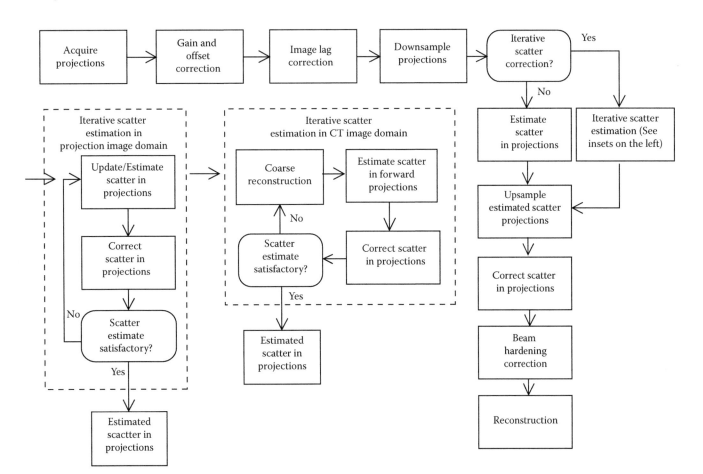

Figure 10.5 Simplified flowchart of projection image corrections.

increase computation speed, only a subset of projections may be used for scatter estimation, and an interprojection interpolation can be performed to estimate scatter in all projections. Because projection images contain both primary and scatter signals, scatter estimation may be performed in an iterative manner to improve the accuracy at the expense of computational speed. The estimated scatter signals in projection domain are interpolated to larger matrix size to match the resolution of the projection images, and the scatter correction is performed. Beam hardening is typically corrected after scatter correction.

10.5.1 MEASUREMENT-BASED CORRECTIONS

Direct measurement methods are commonly used for studying the effects of scattered radiation, and they also have been proposed for patient-specific scatter correction in CBCT. Unlike software-based corrections, measurement-based methods do not require a model for scatter measurement. Scatter is directly measured; therefore, extraction of scatter signal from total image signal is not needed. Beam-stop array is generally used for scatter measurements, and it is made of radio-opaque BBs (i.e. round shaped pellets) or disks, and arranged into a two-dimensional (2D) array. The beam-stop array is placed between the source and the imaged object, and the image signals in the beam-stop shadow are used to estimate the scatter signal. Due to the slowly varying nature of scatter signal in a projection image, beams stops are placed sparsely, and the 2D scatter distribution for each image pixel may be obtained by interpolation. Conventional scatter measurements require two sets of CBCT projections to be acquired: one set with and one set without beam-stop array to restore primary beam signal at beam-stop locations (Ning et al. 2002). Although acquisition of two sets of

CBCT projections can facilitate accurate scatter estimation and corrections, it is not desirable mostly due to excess patient dose and increased image acquisition time.

The majority of measurement-based methods focus on reducing or eliminating the redundant projections needed for scatter measurements. To reduce the number of redundant projections, scatter measurements can be performed on coarsely sampled projection views, and scatter signals for projections in between can be estimated by interprojection interpolation (Ning et al. 2004; Cai et al. 2008). To fully avoid additional projections needed to measure scatter, beam-stop array can be moved during the image acquisition. Due to translational motion of beam-stop array, beam-stops are positioned at different positions in each projection, and total image signal in the beam-stop shadow may be recovered by interpolation. By moving the array, interpolation is avoided at stationary locations in projections, and interpolation artifacts are eliminated (Liu et al. 2005b; Zhu et al. 2005; Wang et al. 2010). If the scatter intensity is expected to be largely uniform within the projection view, it also can be estimated by interpolating image signals in the collimator shadows that may allay the need for beam-stop arrays (Siewerdsen et al. 2006). Also, beam stops can be placed in sections of the projection images that are not used during filtered backprojection (Niu and Zhu 2011). In Figure 10.6a, a schematic representation of beam-stop array method is shown. Horizontal signal profiles for the total, primary, scatter, and SPRs are shown in Figure 10.6b. The transverse image of a uniform phantom before and after scatter correction is shown in Figure 10.6c and 10.6d.

The design of beam-stop array has an implication on the accuracy in sampling and estimating the 2D scatter distribution. The image signals in the beam-stop shadow may not only

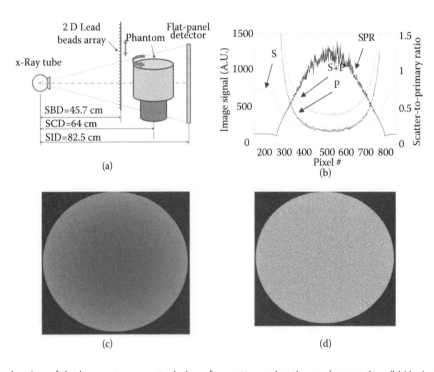

Figure 10.6 (a) Schematic drawing of the beam-stop array technique for scatter estimation and correction. (b) Horizontal profiles from a water phantom are displayed: total image signal (S + P), primary-only signal (P), scatter-only signal (S), and scatter-to-primary ratio (SPR). Reconstructed axial slices before and after scatter correction are shown in panels (c) and (d), respectively. (Reprinted from Liu, X. et al., *SPIE* 614234, 2006. With permission.)

originate from scatter from the imaged object but also from contaminant signal sources, such as detector glare and off-focal spot radiation. Contaminant signal contribution to beam-stop signal is 6% to 15% of primary intensity adjacent to beam stop, and it varies at higher spatial frequencies than object scatter signal (Chen et al. 2009; Bootsma et al. 2011; Lazos and Williamson 2012). In areas of low SPR and high primary intensity gradients, such as tissue–air boundaries, contaminant signal intensity may dominate the image signal at beam-stop shadows and lead to overestimation of scatter signal amplitude. Furthermore, as the contaminant image signal varies at higher spatial frequency, it may be spatially undersampled by sparsely placed beam stops. To address these issues, contaminant signals can be reduced using deconvolution techniques before scatter estimation, or if the contaminant signal is going to be treated as part of scatter signal, radio-opaque bars rather than BBs can be used to increase sampling frequency (Liu et al. 2006b).

As an alternative to beam-stop arrays, a primary signal modulation method has been proposed to exploit the low frequency nature of scatter in Fourier domain. Modulation is achieved by placing a beam filter with high-frequency thickness pattern between the source and the object (Bani-Hashemi et al. 2005; Zhu et al. 2006; Gao et al. 2010). The separation between scatter and modulated primary in Fourier domain can be used to obtain primary-only projections by filtering techniques. Primary modulation method addresses some of the issues associated with the beam-stop array method; it does not require additional projections to be acquired and the beam modulating filter may remain stationary.

10.5.2 SCATTER KERNEL SUPERPOSITION METHODS

Scatter kernel superposition (SKS) is one of the earliest software-based techniques used for both detector scatter-glare and object scatter correction in projections. The main advantage of SKS over more accurate scatter correction algorithms is the computational efficiency that makes real-time scatter corrections feasible. The key element of a SKS method is the scatter kernel, which can be thought as a transfer function that relates normalized primary fluence at a point within the object to its corresponding scatter intensity distribution in the detector plane. In an ideal scenario, if the primary x-ray fluence and the scatter kernel are known for each point within the volume of an object, superposition of scatter signals generated by each point will yield the object's scatter signal distribution in the projection image. In practice, such an implementation requires *a priori* knowledge about the physical properties of the object in 3D and knowledge of point scatter kernels, and superposition operation has to be carried out in 3D. To reduce computational complexity, SKS methods can be implemented in projection image domain, that is, scatter kernel is obtained for a pencil x-ray beam that transverses the object, and falls on the detector. To obtain the scatter signal $S(u,v)$ at point (u, v) in the detector plane, the pencil beam scatter kernel (PBSK) contributions are scaled by the primary intensity, $P(u',v')$, at each detector location (u',v'), and are summed up

$$S(u,v) = \iint_D P(u',v') \cdot H(u-u', v-v') \cdot du' \, dv' \quad (10.18)$$

where $H(u',v')$ is the PBSK for a primary beam at (u', v'). If $H(u', v')$ is stationary, this operation can be performed in Fourier domain. In short notation, the total image signal becomes

$$I(u,v) = P(u,v) + [P * H](u,v) \quad (10.19)$$

where $*$ is the convolution operator.

In early implementations of SKS, the PBSK was assumed to be rotationally symmetric and stationary. This approach is often used for detector glare correction that contributes as part of the scatter signal. In brief, detector glare kernel, $H_G(u,v)$, acts as a low-pass filter that blurs $I(u,v)$, and Equation 10.19 becomes

$$I_G(u,v) = [I * H_G](u,v) \quad (10.20)$$

$I(u,v)$ can be extracted either by using filtering techniques in the spatial domain or by deconvolution in the Fourier domain using fast Fourier transform (FFT) (Shaw et al. 1982; Seibert et al. 1985; Kawata et al. 1996; Poludniowski et al. 2011)

$$I(u,v) = \text{FFT}^{-1}\left(\frac{\text{FFT}\left(I_G(u,v)\right)}{\text{FFT}\left(H_G(u,v)\right)} \right) \quad (10.21)$$

(since the object scatter is more dominant than the detector glare, we will omit this effect and assume that $I_G \approx I$ in the rest of the text for clarity).

However, stationary kernel assumption is an oversimplification for estimating the object scatter. For a given CBCT scan geometry, the spatial distribution of scatter intensity generated by a pencil beam depends on the shape of the object. To address this problem, nonstationary PBSKs can be generated as a function of the object thickness, t, and they have the following form:

$$H(u, v, u', v') = W(u',v') \cdot G(u,v) \quad (10.22)$$

$W(u',v')$ is the weighting term that is proportional to total scatter intensity generated by the pencil beam, and $G(u,v)$ is the kernel shape term that characterizes the spatial dispersion of scattered x-rays in the detector plane. In the first-proposed SKS implementations, stationary PBSKs were used and PBSKs were preselected for specific object sizes and imaging tasks (Love and Kruger 1987; Seibert and Boone 1988). In the first nonstationary kernel implementations, Naimuddin et al. (1987) used a lookup table to determine $W(u',v')$ as a function of image signal, and Kruger et al. (1994) proposed regionally variable PBSKs to be used in different anatomical regions. Ohnesorge et al. (1999) introduced a parameterized PBSK that accounts for changes in object thickness to correct scatter in MDCT.

As the primary intensity is not known, scatter calculation using Equation 10.18 is performed using $I(u,v)$ rather than $P(u,v)$, leading to inaccurate estimation of scatter signals. To allay this problem, Hansen et al. (1997) used an iterative scheme to obtain the primary signals. With their technique, the scatter distribution was first estimated by assuming that $P_1(u,v) = I(u,v)$. The calculated scatter, $S_1(u,v)$, was used to update the primary, $P_2(u,v)$, in the second iteration. Convergence to scatter-only and primary-only signals is achieved in several iterations. Similar approaches also were adopted by SKS implementations (Spies et al. 2001; Maltz et al. 2008; Sun and Star-Lack 2010a).

Scatter kernels are obtained using experimental or MC methods. Boone et al. (1986) used a Gaussian distribution to model the kernel shape and iteratively estimated the model parameters that best fitted the measured scatter distribution. Li et al. (2008) measured the scatter point spread function (PSF) directly from an impulse response generated by a scatter edge spread function. In MC calculations, PBSKs are generated by simulating a pencil-ray beam incident on a water slab and registering scatter distribution in the detector plane (Hansen et al. 1997; Maltz et al. 2008; Sun and Star-Lack 2010a). Compared with experimental methods, MC calculation gives more flexibility for generation of PBSK; kernels can be obtained for any arbitrary object shape and imaging geometry with relative ease. The use of analytical approaches has been limited for scatter kernel generation, because only first-order scatter distribution can be estimated using collision cross sections for coherent and incoherent scattering.

To improve computational efficiency, Ohnesorge et al. (1999) and Sun and Star-Lack (2010a) have developed functional forms for both $W(u',v')$ and $G(u,v)$ with free parameters that could be adjusted empirically. To be able to perform the SKS in Fourier domain, they let $W(u',v')$ vary as a function of the object thickness (or primary intensity), while keeping the kernel shape, $G(u,v)$, stationary. Also, I should briefly touch upon how a scatter kernel shape is derived. Although single Gaussian distributions have been frequently used in the past to model scatter kernels, they are suboptimal for parameterization of $G(u,v)$; a scatter kernel in the kilovolt age energy range has long tails that deviate from the Gaussian model. To alleviate this effect, Sun et al. used a double Gaussian model for $G(u,v)$, and model parameters were generated for different object thickness groups. Meyer et al. (2010) also used a double Gaussian model, and kernel parameters were determined from an ellipsoid object model. In an alternative approach, Maltz et al. (2008) stored precalculated PBSKs in a database that were indexed as a function of water equivalent thickness, object size, and imaging geometry. This approach may be computationally less efficient; in contrast, it avoids issues related to inaccurate kernel parameterization and can potentially accommodate nonstationary kernel shapes.

The major drawback of traditional SKS approaches is the relatively simplistic model of the PBSK; rotationally symmetric $G(u,v)$ and thickness-dependent $W(u',v')$ are generated from uniform, semi-infinite water-equivalent slabs. In more realistic objects, variations in the object thickness and heterogeneities lead to spatially nonuniform transport of scattered radiation; thus, $G(u,v)$ would be asymmetrically shaped and nonstationary. Furthermore, the object-thickness-dependent weighting term does not accurately estimate the total scatter intensity for pencil beams close to the object edge; reduced object volume leads to reduced multiple scattering and attenuation of scattered radiation. Without the knowledge of object shape and composition in 3D, accurate and general enough PBSK generation is not trivial. To address some of these issues, Star-Lack et al. (2009, 2010b) developed an asymmetric PBSK model, which involved a kernel stretching scheme and a modified weighting factor. The kernel stretching term appears as a multiplicative function in Equation 10.22,

and it takes into account the difference in thickness between the position of the pencil beam at (u',v') and the scatter estimation point at (u,v). In a further progression of this method, a hybrid kernel was developed that combined both symmetric and asymmetric PBSKs to improve scatter estimation in the presence of patient couch (Sun et al. 2011). Figure 10.7 demonstrates the effect of SKS corrections in a pelvis CBCT image set. In the uncorrected image (Figure 10.7a), the black hole posterior to prostate was mainly induced by the patient couch and half-fan image acquisition geometry. After SKS correction with symmetric PBSKs (Figure 10.7b), shading artifacts are reduced; however, an overestimation of CT numbers is apparent at the center of the image, and the black hole artifact is still present. With the use of hybrid PBSKs, shading artifacts were further reduced (Figure 10.7c), and the improved image quality is comparable to the CBCT image acquired using a slit collimator (Figure 10.7d).

10.5.3 MC-BASED SCATTER ESTIMATION

MC-based characterization of scattered radiation plays a crucial role in developing strategies for scatter correction. Compared with measurement-based methods, MC calculations enable studying effects of underlying physical processes in more detail and evaluation of a wider range of system parameters to characterize scattered radiation in both radiography (Kalender 1981; Chan and Doi 1983) and CBCT imaging (Malusek et al. 2003; Jarry et al. 2006b; Kyriakou and Kalender 2007a; Lazos and Williamson 2010; Bootsma et al. 2011). However, direct use of MC calculations for patient-specific scatter correction has not been considered until recent years due to the vast computational cost. With the ever-increasing computation speed and fast MC algorithms, MC calculations are increasingly considered for patient-specific scatter corrections.

MC-based scatter corrections are performed in CT image domain; scatter distributions are calculated by forward projecting the CBCT image of the object itself (Zbijewski and Beekman 2006; Jarry et al. 2006a; Bertram et al. 2008). Such calculations offer potentially the most accurate software-based scatter corrections, but the challenge lies in performing MC calculations in a clinically acceptable time frame. To accelerate MC calculations of scatter projections, Mainegra-Hing and Kawrakow (2008, 2010) implemented several variance reduction techniques and a locally adaptive denoising filter into the EGSnrc MC package, thereby improving computation speed by several orders of magnitude. Colijn and Beekman (2004) and Zbijewski and Beekman (2006) used Richardson–Lucy image restoration algorithm to replace noisy scatter projections with 2D fits based on Gaussian basis functions. Their denoising method allowed significantly lower number of photon histories to be used for calculations and lead to an increase in the computation speed by two orders of magnitude. In an alternative approach, Kyriakou et al. (2006) combined computationally efficient analytical single scatter calculation with fast MC-based multiple scatter calculation that exploited the smoothly varying nature of multiple scatter. In addition to variance reduction and image denoising algorithms, use of graphics processing units (GPUs) enabled improved parallelization of MC calculations, and significant gains in computation speeds have been reported. Badal and Badano (2009) reported that calculation of projections from CT image

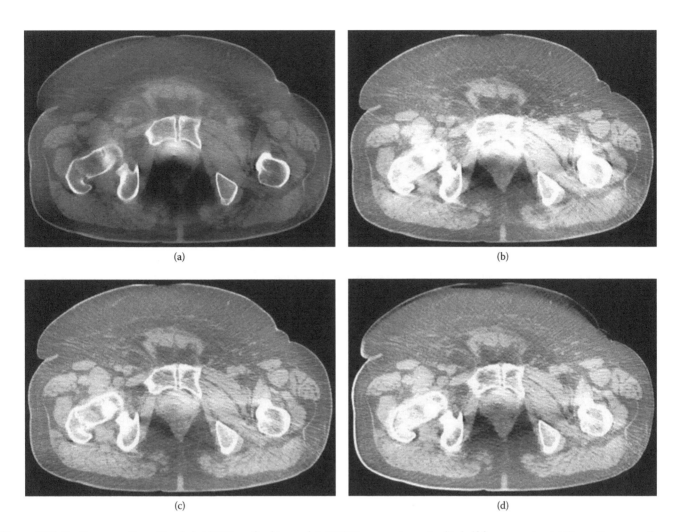

Figure 10.7 Scatter correction with a hybrid SKS method in a pelvis CBCT image set acquired in half-fan geometry. (a) No scatter correction. (b) SKS correction with symmetric kernels. (c) SKS correction using a hybrid method that used both symmetric and asymmetric kernels. (d) Reference image acquired using the slit-scan method. The hybrid SKS method can better account for shape variations due to the patient and the couch. SKS, scatter kernel superposition. (Reprinted from Sun, M. et al., *Med Phys*, 2058–73, 2011. With permission.)

sets was more than an order of magnitude faster with GPUs, compared with calculations based on single central processing unit (CPU). Jia et al. (2102) calculated a single forward projection in less than 2 min using GPU acceleration with image quality comparable to a digitally reconstructed radiograph. Although hardware methods appear to be promising tools for MC acceleration, their use has not been extensively studied yet in the context of patient-specific scatter corrections.

10.5.4 ANALYTICAL METHODS

Klein–Nishina and Rayleigh differential cross sections allow accurate calculation of first-order incoherent and coherent scatter signals in projections. Such calculations are considered a faster alternative to MC calculations for patient-specific scatter corrections. Because multiple scatter is not accounted for with this approach, analytical methods are used in combination with other scatter estimation and correction methods. Yao and Leszczynski (2009a,b) used an iterative approach to calculate multiple scatter and assumed that it is constant or proportional to the first-order scatter intensity. Also, Wiegert et al. (2005) used a simple parametric model that estimated multiple scatter as a constant fraction of analytically calculated single scatter.

Ingeleby et al. (2009) used a similar assumption and estimated the multiple scatter signal within the FOV from the signal behind the x-ray collimator. As mentioned in Section 10.5.3, Kyriakou et al. (2006) used a fast MC algorithm to estimate the multiple scatter component in projections. Rinkel et al. (2007) developed a hybrid method that combined empirical scatter estimation with analytical scatter corrections. They used premeasured scatter intensity maps of phantoms as the initial estimate for scatter distribution of the imaged object and rearranged the scatter distribution spatially using an analytical kernel model.

10.5.5 IMAGE PROCESSING METHODS AND EMPIRICAL MODEL–BASED CORRECTIONS

Scatter correction methods discussed so far use the principles of physics to estimate and correct scatter. CT number accuracy also may be improved by means of image processing rather than physics-based approaches. The performance of image processing methods depends on the detection of scatter-induced image artifacts. For example, cupping and shading artifacts may be detected and corrected in anatomical regions with relatively uniform tissue density, such as brain and breast (Altunbas et al. 2007; Wiegert et al. 2008; Kyriakou et al. 2010b). However, in most imaging tasks, detection of CT number degradation is not trivial, and even if the artifacts are detected, they cannot be fully corrected without the knowledge of ground truth CT numbers. If a prior MDCT image of the patient is available, as in radiation therapy, it can serve as the ground truth to correct CBCT image sets. Using this approach, Marchant et al. (2008) developed a technique to scale CBCT CT numbers with respect to a MDCT image of the same patient. Niu et al. (2010) also used a similar method, but image corrections were made in forward projections of coregistered MDCT and CBCT images.

Besides methods based on image processing, model-based corrections have been proposed to bridge the gap between the computational speed and the accuracy of physics-based scatter correction methods. In this approach, a physics-based method is not directly used to calculate patient- or object-specific scatter calculation, but an empirical model of scatter distribution is established by using previously discussed scatter estimation methods. For example, Bertram et al. (2006) used an ellipsoid phantom in MC simulations to generate parametric lookup tables for scatter distribution and corrected scatter in head CBCT images by approximating head as an ellipsoid. In a similar approach, Kachelriess et al. (2006) developed a model for nonlinearities in projection image signals of cylindrical water phantoms. The model was used to correct both scatter and beam hardening in small animals that were comparable in size to the water phantoms.

In simpler models, spatially smooth nature of scatter distribution was exploited; scatter was assumed to be a constant fraction of projection image signal, and a constant bias was subtracted from image signal to correct for scatter (Suri et al. 2006; Wiegert and Bertram 2006). The major appeal of image processing and model-based corrections is their relatively straightforward implementation and computational efficiency. However, their scope and scatter correction performance is limited to a narrow range of imaging conditions.

10.6 SUMMARY AND DISCUSSION

Increased scatter fraction associated with the large FOV of FPDs is one of the leading sources of image quality degradation problems in FPD-based CBCT imaging. Scattered radiation also is becoming an important image quality issue in MDCT systems due to increased number of detector rows and cross-scatter

contamination in dual x-ray source detectors (Endo et al. 2006; Engel et al. 2008; Petersilka et al. 2010). Nevertheless, the extent of scatter-induced image deterioration is much larger in FPD-based CBCT scanners due to larger FOV and the lack of efficient scatter rejection devices. A rough comparison of SPR values in CBCT and MDCT would be helpful to put the severity of the problem into perspective; in a MDCT projection, maximum SPR is below 0.2 for a phantom with a diameter of 30 cm, whereas in a CBCT projection (with ASG in place), maximum SPR is around 1 for a phantom with similar dimensions (Endo et al. 2006; Kyriakou and Kalender 2007a; Vogtmeier et al. 2008).

Although scattered radiation dominates the errors in attenuation measurements in CBCT, effects of image lag and beam hardening are nonnegligible. Image lag associated with increased charge trapping in a-Si introduces a bias to log-attenuation measurements in FPD-based CBCT systems. Artifacts can be pronounced depending on the imaged object and imaging system parameters. Software-based recursive filtering techniques and hardware-based methods to reduce charge trapping have been proposed to reduce the effects of image lag. Beam hardening in CBCT is also a problem as in MDCT imaging, and first-order (i.e., water equivalent thickness based) corrections have been proposed.

The task of finding a robust scatter rejection or correction method requires a balancing act that leads to a broad range of proposed solutions. The ideal solution to this problem would be a scatter rejection method that reduces scatter intensity before it is "injected" into the image signal. As a result, both CT number accuracy and CNR would be improved. Scatter rejection methods in radiography, such as air gap and ASG, have been used in CBCT; they reduce image artifacts and improve CT number accuracy. Also, bow-tie filters reduce the SPR by 30% to 50% and improve CNR, particularly in the central section of the object. Unfortunately, their performance is still not sufficient to reduce the scatter fraction to levels observed in MDCT systems. To reduce scatter fractions and improve the CNR further, the FOV can be limited to a small, clinically relevant region. The VOI imaging concept exploits this approach, and it uses two CBCT scans, one scan visualizes a small region with high-contrast sensitivity and the other scan visualizes the anatomy surrounding the VOI in a composite CBCT image. This is still a relatively new technique, and its clinical implementation remains to be investigated. A summary of scatter rejection methods is listed in Table 10.1.

Table 10.1 Summary of scatter rejection methods

SCATTER REJECTION METHOD	ADVANTAGES	CONCERNS
Air gap	Simpler to implement in CBCT. Both artifact reduction and increased CNR are achieved.	Has limited efficiency for >50-cm air gap. Other system parameters limit the use of air gaps.
Antiscatter grids	Useful for reducing shading artifacts. Improved CNR in high SPR imaging conditions.	Attenuates the primary beam. Does not improve CNR in low-to-moderate SPR environments.
Bow-tie filters	Already used in CBCT scanners to reduce the detector exposure range and skin dose. Can be further optimized to reduce SPR and improve CNR.	The ideal shape of the bow-tie filter depends on the patient specific imaging geometry.
VOI techniques	An order of magnitude reduction in SPR and improvement in CNR can be achieved in a small region.	Requires VOI images to be combined with large FOV CBCT images. Works only in a limited FOV.

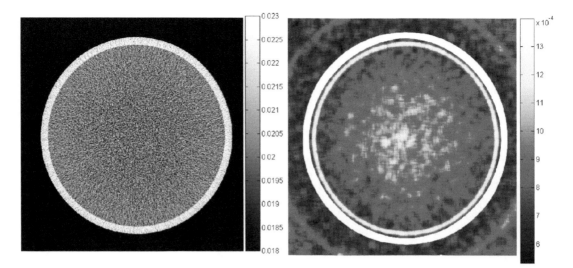

Figure 4.2 (Left) Axial CBCT image of a uniform water phantom. (Right) Map of the noise (standard deviation in voxel values). Note the variation in noise magnitude and correlation within the cylindrical enclosure between the center and edge of the phantom. In this case, the phantom was imaged without a bow-tie filter, so the nonstationarity in quantum noise (i.e., higher noise at the center of the image) is attributed to variation in fluence incident on the detector between the center and edge. The colorbar shows voxel value (or noise) in units of attenuation coefficient (mm^{-1}).

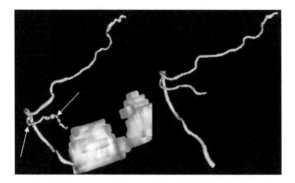

Figure 7.17 Normalization of vessel shape during radiation treatment. (Left) 3D rendering of the segmented tumors (gray) and of the basilar, left superior cerebellar, and left posterior cerebral arteries at baseline as shown from a lateral point of view. Note the abnormal tortuosity of the left superior cerebellar artery (arrows). The posterior cerebral artery is also abnormal, although to a lesser degree. (Right) Rendering of the same vessels from the same point of view at month 2. Note the arterial "straightening." (From figure 2 of Bullitt, E., et al., *Radiology*, 245, 824–30, 2007; Burger, W. and Burge, M.J., *Digital Image Processing: An Algorithmic Introduction Using Java*, Springer, Berlin, 2008. With permission; Courtesy of Elizabeth Bullitt, University of North Carolina.)

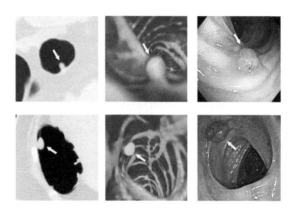

Figure 7.19 Example of identification of polyps in CT colonographic data. In each row, the left image shows an axial CT image of a polyp (arrow), the middle image shows a computer-generated 3D endoluminal view that is color-coded based on the shape index values in Figure 7.18, and the right image shows a colonoscopy image of the polyp. Top row: 7-mm polyp in the transverse colon. Bottom row: 11-mm sessile polyp in the hepatic flexure. The polyp (green) is clearly differentiated from the folds (pink) and the colonic wall (brown). (Courtesy of Hiro Yoshida, Harvard University.)

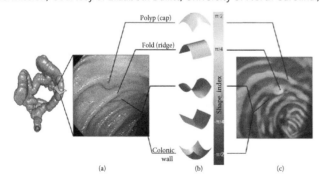

Figure 7.18 Schematic illustration of the morphologic characterization of the structures in the colonic lumen. (a) Colonoscopy image of a 6-mm polyp from endoscopy. (b) Morphologic characterization of the structures in the colonic lumen based on the shape index. (c) Color coding of the colonic lumen in a computer-generated endoscopic view based on the shape index value, in which polyps, folds, and the colonic wall are shown in green, pink, and brown, respectively. (From Perumpillichira, J.J., et al., *Canc Imag*, 5, 11–16, 2005. With permission; Courtesy of Hiro Yoshida, Harvard University.)

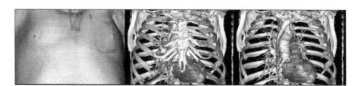

Figure 7.28 Volume rendering of the thorax studies shown in Figure 7.26. (Left) Opacity was chosen so that tissue was opaque, allowing visualization of the skin. (Middle) high voxel values are opaque, allowing visualization of the bone [note the difference between the volume-rendered and MIP-rendered image (see Figure 7.26)]. (Right) LUT of the middle image was used but a cut-plane was placed so as to allow visualization of the inside ribcage.

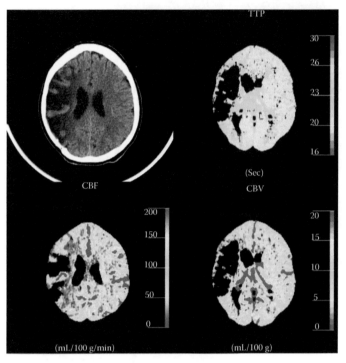

Figure 7.32 Evaluation of cerebral perfusion. In the color scale for the images, red indicates good perfusion and blue indicates poor perfusion. (Top left) CT slice. (Top right) Time to peak (TTP). (Bottom left) Cerebral blood flow. (Bottom right) Cerebral blood volume. The regions of deficit correlate fairly well with each parameter providing additional information to the clinician regarding the perfusion of the brain.

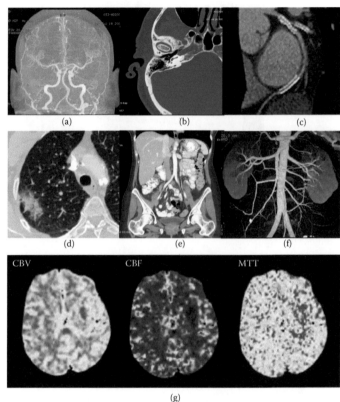

Figure 11.22 Typical clinical application of MDCT imaging. (a) Head CT angiography. (b) Temporal bone. (c) Coronal artery stent. (d) Lung cancer. (e) Abdominal/pelvic. (f) Renal angiography. (g) CT perfusion for the evaluation of acute stroke (images in g are courtesy of GE Healthcare, Buckinghamshire, UK, http://www.gehealthcare.com/euen/ct/pdf/CTClarity2009_Spring.pdf, accessed on 09/28/2011.)

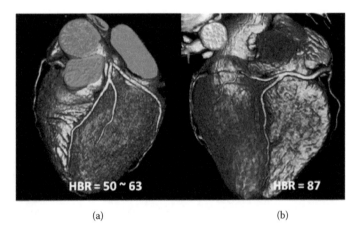

Figure 11.19 3D surface rendering of the heart generated by a single-source single-detector MDCT (a) and a dual-source dual-detector MDCT (b) (Courtesy of Siemens Healthcare, Malvern, PA.)

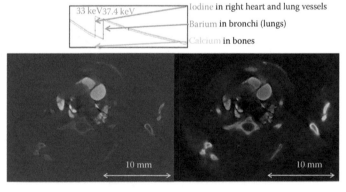

Figure 12.6 (Left) CT image of a transaxial cross section of a mouse thorax. The bronchial tree had barium sulfate infused, and the pulmonary artery had an I-based contrast injected. These different materials and the skeletal features cannot be distinguished unambiguously on the basis of their CT gray scale values. (Right) Use of principal component analysis by virtue of the ability to extract the different x-ray photon energy components from the bremsstrahlung x-ray exposure allowed identification and quantitation of the three elements by virtue of their different attenuation-to-photon energy relationships, as illustrated in the top panel (From Anderson, N.G. et al., *Eur Radiol B*, 20, 2126–2134. With permission.)

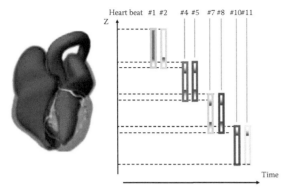

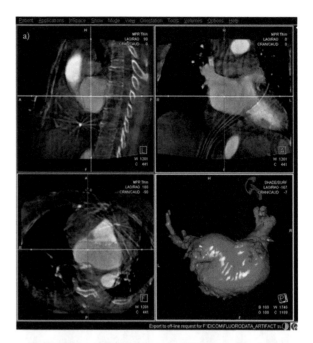

Figure 13.6 A dual-sector method as opposed to a single-sector method shown in Figure 13.2. Two sectors are acquired at the same patient table position with about a half of the time window width for each sector.

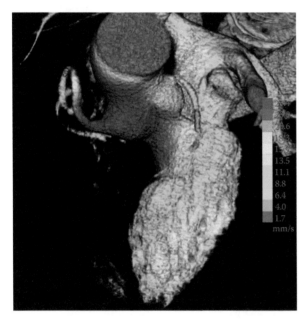

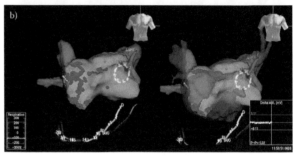

Figure 14.7 (a) Segmentation of the volume of the left atrium from a single sweep C-arm CT acquisition during injection of iodinated contrast into the main pulmonary artery. (b) Overlay of C-arm CT-derived segmentation (red/gray), a segmentation from a prior clinical CT (blue), and points from the electromagnetic tracking system (red dots), indicating locations where radiofrequency ablation has been applied. Continuity of ablation points around the pulmonary veins can be monitored during the procedure.

Figure 13.12 Velocity map fused on the surface of endocardium. (Courtesy of Ziosoft, Rodwood City, CA.)

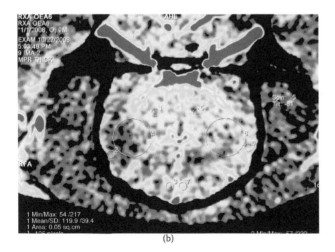

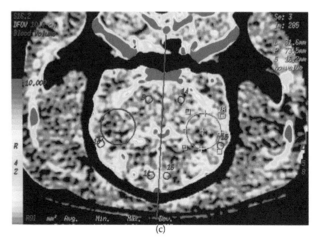

Figure 14.3 (b) C-arm CT CBV image acquired during steady-state injection of iodinated contrast at 1.5 mL per s shows ~30% difference in CBV between right and left sides of the brain. (c) Clinical CT CBV image shows ~45% difference in CBV between right and left sides of the brain. (Images courtesy of Dr. Charles M. Strother, Department of Radiology, University of Wisconsin–Madison.)

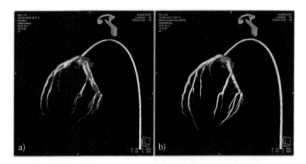

Figure 14.11 Reconstructions using (a) standard ECG-gating and (b) FDK-based reconstruction coupling 3D motion estimation and reconstruction (5-s single sweep acquisition, 4 × 4 binning, 60 fps, coronary injection with a 3-s delay before start of acquisition). (Images courtesy of Dr. Christopher Rohkohl, Medical Image Reconstruction group, Pattern Recognition Lab, Friedrich-Alexander University Erlangen-Nuremberg, Germany.)

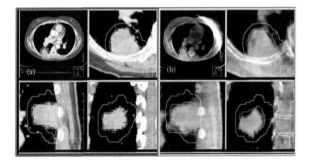

Figure 15.12 Axial and coronal planes of the tumor for a case of lung SBRT in the planning CT (a, red) and from the daily CBCT (b, yellow). Despite the motion artifacts in the CBCT, the information can be used to for accurately positioning the tumor with respect to the PTV (green).

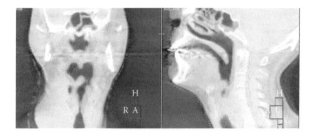

Figure 15.13 Example where alignment of the nasopharyngeal structures results in misalignment of the cervical spine and variations in the skin contour in a head-and-neck cancer treatment.

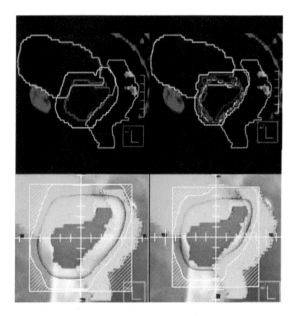

Figure 15.15 PTV prescribed at treatment planning (left) is modified (right) during the course of prostate cancer treatment.

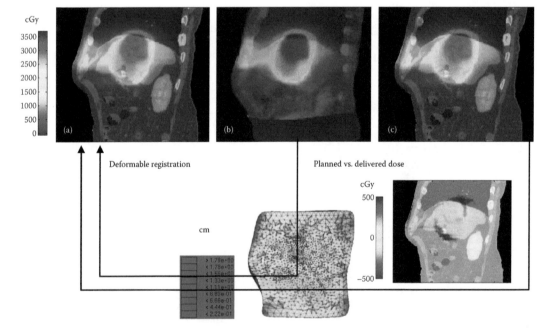

Figure 15.14 Ability to deform anatomic structures is critical for accurate dose estimation during adaptation of treatment plans. Shown is a panel of treatment plans in the sagittal plane representing the prescribed dose distribution on the planning CT (a), the delivered dose distribution based on the CBCT of the day (b), and the delivered dose distribution deformed onto the original planning CT (c) (Courtesy of K. Brock, M. Velec, and L. Dawson, Princess Margaret Hospital).

As opposed to scatter rejection, scatter correction refers to suppressing the effects of scattered radiation after its detection. Scatter correction methods primarily aim to improve quantitative accuracy and reduce image artifacts rather than improve CNR. An ideal scatter correction method should be able to correct scatter accurately, preferably in real time, and its performance should be consistent under a wide range of imaging conditions. In practice, a balance among these requirements is achievable and leads to a spectrum of proposed scatter correction algorithms. A summary of scatter correction methods is given in Table 10.2.

Scatter measurement methods, such as beam stop arrays and multiple-slit assemblies, have been widely investigated for patient-specific scatter corrections. They are accurate and consistent under a broad range of imaging situations. However, their clinical implementation is challenging; typical scatter measurement methods require acquisition of redundant projections that may increase scan time and patient dose. Furthermore, development and integration of scatter measurement hardware may complicate the CBCT system design. Several novel methods, such as primary modulation method, have been proposed to eliminate the shortcomings of typical measurement-based methods. Integration of measurement-based methods in CBCT systems still remains to be demonstrated. However, such methods will remain as the standard to benchmark other scatter correction methods and to study the effects of scattered radiation.

Software-based scatter corrections can potentially overcome the shortcomings of measurement-based corrections; they do not require hardware integration, and corrections can be performed without redundant image sets. The challenge in software-based corrections lies in finding a balance between scatter correction accuracy and computational cost. Among such corrections, image processing and empirical model–based methods lean toward increasing the computational efficiency at the expense of reduced correction accuracy. The former method mostly relies on the detection of scatter-induced image artifacts in CT image domain. The latter method, empirical model–based corrections, aim to characterize scatter distribution using simple parameterized models, such as handling of scatter signal as constant fraction of total image signal. Empirical models have the same pitfall as image processing methods; a simple model of scatter distribution leads to under- or overcorrection of scatter artifacts under a range of imaging conditions. Both image processing and empirical model–based corrections serve as interim solutions, until a more accurate and comprehensive method is developed for scatter correction.

Currently, pencil beam–based SKS algorithms bring together the best compromise between robust, physics-based scatter corrections and computation speed. SKS algorithms are implemented in the projection image domain, and scatter kernels are based on pencil primary beams that traverse the object. SKS algorithms are also suitable for detector glare correction, which is a form of scatter originating from FPD. Many proposed SKS algorithms for object scatter correction use parameterized, stationary kernels to take advantage of fast convolution operations in Fourier domain. Although stationary kernels can greatly improve computation speed and can address the effect of thickness variations to some degree, they do not accurately handle the effect of 3D variations in object shape. Now, parameterized and nonstationary scatter kernels have been developed to address these issues. Alternatively, scatter kernels can be stored in lookup tables that can offer better selection of kernels based on object shape and composition. However, their suitability for real-time

Table 10.2 Summary of scatter correction methods

SCATTER CORRECTION METHOD	ADVANTAGES	CONCERNS
Measurements based on beam-stop arrays or multihole collimators	Direct measurement of patient-specific scatter signal. Computationally not demanding.	Generally requires acquisition of redundant images and may increase image acquisition time. Also requires development and integration of scatter measurement hardware.
Pencil beam based kernel superposition in projection image domain	Applicable to a wide range of imaging conditions. Currently, it offers the best compromise between computation speed and scatter correction accuracy.	Does not fully account for variation in scatter due to object shape and 3D spatial distribution of heterogeneities.
Patient-specific MC calculations	Reconstructed object image is used for scatter calculation that accounts for 3D variations in heterogeneities and object shape. Potentially very accurate.	Requires iterative calculations in CT image domain. It is still computationally very expensive.
Analytical scatter calculations	Accurate calculation of single scatter signal using scattering differential cross sections. General enough and faster than MC-based methods.	Correction of multiple scatter has to be handled using a separate method.
Postreconstruction image processing	Fast correction of image artifacts. May not require measurements or calibration data.	Applicable to specific imaging tasks. True CT numbers may not be fully recovered.
Empirical model–based corrections	Simple and fast scatter correction, using parameterized object models, scatter models, or both.	Applicable to specific imaging tasks. Scatter correction performance is limited due to relatively simple scatter models.

Advanced techniques

corrections remains unaddressed. Even with improved choice of scatter kernels, performance of projection image domain–based corrections would be limited due to lack of 3D information about object shape and density.

On the high end of the scatter correction spectrum, CT image domain–based methods, such as MC calculations, offer further improvement in scatter correction. Because CBCT image itself is used as object model to account for the effect of 3D density and object shape variations, they are potentially the most accurate software-based correction methods. However, CT image domain-based methods are inherently slower; scatter distribution is calculated in forward projections of the 3D object model, and an updated object model is reconstructed from scatter-corrected projections in an iterative way. MC-based calculations are being increasingly used for CT image domain corrections due to the accurate handling of scatter physics. MC corrections are not suitable for real-time applications as yet, but the development of variance reduction algorithms and advances in hardware-based acceleration will eventually make their use viable for practical applications. Also, analytical calculations are powerful in speed and accuracy in estimating the spatial distribution of single scatter. However, their usefulness depends on whether multiple scatter can be accurately estimated using alternative methods.

The disadvantage of scatter correction methods is the lack of improvement in CNR, which particularly hampers the visibility of low-contrast objects. As discussed in Section 10.3, correction of scatter restores the degraded contrast, but at the same time, it enhances the image noise. Suppression of noise plays an important role in improving CNR, because enhancement of contrast is limited by the true contrast in the imaged object. The use of typical linear image processing filters is not preferred as high-frequency details in the image, such as edges, are also smoothed along with image noise. Wavelet-based filters can preserve edges and suppress image noise (Zhong et al. 2004; Borsdorf et al. 2008), but such methods require manual tuning of the smoothing algorithm based on the features of the imaged object. Another possible solution to the noise suppression problem is to use statistical image estimation methods. Zhu et al. (2009) developed a penalized weighted least squares algorithm to suppress noise and predict ideal primary image signal after scatter correction, while preserving spatial resolution at high frequencies. In radiography, maximum likelihood and Bayesian image estimation algorithms that constrained noise during scatter correction were developed, and significant improvements in CNR were reported. Ruhrnschopf et al. suggested that these approaches can be extended into CBCT imaging (Floyd et al. 1993; 1994; Baydush and Floyd 1995; Ogden et al. 2002; Ruhrnschopf and Klingenbeck 2011b). On the downside, statistical methods are computationally expensive due to iterative image estimation and computational complexity, require tuning of free parameters, and an accurate scatter estimation model is still required. The future of CBCT scatter corrections may see more progress in statistical image estimation methods; they potentially offer scatter physics–based suppression of noise and preservation of spatial resolution as opposed to an approach involving brute force in typical subtractive or multiplicative scatter correction methods that amplify image noise.

10.7 CONCLUSIONS

The major advantage of FPD-based CBCT scanners is their ability to reconstruct relatively large imaging volumes with a single circular gantry rotation. At the same time, this is the root cause behind the image quality issues with such systems. High fraction of scatter in projections is one of the leading problems to be solved to improve the image quality of CBCT. Scatter rejection methods are, in principle, the ultimate solution to this problem, because they can potentially improve both low-contrast sensitivity and restore the true CT numbers. Conventional scatter rejection techniques adopted from projection radiography have been investigated extensively for use with CBCT systems. Unless a revolutionary new approach is introduced, such techniques do not provide sufficient scatter rejection to reduce image artifacts and to improve low-contrast sensitivity.

The lack of adequate scatter rejection methods may have motivated the development of various scatter correction methods. These methods aim to recover true CT numbers and reduce image artifacts after detection of scattered radiation. At the postdetection stage, stochastic noise from scatter detection cannot be removed; hence, the CNR improvement is limited with the scatter correction methods. Currently, scatter physics–based corrections in projection image domain, such as pencil beam scatter kernel superposition techniques, offer a balance between accuracy and ease of practical implementation. However, scatter corrections in CT image domain, such as MC calculations, offer potentially the most accurate software-based solutions. These methods suffer from low computational efficiency at this point.

For now, scatter in CBCT remains a challenging problem in image quality. Although it may not be possible for FPD-based CBCT to reach the low-contrast sensitivity of conventional MDCT systems, advances in scatter rejection/correction techniques would help to narrow the gap in image quality and quantitative accuracy between the two modalities.

ACKNOWLEDGMENTS

I thank Dimitrios Lazos (Beth Israel Medical Center, NY) for valuable suggestions during the preparation of this chapter.

REFERENCES

Abella, M., Vaquero, J.J., Sisniega, A., et al. 2012. Software architecture for multi-bed FDK-based reconstruction in X-ray CT scanners. *Comput Meth Programs Biomed* 107: 218–32.

Altunbas, C., Shaw, C.C., Chen, L., et al. 2007. A post-reconstruction method to correct cupping artifacts in cone beam breast computed tomography. *Med Phys* 34: 3109–18.

Altunbas, M., Shaw, C. & Chen, L. 2006. TH-E-330A-03: reduction of ring artifacts in cone beam CT: artifact detection and correction for flat panel imagers. *Med Phys* 33: 2287–8.

Anas, E.M.A., Kim, J.G., Lee, S.Y. and Hasan, M.K. 2011. High-quality 3D correction of ring and radiant artifacts in flat panel detector-based cone beam volume CT imaging. *Phys Med Biol* 56: 6495.

Badal, A. and Badano, A. 2009. Accelerating Monte Carlo simulations of photon transport in a voxelized geometry using a massively parallel graphics processing unit. *Med Phys* 36: 4878–80.

Bani-Hashemi, A., Blanz, E., Maltz, J., Hristov, D. and Svatos, M. 2005. TU-D-I-611-08: cone beam X-ray scatter removal via image frequency modulation and filtering. *Med Phys* 32: 2093.

Baydush, A.H. and Carey E. Floyd, J. 1995. Bayesian image estimation of digital chest radiography: interdependence of noise, resolution, and scatter fraction. *Med Phys* 22: 1255–61.

Benitez, R.B. and Ning, R. 2010. Development of a beam hardening filter and characterization of the spatial resolution for a cone beam CT imaging system. In: Ehsan, S. and Norbert, J.P. (eds.), *SPIE Proceedings* 762237, Bellingham, WA.

Bertram, M., Sattel, T., Hohmann, S. and Wiegert, J. 2008. Monte-Carlo scatter correction for cone-beam computed tomography with limited scan field-of-view. In: Jiang, H. and Ehsan, S. (eds.) *SPIE Proceedings* 69131Y, Bellingham, WA.

Bertram, M., Wiegert, J. and Rose, G. 2006. Scatter correction for cone-beam computed tomography using simulated object models. In: Flynn, M. and Hsieh, J. (eds.). Bellingham, WA: SPIE, 61421C.

Boone, J.M., Arnold, B.A. and Seibert, J.A. 1986. Characterization of the point spread function and modulation transfer function of scattered radiation using a digital imaging system. *Med Phys* 13: 254–6.

Boone, J.M., Makarova, O.V., Zyryanov, V.N. et al., 2002. Development and Monte Carlo analysis of antiscatter grids for mammography. *Technol Cancer Res Treat* 1(6): 441–7.

Boone, J.M., Nelson, T.R., Lindfors, K.K. and Seibert, J.A. 2001. Dedicated breast CT: radiation dose and image quality evaluation. *Radiology* 221: 657–67.

Bootsma, G.J., Verhaegen, F. and Jaffray, D.A. 2011. The effects of compensator and imaging geometry on the distribution of x-ray scatter in CBCT. *Med Phys* 38: 897–914.

Borsdorf, A., Raupach, R., Flohr, T. and Hornegger, J. 2008. Wavelet based noise reduction in CT-images using correlation analysis. *IEEE Trans Med Imaging* 27: 1685–703.

Brooks, R.A. and DI Chiro, G. 1976. Beam hardening in x-ray reconstructive tomography. *Phys Med Biol* 21: 390–8.

Bucky, G. 1913. Über die Ausschaltung der im Objekt entstehenden Sekundärstrahlen bei Röntgenstrahlen. *Verh d Dt Röntgengesellschaft* 9: 30–2.

Buzug, T. 2008. *Computed Tomography: From Photon Statistics to Modern Cone-Beam CT*. Berlin: Springer.

Cai, W., Ning, R. and Conover, D. 2008. Simplified method of scatter correction using a beam-stop-array algorithm for cone-beam computed tomography breast imaging. *Opt Eng* 47: 97003.

Chan, H.P. and Doi, K. 1982. Investigation of the performance of antiscatter grids: Monte Carlo simulation studies. *Phys Med Biol* 27: 785–803.

Chan, H.P. and Doi, K. 1983. The validity of Monte Carlo simulation in studies of scattered radiation in diagnostic radiology. *Phys Med Biol* 28: 109–29.

Chen, L., Shaw, C.C., Altunbas, M.C., et al. 2008. Feasibility of volume-of-interest (VOI) scanning technique in cone beam breast CT—a preliminary study. *Med Phys* 35: 3482–90.

Chen, Y., Liu, B., O'connor, J.M., Didier, C.S. and Glick, S.J. 2009. Characterization of scatter in cone-beam CT breast imaging: comparison of experimental measurements and Monte Carlo simulation. *Med Phys* 36: 857–69.

Chityala, R., Hoffmann, K.R., Bednarek, D.R. and Rudin, S. 2004. Region of interest (ROI) computed tomography. *Proc SPIE* 5368: 534–41.

Colijn, A.P. and Beekman, F.J. 2004. Accelerated simulation of cone beam x-ray scatter projections. *IEEE Trans Med Imaging* 23: 584–90.

Crotty, D.J., Mckinley, R.L. and Tornai, M.P. 2006. Experimental spectral measurements of heavy K-edge filtered beams for x-ray computed mammotomography. *Proc SPIE*: 61421V.

Endo, M., Mori, S., Tsunoo, T. and Miyazaki, H. 2006. Magnitude and effects of x-ray scatter in a 256-slice CT scanner. *Med Phys* 33: 3359–68.

Engel, K.J., Herrmann, C. and Zeitler, G. 2008. X-ray scattering in single- and dual-source CT. *Med Phys* 35: 318–32.

Feldkamp, L., Davis, L. and Kress, J. 1984. Practical cone-beam algorithm. *J Opt Soc Am A* 1: 612–19.

Floyd, Jr., C.E., Baydush, A.H., Lo, J.Y., Bowsher, J.E. and Ravin, C.E. 1993. Scatter compensation for digital chest radiography using maximum likelihood expectation maximization. *Invest Radiol* 28: 427–33.

Floyd, Jr., C.E., Baydush, A.H., Lo, J.Y., Bowsher, J.E. and Ravin, C.E. 1994. Bayesian restoration of chest radiographs. Scatter compensation with improved signal-to-noise ratio. *Invest Radiol* 29: 904–10.

Gao, H., Fahrig, R., Bennett, N.R., et al. 2010. Scatter correction method for x-ray CT using primary modulation: phantom studies. *Med Phys* 37: 934–46.

Glover, G.H. 1982. Compton scatter effects in CT reconstructions. *Med Phys* 9: 860–7.

Graham, S.A., Moseley, D.J., Siewerdsen, J.H. and Jaffray, D.A. 2007. Compensators for dose and scatter management in cone-beam computed tomography. *Med Phys* 34: 2691–703.

Granfors, P.R., Aufrichtig, R., Possin, G.E., et al. 2003. Performance of a 41 x 41 cm[sup 2] amorphous silicon flat panel x-ray detector designed for angiographic and R&F imaging applications. *Med Phys* 30: 2715–26.

Grimmer, R. and Kachelriess, M. 2011. Empirical binary tomography calibration (EBTC) for the precorrection of beam hardening and scatter for flat panel CT. *Med Phys* 38: 2233–40.

Hansen, V.N., Swindell, W. and Evans, P.M. 1997. Extraction of primary signal from EPIDs using only forward convolution. *Med Phys* 24: 1477–84.

Herman, G.T. 1979. Correction for beam hardening in computed tomography. *Phys Med Biol* 24: 81–106.

Hsieh, J. 2009. *Computed Tomography: Principles, Design, Artifacts, and Recent Advances*. Bellingham, WA: SPIE.

Hsieh, J., Gurmen, O.E. and King, K.F. 2000a. Recursive correction algorithm for detector decay characteristics in CT. In: James, D. and John, B. (eds.), *SPIE Proceedings* 614254, April 25, Bellingham, WA, 298–305.

Hsieh, J., Molthen, R.C., Dawson, C.A. and Johnson, R.H. 2000b. An iterative approach to the beam hardening correction in cone beam CT. *Med Phys* 27: 23–9.

Ingleby, H.R., Elbakri, I.A., Rickey, D.W. and Pistorius, S. 2009. Analytical scatter estimation for cone-beam computed tomography. In: Samei, E. and Jiang, H. (eds.), *SPIE Proceedings* 725839, Bellingham, WA.

Jaffray, D.A., Siewerdsen, J.H., Wong, J.W. and Martinez, A.A. 2002. Flat-panel cone-beam computed tomography for image-guided radiation therapy. *Int J Radiat Oncol Biol Phys* 53: 1337–49.

Jarry, G., Graham, S.A., Jaffray, D.A., Moseley, D.J. and Verhaegen, F. 2006a. Scatter correction for kilovoltage cone-beam computed tomography (CBCT) images using Monte Carlo simulations. In: Michael, J.F. and Jiang, H. (eds.), *SPIE Proceedings* 614254, Bellingham, WA.

Jarry, G., Graham, S.A., Moseley, D.J., et al. 2006b. Characterization of scattered radiation in kV CBCT images using Monte Carlo simulations. *Med Phys* 33: 4320–9.

Jia, X., Yan, H., Cervino, L., Folkerts, M. and Jiang, S.B. 2012. A GPU tool for efficient, accurate, and realistic simulation of cone beam CT projections. *Med Phys* 39: 7368–78.

Johns, P.C. and Yaffe, M.J. 1983. Coherent scatter in diagnostic radiology. *Med Phys* 10: 40–50.

Kachelriess, M., Sourbelle, K. and Kalender, W.A. 2006. Empirical cupping correction: a first-order raw data precorrection for cone-beam computed tomography. *Med Phys* 33: 1269–74.

Kak, A. and Slaney, M. 1988. *Principles of Computerized Tomographic Imaging*. New York: IEEE Press.

Kalender, W. 1981. Monte Carlo calculations of x-ray scatter data for diagnostic radiology. *Phys Med Biol* 26: 83–549.

Kalender, W.A. 1982. Calculation of x-ray grid characteristics by Monte Carlo methods. *Phys Med Biol* 27: 353–61.

Kawata, Y., Niki, N. and Kumazaki, T. 1996. 3-D image reconstruction with veiling glare correction to improve the contrast of 3-D reconstructed vascular images. *IEEE Trans Nucl Sci* 43: 304–9.

Kijewski, P.K. and Bjarngard, B.E. 1978. Correction for beam hardening in computed tomography. *Med Phys* 5: 209–14.

Kolditz, D., Kyriakou, Y. and Kalender, W.A. 2010. Volume-of-interest (VOI) imaging in C-arm flat-detector CT for high image quality at reduced dose. *Med Phys* 37: 2719–30.

Kruger, D.G., Zink, F., Peppler, W.W., ERGUN, D.L. and Mistretta, C.A. 1994. A regional convolution kernel algorithm for scatter correction in dual-energy images: comparison to single-kernel algorithms. *Med Phys* 21: 175–84.

Kwan, A.L., Boone, J.M. and Shah, N. 2005. Evaluation of x-ray scatter properties in a dedicated cone-beam breast CT scanner. *Med Phys* 32: 2967–75.

Kyriakou, Y. and Kalender, W. 2007a. Efficiency of antiscatter grids for flat-detector CT. *Phys Med Biol* 52: 6275–93.

Kyriakou, Y. and Kalender, W.A. 2007b. X-ray scatter data for flat-panel detector CT. *Phys Med Biol* 23: 3–15.

Kyriakou, Y., Meyer, M., Lapp, R. and& Kalender, W. A. 2010b. Histogram-driven cupping correction (HDCC) in CT. *Proc. SPIE* 7622: 76221S.

Kyriakou, Y., Meyer, E., Prell, D. and Kachelriess, M. 2010a. Empirical beam hardening correction (EBHC) for CT. *Med Phys* 37: 5179–87.

Kyriakou, Y., Riedel, T. and Kalender, W.A. 2006. Combining deterministic and Monte Carlo calculations for fast estimation of scatter intensities in CT. *Phys Med Biol* 51: 4567–86.

Lai, C.J., Chen, L., Zhang, H., et al. 2009. Reduction in x-ray scatter and radiation dose for volume-of-interest (VOI) cone-beam breast CT—a phantom study. *Phys Med Biol* 54: 6691–709.

Lazos, D. and Williamson, J.F. 2010. Monte Carlo evaluation of scatter mitigation strategies in cone-beam CT. *Med Phys* 37: 5456–70.

Lazos, D. and Williamson, J.F. 2012. Impact of flat panel-imager veiling glare on scatter-estimation accuracy and image quality of a commercial on-board cone-beam CT imaging system. *Med Phys* 39: 5639–51.

Li, H., Mohan, R. and Zhu, X.R. 2008. Scatter kernel estimation with an edge-spread function method for cone-beam computed tomography imaging. *Phys Med Biol* 53(23): 6729–48.

Liu, B., Glick, S.J. and Groiselle, C. 2005a. Characterization of scatter radiation in cone beam CT mammography. In: Michael, J.F. (ed.), *SPIE Proceedings*, Bellingham, WA, 818–27.

Liu, X., Shaw, C., Altunbas, M. and Wang, T. 2005b. TU-D-I-611-07: a scanning sampled measurement (SSM) technique for scatter measurement and correction in cone beam breast CT. *Med Phys* 32: 2093.

Liu, X., Shaw, C.C., Altunbas, M.C. and Tianpeng, W. 2006a. An alternate line erasure and readout (ALER) method for implementing slot-scan imaging technique with a flat-panel detector-initial experiences. *IEEE Trans Med Imaging* 25: 496–502.

Liu, X., Shaw, C.C., Wang, T., et al. 2006b. An accurate scatter measurement and correction technique for cone beam breast CT imaging using scanning sampled measurement (SSM) technique. *Proc SPIE* 6142: 6142341–7.

Love, L.A. and Kruger, R.A. 1987. Scatter estimation for a digital radiographic system using convolution filtering. *Med Phys* 14: 178–85.

Mail, N., Moseley, D.J., Siewerdsen, J.H. and Jaffray, D.A. 2008. An empirical method for lag correction in cone-beam CT. *Med Phys* 35: 5187–96.

Mail, N., Moseley, D.J., Siewerdsen, J.H. and Jaffray, D.A. 2009. The influence of bowtie filtration on cone-beam CT image quality. *Med Phys* 36: 22–32.

Mainegra-Hing, E. and Kawrakow, I. 2008. Fast Monte Carlo calculation of scatter corrections for CBCT images. *J PhysConf* 102: 12017.

Mainegra-Hing, E. and Kawrakow, I. 2010. Variance reduction techniques for fast Monte Carlo CBCT scatter correction calculations. *Phys Med Biol* 55: 4495–507.

Maltz, J.S., Gangadharan, B., Bose, S., et al. 2008. Algorithm for x-ray scatter, beam-hardening, and beam profile correction in diagnostic (kilovoltage) and treatment (megavoltage) cone beam CT. *IEEE Trans Med Imaging* 27: 1791–810.

Malusek, A., Sandborg, M.P. and Carlsson, G.A. 2003. Simulation of scatter in cone beam CT: effects on projection image quality. *Proc SPIE* 5030: 740–51.

Marchant, T.E., Moore, C.J., Rowbottom, C.G., Mackay, R.I. and Williams, P.C. 2008. Shading correction algorithm for improvement of cone-beam CT images in radiotherapy. *Phys Med Biol* 53(20): 5719–33.

Matsinos, E. 2005. Current status of the CBCT project at Varian Medical Systems. In: Michael, J.F. (ed.), *SPIE Proceedings*, Bellingham, WA. 340–51.

Menser, B., Wiegert, J., Wiesner, S. and Bertram, M. 2010. Use of beam shapers for cone-beam CT with off-centered flat detector. In: Ehsan, S. and Norbert, J.P. (eds.), *SPIE Proceedings* 762233, Bellingham, WA.

Meyer, M., Kalender, W.A. and Kyriakou, Y. 2010. A fast and pragmatic approach for scatter correction in flat-detector CT using elliptic modeling and iterative optimization. *Phys Med Biol* 55: 99–120.

Naimuddin, S., Hasegawa, B. and Mistretta, C.A. 1987. Scatter-glare correction using a convolution algorithm with variable weighting. *Med Phys* 14: 330–4.

Neitzel, U. 1992. Grids or air gaps for scatter reduction in digital radiography: a model calculation. *Med Phys* 19: 475–81.

Niklason, L.T., Sorenson, J.A. and Nelson, J.A. 1981. Scattered radiation in chest radiography. *Med Phys* 8: 677–81.

Ning, R., Tang, X. and Conover, D. 2004. X-ray scatter correction algorithm for cone beam CT imaging. *Med Phys* 31: 1195–202.

Ning, R., Tang, X. and Conover, D.L. 2002. X-ray scatter suppression algorithm for cone-beam volume CT. In: Larry, E.A. and Martin, J.Y. (eds.), *SPIE Proceedings*, Bellingham, WA. 774–81.

Niu, T., Sun, M., Star-Lack, J., et al. 2010. Shading correction for on-board cone-beam CT in radiation therapy using planning MDCT images. *Med Phys* 37: 5395–406.

Niu, T. and Zhu, L. 2011. Single-scan scatter correction for cone-beam CT using a stationary beam blocker: a preliminary study. In: Norbert, J.P., Ehsan, S. and Robert, M.N. (eds.), *SPIE Proceedings* 796126, Bellingham, WA.

Ogden, K.M., Wilson, C.R. and Cox, R.W. 2002. Contrast improvements in digital radiography using a scatter-reduction processing algorithm. *Proc. SPIE* 4684: 1034–47.

Ohnesorge, B., Flohr, T. and Klingenbeck-Regn, K. 1999. Efficient object scatter correction algorithm for third and fourth generation CT scanners. *Eur Radiol* 9: 563–9.

Petersilka, M., Stierstorfer, K., Bruder, H. and Flohr, T. 2010. Strategies for scatter correction in dual source CT. *Med Phys* 37: 5971–92.

Petrov, I.E., Nikolov, H.N., Holdsworth, D.W. and Drangova, M. 2011. Image performance evaluation of a 3D surgical imaging platform. *Proc SPIE* 7961: 7961501--6.

Poludniowski, G., Evans, P.M., Kavanagh, A. and Webb, S. 2011. Removal and effects of scatter-glare in cone-beam CT with an amorphous-silicon flat-panel detector. *Phys Med Biol* 56(6): 1837–51. England.

Prell, D., Kyriakou, Y. and Kalender, W.A. 2009. Comparison of ring artifact correction methods for flat-detector CT. *Phys Med Biol* 54(12): 3881–95. England.

Advanced techniques

Rinkel, J., Gerfault, L., Esteve, F. and Dinten, J.M. 2007. A new method for x-ray scatter correction: first assessment on a cone-beam CT experimental setup. *Phys Med Biol* 52: 4633–52.

Ruhrnschopf, E.P. and Klingenbeck, A.K. 2011a. A general framework and review of scatter correction methods in cone beam CT. Part 2: scatter estimation approaches. *Med Phys* 38: 5186–99.

Ruhrnschopf, E.P. and Klingenbeck, K. 2011b. A general framework and review of scatter correction methods in x-ray cone-beam computerized tomography. Part 1: scatter compensation approaches. *Med Phys* 38: 4296–311.

Schafer, S., Stayman, J.W., Zbijewski, W., et al. 2012. Antiscatter grids in mobile C-arm cone-beam CT: effect on image quality and dose. *Med Phys* 39: 153–9.

Schmidgunst, C., Ritter, D. and Lang, E. 2007. Calibration model of a dual gain flat panel detector for 2D and 3D x-ray imaging. *Med Phys* 34: 3649–64.

Seibert, J.A. and Boone, J.M. 1988. X-ray scatter removal by deconvolution. *Med Phys* 15: 567–75.

Seibert, J.A., Boone, J.M. and Lindfors, K.K. 1998. *Flat-Field Correction Technique for Digital Detectors*. Bellingham, WA: SPIE, 348–54.

Seibert, J.A., Nalcioglu, O. and Roeck, W. 1985. Removal of image intensifier veiling glare by mathematical deconvolution techniques. *Med Phys* 12: 281–8.

Shaw, C.G., Ergun, D.L., Myerowitz, P.D., et al. 1982. A technique of scatter and glare correction for videodensitometric studies in digital subtraction videoangiography. *Radiology* 142: 209–13.

Siewerdsen, J.H., Chan, Y., Rafferty, M.A., et al. 2005. Cone-beam CT with a flat-panel detector on a mobile C-arm: preclinical investigation in image-guided surgery of the head and neck. *Proc SPIE*: 789–97.

Siewerdsen, J.H., Daly, M.J., Bakhtiar, B., et al. 2006. A simple, direct method for x-ray scatter estimation and correction in digital radiography and cone-beam CT. *Med Phys* 33: 187–97.

Siewerdsen, J.H. and Jaffray, D.A. 1999a. Cone-beam computed tomography with a flat-panel imager: effects of image lag. *Med Phys* 26: 2635–47.

Siewerdsen, J.H. and Jaffray, D.A. 1999b. A ghost story: spatio-temporal response characteristics of an indirect-detection flat-panel imager. *Med Phys* 26: 1624–41.

Siewerdsen, J.H. and Jaffray, D.A. 2000. Optimization of x-ray imaging geometry (with specific application to flat-panel cone-beam computed tomography). *Med Phys* 27: 1903–14.

Siewerdsen, J.H. and Jaffray, D.A. 2001. Cone-beam computed tomography with a flat-panel imager: magnitude and effects of x-ray scatter. *Med Phys* 28: 220–31.

Siewerdsen, J.H., Moseley, D.J., Bakhtiar, B., Richard, S. and Jaffray, D.A. 2004. The influence of antiscatter grids on soft-tissue detectability in cone-beam computed tomography with flat-panel detectors. *Med Phys* 31: 3506–20.

Sijbers, J. and Andrei, P. 2004. Reduction of ring artefacts in high resolution micro-CT reconstructions. *Phys Med Biol* 49: N247.

Smith, B.D. 1985. Image reconstruction from cone-beam projections: necessary and sufficient conditions and reconstruction methods. *IEEE Trans Med Imaging* 4: 14–25.

Sorenson, J.A. and Floch, J. 1985. Scatter rejection by air gaps: an empirical model. *Med Phys* 12: 308–16.

Spies, L., Ebert, M., Groh, B.A., Hesse, B.M. and Bortfeld, T. 2001. Correction of scatter in megavoltage cone-beam CT. *Phys Med Biol* 46: 821–33.

Star-Lack, J., Starman, J., Munro, P., et al. 2006. SU-FF-I-04: a fast variable-intensity ring suppression algorithm. *Med Phys* 33: 1997.

Star-Lack, J., Sun, M., Kaestner, A., et al. 2009. Efficient scatter correction using asymmetric kernels. In: Ehsan, S. and Jiang, H. (eds.). Bellingham, WA: SPIE, 72581Z.

Starman, J., Star-Lack, J., Virshup, G., Shapiro, E. and Fahrig, R. 2011. Investigation into the optimal linear time-invariant lag correction for radar artifact removal. *Med Phys* 38: 2398–411.

Starman, J., Tognina, C., Partain, L. and Fahrig, R. 2012. A forward bias method for lag correction of an a-Si flat panel detector. *Med Phys* 39: 18–27.

Sukovic, P. 2003. Cone beam computed tomography in craniofacial imaging. *Orthod Craniofac Res* 6: 31–6.

Sun, M., Nagy, T., Virshup, G., et al. 2011. Correction for patient table-induced scattered radiation in cone-beam computed tomography (CBCT). *Med Phys* 38: 2058–73.

Sun, M. and Star-Lack, J.M. 2010. Improved scatter correction using adaptive scatter kernel superposition. *Phys Med Biol* 55: 6695–720.

Suri, R.E., Virshup, G., Zurkirchen, L. and Kaissl, W. 2006. Comparison of scatter correction methods for CBCT. In: Michael, J.F. and Jiang, H. (eds.), *SPIE Proceedings* 614238, Bellingham, WA.

Tuy, H.K. 1983. An inversion formula for cone-beam reconstruction. *SIAM J Appl Math* 43: 546–52.

Vogtmeier, G., Dorscheid, R., Engel, K.J., et al. 2008. Two-dimensional anti-scatter grids for computed tomography detectors. In: Jiang, H. and Ehsan, S. (eds.), *SPIE Proceedings* 691359, Bellingham, WA.

Wang, J., Mao, W. and Solberg, T. 2010. Scatter correction for cone-beam computed tomography using moving blocker strips: a preliminary study. *Med Phys* 37: 5792–800.

Wiegert, J. and Bertram, M. 2006. Scattered radiation in flat-detector based cone-beam CT: analysis of voxelized patient simulations. In: Michael, J.F. and Jiang, H. (eds.), *SPIE Proceedings* 614235, Bellingham, WA

Wiegert, J., Bertram, M., Rose, G. and Aach, T. 2005. Model based scatter correction for cone-beam computed tomography. 271–82.

Wiegert, J., Bertram, M., Schaefer, D., et al. 2004. Performance of standard fluoroscopy antiscatter grids in flat-detector-based cone-beam CT. In: Martin, J.Y. and Michael, J.F. (eds.), *SPIE Proceedings*, Bellingham, WA, 67–78.

Wiegert, J., Hohmann, S. and Bertram, M. 2008. Iterative scatter correction based on artifact assessment. In: Jiang, H. and Ehsan, S. (eds.), *SPIE Proceedings* 69132B, Bellingham, WA.

Yao, W. and Leszczynski, K.W. 2009a. An analytical approach to estimating the first order scatter in heterogeneous medium. II. A practical application. *Med Phys* 36: 3157–67.

Yao, W. and Leszczynski, K.W. 2009b. An analytical approach to estimating the first order x-ray scatter in heterogeneous medium. *Med Phys* 36: 3145–56.

Zbijewski, W. and Beekman, F.J. 2006. Efficient Monte Carlo based scatter artifact reduction in cone-beam micro-CT. *IEEE Trans Med Imaging* 25: 817–27.

Zellerhoff, M., Scholz, B., Ruehrnschopf, E.P. and Brunner, T. 2005. Low contrast 3D reconstruction from C-arm data. *Proc. SPIE* 5745: 646–55.

Zhong, J., Ning, R. and Conover, D. 2004. Image denoising based on multiscale singularity detection for cone beam CT breast. *IEEE Trans Med Imaging* 23: 696–703.

Zhu, L., Bennett, N.R. and Fahrig, R. 2006. Scatter correction method for X-ray CT using primary modulation: theory and preliminary results. *IEEE Trans Med Imaging* 25: 1573–87.

Zhu, L., Strobel, N. and Fahrig, R. 2005. X-ray scatter correction for cone-beam CT using moving blocker array. In: Michael, J.F. (ed.), *SPIE Proceedings*, Bellingham, WA, 251–8.

Zhu, L., Wang, J. and Xing, L. 2009. Noise suppression in scatter correction for cone-beam CT. *Med Phys* 36: 741–52.

Advanced techniques

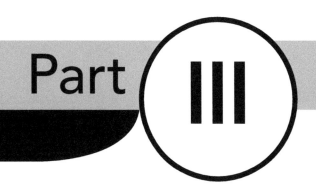

Part III

Applications

11 Multidetector row CT

Xiangyang Tang

Contents

11.1 INTRODUCTION

Since its advent in the early 1970s, x-ray computed tomography (CT) has advanced substantially in every aspect of its capability for clinical applications, with the most remarkable advancement being made in its speed of data acquisition and image generation. In the early days, approximately 5 min was needed in a first-generation CT scanner to acquire a full set of data for the generation of one single image slice. Nowadays, on average, fewer than 5 ms is needed in state-of-the-art multidetector row CT (MDCT) scanners to acquire the data for generating one image slice. Note that this is a 60,000 [(5 × 60)/(5/1000) = 60,000]–fold increase in speed. Thus far, at least three major milestones have been passed in the advancement of CT technology. The first milestone is the evolution from the first- and second-generation geometry to the third- and fourth-generation geometry. The narrow pencil or small fan beam has expanded into a fan beam that can accommodate the entire body of a patient, and the rotation speed of CT gantry has increased significantly, speeding up the data acquisition substantially. The second milestone is the availability of spiral/helical CT enabled by the slip-ring technology in 1990 (Kalender et al. 1989, 1990;

Crawford and King 1990). The elimination of the step-and-shoot scan mode and the resultant interscan delay marked the entrance of CT technology and application into a new era, resulting in remarkable advantages in the clinic, for example, faster patient throughput, less contrast agent, improvement in patient comfort, and resultant reduction of motion artifact or spatial misregistration. The clinical community acclaimed the overwhelming success of spiral/helical CT, driving all major CT manufacturers to deliver their spiral/helical CT products within a short time in the beginning of the 1990s.

The third major milestone is the MDCT enabled by the multidetector row technology. The initial attempt to transition from a single detector row CT (SDCT) to MDCT was the twin-slice CT scanner offered by Elscint (Elscint TWIN) in 1992 (Liang and Kruger 1996; http://www.medcompare.com/spotlight.asp?spotlightid=147). Six years later, all major vendors unveiled their 4-detector row CT scanners (Taguchi and Aradate 1998; Hu 1999) in the Radiological Society of North America (RSNA) Exhibition Hall at the McCormick Place in Chicago, IL. Historically, one significant thing occurred with the introduction of the four-slice CT scanner–the CT technology based on the fourth-generation geometry was forced to phase out because the cost for deploying a two-dimensional (2D) detector array along the entire CT gantry made the MDCT based on this geometry competitively impotent against those based on the third-generation geometry. In 2002, all major CT manufacturers launched their 16-detector row flagship scanners (Flohr et al. 2003) in which the submillimeter craniocaudal spatial resolution and three-dimensional (3D) isotropic spatial resolution became true the first time, enabling numerous advanced applications in the clinic, such as the imaging of temporal bone and coronary artery angiographies. Note that the leap from 4 to 16 detector rows took only about 4 years, whereas about 8 years elapsed from 1 to 4 detector rows. In 2005, all major CT manufactures launched their flagship 64-detector row CT scanner (Flohr et al. 2005), an even larger leap in the number of detector rows in just 3 years. Since then, the major CT manufacturers have competed fiercely by launching their flagship products at a variety of detector rows, for example, the 128-detector row scanner in 2007, 256-detector row scanner in 2007, and 320-detector row scanner (Rybicki et al. 2008) in 2008.

There has been a slice war since the mid-1990s, driven by the desire to scan a patient's entire heart and other large organs without table movement. As a result, the x-ray radiation dose, contrast agent dose, and interslab artifact can be reduced substantially, in addition to the efficiency in x-ray tube power use. The dual-source dual-detector CT (Flohr et al. 2008; Petersilka et al. 2008) for cardiac applications at almost doubled temporal resolution became available in 2008, followed by the scan mode at dual peak energies to conduct advanced clinical applications for material differentiation with spectral resolution. To meet the challenges imposed by advanced clinical applications, the CT technology is continuing to advance in leaps. In this chapter, I provide an introductory review of MDCT's system architecture, image reconstruction solutions, image qualities and clinical applications, and technological and clinical potential in the foreseeable future.

11.2 FUNDAMENTALS OF PHYSICS IN CT IMAGING

The subject contrast in x-ray CT imaging is generated by the attenuation of x-ray beam while it penetrates human body. In the energy range (20–150 keV) for diagnostic imaging, the x-ray attenuation is mainly determined by photoelectric absorption and Compton scatter. In physics, the mass attenuation coefficient of a material is used to describe the attenuation (Johns and Cunninham 1983; Bushberg et al. 2002):

$$\mu(x, y; E) = \alpha(x, y) f_c(E) + \beta(x, y) f_p(E), \quad (11.1)$$

where $f_p(E) \cong 1/E^{3.2}$ is the energy dependency of photoelectric absorption, $f_C(E)$ is the energy dependency of Compton scatter (Klein–Nishina function), and $\alpha(x, y)$, and $\beta(x, y)$ are characteristic coefficients of the material at location (x, y):

$$\alpha(x, y) \approx K_1 Z^{3.8} \rho / A, \quad (11.2)$$

$$\beta(x, y) \approx K_2 Z \rho / A. \quad (11.3)$$

where Z represents the atomic number, A the mass number, and ρ the mass density; K_1 and K_2 are constants. It is important to note that, given a material, Z/A is virtually constant. Thus, $\alpha(x, y)$ is determined by the atomic number of a material, whereas $\beta(x, y)$ is dominantly determined by its mass or electron density.

CT images are obtained by reconstruction of the 2D linear attenuation distribution from its projection acquired with either energy integration or photon counting detector. In the energy integration mode, an electric current proportional to the total energy carried by the x-ray fluency impinging upon a detector cell is recorded. In the photon counting mode, the electric pulse corresponding to an interaction between an x-ray photon and the detector scintillator at each cell is counted, whereby the pulse height is proportional to the energy deposited by the x-ray photon. Consequently, a threshold and range in the pulse height can be set to suppress electronic noise and endow each detector cell with energy resolution, respectively. Regardless of whether energy integration or photon counting is used for data acquisition, a CT with monochromatic x-ray source can be conceived as to obtain the 2D distribution of linear attenuation coefficient $\mu(x, y; E)$ from its projection:

$$\int_L \mu(x, y, E) \, dl = \int_L [\alpha(x, y) f_c(E) + \beta(x, y) f_p(E)] \, dl, \quad (11.4)$$

where $\int_L dl$ represents line integrals along L, a family of lines passing through point (x, y) at various orientations. As long as the data sufficiency condition is satisfied, numerous algorithms can be used to reconstruct $\mu(x, y; E)$, although the algorithms in the fashion of filtered backprojection (FBP) have been preferably adopted by all major CT vendors because of its efficient data flow and the capability to reach the most achievable spatial resolution determined by detector cell dimension.

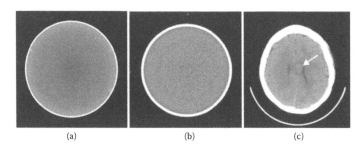

Figure 11.1 Artifacts caused by the polychromatics of x-ray source in x-ray MDCT: (a) Cupping artifacts in a cylindrical water phantom. (b) Spectral artifacts in a cylindrical water phantom. (c) Bone (skull)-induced spectral artifacts in a clinical head scan. (Images in (b) and (c) adopted from Cody, D.D. et al., *Radiology* 236, 756–61, 2005. With permission.)

Although the pursuit of a monochromatic x-ray source continues, no viable technology that can provide a monochromatic x-ray source with sufficient intensity for diagnostic imaging is currently available. In current practice, a polychromatic x-ray source is used, in which the energy of x-ray photons distributes over a spectrum up to the peak voltage (E_{kVp}) applied to the x-ray tube's anode. By taking all x-ray photons at various energies into account, Equation 11.4 becomes

$$\int_E S_{kVp}(E) \left\{ \int_L (x,y;E)\, dl \right\} dE$$

$$= \int_E S_{kVp}(E) \left\{ \int_L \left[\alpha(x,y) f_c(E) + \beta(x,y) f_p(E) \right] dl \right\} dE, \quad (11.5)$$

where $\int_E S_{kVp}(E)\{\circ\}\, dE$ denotes the integration over the energy spectrum from 0 to E_{kVp}. Note that E represents a single energy level in Equation 11.4, whereas it becomes a variable in Equation 11.5 within the energy range from 0 to E_{kVp}. All existing image reconstruction algorithms assume Equation 11.4, rather than Equation 11.5. Hence, the x-ray polychromatics underlying Equation 11.5 may result in beam-hardening effects (Cody et al. 2005; Ertl-Wagner et al. 2008) in CT images, such as the severe cupping artifacts shown in Figure 11.1a or subtle spectral artifacts shown in Figure 11.1b and 11.1c, that necessitates the use of empirical approaches for image correction in state-of-the-art MDCT scanners.

11.3 SYSTEM ARCHITECTURE OF MDCT

The 3D effect display of an x-ray CT scanner is illustrated in Figure 11.2a, and a schematic of its imaging chain is shown in Figure 11.2b. The seven major components or subsystems of an MDCT scanner are as follows: (1) x-ray source generating the x-ray fluency to penetrate a patient; (2) x-ray filtration removing low-energy x-ray photons and shaping the beam's intensity to conform patient's body contour for radiation dose reduction; (3) postpatient collimator removing the Compton scattering that degrades image contrast and CT number (Hounsfield unit) accuracy; (4) detector array made of scintillator converting x-ray photons into light

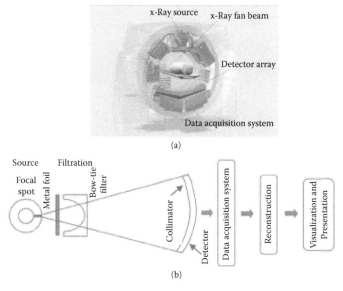

Figure 11.2 Diagrams showing the 3D effect display of an x-ray CT scanner for diagnostic imaging (a) and schematic of its imaging chain (b) (Picture in (a) courtesy Analogic Corporation, Peabody, MA, http://www.analogic.com/products-medical-computer-tomography.htm.)

photons; (5) data acquisition system (DAS) collecting the current generated by diodes and converting it into digital data and transferring for data storage; (6) image reconstruction engine for data preprocessing and generating transverse image slices; and (7) computation engine for image presentation, such as coronal and sagittal multiplanar reformatting, maximum intensity projection (MIP), and volume and surface rendering. Every component plays an important role, no matter if its implementation is costly or cheap. For example, the x-ray filtration is just a thin layer of Al, Cu, or Mo on the top of the bow-tie filter's graphite substrate, but it is critical to determine the low-contrast detectability and dose efficiency of an MDCT for diagnostic imaging. Similarly to the strength of a chain being determined by its weakest link, the overall image quality of a CT scanner is determined by the component in the imaging chain with the poorest performance. Thus, an adequate balance and trade-off over spatial, contrast, temporal, and spectral resolutions is the key to reach the best possible imaging performance.

As schematically illustrated in Figure 11.3, the major difference between an SDCT and the MDCT is the use of a multirow detector for data acquisition. The full cone angle α_m spanned by the detector is proportional to the number of detector rows. By convention, MDCT also has been called multislice or multisection CT (MSCT). Due to the rationale that will be elucidated later in this chapter, an MDCT may not simultaneously generate a number of image slices with the number of slices equal to the number of detector rows. Hence, unless otherwise specified, I refer to the multislice, multisection, and multidetector row CT as MDCT in this chapter.

11.4 DATA ACQUISITION IN MDCT

In an SDCT, the geometries of both data acquisition and image reconstruction are 2D, that is, in fan beam geometry (Figure 11.4a), wherein a ray is uniquely determined by its view angle β and fan angle γ. However, once evolved into MDCT, the

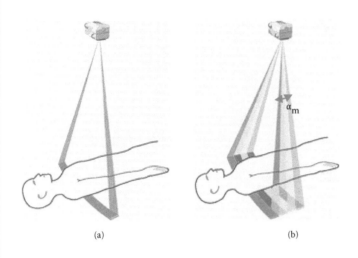

(a)   (b)

Figure 11.3 Exaggerated schematic diagrams showing the scan of single detector row CT (a) and multidetector row CT (b) (Adopted and modified from Rydberg, J. et al., *Radiographics*, 20, 1787–806, 2000. With permission.)

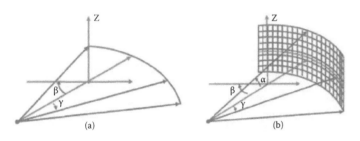

(a)   (b)

Figure 11.4 Schematic diagrams showing the geometries of fan beam (a) and cone beam (b) for either data acquisition or image reconstruction.

geometry of data acquisition is of course cone beam, that is, 3D (Figure 11.4b) but that for image reconstruction is still in fan beam for the number of detector rows up to 16. This is because the cone angle corresponding to detector rows up to 16 is still relatively small; thus, each of the images can be treated as slices stacked parallel to each other and orthogonal to the rotation axis of CT gantry. Similar to the scenario in the SDCT, as required by clinical procedures, the patient table can remain motionless or proceed in data acquisition, corresponding to the axial and spiral/helical scan modes, respectively. Under either mode, the angular range of the projection data used for image reconstruction can be equal to 360° (full-scan) (Crawford and King 1990), larger than 360° (over-scan) (Crawford and King 1990), equal to 180°+γ_m [half-scan (Parker 1982), where γ_m is the full fan angle of x-ray beam], or between 180°+γ_m and 360° (partial scan) (Silver 2000). The full- and over-scan is usually used in noise-critical applications of detecting pathologic lesions in low contrast, whereas the half- or partial scan is used for applications wherein temporal resolution is of essence, for example, cardiovascular CT imaging, pulmonary CT imaging, or a combination. In practice, the over-scan and partial scan have advantages in suppressing artifact caused by the patient's voluntary and involuntary motion, such as the head's rotation in scanning pediatric or unconscious adult patients. No all-in-one solution can meet all

the requirements imposed by various clinical applications. As illustrated in Section 11.6, the variety of scan modes and number of detector rows (and resultant cone angle) makes the design and optimization of image reconstruction solutions in MDCT very challenging.

11.5 IMAGING PERFORMANCE IN MDCT

In general, the major image qualities to evaluate the performance of an MDCT are contrast, spatial, and temporal resolution, with the recent addition of energy or spectral resolution implemented in state-of-the-art MDCT via dual peak energies (kVp) scanning.

11.5.1 CONTRAST RESOLUTION

Contrast resolution is also called low contrast detectability (LCD) and is defined as the capability of identifying low-contrast (0.1%–0.5%) targets at various dimensions (1–5 mm), given a radiation dose quantified as computed tomography dose index (CTDI). The contrast resolution is dependent on the CT detector's absorption and conversion efficiency, in addition to its geometrical efficiency determined by the postpatient collimator and active area of each detector cell. The LCD is critical in identifying low-contrast pathology over patient body habitus. For example, in the scanning of a large size patient the noise level is usually high; high noise levels also occur when scanning pediatric patients, because the radiation dose has to be compromised to accommodate the pediatric patient tissue or organ's sensitivity to radiation. Figure 11.5a is the drawing of the CTP515 LCD module in the CatPhan600 phantom (http://www.phantomlab.com/library/pdf/catphan500-600manual.pdf); the corresponding CT image is in Figure 11.5b, in which the LCD at given radiation dose can be evaluated. The contrast resolution is the differentiator between the CT for diagnostic imaging and that for other special purposes, such as the cone beam CT (CBCT) for image-guided radiation therapy and micro-CT for animal or specimen imaging in preclinical research. To make use of the x-ray photons that have penetrated the patient's body as much as possible, the scintillator in diagnostic MDCT's detector is approximately 3.0 mm, substantially thicker than that of the flat panel used in CBCT (~0.5 mm).

11.5.2 SPATIAL RESOLUTION

Spatial resolution is quantitatively defined by the modulation transfer function (MTF) and serves to evaluate the MDCT's capability of differentiating two objects that are in high contrast and stay close to each other. The spatial resolution of an MDCT is primarily determined by the dimension of its detector cell, but resolution can be boosted to approach twice the Nyquest frequency determined by the detector cell dimension (Flohr et al. 2007; Tang et al. 2010). The typical detector cell size in MDCT is approximately 0.5 mm, corresponding to a Nyquest frequency of 10.0 lp/cm. However, almost all MDCT offers the highest spatial resolution beyond 15.0 lp/cm. For example, presented in Figure 11.6a is the MTF corresponding to the standard kernel (STAND) used in an MDCT, in which the 10% cut-off frequency is well below the Nyquest frequency. With sophisticated boosting techniques (Figure 11.6b), the

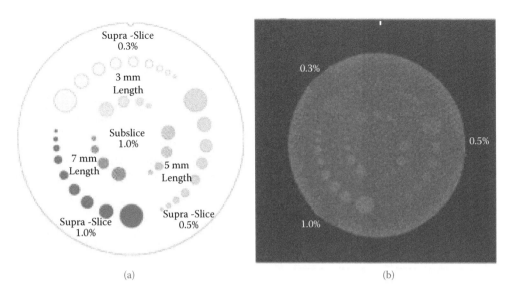

(a)   (b)

Figure 11.5 Schematic diagram showing the CTP515 LCD module of the CatPhan-600 phantom (a) and an example of its transverse MDCT image (b). (image in (b) adopted from Thilander-Klang, A. et al., *Radiat Prot Dosimetry*, 139, 449–54, 2010. With permission.)

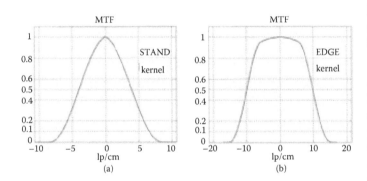

Figure 11.6 MTF corresponding to the STANDARD (a) and EDGE (b) reconstruction kernels in a typical MDCT scanner.

10% cut-off frequency of the edge kernel (EDGE) of the same MDCT can be readily beyond the Nyquest frequency. Aliasing artifacts may appear when the Nyquest frequency is exceeded. However, the so-called quarter-offset technique (Tang et al. 2010) can be effectively applied to improve the sampling rate substantially, if not double it, thereby avoiding the occurrence of aliasing artifacts in clinical applications demanding high spatial resolution.

11.5.3 TEMPORAL RESOLUTION

Temporal resolution, determined by the period of time during which the projection data to generate the CT images are acquired, aims to evaluate MDCT's capability of imaging the organs and tissues in motion, for example, heart or lung in cardiac or respiratory motion, respectively. In practice, given an MDCT gantry rotation speed, the short scan mode is used to attain the best possible temporal resolution. The temporal resolution of a short scan is defined as $T \times (180° + \gamma_m)/360°$, where T is the period of time for the CT gantry to rotate one full circle. With the increasing number of detector rows, MDCT is becoming a routine modality in the clinic for cardiovascular imaging wherein

high temporal resolution is of essence. Only a brief introduction on temporal resolution is given here; details can be found in Section 11.7.2.

11.5.4 ENERGY RESOLUTION

Energy resolution implemented with dual-kVp scan is a new addition to the potency of MDCT. In single kVp CT scan, the pixel intensity in a reconstructed image is the mass attenuation coefficient that is jointly determined by the effective atomic number and mass density of the material. Consequently, a material, for example, I, with higher atomic number but lower mass density, may happen to have approximately the same mass attenuation as that of another material, for example, Ca, with lower atomic number but higher mass density. However, the mass attenuation coefficient of a material varies over x-ray photon energy and that of various materials vary at different rate. It is apparent, as is elucidated in Section 11.7.3, that such a dependence on x-ray photon energy can be used to differentiate materials that generate no contrast in a single peak voltage scan.

Enormous effort has been devoted by the scientists and researchers in the CT industry to make MDCT more potent for clinical excellence. Generally, each aspect of MDCT's imaging performance may not be the best in the clinic in comparison with other imaging modalities. For example, the contrast resolution of MDCT is not as high as that of positron emission tomography (PET),single-photon emission computed tomography (SPECT), or magnetic resonance imaging (MRI); the temporal resolution of MDCT may be inferior to that of MRI when special pulse sequences, for example, echo planar imaging (EPI), are used. Furthermore, the spatial resolution of CT is not as good as that of ultrasound when only a small and shallow region of interest (ROI) is to be imaged. However, putting all the resolution together, it is quite fair to say that MDCT is the best and most robust imaging modality to fulfill the requirements imposed by the majority of clinical applications.

As is illustrated in the next section, the geometry of both data acquisition and image reconstruction in MDCT with detector row number larger than 16 is 3D, that is, it is in cone beam or volumetric geometry. Nevertheless, although they are still being used for imaging performance evaluation in MDCT, almost all the phantoms used for image performance evaluation and verification, for example, the LCD phantom displayed in Figure 11.5a, are designed for the SDCT working at fan beam or slice mode. The targets in these phantoms are cylindrical and required to be placed in parallel with the gantry's rotation axis, that is, no variation along the craniocaudal direction. These cylindrical targets work well in the SDCT or MDCT with the fan beam geometry for image reconstruction, but they may result in at least two consequences in the MDCT with the cone beam geometry for image reconstruction. First, in general, a cylindrical target cannot detect cone beam artifacts (see Section 11.6.3 for details on cone beam artifacts). Second, one may take advantage of the fact that there is no variation along the cylindrical targets to attain imaging performance that is not real. For example, the LCD (Figure 11.5b) measured with the LCD phantom shown in Figure 11.5a may falsely appear better than what it actually is, when certain filtering along the longitudinal direction is applied. Hence, new phantoms with adequate longitudinal variation to ensure the accuracy of imaging performance evaluation in MDCT are anticipated to be defined by federal or state regulatory agencies. The availability of such phantoms may not only benefit the patients and physicians with diagnosis accuracy in clinical practice but also help identify the front-runner among the major MDCT vendors in their technological race.

11.6 IMAGE RECONSTRUCTION IN MDCT

Image reconstruction plays a central role in CT imaging (Kak and Slaney 1988). As indicated earlier, the algorithms in the fashion of FBP have been preferably adopted by all major CT vendors because of the efficient data flow and the capability to reach the most achievable spatial resolution determined by detector cell dimension. In the following is a description of the typical image reconstruction solutions used in MDCT scanners for diagnostic imaging.

11.6.1 IMAGE RECONSTRUCTION SOLUTIONS IN 4-DETECTOR ROW CT

11.6.1.1 Axial scan

As indicated earlier, the geometry for image reconstruction in 4-detecor row CT scanner is assumed as 2D or fan beam, even though the data acquisition is in fact carried out in 3D or cone beam. In an axial scan, the mismatch between data acquisition and image reconstruction geometries may result in inaccuracy in reconstructed images. However, corresponding to the typical 20-mm longitudinal beam aperture that can be implemented in 4-detector row CT scanner by 5 mm × 4 or 10 mm × 2 mode, the cone angle of the outmost image slice is $\frac{1}{2}\alpha_m = \sim 0.79°$ or $\frac{1}{2}\alpha_m = \sim 0.53°$, respectively, which is quite small. The resultant inaccuracy or artifacts in reconstructed images is almost undetectable when the cone beam at such a small cone angle is assumed as four

fan beams stacked parallel to each other along the longitudinal direction. This means that each image slice in the 4-detector row CT scanner in axial scan mode is treated exactly the same as that in a SDCT. Moreover, it should be pointed out that the backprojector used by all the major CT vendors in 4-detector row CT for image reconstruction is one-dimensional (1D), which is exactly the same as those used in SDCT scanners.

11.6.1.2 Spiral/helical scan

A brief review of the image reconstruction in spiral/helical SDCT would be beneficial for readers to understand the spiral/helical image reconstruction algorithms used in MDCT. In a single slice spiral/helical scan, the artifact is mainly owing to the data inconsistency, because, given an image at specified location, its projection can be recorded only with full fidelity by the 1D detector array, while the spiral/helical source trajectory exactly intercepts the image slice (namely, midway). At other angular locations at which the image slice does not intercept the source trajectory, interpolation, either in the 180° or 360° fashion, has to be exercised to obtain the corresponding projection (Kalender et al. 1989, 1990; Crawford and King 1990). In geometry, this is to approximately obtain the desired projection via view-wise (360° interpolation) or ray-wise (180° interpolation) interpolation of two corresponding projections based on the longitudinal distance. Apparently, only the projection at the midway is identical to or consistent with the true projection of the image slice, but every other projection obtained via the interpolation is not identical to or inconsistent with the true projection. The inconsistence causes inaccuracy in reconstructed images, and this is the underlying reason that the spiral/helical artifacts are called inconsistency artifact. It should be indicated that the slice sensitivity profile (SSP) is dependent on the interpolation method used. In addition, the SSP is dependent on spiral/helical pitch that is usually defined as the ratio of the distance proceeded by the patient table within one helical turn over the longitudinal beam aperture of the x-ray detector used in the scan.

In spiral/helical MDCT scan, one is no longer bothered by the data inconsistence problem, because, in principle, the wider longitudinal dimension of the 2D detector keeps intercepting the x-ray flux that have penetrated the image slice at the midway position, that is, recording the projection, as long as the orthogonal distance between the x-ray focal spot to the image slice at the midway position is not too far. Thus, with resort to adequate ray tracking and view weighting techniques, the projection data over the angular positions of the image slice at a specified position can be obtained via cross-detector row interpolation (Taguchi and Aradate 1998; Hu 1999). It should be pointed out that the cross-row interpolation in MDCT differs from that in the spiral/helical SDCT. This can be better understood if the reader realizes that the interpolation in MDCT can be eliminated if the longitudinal sampling rate of the multidetector row detector is sufficient and aligned to record the projection at each angular position, whereas the interpolation in the spiral/helical SDCT is always necessary. Because the interpolation in MDCT is conducted across detector row, rather than across views (Kalender et al. 1989, 1990; Crawford and King 1990) in the spiral/helical SDCT, the SSP in MDCT in principle is no longer dependent on the

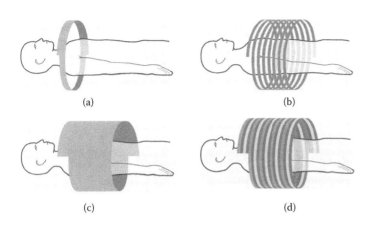

Figure 11.7 Schematic diagrams showing the scanning of SDCT at helical pitch 1:1 (a), SDCT at helical pitch 4:1 (b), SDCT at helical pitch 1:1 but four times thicker image slice (c), and 4-detector row CT at helical pitch 1:1 (d). (all drawings adopted from Rydberg, J. et al. *Radiographics*, 20, 1787–806, 2000. With permission.)

spiral/helical pitch. Once the projection data are obtained, ramp filtering and 1D backprojection are used to generate tomographic images.

The most remarkable benefit brought about by the 4-detector row CT to clinical applications is the speeding-up of data acquisition (Rydberg et al. 2000). In the step-and-shoot axial scan, it is quite intuitive to understand that each step of patient table proceeding is equal to four times that of an SDCT. The speeding up of helical/spiral scan is schematically illustrated in Figure 11.7. Figure 11.7a shows that a helical/spiral SDCT scans the patient at pitch 1:1. If the scan speed needs to be increased by a factor of 4, the SDCT may increase either the pitch or slice thickness by four times (Figure 11.7b and 11.7c), resulting in substantial interhelix gap or degradation in the longitudinal spatial resolution, respectively. Note that a spiral/helical scan at pitch larger than 1:1 does exist in clinical applications, but a pitch as large as 4:1 definitely makes high-quality image reconstruction impossible. However, if there are four detector rows in the scanner, a helical/spiral scan at pitch 1:1 can scan the patient four times faster and without interhelix gap, and thin slice thickness can be maintained (Figure 11.7d). In general, with recourse to the multidetector row technology, the upper limit of spiral/helical pitch is approximately 1.5:1 but may vary in practice, depending on the gantry geometry and the field of view (FOV) of scan and image reconstruction. It should be noted that an increase in spiral/helical scan reduces the radiation dose to the patient proportionally, whereas the noise index in a CT image deteriorates in a manner of square root.

11.6.2 IMAGE RECONSTRUCTION SOLUTIONS IN 16-DETECTOR ROW CT

11.6.2.1 Axial scan

Although other numbers of detector rows, such as 8, 10, or 12, exist in MDCT, every major CT vendor positions their 16-detector row CT scanner as the flagship product. Despite the number of detector rows being increased by fourfold, the typical longitudinal beam aperture is still 20 mm in 16-detector row CT, which can

be implemented by 1.25 mm × 16, 2.5 mm × 8, 5 mm × 4, and 10 mm × 2 via adequate row combination. The maximum half cone angle corresponding to the outmost slice at 1.25 mm × 16 mode is $\frac{1}{2}\alpha_m \cong 0.99°$ and that of the outmost slice in the 5 mm × 4 mode in 4-detector row CT scanner is $\frac{1}{2}\alpha_m \cong 0.79°$. Obviously, the maximum full cone angle in 16-detector row CT scanner is approximately the same as that of the 4-detector row CT scanner. Consequently, the geometry of stacked fan beams is still assumed for image reconstruction in the axial scan of 16-detector row CT.

11.6.2.2 Spiral/helical scan

The leap from 4 to 16 detector rows actually has provided the opportunity to design the image reconstruction solution in 3D geometry wherein a 2D detector is used. However, rather than taking this opportunity, the image reconstruction solution developers of almost all the major CT vendors still constrain themselves to what they have done in the single- or 4-detector row CT scanner—converting the 3D geometry into 2D geometry wherein the 1D backprojector can still be used. The main reason behind this choice is business strategy for cost savings, because the 1D backprojector implemented with a specially designed array processor is still fast enough to meet the requirements for image generation speed in the clinic. This constraint makes the spiral/helical image reconstruction in 16-detector row CT extremely difficult. Figure 11.8a shows projections of an orthogonal disc with its height equal to that of a detector row (Figure 11.8a) when the x-ray source focal spot is at view angle β = −90°, −45°, 0°, 45° and 90°, respectively. It is observed that, except at the midway position (β = 0°), the projection of a thin disc in the multirow detector occupies a variable number of detector rows. The larger the magnitude of the viewing angle, the greater the number of detector rows that are intercepted by the projection of the thin disc. It is not hard to understand that, if a 1D backprojector is used, all the projection data must be fitted into one detector row. Consequently, data loss occurs with increasing view angle β. In contrast, if the thin disc is tilted to conform to the spiral/helical source trajectory as illustrated in Figure 11.8b, its projection at various angular positions (Figure 11.8b′) can fit into an oblique 1D detector, that is, the loss of projection data can be mitigated substantially in comparison with the case of the orthogonal thin disc (Larson et al. 1998; Bruder et al. 2000; Kachelrieß et al. 2000; Heuscher 2002; Tang 2003). In reality, no oblique 1D detector is needed, because the projection of the tilted thin disc can be obtained with cross-row interpolation. In such a way, the tilted thin disc can be well reconstructed using a 1D backprojector from the projection data obtained through across-row interpolation. Subsequently, the entire 3D Cartesian coordinate system needs to be exhaustively covered by a nutation of tilted thin discs. Any image corresponding to the orthogonal thin disc in the Cartesian coordinate system can be readily obtained via 1D interpolation along the z-axis. An inspection of the images presented in Figure 11.9a and 11.9b shows that the image reconstruction through a nutation of tilted thin discs outperforms the reconstruction with orthogonal thin discs in terms of reducing the artifacts caused by the spiral/helical inconsistency. However, three side effects are attributed to the nutation of tilted thin discs: (1) the spatial sampling by

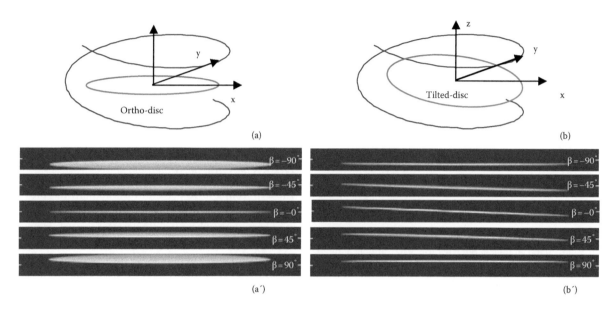

Figure 11.8 Schematic diagram showing the data acquisition geometry in MDCT with a disc orthogonal (a) or tilted (b) to its rotation axis, and the projection at view angle β = –90°, –45°, 0°, 45°, and 90° of the orthogonal (a′) and tilted (b′) discs.

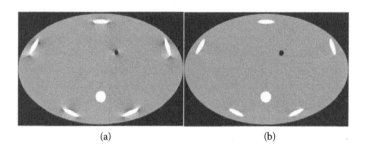

Figure 11.9 Transverse images of the helical body phantom reconstructed from the simulated projection data acquired by a 16-detector row CT at spiral/helical pitch 25/16:1 = 1.5265:1, using view weighted algorithm with orthogonal (a) and tilted (b) image slices without view weighting.

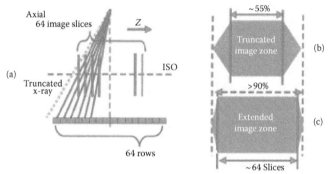

Figure 11.10 Schematic diagram showing the data acquisition in the axial scan (a), the image zone truncation due to the cone angle (b), and the extension of the image zone by cone angle–dependent weighting (c).

the tilted thin disc is not uniform, (2) the 1D interpolation along the z-axis may slightly broaden the SSP, and (3) a larger beam over-range at the starting and finishing ends of the spiral/helical scan (Tzedakis et al. 2005; Molen and Geleijns 2006) in comparison with that without tilting the thin disc given an identical imaging zone.

11.6.3 IMAGE RECONSTRUCTION SOLUTIONS IN 64-SLICE CT AND BEYOND

When the number of detector rows increases to 64, the half cone angle $\frac{1}{2}\alpha_m$ typically becomes larger than 2°. Consequently, no matter how the projection data is cleverly manipulated, there is no choice but to use the 2D detector or 3D geometry for image reconstruction. This means that there is no geometric mismatch between image reconstruction and data acquisition anymore, but the cone angle becomes a troublemaker now, manifesting itself as artifacts through three mechanisms: (1) longitudinal truncation, (2) shift-variant spatial sampling rate, and (3) cone angle.

11.6.3.1 Axial scan

A 2D sectional view of the axial data acquisition geometry is illustrated in Figure 11.10a, whereby 64 slices of images are to be reconstructed from the data acquired with a 64-row detector. Owing to the cone angle, truncation occurs unavoidably and indents the image zone to be just about 55% of the detector's longitudinal dimension, if the original FDK reconstruction algorithm (Feldkamp et al. 1984) is used. However, in a full axial scan, the data redundancy of the majority of the voxels in the volume to be reconstructed is either one or two, whereas only a data redundancy of one is sufficient for image reconstruction. Illustrated in Figure 11.11 is the data redundancy in the three outmost image slices in an axial scan of 64-detector row CT, in which the FOV is assumed 500 mm. It is clearly observed that almost all the voxels in the third outmost image slice are of a data redundancy larger than 1 and thus can be reconstructed appropriately. This means that in the 64 image slices corresponding to each detector row in the detector array, all but the two outer slices at the upper and lower ends of the detector have enough projection data for image reconstruction. However, as further illustrated in Figure 11.12a, given a voxel P with the data redundancy larger than 1, there exists a pair of conjugate rays SP and $S'P$ that may contribute to the reconstruction.

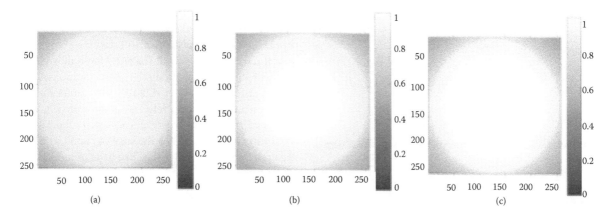

Figure 11.11 Pictures showing the data redundancy in the outmost (a), second (b), and third (c) outmost image slices in the axial scan of 64-detector rows [detector dimension, 64 × 0.625 mm; source-to-imager distance (SID), 541 mm].

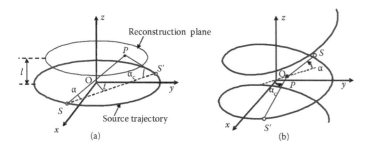

Figure 11.12 Schematic diagram showing the rationale of cone angle–dependent weighting to deal with the data redundancy in axial (a) and spiral/helical scan (b).

Intuitively, the contribution from the ray with a smaller cone angle, for example, ray SP with cone angle α in Figure 11.12 a, should be more trustworthy (Patch 2004; Taguchi et al. 2004; Tang et al. 2005, 2008) than the ray with a larger cone angle, for example, ray $S'P$ with cone angle α' in Figure 11.12a, from the perspective of image reconstruction. Based on this insightful understanding, a cone angle–dependent weighting scheme is proposed to suppress the artifacts caused by the inconsistency between the rays of the conjugate pair (Patch 2004; Tang et al. 2005, 2008). Figure 11.13 shows the performance of the cone angle–dependent weighting scheme, whereby the artifacts in the helical body phantom (Figure 11.13a and 11.13a′) and the humanoid head phantom (Figure 11.13b and 11.13b′) are reduced significantly.

11.6.3.2 Spiral/helical scan

As illustrated in Figure 11.12b, the cone angle–dependent weighting scheme also can be used in spiral/helical scan, whereby the calculation of the cone angle corresponding to each conjugate pair is a little bit more complicated, because the movement of patient table during scan has to be taken into account (Heuscher et al. 2004; Stierstorfer et al. 2004; Tang et al. 2006; Tang and Hsieh 2007; Wang et al. 1993). Figure 11.14 presents typical clinical images in the transverse and coronal review, respectively, in which the superior image quality provided by the spiral/helical scan in state-of-the-art MDCT scanners for clinical applications can be appreciated.

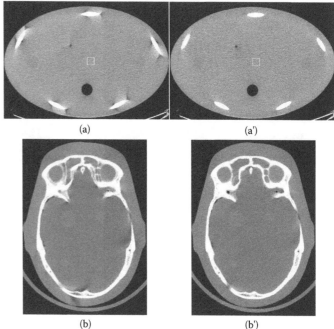

Figure 11.13 Images of the helical body phantom reconstructed by the FDK algorithm (a) and the algorithm with cone angle–dependent weighting (Tang, X. et al. *Phys Med Biol*, 50, 3889–905, 2005; Tang, X. et al., *Med Phys*, 35, 3232–8, 2008.) (a′), and the images of the humanoid head phantom reconstructed by the FDK algorithm (b) and the algorithm with cone angle–dependent weighting (Tang, X. et al., *Phys Med Biol*, 50, 3889–905, 2005; Tang, X. et al., *Med Phys*, 35, 3232–8, 2008.) (b′).

11.7 RECENT ADVANCEMENTS IN MDCT TECHNOLOGY

11.7.1 UP-SAMPLING TO SUPPRESS CRANIOCAUDAL ALIASING ARTIFACTS

With the advent of MDCT, the radiology community is bothered by an annoying artifact called windmill, pinwheel, or even "bear claw" (referred to as windmill artifact hereafter), because of its spoke-like pattern surrounding bony structures, as exemplified by Figure 11.15a. The windmill artifact frequently occurs in

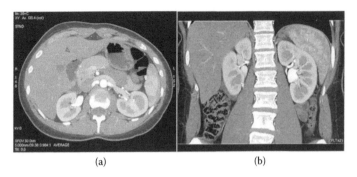

(a)   (b)

Figure 11.14 Typical transverse images (a) reconstructed from the projection data of a helical/spiral scan of 64-detector row using cone angle–dependent weighting scheme and the coronal view of multiplanner reformatted image (b).

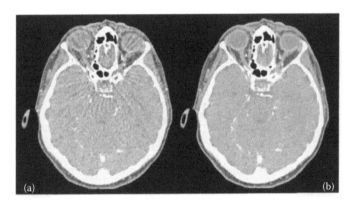

Figure 11.15 Transverse head images reconstructed from the projection data acquired without (a) and with (b) the focal spot wobbling (z-sharp technique) along the z-direction, respectively (Courtesy of Siemens Healthcare, Malvern, PA).

neurological scans if a high-contrast bony structure gets involved, for example, the circle of Willis or the spinal cord. The root cause of this artifact is because a bony structure may possess the spatial frequency beyond the Nyquest frequency that is determined by MDCT's detector cell dimension. In other words, the abrupt variation along the craniocaudal direction is too severe to be sampled adequately by the MDCT's detector; hence, the windmill artifact is actually an aliasing artifact that in principle can be suppressed through two approaches: (1) reducing the highest frequency of the bony structure by smooth filtering along the craniocaudal direction to make sure no frequency component exceeds the Nyquest frequency; or (2) increasing the sampling rate in the craniocaudal direction to increase MDCT's Nyquest frequency so that the projection data of the bony structure can be acquired by MDCT's detector without aliasing. Both of them work effectively in terms of suppressing aliasing artifact, but the latter outperforms the former in maintaining spatial resolution along the craniocaudal direction and thus is preferable in clinical applications wherein a thinner image slice is desirable.

The "z-sharp" technique (Flohr et al. 2005) offered by one of the major CT vendors is intended to increase the spatial sampling rate along the craniocaudal direction, whereas those by other vendors are aimed at reducing the highest frequency component via smooth filtering. The z-sharp technique is implemented by wobbling the focal spot along the craniocaudal direction in

data acquisition, which is actually an extension of the focal spot wobbling in the lateral direction that is initially used in SDCT for suppression of aliasing artifact and the enhancement of in-plane spatial resolution. Figure 11.16a and 11.16b illustrates the focal spot wobbling schemes along the lateral (Sohval and Freundlich 1987; Lonn 1992; Tang et al. 2010) and craniocaudal (Flohr et al. 2005) directions, respectively. As demonstrated by Figure 11.15b, the z-sharp technique is very efficacious in suppressing the windmill artifact caused by the stephoid bone at the bottom of the brain, while the image slice can be maintained thin.

11.7.2 DUAL-SOURCE DUAL-DETECTOR TO DOUBLE TEMPORAL RESOLUTION FOR CARDIOVASCULAR IMAGING

With the increasing number of detector rows, MDCT is becoming one of the most popular modalities for cardiac imaging, for example, the diagnosis of stenosis in coronary arteries, in addition to the standard of fluoroscopy-guided catheterization. To take a snapshot of the heart that is in cyclic motion, the temporal resolution becomes the most important imaging performance (Flohr and Ohnesorge 2000, 2008; Ohnesorge et al. 2000; Vembar et al. 2003; Tang and Pan 2004; Hsieh et al. 2006; Taguchi et al. 2006; Tang et al. 2008). The temporal resolution of an MDCT scanner is dependent on the duration of time during which the projection data are acquired (Flohr and Ohnesorge 2000, 2008; Ohnesorge et al. 2000; Vembar et al. 2003; Tang and Pan 2004; Hsieh et al. 2006; Taguchi et al. 2006; Tang et al. 2008). Accordingly, the short scan mode mentioned in Section 11.4 is usually used for cardiovascular imaging. If, for instance, the time for an MDCT gantry to rotate one circle is 0.3 s., the temporal resolution is $0.3 \times (55° + 180°)/360°$ s ≈ 196 ms, a sufficient time for imaging a heart that beats fewer than 65 times in a minute, that is, 65 beats per minute (bpm). For patients with a heartbeat rate (HBR) higher than 65 bpm, an HBR that occurs frequently in the clinic, beta blocker is usually administered to decrease the HBR until it is stably lower than 65 bpm. However, the avoidance of beta blocker injection is of clinical relevance, especially for the patients with suspected myocardial infarction. Therefore, in addition to short scan, more methods to improve the temporal resolution for clinical excellence are needed. A straightforward way to do so is to increase the rotation speed of MDCT's gantry. For instance, if the gantry rotation speed can be increased to 0.2 s per rotation (s/r), the temporal resolution would be $0.2 \times (55° + 180°)/360°$ s ≈ 130 ms. However, to reach a gantry speed of 0.2 s/r, the G-force in a typical MDCT would be larger than 70 g, making the fabrication of an MDCT gantry extremely challenging and costly, if not impossible.

An alternative way is to acquire the projection data in an intercycle multisector manner, as illustrated in Figure 11.17 (Taguchi et al. 2006; Flohr and Ohnesorge 2008; Tang et al. 2008). Because a heart physiologically repeats itself, the required projection data can be acquired over multicycles at an appropriate phase gated by the electrocardiogram (ECG) signal. Figure 11.17a illustrates an ideal case in the two-sector data acquisition in which half of the data come from cardiac cycle I and the rest from cycle II. It is not hard to imagine, however, that the ideal case rarely occurs in reality, because the temporal relationship between the two sectors is jointly determined by

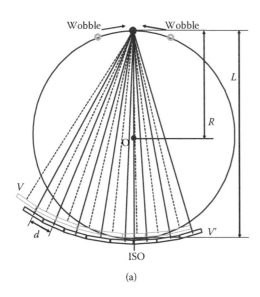
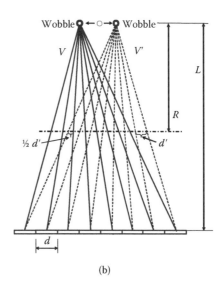

(a)   (b)

Figure 11.16 The schematic diagrams showing the lateral focal spot wobbling (a) for enhancing in-plane spatial resolution and craniocaudal focal spot wobbling (b) for enhancing longitudinal resolution, where L and R are the distances from the focal spot to iso and detector, respectively.

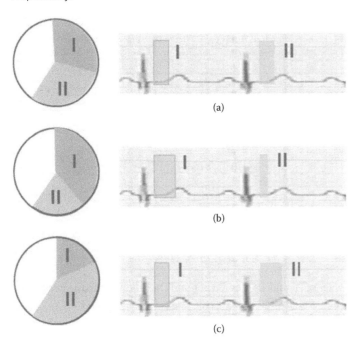

(a)

(b)

(c)

Figure 11.17 Schematic diagram showing the variation of sector width in the two-sector data acquisition and image reconstruction for cardiac imaging. (a) Ideal case with equal sectors I and II. (b) A nonideal case with the width of cycle I larger than that of cycle II. (c) Another nonideal case with the width of cycle II larger than that of cycle I.

MDCT gantry rotation speed and patient's heart beat rate and initial phase, which seldom guarantees a perfect timing for the ideal case, not mention the fact that the patient's HBR variation may further complicate the situation. Actually, the cases illustrated in Figure 11.17b and 11.17c occur the majority of the time in practice. In principle, the effective temporal resolution T_{eff} of a two-sector data acquisition and image reconstruction can be defined as $T_{eff} = \text{maximum}(T_I, T_{II})$, where T_I and T_{II} are the duration of time to acquire the data in cycle I and II, respectively, and $\max(\cdot, \cdot)$ denotes an operation to select the larger of the two variables. Consequently, only the ideal case can assure a doubled temporal resolution, and all other cases are between the best (doubled temporal resolution) and worst (no gain in temporal resolution) scenarios (Tang et al. 2008). In general, the larger the difference between the two sectors, the less the gain in temporal resolution. It also should be realized that although the ECG repeats itself, the mechanical state of the heart never repeats exactly, particularly for MDCT imaging at a spatial resolution that is significantly better than that in SPECT or PET, whereby the heart is assumed to be mechanically repeating itself.

Fortunately, the shortcomings of the data acquisition in the intercycle two-sector manner can be overcome by acquiring the projection data in an intracycle two-sector manner (Flohr et al. 2008; Petersilka et al. 2008) that can be implemented with the dual-source dual-detector technology, as illustrated in Figure 11.18 (Flohr and Ohnesorge 2008). The data corresponding to each sector come from the identical cardiac cycle with an equal period of time for data acquisition and thus guarantee a doubled temporal resolution. It should be emphasized that there is no chance for the heart rate arrhythmia to degrade the temporal resolution, because all the data come from the same single cardiac cycle. Using a dual-source dual-detector MDCT, the HBR of a patient can readily exceed 65 bpm, as demonstrated by the images of the coronary arteries presented in Figure 11.19.

11.7.3 DUAL PEAK VOLTAGE (DUAL-KVP) SCAN FOR MATERIAL DIFFERENTIATION WITH ENERGY RESOLUTION

As demonstrated by Equations 11.1–11.3 the mass attenuation coefficient $\mu(x, y; E)$ of a material is jointly dependent on its atomic number and mass density (Johns and Cunninham 1983; Bushberg et al. 2002). There exist situations in practice where two different materials are not differentiable in an MDCT image acquired at single peak voltage, because the material with the lower atomic

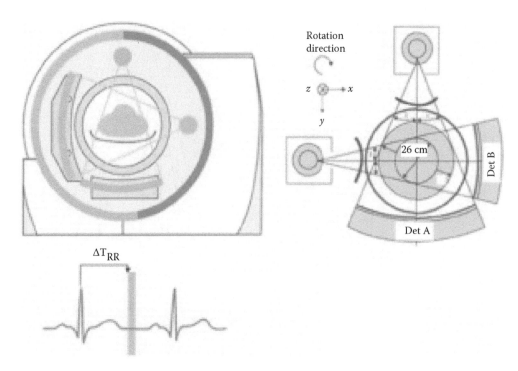

Figure 11.18 The diagram showing the schematic of data acquisition in the dual-source-dual-detector CT to make sure the ideal case always occurs while the data corresponding to both sectors came from the identical cycle (Adopted from Flohr, T.G. and Ohnesorge, B.M. *Basic Res Cardiol* 103, 161–73, 2008. With permission).

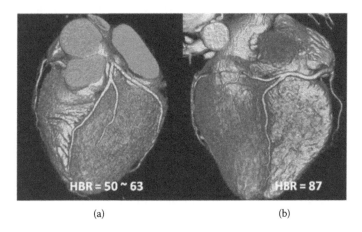

(a)   (b)

Figure 11.19 (See color insert.) 3D surface rendering of the heart generated by a single-source single-detector MDCT (a) and a dual-source dual-detector MDCT (b) (Courtesy of Siemens Healthcare, Malvern, PA.)

number may possess a higher mass density. It occurs often in the diagnosis of stenosis with CT angiography that the iodine-contrast in vessel lumen may not be differentiable from the calcified plaques attached to vessel wall. However, the difficulty in such situations can be overcome using the dual-kVp scan capability that is newly available in state-of-the-art MDCT scanners.

11.7.3.1 Separation between material atomic number and mass density

A brief review of Equation 11.4 tells us that, if a pair of scans at high and low monochromatic energies can be made, respectively, one has (Alvarez and Macovski 1976; Alvarez and Seppi 1979)

$$I_{\text{low}} = \int_L \alpha(x,y)\,dl\,f_c(E_{\text{low}}) + \int_L \beta(x,y)\,dl\,f_p(E_{\text{low}})$$

$$\cong A_\alpha f_c(E_{\text{low}}) + A_\beta f_p(E_{\text{low}}), \tag{11.6}$$

$$I_{\text{high}} = \int_L \alpha(x,y)\,dl\,f_c(E_{\text{high}}) + \int_L \beta(x,y)\,dl\,f_p(E_{\text{high}})$$

$$\cong A_\alpha f_c(E_{\text{high}}) + A_\beta f_p(E_{\text{high}}), \tag{11.7}$$

where E_{low} and E_{high} correspond to the monochromatic energy at high and low levels, and

$$A_\alpha \equiv \int_L \alpha(x,y)\,dl \tag{11.8}$$

$$A_\beta \equiv \int_L \beta(x,y)\,dl. \tag{11.9}$$

Both $f_c(E_{\text{low}})$ and $f_p(E_{\text{high}})$ can be calculated according to Equations 11.2 and 11.3. Equations 11.6 and 11.7 are actually two simultaneous linear equations; thus, $A\alpha$ and $A\beta$ can be analytically solved given the intensity measurements I_{low} and I_{high}. Subsequently, according to Equations 11.8 and 11.9, $\alpha(x,y)$ that is determined by the atomic number and $\beta(x,y)$ by the mass density can be reconstructed via numerous image reconstruction algorithms.

Because only polychromatic x-ray sources are currently available in practice, the dual energy MDCT imaging can be implemented only via dual-kVp CT scans. Starting from Equation 11.5 and exercising the same logic in getting Equations 11.6 and 11.7, we have (Alvarez and Macovski 1976; Alvarez and Seppi 1979; Lehmann et al. 1981)

$$
\begin{aligned}
I_{\text{low}} &= \int_E S_{\text{low}}(E)\{\int_L \alpha(x,y)\,dl\}f_c(E)\,dE \\
&+ \int_E S_{\text{low}}(E)\{\int_E \beta(x,y)\,dl\}f_p(E)\,dE \\
&\equiv \int_E S_{\text{low}}(E)A_\alpha f_c(E)\,dE + \int_E S_{\text{low}}(E)A_\beta f_p(E)\,dE \quad (11.10)
\end{aligned}
$$

$$
\begin{aligned}
I_{\text{high}} &= \int_E S_{\text{high}}(E)\{\int_L \alpha(x,y)\,dl\}f_c(E)\,dE \\
&+ \int_E S_{\text{high}}(E)\{\int_E \beta(x,y)\,dl\}f_p(E)\,dE \\
&\equiv \int_E S_{\text{high}}(E)A_\alpha f_c(E)\,dE + \int_E S_{\text{high}}(E)A_\beta f_p(E)\,dE. \quad (11.11)
\end{aligned}
$$

Equations 11.10 and 11.11 are no longer simultaneous linear equations; thus, $A\alpha$ and $A\beta$ have to be obtained via data fitting. For example, through a third-order polynomial data fitting, one has

$$
\begin{aligned}
I_{\text{low}} &\equiv \lambda_0 + \lambda_1 A_\alpha + \lambda_2 A_\beta + \lambda_3 A_\alpha^2 + \lambda_4 A_\alpha A_\beta + \lambda_5 A_\beta^2 \\
&+ \lambda_6 A_\alpha^3 + \lambda_7 A_\alpha^2 A_\beta + \lambda_8 A_\alpha A_\beta^2 + \lambda_9 A_\beta^3 \quad (11.12)
\end{aligned}
$$

$$
\begin{aligned}
I_{\text{high}} &\equiv \chi_0 + \chi_1 A_\alpha + \chi_2 A_\beta + \chi_3 A_\alpha^2 + \chi_4 A_\alpha A_\beta + \chi_5 A_\beta^2 \\
&+ \chi_6 A_\alpha^3 + \chi_7 A_\alpha^2 A_\beta + \chi_8 A_\alpha A_\beta^2 + \chi_9 A_\beta^3. \quad (11.13)
\end{aligned}
$$

Coefficients $\lambda_0, \lambda_1, \lambda_2, \ldots, \lambda_9$ and $\chi_0, \chi_1, \chi_2, \ldots, \chi_9$ can be attained either analytically or experimentally, and such a process is termed as system calibration (Alvarez and Macovski 1976; Alvarez and Seppi 1979; Lehmann et al. 1981; Kalender et al. 1986; Chuang and Huang 1988; Heismann et al. 2003; Walter et al. 2004; Liu et al. 2008; Zou and Silver 2008, 2009; Liu et al. 2009; Yu et al. 2009). Once these coefficients are obtained, $A\alpha$ and $A\beta$ can be obtained from I_{low} and I_{high} with algorithms to solve the nonlinear simultaneous equations. This means that an MDCT image corresponding to the distribution of mass attenuation coefficient at a sectional slice of patient can be separated into two images corresponding to the distribution of atomic number and mass density, respectively, and the clinical relevance of such a separation cannot be over appreciated.

11.7.3.2 Material decomposition

The separation of atomic and mass density images is a straightforward application of Equations 11.2 through 11.5. A further development in dual-kVp MDCT imaging is the material decomposition (Kalender et al. 1986) illustrated below, which is of even more relevance in the clinic.

Suppose two materials are given and their mass attenuation coefficients at pixel (x, y) and x-ray photon energy can be represented as

$$
\mu_1(x,y;E) = \alpha_1(x,y)f_c(E) + \beta_1(x,y)f_p(E) \quad (11.14)
$$

$$
\mu_2(x,y;E) = \alpha_2(x,y)f_c(E) + \beta_2(x,y)f_p(E). \quad (11.15)
$$

From Equations 11.14 and 11.15, it is not hard to attain

$$
\begin{aligned}
f_c(E) &= \frac{\mu_1(x,y)\beta_2(x,y) - \mu_2(x,y)\beta_1(x,y)}{\alpha_1(x,y)\beta_2(x,y) - \alpha_2(x,y)\beta_1(x,y)} \\
&= \frac{\beta_2(x,y)}{\alpha_1(x,y)\beta_2(x,y) - \alpha_2(x,y)\beta_1(x,y)}\mu_1(x,y) \\
&- \frac{\beta_1(x,y)}{\alpha_1(x,y)\beta_2(x,y) - \alpha_2(x,y)\beta_1(x,y)}\mu_2(x,y) \quad (11.16)
\end{aligned}
$$

$$
\begin{aligned}
f_p(E) &= \frac{\mu_2(x,y)\alpha_1(x,y) - \mu_1(x,y)\alpha_2(x,y)}{\alpha_1(x,y)\beta_2(x,y) - \alpha_2(x,y)\beta_1(x,y)} \\
&= \frac{\alpha_2(x,y)}{\alpha_1(x,y)\beta_2(x,y) - \alpha_2(x,y)\beta_1(x,y)}\mu_1(x,y) \\
&- \frac{\alpha_1(x,y)}{\alpha_1(x,y)\beta_2(x,y) - \alpha_2(x,y)\beta_1(x,y)}\mu_2(x,y). \quad (11.17)
\end{aligned}
$$

Hence, as for another material, its mass attenuation coefficient at pixel (x, y) can be represented by

$$
\begin{aligned}
\mu(x,y;E) &= \alpha(x,y)f_c(E) + \beta(x,y)f_p(E) \\
&\cong \alpha_1(x,y)\mu_1(x,y;E) + \alpha_2(x,y)\mu_2(x,y;E), \quad (11.18)
\end{aligned}
$$

where

$$
a_1(x,y) = \frac{\alpha(x,y)\beta_2(x,y) - \beta(x,y)\alpha_2(x,y)}{\alpha_1(x,y)\beta_2(x,y) - \alpha_2(x,y)\beta_1(x,y)} \quad (11.19)
$$

$$
a_2(x,y) = \frac{\beta(x,y)\alpha_1(x,y) - \alpha(x,y)\beta_1(x,y)}{\alpha_1(x,y)\beta_2(x,y) - \alpha_2(x,y)\beta_1(x,y)}. \quad (11.20)
$$

Note that $a_1(x, y)$ and $a_2(x, y)$ are linear combinations of $\alpha(x, y)$ and $\beta(x, y)$, respectively, which can be obtained via reconstruction (Equations 11.8 and 11.9) from the data A_α and A_β obtained from Equations 11.12 and 11.13. Let us have a comparison between Equation 11.1 and 11.8 in detail. Conceptually, Equation 11.1 implies that the mass attenuation of a material is a function in the functional space spanned by the two base functions $f_c(E)$ and $f_p(E)$, whereas Equation 11.18 implies that the mass attenuation of a material is a function in the functional space spanned by the two base functions $\mu_1(x, y; E)$ and $\mu_2(x, y; E)$ corresponding to

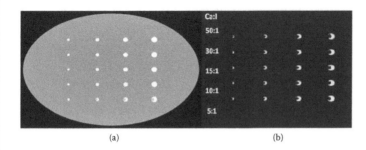

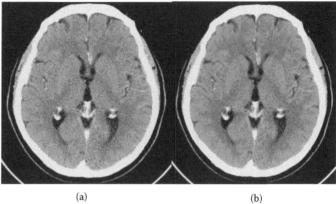

Figure 11.20 Transverse image of an I-Ca phantom consisting of cylinders made of I and Ca scanned at single peak voltage (a) and dual peak voltage (b).

Figure 11.21 Images generated without IRIS (a) and with IRIS (b) in which the noise is reduced significantly while the neurological detail is maintained (images in g are courtesy of Siemens Healthcare, Malvern, PA, http://www.medical.siemens.com/siemens/en_US/gg_ct_FBAs/files/Case_Studies/CT_IRIS_final.pdf.)

the two different materials. This means that any material can be decomposed into two materials that are the projections on the two base materials (Kalender et al. 1986; Chuang and Huang 1988; Heismann et al. 2003; Walter et al. 2004; Liu et al. 2008; Zou and Silver 2008, 2009; Boll et al. 2009; Graser et al. 2009; Liu et al. 2009; Yu et al. 2009). It is important to note that $a_1(x, y)$ and $a_2(x, y)$ in Equation 11.18 have no dependence on x-ray photon energy, because the x-ray energy dependence has been taken into account by the mass attenuation coefficients $\mu_1(x, y; E)$ and $\mu_2(x, y; E)$ of the two base materials (Lehmann et al. 1981; Kalender et al. 1986).

Figure 11.20 is an example of the material decomposition in dual-kVp MDCT imaging in which an I-Ca phantom is used. Each cylindrical rod in Figure 11.20a consists of Ca and I, respectively, and their mass densities are deliberately manipulated to make them nondifferentiable from each other at single-kVp MDCT imaging (Figure 11.20a). However, as demonstrated in Figure 11.20b, via material decomposition with Ca and I as the base materials, the I at low mass density can be readily differentiated from the Ca at high mass density. With such a capability of differentiating Ca from I, a physician can diagnose the stenosis in carotid or coronary arteries with much higher confidence and accuracy. The application of dual-kVp MDCT imaging is quickly growing in the clinic, and interested readers are referred to other literature covering its current and future application (Boll et al. 2009; Graser et al. 2009).

11.7.4 REDUCTION OF NOISE AND RADIATION DOSE

It is always desired in the clinic to detect pathological lesions at high spatial resolution and low radiation noise, with resort to certain imaging processing methods. In general, however, if linear image processing methods are used, this desire can never be fulfilled, because there are always trade-offs between the spatial resolution and noise in CT imaging (Chesler et al. 1977; Hanson 1979, 1981; Ritman 2008), as one may have experienced in the situations wherein the so-called "STAND" or "BONE" filter kernels are used for image reconstruction. In practice, the techniques of modulating x-ray tube current according to the angular and longitudinal variation in patient's body habitus have been used to significantly reduce the radiation dose (Klara et al. 2004a, 2004b). Several nonlinear shift-variant approaches in image space to significantly reduce noise while maintaining spatial resolution have been proposed and implemented in MDCT for neurological, body, and cardiovascular applications.

These image space-based methods vary in implementations but have the following features in common: (1) noise map–guided anisotropic diffusion (Perona and Malik 1990; Gerig et al. 1992; Black et al. 1998), (2) preservation and even boosting of edge, and (3) blending of the nonlinear processed image with the original image reconstructed by the FBP algorithm to make the appearance of the finally obtained images similar to that of conventional CT images. Figure 11.21 (right) is an MDCT image of the basal ganglia with the application of such a nonlinear method called iterative reconstruction in image space (IRIS) (Yang et al. 2011; http://www.medical.siemens.com/siemens/en_US/gg_ct_FBAs/files/Case_Studies/CT_IRIS_final.pdf) and that of the original image (Figure 11.21, left) for comparison. It is interesting to note that because the anisotropic diffusion is usually carried out in the manner of iteration, these nonlinear approaches have been claimed as iterative image reconstruction by MDCT vendors, even though all these nonlinear approaches are confined to be carried out in image space only. In light of the widely accepted concept of iterative image reconstruction wherein the back-and-forth operations between the projection and image spaces are essential (Shepp and Vardi 1982; Lange and Carson 1984; Bouman and Sauer 1993, 1996; Barret et al. 1994; Wilson et al. 1994; Lange and Fessler 1995; Fessler 1996; Fessler and Rogers 1996; Saquib et al. 1996; Wang and Gindi 1997; Fessler and Booth 1999; Fessler 2000; Qi and Leathy 2000; Qi 2003, 2005; De Man et al. 2005; Thibault et al. 2007; Xu et al. 2009), these controversial claims have triggered debate in the community of CT imaging.

One may intuitively think that a reduction of noise in an MDCT image can result in radiation dose savings by observing the "square-root" rule, that is, a k times reduction in noise result in k^2 times saving in radiation dose, and vice versa. Nevertheless, it is important to clarify that this intuitive logic works only in the case in which linear methods are used. If nonlinear methods are used, this square-root rule may not hold anymore. In addition, these nonlinear approaches are usually shift-variant from the perspective of image processing. Hence, one has to be cautious about the appealing claims in radiation dose savings made by the vendors whenever nonlinear shift-variant methods are used to support such claims.

11.8 CLINICAL APPLICATIONS OF MDCT

As one of the most popular imaging modalities, MDCT is playing a significant role in routine clinical practice (Rogalla et al. 2009). Numerous investigations have been conducted to evaluate and verify MDCT's sensitivity and specificity in cardiovascular, thoracic, abdominal, and neurologic applications and the imaging of extremities. A detailed discussion about MDCT's clinical applications is beyond the scope of this chapter; interested readers are referred to the large body of introductory, review, and research papers published in the literature (Rydberg et al. 2000). For readers to have a broader impression about the significant role that is being played by MDCT in the clinic, several important clinical applications in addition to the examples that have already been presented earlier are provided in Figure 11.22.

11.9 RADIATION DOSE IN MDCT

The metric of radiation dose in CT is defined as the CTDI (U.S. Nuclear Regulatory Commission), a value that is theoretically the integral of dose profile corresponding to the aperture of x-ray beam from negative to positive infinite. Apparently, such

a definition is not feasible in practice (see Equation 11.21). In the early days of CT technology, the Food and Drug Administration (FDA) specified a more feasible definition given in Equation 11.22 (FDA 1980). Nowadays, the $CTDI_{100}$ is the definition (Equation 11.23) that has been widely accepted, in which a 100-mm-long pencil ion chamber is used to measure the exposure that is then converted to the radiation dose (air kerma) to soft tissue (Bushberg et al. 2002). Considering the human body's attenuation, the weighted CT dose index $CTDI_W$ defined in Equation 11.24 has become routinely used in the clinic and has been extended for spiral/helical CT scan by taking the spiral/helical pitch into account (see Equation 11.25) (McCollough et al. 2008). Note that the length of the pencil ion chamber to measure $CTDI_{100}$ is only 100 mm along the longitudinal direction. However, as one has already experienced in the 320-detector row CT, the longitudinal beam aperture in the clinic can be up to 160 mm, which exceeds the longitudinal range defined by $CTDI_{100}$ and its derivatives $CTDI_W$ and $CTDI_{vol}$. Hence, immediate actions by the federal or state regulatory agencies to define new radiation dose phantoms and metrics that can accommodate the MDCT scanner with the x-ray beam aperture larger than 100 mm are anticipated.

$$CTDI = \frac{1}{NT} \int_{-\infty}^{\infty} D(z)\,dz \qquad (11.21)$$

$$CTDI = \frac{1}{NT} \int_{-7T}^{7T} D(z)\,dz \qquad (11.22)$$

$$CTDI_{100} = \frac{1}{NT} \int_{-50mm}^{50mm} D(z)\,dz \qquad (11.23)$$

$$CTDI_w = \frac{1}{3} CTDI_{100,\,center} + \frac{2}{3} CTDI_{100,\,peripheral} \qquad (11.24)$$

$$CTDI_{vol} = \frac{1}{pitch} CTDI_w. \qquad (11.25)$$

The ever-increasing radiation dose rendered by CT, particularly MDCT, to the population has been drawing concerns in the public (FDA 1980; International Commission on Radiological Protection 1991; McCollough et al. 2006, 2008; American College of Radiology 2008; National Council on Radiation Protection and Measurement 2008; Yu et al. 2009). According to Report 160 of the National Council on Radiological Protection (NCRP), up to 2006, the effective radiation dose contributed by all medical imaging modalities to an individual in the U.S. population accounts for 48% (3.0 mSv) of that from all natural and artificial sources (6.1 mSv), of which the contribution from CT alone is 24% (1.5 mSv). Hence, the importance of accurately measuring the radiation dose rendered by MDCT with large beam aperture can never be overstated. However, detailed coverage on the radiation dose of MDCT is beyond the scope of this chapter. Interested readers are referred to Chapter 5 of this book and numerous references in the literature (FDA 1980;

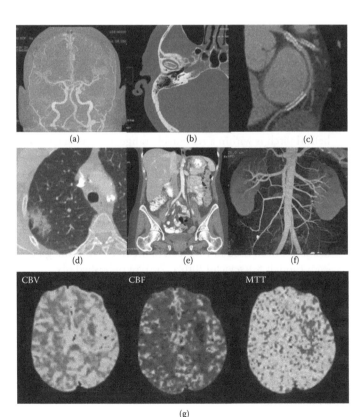

Figure 11.22 (See color insert.) Typical clinical application of MDCT imaging. (a) Head CT angiography. (b) Temporal bone. (c) Coronal artery stent. (d) Lung cancer. (e) Abdominal/pelvic. (f) Renal angiography. (g) CT perfusion for the evaluation of acute stroke (images in g are courtesy of GE Healthcare, Buckinghamshire, UK, http://www.gehealthcare.com/euen/ct/pdf/CTClarity2009_Spring.pdf, accessed on 09/28/2011.)

International Commission on Radiological Protection 1991; McCollough et al. 2006, 2008; American College of Radiology; National Council on Radiation Protection and Measurement 2008; Yu et al. 2009; U.S. Nuclear Regulatory Commission 2013).

11.10 DISCUSSION

An introductory review on MDCT imaging provided in this chapter covers its physics, system architecture, data acquisition modes, imaging performance evaluation, image reconstruction solutions, typical clinical applications, and recent technological advancement. Before ending this chapter, I discuss the future of MDCT technology, from a similar and also expanded perspective in to what has been discussed in the literature (Pan et al. 2008; Wang et al. 2008).

First, I speculate how many detector rows would eventually be available in MDCT. The number of detector rows is driven by the clinical desire to cover large organs in human body with one gantry rotation and the fabrication cost of CT detector. Displayed in Figure 11.23 are the typical longitudinal ranges corresponding to the major organs in human body. Most likely, the ultimate goal of MDCT is to cover the entire heart in one gantry rotation, so that the interslab discontinuity caused by the inconsistency in cardiac motion or contrast agent circulation can be avoided. The longitudinal range of the heart for the majority of the population is approximately 160 mm. Hence, the number of detector rows is 320, if the detector row width is 0.5 mm as we have already seen in the market; or 256, if the detector row width is 0.625 mm, as we may see very soon in the market. All other organs with their longitudinal range larger than that of the heart would be scanned by the spiral/helical modes of MDCT, as we are conducting as a routine in the clinic.

Second, I discuss the accuracy of image reconstruction solutions in MDCT, especially its prognosis with increasing number of detector rows. Theoretically, only the image reconstruction of the SDCT at axial scan is accurate. All other reconstruction solutions, starting from the spiral/helical scan

in SDCT and the axial scan in MDCT with the number of detector row more than one, are all approximate. This fact may be surprising but is what has happened so far in the SDCT and MDCT and most likely will continue in the future. One may have to be cautious about the reconstruction accuracy that can be achieved by upcoming state-of-the-art CT scanners with an increasing number of detector rows. The following points may be informative for reader's scrutiny about the reconstruction accuracy:

1. In the axial scan, owing to the cone angle spanned by detector rows that are not located within the central plane determined by the source trajectory, even an MDCT with only two detector rows in principle does not satisfy the so-called data sufficiency condition (DSC) (Tuy 1983). The greater the number of detector rows, the more severe the cone beam artifact, as demonstrated in Figure 11.24, wherein a phantom consisting of seven identical discs stacked parallel to each other along the craniocaudal direction is used to highlight the cone beam artifacts. The root cause of cone beam artifact is the violation of the DSC, and it may manifest itself as (1) streak-like shading or glaring adjacent to high contrast structures, (2) dropping of CT number (or Hounsfield unit) at the pixels that are not located within the central plane, and (3) geometric distortion. Artifacts 1 and 2 may be correctable with empirical approaches (Forthmann et al. 2009), but artifact 3—the geometric distortion—may result in distorted shape of organs and is much more difficult, if not impossible, to correct. It may be argued that no anatomic structure like the discs shown in Figure 11.24 exists in human body. However, the cone angle and the artifacts caused by it are indeed an open problem to be overcome in MDCT technology.

2. The spiral/helical scan of MDCT actually satisfies the DSC (Forthmann et al. 2009), as long as the effective spiral/helical pitch is within a reasonable range. For instance, the allowable spiral/helical pitch in MDCT scan is dependent on gantry geometry and detector deployment, and a reasonable spiral/helical pitch up to 1.5:1 is routinely used in the clinic (Heuscher et al. 2004; Stierstorfer et al. 2004; Taguchi et al. 2004; Tang et al. 2006, 2008; Tang and Hsieh 2007). However, no theoretically accurate image reconstruction has so far been used in this scan mode in MDCT imaging, even though Katsevich (2002a, 2002b) published his breakthrough accurate reconstruction algorithm for spiral/

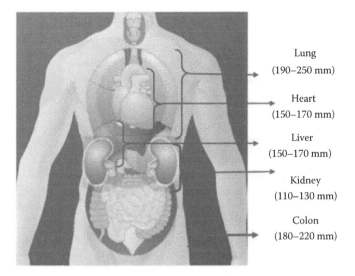

Lung
(190–250 mm)

Heart
(150–170 mm)

Liver
(150–170 mm)

Kidney
(110–130 mm)

Colon
(180–220 mm)

Figure 11.23 Diagram showing the longitudinal range of the major organs in human body.

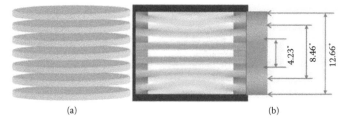

(a)   (b)

Figure 11.24 Defrise phantom (a) and the tomographic images in coronal view reconstructed from the projection acquired along a circular source trajectory (b), whereby the relationship between artifact severity and cone angle is illustrated.

helical scan right before the launching of the 16-detector row CT in the market by all major CT vendors. The most distinct feature of Katsevich's algorithm and its derivatives is the conducting of filtering along the white curves shown in Figure 11.25. Another important feature is the handling of data redundancy with the Tam–Danielsson (Tam 1995; Danielsson et al. 1997) window that also is shown in Figure 11.25 by the curves in red (indicated by red solid arrows). The DSC is satisfied, as long as the boundary of the Tam–Danielsson window, which is dependent on the spiral/helical pitch, is within the dimension of an MDCT detector. It has been experimentally evaluated and verified that at a cone angle up to 4.5°, an angle that approximately corresponds to that spanned in a 64-detector row CT, there is no dominant advantage in reconstruction accuracy by Katsevich's algorithm over the approximate solutions that use various weighing schemes to suppress the artifacts caused by data truncation and inadequate handling of the data redundancy (Tang et al. 2005, 2008).

3. Even though the number of detector rows in MDCT continues to increase, the number of detector rows used for spiral/helical scanning in the clinic may not exceed 64 or 128; thus, the increase in the number of detector rows is mainly to benefit the axial scan for covering a large organ within one gantry rotation. One primary reason accountable for this limitation is that, at given spiral/helical pitch, detector row width and gantry rotation speed, for example, 1:1, 0.625 mm and 0.5 s/r, respectively, an x-ray beam aperture larger than 128 mm × 0.625 = 80 mm results in a motion of patient table at a speed of 160 mm/s, a speed that may cause unacceptable patient discomfort due to the acceleration at the start and deceleration at the end of the scan. Moreover, a patient table proceeding at such a high speed may substantially advance the contrast agent circulation, making the bolus chasing routinely conducted in the clinic no longer feasible.

The image space–based nonlinear shift-variant noise reduction methods (Fan et al. 2010; Yang et al. 2011; http://www.medical.

siemens.com/siemens/en_US/gg_ct_FBAs/files/Case_Studies/ CT_IRIS_final.pdf) are playing an increasingly important role in the clinic. However, in comparison with the statistical iterative or optimization-based image reconstruction solutions, the efficacy of these image space-based solutions from the perspective of noise reduction or dose saving is limited. As for the statistical iterative image reconstruction solutions, encouraging data have been demonstrated for certain clinical applications (https://www.medical.siemens.com/siemens/it_IT/gg_ct_FBAs/files/brochures/SAFIRE_Brochure.pdf; http://www.gehealthcare.com/euen/ct/pdf/CT-Clarity-Spring-2011.pdf). Fairly speaking, the statistical iterative reconstruction solution needs more intensified computation and thus a much more powerful computation engine than that of analytic image reconstruction solutions, for example, the algorithms in the manner of FBP. But this may not be the real cause of the delayed availability of statistical iterative image reconstruction solutions for routine applications in the clinic. The real root cause is more likely its robustness over clinical applications and patients. The image quality of the MDCT provided by the existing image reconstruction solution has already been superior. To make the image quality even better, aggressive regularization schemes have to be exercised by the statistical iterative image reconstruction solutions, but they may result in unexpected artifacts over anatomic areas or patients. In principle, the statistical iterative image reconstruction is an optimization-based solution in which an accurate modeling of the imaging chain of MDCT at high fidelity is critical to its success. However, in practice, it would be very hard, if not impossible, to accurately model an imaging system. Moreover, a patient is actually a central component in the modeling of an imaging system when the optimization-based statistical iterative reconstruction is used. Recognizing the variety of anatomic structures over patients, tremendous effort may still be needed to make the statistical iterative image reconstruction solution routinely and reliably running in the clinic.

The energy resolution implemented by dual-kVp scan is the latest major addition to MDCT's capability for clinical applications. However, the potential of energy resolution is limited by the technologies that are currently available in MDCT. If more advanced technologies, such as the photon counting (Taguchi et al. 2010; Wang et al. 2011) detector with high counting rate, energy resolution, and spatial resolution are available, the energy resolution of MDCT may lead to breakthrough advancement for advanced clinical applications. For example, if its potential were fully realized, the capability of material differentiation at high spatial resolution may enable MDCT to substantially improve its contrast sensitivity, a feature paramount importance in the early detection of tumor.

A frequently asked question related to MDCT imaging is, Would the MDCT for general diagnostic imaging merge in the future with the flat panel imager–based CBCT aimed at special applications? Two prerequisites are mandatory to fulfill if such a merge can eventually become a reality: (1) the absorption and conversion efficiency of the sodium iodine (NaCl) or other scintillator based flat panel imager needs to be substantially improved to reach that of the x-ray detectors used in MDCT and (2) the data acquisition speed and transferring bandwidth of thin-film transistor (TFT) in the flat panel imager need to be

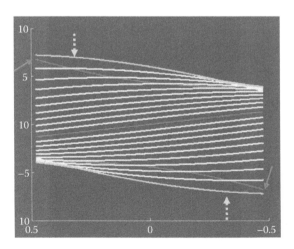

Figure 11.25 Schematic diagram showing the curves (white, no arrow) along which the filtering required by Katsevich-type algorithms is carried out, the Tam–Danielson window indicated by the two red curves (solid arrows), and the boundary of data detection indicated by the outmost curves (dashed arrow).

improved substantially. It should be noted that as an imaging device for x-ray radiography and fluoroscopic procedures, the detection efficiency and data transferring bandwidth of the flat panel imager are sufficient to replace the screen/film radiography or the image intensifier and TV-based fluoroscopy. Nevertheless, many more projection views are needed in CT because it demands a high x-ray quantum detection efficiency to reduce radiation dose. Unless the two prerequisites are fulfilled, the flat panel imager–based CBCT would remain as an imaging modality for special-purpose applications, such as dental, image-guided radiation therapy, and image-guided surgery.

Finally, I sketch a landscape of the technological advancement that is occurring in MDCT (Figure 11.26). The slice war in the past decade has driven the major MDCT vendors to not only pass the milestones in the number of detector rows but also make the imaging performance of MDCT better in contrast, spatial, temporal, and the very recently added energy resolution. Both hardware and software are the enablers of the technological advancement, but the hardware-based methods, such as the dual-source dual-detector MDCT system for improving the temporal resolution of cardiovascular imaging, are the cornerstone. In addition, one should pay close attention to the technological advancement in biomarker-targeted contrast agent (Hainfeld et al. 2006; Hyafil et al. 2007; Desai and Schoenhagen 2009; Chithrani et al. 2010a, 2010b; Hallouard et al. 2010; Lee et al. 2010). At present, almost all contrast agents used in the MDCT imaging are I-based organic compounds based on the mechanism of blood compartment retention (Idee et al. 2002). The molecular size of the iodine-contrast agent is relatively small; thus, the contrast agents are removed from circulation very quickly (in seconds) via renal excretion. Also, the nanoparticulation of contrast agent has been the subject of research to enable the retention of contrast agents in human body for a substantially prolonged period by escaping the renal excretion and reticuloendothelial clearance. Moreover, via biomarker-targeted delivery, the subject contrast of pathological lesions, such as tumor and vulnerable plaque in atherosclerosis, can be substantially improved. All these technological advancements inspire us to anticipate that the MDCT will play a more significant role in routine clinical practice in the future,

and even a significant role in molecular imaging (Weissleder and Mahmood 2001; Czernin et al. 2006) wherein the subject contrast is of essence.

ACKNOWLEDGMENTS

I thank Shaojie Tang, PhD, for generating the diagrams in Figure 11.11 and Ms. Jessica Paulishen for proofreading this chapter.

REFERENCES

Alvarez, R. and Seppi, E. 1979. A comparison of noise and dose in conventional and energy selective computed tomography. *IEEE Trans Nucl Sci* 26: 2853–6.

Alvarez, R.E. and Macovski, A. 1976. Energy-selective reconstruction in x-ray computerized tomography. *Phys Med Biol* 21: 733–44.

American College of Radiology. 2008. *ACR Practice Guideline for Diagnostic Reference Levels in Medical X-ray Imaging.*

Barret, H.H., Wilson, D.W. and Tsui, B. 1994. Noise properties of the EM algorithm: I. Theory. *Phys Med Biol* 39: 833–46.

Black, M.J., Sapiro, G., Marimont, D.H. and Heeger, D. 1998. Robust anisotropic diffusion. *IEEE Trans Image Process* 7(3): 421–32.

Boll, D.T., Patil, N.A., Paulson, E.K., Merkle, E.M., Simmons, W.N., Pierre, S.A. and Preminger, G.M. 2009. Renal stone assessment with dual-energy multidetector CT and advanced postprocessing techniques: improved characterization of renal stone composition – pilot study. *Radiology* 250: 813–20.

Bouman, C. and Sauer, K. 1993. A generalized Gaussian image model for edge-preserving MAP estimation. *IEEE Trans Image Process* 2: 296–310.

Bouman, C. and Sauer, K. 1996. A unified approach to statistical tomography using coordinate descent optimization. *IEEE Trans Image Process* 5: 480–92.

Bruder, H., Kachelrieß, M., Schaller, S., Stierstorfer, K. and Kalender, W.A. 2000. Single-slice rebinning reconstruction in spiral cone-beam computed tomography. *IEEE Trans Med Imaging* 19(9): 873–87.

Bushberg, J.T., Seibert, J.A., Leidholdt, E.M., Jr. and Boone, J.M. 2002. *The Essential Physics of Medical Imaging*, 2nd edn. Philadelphia, PA: Lippincott Williams & Wilkins.

Chesler, D.A., Riederer, S.J. and Norbert, N.J. 1977. Noise due to photon counting statistics in computed x-ray tomography. *J Comput Assist Tomogr* 1: 64–74.

Chithrani, D.B., Dunne, M., Stewart, J., Allen, C. and Jaffray, D.A. 2010. Cellular uptake and transport of gold nanoparticles incorporated in a liposomal carrier. *Nanomedicine* 6: 161–9.

Chithrani, D.B., Jelveh, S., Jalali, F., Prooijen, M.V., Allen, C., Bristow, R.G., Hill, R.P. and Jaffray, D.A. 2010. Gold nanoparticles as radiation sensitizer in cancer therapy. *Radiat Res* 173: 719–28.

Chuang, K. and Huang, H.K. 1988. Comparison of four dual energy image decomposition methods. *Phys Med Biol* 33(4): 455–66.

Cody, D.D., Stevens, D.M. and Glnsberg, L.E. 2005. Multi-detector row CT artifacts that mimic diseases. *Radiology* 236: 756–61.

Crawford, C.R. and King, K.F. 1990. Computed tomography scanning with simultaneous patient translation. *Med Phys* 17: 967–82.

Czernin, J., Weber, W.A. and Herschman, H.R. 2006. Molecular imaging in the development of cancer therapeutics. *Annu Rev Med* 57: 99–118.

Danielsson, P.E., Edholm, P., Eriksson, J. and Magnusson-Seger, M. 1997. Towards exact 3D-reconstruction for helical cone-beam scanning of long objects: a new arrangement and a new completeness condition. *International Meeting on Fully Three-dimensional Image Reconstruction in Radiology and Nuclear Medicine*, (Pittsburgh) ED Townsend and Kinahan P E 25-8.

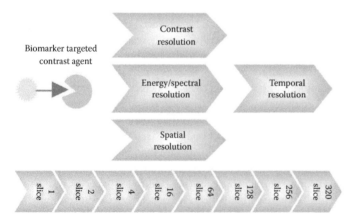

Figure 11.26 The diagram showing the landscape of technologic advancement in MDCT.

De Man, B., Basu, S., Thibault, J., Hsieh, J., Fessler, J.A., Sauer, K. and Bouman, C. 2005. A study of different minimization approaches for iterative reconstruction in x-ray CT. *Proceedings of the IEEE Nuclear Science Symposium and Medical Imaging Conference*, v. 3, pp. 2708-2710, San Juan, Puerto Rico, Oct. 23-29, 2005.

Desai, M.Y. and Schoenhagen, P. Emergence of targeted molecular imaging in atherosclerotic cardiovascular disease. *Expert Rev Cardiovasc Ther* 7(2): 197–203.

Ertl-Wagner, B., Eftimov, L., Blume, J., Bruening, R., Becker, C., Cormack, J., Brueckmann, H. and Reiser, M. 2008. Cranial CT with 64-, 16-, 4- and single-slice CT systems–comparison of image quality and posterior fossa artifacts in routine brain imaging with standard protocols. *Eur Radiol* 18: 1720–26.

Fan, J., Hsieh, J., Sainath, P. and Crandall, P. 2010. Evaluation of the low dose cardiac CT imaging using ASIR technique. *SPIE Proc* 7622 (76222U): 9p.

FDA dose definition: Shope, T.B., Gagne, R.M. and Johnson, G.C. 1980. A method for describing the doses delivered by transmission x-ray computed tomography. *Med Phys* 8: 488–95.

Feldkamp, L.A., Davis, L.C. and Kress, J.W. 1984. Practical cone-beam algorithm. *J Opt Soc Am A* 1:612–9.

Fessler, J. 2000. Statistical image reconstruction methods for transmission tomography. In: Sonka, M. and Fitzpatrick, J.M. (eds.). *Handbook of Medical Imaging, Vol. 2, Medical Imaging Processing and Analysis*. Bellingham, WA: SPIE, 1–70.

Fessler, J.A. 1996. Mean and variance of implicitly defined biased estimators (such as penalized maximum likelihood): application to tomography. *IEEE Trans Med Imaging* 5: 493–506.

Fessler, J.A. and Booth, S.D. 1999. Conjugate-gradient preconditioning methods for shift-variant reconstruction. *IEEE Trans Med Imaging* 8: 688–99.

Fessler, J.A. and Rogers, W.L. 1996. Spatial resolution properties of penalized-likelihood image reconstruction: space-invariant tomographs. *IEEE Trans Image Process* 5: 1346–58.

Flohr, T., Bruder, H., Stierstorfer, K., Petersilka, M., Schmidt, B. and McCollough, C.H. 2008. Image reconstruction and image quality evaluation for a dual source CT scanner. *Med Phys* 35: 5882–97.

Flohr, T. and Ohnesorge, B. 2000. Heart rate adaptive optimization of spatial and temporal resolution for electrocardiogram-gated multislice spiral CT of the heart. *J CAT* 27(6): 907–23.

Flohr, T., Stierstorfer, K., Bruder, H., Simon, J., Polacin, A. and Schaller, S. 2003. Image reconstruction and image quality evaluation for a 16-slice CT scanner. *Med Phys* 30: 2650–62.

Flohr, T., Stierstorfer, K., Ulzheimer, S., Bruder, H., Primak, A.N. and McCollough, C.H. 2005. Image reconstruction and image quality evaluation for a 64-slice CT scanner with z-flying focal spot. *Med Phys* 32: 2536–47.

Flohr, T.G. and Ohnesorge, B.M. 2008. Imaging of the heart with computed tomography. *Basic Res Cardiol* 103: 161–73.

Flohr, T.G., Stierstortfer, K., Suss, C., Schmidt, B., Primak, A.N. and McCollough, C.H. 2007. Novel ultrahigh resolution data acquisition and image reconstruction for multi-detector row CT. *Med Phys* 34: 1712–23.

Forthmann, P., Grass, M. and Proksa, R. 2009. Adaptive two-pass cone-beam artifact correction using a FOV-preserving two-source geometry: a simulation study. *Med Phys* 36: 4440–50.

Gerig, G., Kübler, O., Kikinis, R. and Jolesz, F.A. 1992. Nonlinear anisotropic filtering of MRI data. *IEEE Trans Med Imaging* 11(2): 221–32.

Graser, A., Johnson, T.R.C., Chandarana, H. and Macari, M. 2009. Dual energy CT: preliminary observations and potential clinical applications in the abdomen. *Eur Radiol* 19: 13–23.

Hainfeld, J.F., Slatkin, D.N., Focella, T.M. and Smilowitz, H.M. 2006. Gold nanoparticles: a new x-ray contrast agent. *Br J Radiol* 79: 248–53.

Hallouard, F., Anton, N., Choquet, P., Constantinesco, A. and Vandamme, T. 2010. Iodinated blood pool contrast media for preclinical x-ray imaging applications – a review. *Biomaterials* 31: 6249–68.

Hanson, K.M. 1979. Detectability in computed tomography images. *Med. Phys.* 6: 441–51.

Hanson, K.M. 1981. Noise and contrast discrimination in CT. In: Newton, T.H. and Potts, D.G. (eds.). *Ridiology of the Skull and Brain, vol. V: Technical Aspects of Computed Tomography.* 3941–55.

Heismann, B.J. Leppert, J. and Stierstorfer, K. 2003. Density and atomic number of measurements with spectral x-ray attenuation method. *J Appl Phys* 94: 2074–9.

Heuscher, D. 2002. Cone beam scanner using oblique surface reconstructions. US Patent Pub. No. 2002/0122529 A1.

Heuscher, D., Brown, K. and Noo, F. 2004. Redundant data and exact helical cone-beam reconstruction. *Phys Med Biol* 49: 2219–38.

Hsieh, J., Londt, J., Vass, M., Li, J., Tang, X. and Okerlund, D. 2006. Step-and-shoot data acquisition and reconstruction for cardiac x-ray computed tomography. *Med Phys* 33(11): 4236–48.

Hu, H. 1999. Multi-slice helical CT: scan and reconstruction. *Med Phys* 26: 5–18.

Hyafil, F., Cornily, J.C., Feig, J.E., Gordon, R., Vucic, E., Amirbekian, V., Fisher, E.A., Fuster, V., Feldman, L.J. and Fayad, Z.A. 2007. Noninvasive detection of macrophages using a nanoparticulate contrast agent for computed tomography. *Nat Med* 13(5): 636–41.

Idee, J., Nachman, I., Port, M., Petta, M., Lem, G.L., Greneur, S.L., et al. 2002. *Contrast Agents II: Iodinated Contrast Media – From Non-Specific to Blood-Pool Agent*. Paris: Springer.

International Commission on Radiological Protection. 1991. *1990 Recommendations of the International Commission on Radiological Protection (*Report 60), Annals of the ICRP, 21(1–3).

Johns, H.E. and Cunninham, J.R. 1983. *The Physics of Radiology,* 4th edn. Springfield, IL: Charles C Thomas.

Kachelrieβ, M., Schaller, S. and Kalender, W.A. 2000. Advanced single-slice rebinning in cone-beam spiral CT. *Med Phys* 27(4): 754–72.

Kak, A.C. and Slaney, M. 1988. *Principles of Computerized Tomographic Imaging*. New York: IEEE Press.

Kalender, W.A., Perman, W.H., Vetter, J.R. and Klotz, E. 1986. Evaluation of a prototype dual-energy computed tomographic apparatus. *Med Phys* 13(5): 334–9.

Kalender, W.A., Seissler, W., Klotz, E. and Vock, P. 1990. Spiral volumetric CT with single-breathhold technique, continuous transport, and continuous scanner rotation. *Radiology* 176: 181–83.

Kalender, W.A., Seissler, W. and Vock, P. 1989. Single-breathhold spiral volumetric CT by continuous patient translation and scanner rotation. *Radiology* 173(P): 414.

Kalra, M.K., Maher, M.M., Toth, T.L., Kamath, R.S., Halpern, E.F. and Saini, S. 2004b. Comparison of Z-axis automatic tube current modulation technique with fixed tube current CT scanning of abdomen and pelvis. *Radiology* 232: 347–53.

Kalra, M.K., Maher, M.M., Toth, T.L., Schmidt, B., Westerman, B.L., Morgan, H.T. and Saini, S. 2004a. Techniques and applications of automatic tube current modulation for CT. *Radiology* 233: 649–57.

Katsevich, A. 2002a. Analysis of an exact inversion algorithm for spiral cone-beam CT. *Phys Med Biol* 47: 2583–98.

Katsevich, A. 2002b. Theoretically exact filtered backprojection-type inversion algorithm for spiral CT. *SIAM J Appl Math* 62: 2012–26.

Lange, K. and Carson, R. 1984. EM reconstruction algorithms for emission and transmission tomography. *J Comput Assist Tomogr* 8: 306–16.

Lange, K. and Fessler, J.A. 1995. Globally convergent algorithms for maximum a posteriori transmission tomography. *IEEE Trans Image Process* 4: 1430–50.

Larson, G.L., Ruth, C.C. and Crawford, C.R. 1998. Nutating slice CT image reconstruction. US Patent No. 5802134.

Lee, H., Hoang, B., Fonge, H., Reilly, R.M. and Allen, C. 2010. In vivo distribution of polymeric nanoparticles at the whole-body, tumor, and cellular levels. *Pharmaceut Res* 27: 2343–55.

Lehmann, L.A., Alvarez, R.E., Macovski, A., Brody, W.R., Pelc, N.J., Riederer, S.J. and Hall, A.L. 1981. Generalized image combinations in dual KVP digital radiography. *Med Phys* 8(5): 659–67.

Liang, Y. and Kruger, R.A. 1996. Dual-slice spiral versus single-slice spiral scanning: comparison of the physical performance of two computed tomography scanners. *Med Phys* 23(2): 205–20.

Liu, X., Primak, A.N., Yu, L., McCollough, C.H. and Morin, R.L. 2008. Quantitative imaging of chemical composition using dual-energy, dual-source CT. *SPIE Proc* 6913(69134Z-1): 8 pages (doi: 10.1117/12.773042).

Liu, X., Yu, L., Primak, A.N. and McCollough, C.H. 2009. Quantitative imaging of element composition and mass fraction using dual-energy CT: three-material decomposition. *Med Phys* 36(5): 1602–3.

Lonn, A.H. 1992. Computed tomography system with translatable focal spot. US Patent No. 5173852.

McCollough, C.H., Bruesewitz, M.R. and Kofler, J.M. 2006. CT dose reduction and dose management tools: overview of available options. *RadioGraphics* 26: 503–12.

McCollough, C.H., Cody, D., Edyvean, S., Geise, R., Gould, B., Keat, N., Huda, W., Judy, P., Kalender, W., McNitt-Gray, M., Morin, R., Payne, T., Stern, S., Rothenberg, L., Shrimpton, P., Timmer, J. and Wilson, C. 2008. The measurement, reporting, and management of radiation dose in CT. *AAPM Rep* 96: 1–28.

Molen, A.J. and Geleijns, J. 2006. Overranging in multisection CT: quantification and relative contribution to dose – comparison of four 16-section CT scanners. *Radiology* 242: 208–16.

National Council on Radiation Protection and Measurement. 2008. *Ionizing Radiation Exposure of the Population of the United States*, NCRP Report No. 160, Bethesda, MD: National Council on Radiological Protection and Measurement.

Ohnesorge, B., Flohr, T., Becker, C., et al. 2000. Cardiac imaging by means of eletrocardiagraphically gated multisection spiral CT: initial experience. *Radiology* 217 (2): 564–71.

Pan, X., Siewerdsen, J., La Riviere, P.J. and Kalender, W. 2008. Anniversary paper: development of x-ray computed tomography: the role of medical physics and AAPM from the 1970s to presentanalytical model of the effects of pulse pileup on the energy spectrum recorded by energy resolved photon counting x-ray detectors. *Med Phys* 35: 3728–39.

Parker, D. 1982. Optimal short scan convolution reconstruction for fan beam CT. *Med Phys* 9: 254–7.

Patch, S.K. 2004. Methods and apparatus for weighting of computed tomography data. US Patent No. 6754299.

Perona, P. and Malik, J. 1990. Scale-space and edge detection using anisotropic diffusion. *IEEE Trans Pattern Anal Mach Intell* 12(7): 629–39.

Petersilka, M., Bruder, H., Krauss, B., Stierstorfer, K. and Flohr, T.G. 2008. Technical principles of dual source CT. *J Eur Radiol* 68: 362–68.

Qi, J. 2003. A unified noise analysis for iterative image estimation. *Phys Med Biol* 48: 3505–19.

Qi, J. 2005. Noise propagation in iterative reconstruction algorithms with line searches. *IEEE Trans Nucl Sci* 52: 57–62.

Qi, J. and Leathy, R.M. 2000. Resolution and noise properties of MAP reconstruction for fully 3D PET. *IEEE Trans Med Imaging* 19: 493–506.

Ritman, E.L. 2008. Vision 20/20: increased image resolution versus reduced radiation exposure. *Med Phys* 35(6): 2502–12.

Rogalla, P., Kloeters, C. and Hein, P. 2009. CT technology overview: 64-slice and beyond. *Radiol Clin N Am* 47: 1–11.

Rybicki, F.J., Otero, H.J. Steigner, M.L., Vorobiof, G., Nallamshetty, L., Mitsouras, D., Ersoy, H., Mather, R.T., Judy, P.F., Cai, T., Coyner, K., Schultz, K., Whitmore, A.G. and Di Carli, M.F. 2008. Initial evaluation of coronary images from 320-detetor row computed tomography. *Int J Cardiovasc Imag* 24: 535–46.

Rydberg, J., Buckwalter, K.A., Caldemeyer, K.S., Philips, M.D., Conces, D.J., Aisen, A.M., Persohn, S.A. and Kopecky, K.K. 2000. Multisection CT: scanning techniques and clinical applications. *Radiographics* 20: 1787–806.

Saquib, S.S., Bouman, C. and Sauer, K. 1996. ML parameter estimation for Markov random fields with applications to Bayesian tomography. *IEEE Trans Image Process* 7: 480–92.

Shepp, L. and Vardi, Y. 1982. Maximum likelihood reconstruction for emission tomography. *IEEE Trans Med Imaging* 1: 113–22.

Silver, M. 2000. A method for including redundant data in computed tomography. *Med Phys* 27: 773–4.

Sohval, A.R. and Freundlich, D. 1987. Plural source computerized tomography device with improved resolution. US Patent No. 4637040.

Stierstorfer, K., Rauscher, A., Boese, J., Bruder, H., Schaller, S. and Flohr, T. 2004. Weighted FBP— a simple approximate 3D FBP algorithm for multislice spiral CT with good dose usage for arbitrary pitch. *Phys Med Biol* 49: 2209–18.

Taguchi, K. and Aradate, H. 1998. Algorithm for image reconstruction in multi-slice CT system. *Med Phys* 25: 550–65.

Taguchi, K., Chiang, B. and Silver, M. 2004. New weighting scheme for cone-beam helical CT to reduce the image noise. *Phys Med Biol* 49: 2351–64.

Taguchi, K., Chiang, B.S. and Hein, I.A. 2006. Direct cone-beam cardiac reconstruction algorithm with cardiac banding artifact correction. *Med Phys* 33: 521–39.

Taguchi, K., Frey, EC., Wang, X., Iwanczyk, J.S. and Barber, W.C. 2010. An analytical model of the effects of pulse pileup on the energy spectrum recorded by energy resolved photon counting x-ray detectors. *Med Phys* 37: 3957–69.

Tam, K.C. 1995. Three-dimensional computerized tomography scanning methods and system for large objects with smaller area detectors. US Patent No. 5390112.

Tang, X. 2003. Matched view weighting in tilted-plane-based reconstruction algorithms to suppress helical artifacts and optimize noise characteristics. *Med Phys* 30: 2912–18.

Tang, X., and Hsieh, J. 2007. Handling data redundancy in helical cone beam reconstruction using a cone-angle-based window function and its asymptotic approximation. *Med Phys* 34 1989–98.

Tang, X., Hsieh, J., Dong, F., Fan, J. and Toth, T.L. 2008. Minimization of over-ranging in helical CT scan via hybrid cone beam image reconstruction – benefits in dose efficiency. *Med Phys* 35: 3232–8.

Tang, X., Hsieh, J., Nilsen, R.A., Dutta, S., Samsonov, D., Hagiwara, A., Shaughnessy, C. and Drapkin, E. 2006. A three-dimensional weighted cone beam filtered backprojection (CB-FBP) algorithm for image reconstruction in volumetric CT –helical scanning. *Phys Med Biol* 51: 855–74.

Tang, X., Hsieh, J., Nilsen, R.A., Hagiwara, A., Thibault, J. and Drapkin, E. 2005. A three-dimensional weighted cone beam filtered backprojection (CB-FBP) algorithm for image reconstruction in volumetric CT under a circular source trajectory. *Phys Med Biol* 50: 3889–905.

Tang, X., Hsieh, J., Seamans, J., Dong, F. and Okerlund, D. 2008. Cardiac imaging in diagnostic volumetric CT using multi-sector data acquisition and image reconstruction: step-and-shoot scan vs. helical scan. *SPIE Proc* 6913(69131H): 11 pages.

Tang, X., Narayana, S., Fan, J., Hsieh, J., Pack, J.D., Nilsen, R.A. and Taha, B. 2010. Enhancement of in-plane spatial resolution in volumetric CT with focal spot wobbling – overcoming the constraint on number of projection view per gantry rotation. *J X-ray Scien Tech* 18: 251–65.

Tang, X. and Pan, T. 2004. CT cardiac imaging: evolution from 2D to 3D backprojection. *SPIE Proc* 5369: 44–50.

Thibault, J., Sauer, K.D., Bouman, C.A. and Hsieh, J. 2007. A three-dimensional statistical approach to improved image quality for multislice helical CT. *Med Phys* 34: 4526–44.

Thilander-Klang, A., Ledenius, K., Hansson, J., Sund, P. and Båth, M. 2010. Evaluation of subkective assessment of the low-contrast visibility in constancy control of computed tomography. *Radiat Prot Dosimetry* 139: 449–54.

Title 10 (Energy) of the code of Federal Regulations (CFR), Part 20: Standards for Protection against Radiation.

Tuy, H. 1983. An inversion formula for cone-beam reconstruction. *SIAM J Appl Math* 43: 546–52.

Tzedakis, A., Damilakis, J., Perisinakis, K. and Stratakis, J. 2005. The effect of z overscanning on patient effective dose from multidetector helical computed tomography examinations. *Med Phys* 32: 1621–9.

Vembar, M., Garcia, M.J., Heuscher, D.J., Haberl, R., Matthews, D., Bohme, G.E. and Greenberg, N.L. 2003. A dynamic approach to identifying desired physiological phase for cardiac imaging using multislice spiral CT. *Med Phys* 30: 1683–93.

Walter, D.J., Wu, X., Du, Y., Tkaczyk, E.J. and Ross, W.R. 2004. Dual KVP material decomposition using flat-panel detectors. *SPIE Proc* 5368: 23.

Wang, G., Lin, T.H., Cheng, P.C. and Shinozaki, D.M. 1993. A general cone-beam reconstruction algorithm. *IEEE Trans Med Imaging* 12:486–96.

Wang, G., Yu, H. and De Man, B. 2008. An outlook on x-ray CT research and development. *Med Phys* 35: 1051–64.

Wang, W. and Gindi, G. 1997. Noise analysis of MAP-EM algorithms for emission tomography. *Phys Med Biol* 42: 2215–31.

Wang, X., Meier, D., Taguchi, K., Wagenaar, D.J., Patt, B.E. and Frey, E.C. 2011. Material separation in x-ray CT with energy resolved photon-counting detectors. *Med Phys* 38: 1534–46.

Weissleder, R. and Mahmood, U. 2001. Molecular imaging. *Radiology* 219: 316–33.

Wilson, D.W., Tsui, B. and Barret, H.H. 1994. Noise properties of the EM algorithm: II. Monte Carlo simulations. *Phys Med Biol* 39: 847–72.

Xu, J., Mahesh, M. and Tsui, B. 2009. Is iterative reconstruction ready for MDCT? *J Am Coll Radiol* 6: 274–6.

Yang, Z., Silver, M.D. and Noshi, Y. 2011. Adaptive weighted anisotropic diffusion for computed tomography denoising. *Proceeding of the 11th International Meeting on Fully Three-dimensional Image Reconstruction in Radiology and Nuclear Medicine*, 210–13, Potsdam, 2011.

Yu, L., Liu, X., Leng, S., Kofler, J.M., Ramirez-Giraldo, J.C., Qu, M., Christner, J., Fletcher, J.G. and McCollough, C.H. 2009. Radiation dose reduction in computed tomography: techniques and future perspective. *Imag Med* 1: 65–84.

Yu, L., Primak, A.N., Liu, X. and McCollough, C.H. 2009. Image quality optimization and evaluation of linearly mixed images in dual-source, dual-energy CT. *Med Phys* 36(5): 1019–24.

Zou, Y. and Silver, M.D. 2008. Analysis of fast kV-switching in dual energy CT using a pre-reconstruction decomposition technique. *SPIE Proc* 6913(691313-1): 10 pages (doi: 10.1117/12.772826).

Zou, Y. and Silver, M.D. 2009. Elimination of blooming artifacts off stents by dual energy CT. *SPIE Proc* 7258(7258116-1): 8 pages.

12 Cone beam micro-CT for small-animal research

Erik L. Ritman

Contents

12.1 INTRODUCTION

Cone beam micro-computed tomography (CT) is a three-dimensional (3D) x-ray imaging method that involves obtaining x-ray projection images at many angles of view around an axis through an object and then applying a tomographic reconstruction algorithm to generate a stack of thin tomographic images of transaxial slices through the object. The transaxial images are made up of voxels (3D pixels).

Micro-CT was first developed in the early 1980s (Elliott and Dover 1982; Flannery et al. 1987; Sasov 1987). In the later 1980s, the use of bench-top micro-CT was greatly facilitated by the development of a cone beam reconstruction algorithm by Feldkamp et al. (1984). The x-ray cone beam has the advantage that it magnifies the x-ray image, but in doing so, it introduces the problem of cone beam geometry which could not be adequately dealt with by representing the cone with a stack of fan beams. Although the Feldkamp algorithm greatly reduced the cone beam artifact, the tomographic images at the upper and lower axial extents of the specimen were still prone to some distortion, thus limiting the axial length of object that could be imaged with a single scan. This effect can be overcome by a "step-and-shoot" method in which the animal is advanced one axial field-of-view length after completing each sequential scan and then "stitching" these individual images together into a single "long" 3D image. The helical CT scanning mode, in which the specimen is translated along the axis of rotation during the scan, allows coverage over a long axial extent but reduces the temporal resolution of the tomographic image data set. This approach greatly reduces the duration of the total scan sequence.

The use and availability of small-animal CT systems has increased markedly over the past decade. It has evolved from custom-made scanners (applied mostly to imaging small-animal bones and segments of larger animal bones) to commercially available scanners designed for *in vivo* imaging of skeletal and soft tissues. Numerous reviews of the development and applications of micro-CT have been published (Paulus et al. 2000; Holdsworth and Thornton 2002; Badea et al. 2008; Ritman 2011). Several commercially marketed micro-CT scanners are now available for *in vivo* small-animal imaging. Because this market is rapidly evolving, performance characteristics are likely to change over the foreseeable future. Nonetheless, because the functional characteristics of these scanners differ and the imaging needs of the potential purchasers also differ, the imaging needs and capabilities have to be carefully matched. Similarly, because some scanners have a range of operational characteristics but others are more suitable for "turn-key" operation, an investigator needs to consider the positives and negatives of the operational flexibility of a scanner. Figure 12.1 is a schematic of a typical small-animal CT scanner.

The gray scale of the CT images is proportional to the attenuation coefficient of the material at the spatial location depicted by the voxel. The voxel is usually on the order of approximately 50–100 μm on-a-side when intact small animals are scanned, perhaps more appropriately called mini-CT because its CT images are scaled so as to provide voxel resolution such

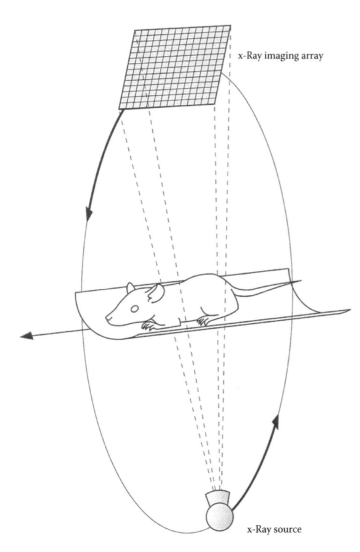

x-Ray imaging array

x-Ray source

Figure 12.1 Schematic of a small-animal CT scanner system. The small animal is anesthetized and lies on an horizontal table. If the animal's electrocardiogram (ECG), thoracic movement, or both are monitored, then either prospective or retrospective gated scans, reconstructions, or both can be performed by incremental recording and selection of the different angles of view required to generate transaxial CT images. The x-ray source and its opposite x-ray imaging array rotate about the cephalocaudal axis of the animal. Some scanners have dual x-ray source–detector arrays arranged at right angles to each other, thereby halving the scan time required. The animal table can be translated axially (i.e., at right angles to the plane described by the x-ray source trajectory), so that the length of the body scanned can be several times the length of animal exposed by the x-ray source (From Ritman, E.L., *Multimodality Molecular Imaging of Small Animals: Instrumentation and Applications*, Springer-Verlag, New York, 2013. With permission.)

that the number of voxels per organ is similar to that obtained in human CT images of those organs. This imaging generally involves clinical level x-ray photon energy. However, because small animals have higher heart and respiratory rates than humans, imaging of the thorax involves scans that provide incremental scan data acquired over many sequential heart cycles, respiratory cycles, or both, so-called gated scanning. To provide CT image signal-to-noise comparable with clinical CT scanners, the x-ray exposure of the animal or specimen should increase by an amount proportional to at least the inverse of the voxel volume (Brooks and Di Chiro 1976; Faulkner and Moores 1984; Ford et al. 2003).

As discussed in Section 12.5, the radiation itself might affect the pathophysiology of interest in, for example, angiogenesis or cancer (Paulus et al. 2000), and a voxel size $<(50\ \mu m)^3$ could result in radiation exposures in living animals that could alter the very pathophysiology of interest if repeated scans are involved.

True micro-CT has voxel resolution in the order of approximately 5–50 µm and is suitable for scanning isolated organs from small animals, tissue biopsies from larger animals, or even intact dead small animals. For isolated specimens, for which higher resolution is often desired, the scanner generally operates at lower x-ray photon energy which is optimally matched to the diameter of the specimen (Grodzins 1983). Several bench-top nano-CT scanners with submicrometer voxel resolutions have been developed and are commercially available. These scanners can provide 3D images at the cellular level of resolution, but they scan only relatively small volumes.

12.2 RATIONALE FOR USE OF SMALL-ANIMAL CT

CT has been used primarily to provide 3D images of anatomic structures and to some extent the function of those structures by virtue of their motion, distribution of contrast agent within the vascular tree lumen, or both. Traditional clinical CT and small-animal CT approaches have rarely been used to generate images of the spatial distribution of specific molecules by virtue of the CT image data itself. The uses of small-animal CT in biology are discussed in the following.

12.2.1 PHENOTYPE CHARACTERIZATION BY ANATOMIC STRUCTURES AND MATERIAL COMPOSITION

Small-animal CT has been used to study organ dimensions (e.g., dimensions of airways or volumes of lung, heart wall, or chambers; Badea et al. 2004; Drangova et al. 2007), bone mineralization (Borah et al. 2005; Cowan et al. 2007), microarchitecture of the cancellous bone and cortical bone thickness (Kinney et al. 1995), blood vessel lumen diameter and branching geometry (Nordsletten et al. 2006; Op den Buijs et al. 2006; Lee et al. 2007), and tumor size and impact on its surrounding tissues (e.g., bone erosion or compression of adjacent blood vessels). Such measurements would be seen to change in response to maturation or disease or via exposure to various pharmacological agents, environmental conditions, or radiation. These dimensions and local CT gray scales can be measured directly from the 3D CT image data and thereby represent the main application of small-animal CT imaging to date.

12.2.2 PHYSIOLOGICAL SPACES AND THEIR CONTENTS

In addition to anatomic structures, especially of entire organs, there are "macroscopic" physiological spaces such as intravascular lumens and lumens of ducts (e.g., renal tubules, ureters, bowel, and bladder and bile ducts that tend to vary with time or pathophysiological conditions) and less well-defined "microscopic" spaces such as the widely distributed, but microscopic, extravascular spaces between vessel endothelium and parenchymal cells. The extravascular spaces swell with accumulation of serous fluid (edema) or with deposition of pathological proteins such as occurs in amyloidosis, or with deposition of lipids such as occurs in atherosclerosis. These spaces can be detected by delineating them by use of contract agents that selectively accumulate (or avoid) those spaces. For the vascular tree, iodinated molecule solutions are used, and bile and renal ducts can be opacified by virtue of intravascular injection of contrast agents that are selectively taken up and excreted by the liver or kidney, respectively. Very transient labeling of those spaces can still be scanned despite the relatively slow micro-CT scans by use of incremental scans acquired from repeated contrast injections (Badea et al. 2007), use of long-duration contrast agent concentration in the bloodstream (Badea et al. 2008), or snap-freezing of the tissue of interest for subsequent cryostatic scanning (Kantor et al. 2002). The volume of these spaces can be estimated by use of the increase in local CT values within those spaces. Developments in pulsed x-ray sources facilitate the scanning of dynamic physiological processes (Cao et al. 2010).

12.2.3 TISSUE PERFUSION, DRAINAGE, AND SECRETION: MOLECULAR TRANSPORT

Tissue perfusion (F) can be estimated from CT scans if they provide images at each heart cycle during the passage of a bolus of intravascular contrast agent (Schmermund et al. 1997). Given the values of F and the extraction (E) of the contrast from the bloodstream into the extravascular spaces, the rate of influx or washout of the contrast agent from a physiological space can be used to estimate the transport into or out of that space from the Crone–Renkin relationship (Crone 1963): $PS = -F.\ln(1 - E)$, where P is the endothelial permeability and S is the surface area of the endothelial surface. The value of S can be estimated from the vascular interbranch segment's lumen diameter and length.

12.2.4 NEED TO SCAN ENTIRE ORGAN AND RESOLUTION

The volume that needs to be scanned is determined by several, sometimes conflicting, needs. Thus, we would need to scan an entire organ if we are looking for a focal lesion, such as early cancer. Conversely, at high voxel resolution, it may technically not be possible to scan an entire organ at that resolution due to, for instance, limits on the x-ray detection system resolution and size. For estimation of organ volume, relatively large voxel sizes can be tolerated; for example, a $(2\text{-cm})^3$ heart needs approximately 4000 voxels of $(30\ \mu m)^3$ if better than a 1% uncertainty is desired. However, if a 200-μm-diameter basic functional unit (BFU; the smallest accumulation of diverse cells that behaves like the organ it is in, for example, an hepatic lobule or a Haversian canal-centered osteome) is of interest, then voxel resolutions of better than $(100\ \mu m)^3$ will be needed just to unambiguously detect it, but a $(3\text{-}\mu m)^3$ voxel would be needed if the volume of the BFU is to be estimated within 10%.

12.3 TYPES OF SMALL-ANIMAL CT APPROACHES

The above-mentioned considerations apply to most current uses of small-animal CT. These applications also can provide some information about atomic content and therefore relate to molecular discrimination and quantitation at only an indirect level. There are, however, other aspects of x-ray–matter interaction that can be used to discriminate and quantitate atom concentration as well as some chemical bonds, that is, a more direct aspect of molecular characteristics.

12.3.1 ATTENUATION-BASED SCANNING

Attenuation-based scanning is the basis for the most common and most technically straightforward mode of CT scanning. The basic mechanism is the generation of a shadowgraph that is quantitated by measurement of the reduction in local x-ray intensity. By use of the Beer–Lambert law, $I = I_o e^{-\mu x}$, where I is the detected x-ray intensity at a detector pixel after passing through an object of thickness x, I_o is the incident x-ray intensity at the same detector pixel, and cumulative μ is the attenuation coefficient of the specimen's material along the x-ray beam joining the x-ray source to the detector pixel. As shown in Figure 12.2, the attenuation coefficient, expressed as "per cm" of matter traversed, decreases exponentially with increasing x-ray photon energy up to an energy of approximately 50 keV due to the photoelectric effect (proportional to Z^3/E^3, where Z is the atomic number and E is the photon energy); beyond 50 keV, μ decreases more slowly with photon energy due to the Compton effect (largely independent of Z).

This image can be converted to a projection of the attenuation × thickness product (i.e., the line integral) along the

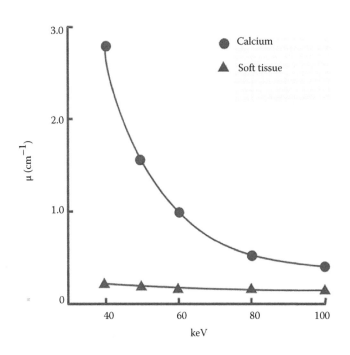

Figure 12.2 Attenuation of x-rays passing through tissues differs depending on the tissue elemental content and the energy of the x-ray photons. In this example of a filtered 80-kVp source (effective 54 keV), the difference between Ca and soft tissue is dramatic. If the Ca is diluted so that its attenuation is comparable with the soft tissue, then by generating images at two energies, for example, 80 and 50 keV, the Ca component will change more rapidly than the soft tissue; hence, a subtraction of the two images will tend to leave a Ca signal but eliminate the tissue signal (From Cann, C.E. et al., *Radiology*, 145, 493–6, 1982. With permission.)

x-ray beam traversing the object. With multiangular projection data, the data can be mathematically converted to the 3D distribution of the local attenuation coefficient at the site of each voxel making up the 3D image data set (Herman 1980). The x-ray beam should be monochromatic if beam hardening is to be avoided and is readily achievable with a synchrotron (Dilmanian 1992; Bonse et al. 1986) by use of a diffraction crystal that can select those x-ray photons within a ±50-eV energy range about some mean value, for example, 20 keV. This also can be achieved with a bench-top x-ray source in part by filtering the x-ray beam before it encounters the specimen. This generally involves use of a layer of aluminium that preferentially removes the lower energy photons, but if the K_α emission of the anode is to be used as the primary source (e.g., 17.5 keV for an Mo anode), then a suitably matched filter with a Kedge absorption energy just greater than the K_α energy also would selectively reduce the photons with energy greater than the K_α energy (e.g., 18-keV K-edge for a zirconium filter to match the Mo K_α) (Ross 1928). This approach is effective but for the bench-top x-ray source it results in a greatly diminished x-ray flux and hence requires long scan periods that are generally incompatible with *in vivo* scanning unless gated scan acquisition can be used to reduce the CT image blurring due to cardiac or respiratory motion.

The signal (i.e., the change in local contrast of the shadowgraph) in all attenuation-based imaging approaches involves local reduction of x-ray intensity that is accompanied by a reduced signal-to-noise ratio due to the reduction in the number of photons impinging on each detector pixel. Noise in this context is the variation of signal between adjacent pixels that should have identical signals due to the line integral of the specimen along the x-ray beam illuminating each pixel being identical. For specimens, higher contrast resolution can be achieved by use of lower energy photons. As shown in Figure 12.3 (Spanne 1989), E has to increase with sample diameter (actually the μx product) if CT image noise is to be kept constant.

Grodzins (1983) showed that the optimal trade-off between signal and contrast resolution occurs when 10% of the incident beam is transmitted. If the duration of the scan is important (especially in living animals), then higher x-ray photon energy is used because of the higher signal (due to less attenuation). However, this is at the "cost" of lower "density" (i.e., μ) resolution.

The absolute value and rate of decrease of the attenuation coefficient differ depending on the element and the density of the material. Thus, the attenuation value of muscle tissue decreases 765-fold from 1 to 10 keV but only 31-fold from 10 to 100 keV, whereas blood decreases 690- and 32-fold over the same ranges of photon energies. Subtracting the image obtained at a low, from that obtained at higher, photon energy would differentiate blood and muscle tissue better than their attenuation coefficient alone would at any one photon energy.

At 10 keV, tissues of different density (e.g., fat, muscle, and bone) show considerable differences in attenuation coefficients (3.1, 5.6, and 54 per cm, respectively) and hence can be distinguished from each other by their attenuation coefficient alone.

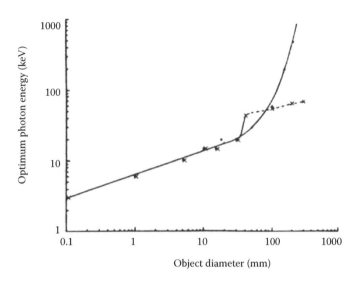

Figure 12.3 Optimal photon energy for the detection of a 1% density difference in circular water phantoms as a function of the diameter of the phantom. The diameter of the contrasting detail is 1/200 of the phantom diameter. Optimization criteria: x, minimum absorbed dose at center of phantom: •, minimum number of incident photons. (From Spanne, P., *Phys Med Biol*, 34, 679–90, 1989. With permission.)

The attenuation coefficient can change dramatically at the so-called Kedge. As illustrated in Figure 12.4, for I, the attenuation coefficient increases abruptly from 6.55 to 35.8 per cm when the photon energy increases by a mere electron volt at 33.1694 keV.

Certain biologically relevant elements (such as I that occurs naturally in the thyroid gland or when purposely bound chemically to biological molecules of interest, as is the case for clinical contrast agents) can be identified and quantitated by subtracting images generated at x-ray photon energy just below and just above the Kedge transition voltage. Unfortunately, none of the common elements that occur naturally in the tissues of the body (e.g., Na, K, Ca, P) have Kedges at sufficiently high keV photon energy that can be used for imaging of even isolated mouse organs, much less intact mice. This is because at these very low photon energies (i.e., <10 keV), the attenuation of the x-ray is so large (i.e., only 0.5% of photons pass through 1 cm of water—the thickness of a mouse abdomen) that useful images cannot be generated at acceptable radiation exposure levels.

This methodology traditionally involved use of two x-ray photon energies of quasi-monochromatic radiation with narrow spectral bandwidths that lie below and above the Kedge energy of the atom of interest. Recently, with the advent of high spatial resolution, energy-selective x-ray detection systems allow use of broad-spectrum x-ray exposure. Narrowing of the spectral bandwidth (i.e., range of x-ray photon energy) down to levels of 50 eV (i.e., ~0.1% bandwidth) can be achieved by use of a diffraction crystal at a synchrotron because even with this great restriction of x-ray flux, there is still adequate x-ray flux to allow rapid imaging. However, two photon energies can be used to discriminate two different elements due to their

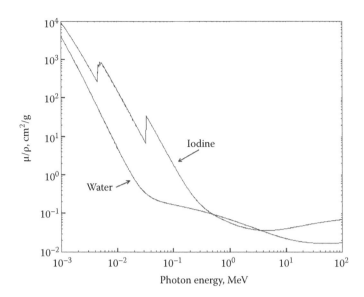

Figure 12.4 Attenuation of x-ray, normalized for gravimetric density, by I decreases with increasing energy, but at approximately 5 and 33 keV, there are step changes in attenuation. These are the so-called Ledge and Kedge of I that correspond to the energy of the electrons in the L and K shells of the I atom. Water, the main component of living tissues, has those discontinuities due to the hydrogen and oxygen at 16 and 500 eV, respectively, well below the x-ray photon energies used in small-animal-CT. Subtraction of two x-ray images involving x-ray photon energies at 32 and 34 keV would show a large difference in I signal but a relatively small change in the water signal, resulting in an essentially I-only image. Micromolar (15 μg per cm³) concentrations of I can be detected by this method (From http://physics.nist.gov/PhysRefData/XrayMassCoef/. With permission.)

different rates of change of attenuation coefficient with photon energy.

With conventional x-ray sources that produce broad-spectrum bremsstrahlung, suitable selection of the anode material for its characteristic K_α emission of the material, combined with a thin metal foil filter that has an absorption K edge just above the K_α photon energy, the spectral bandwidth can be reduced to less than 30% (Figure 12.5).

If an energy discriminating x-ray detector array is used, then those photons falling within selected energies based can be used to form the x-ray image (Gleason et al. 1999; Panetta et al. 2007; Butzer et al. 2008; Anderson et al. 2009; Firsching et al. 2009). Detector arrays with (55-μm)² pixels, energy discrimination, and photon counting (up to 8000 photons per s per pixel) have become available (Butzer et al. 2008) for energies up to 18 keV (Si-based array), 50 keV [gallium arsenide (GaAs) array], and 75 keV [cadmium telluride (CdTe) array] at 50% detector efficiency (Figure 12.6). This approach can be used to select two different photon energy bands so that the K absorption edge falls between these bands. Subtraction of the two images provides Kedge subtraction images of increased contrast for the selected element. The spectral CT approach is in a state of rapid development. A monograph by Heismann et al. (2012) provides a good introduction.

12.3.2 PHASE CONTRAST SCANNING

X-Rays, like light, are refracted by matter, resulting in slight deviations of the x-ray beam from its initial straight-line

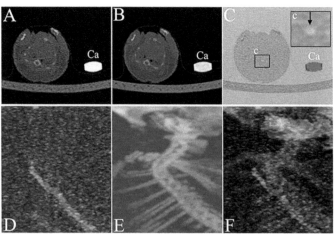

Figure 12.5 Image data from the dual-energy 64-slice CT scanner. After subtraction, a "positive signal" was obtained in animals at the age of 52 weeks (C). Sagittal reconstruction of the descending aorta showing the distribution of the positive signal in the same animal. The anatomic situation becomes apparent if the skeleton (E) is superimposed to the subtracted image, resulting in F (From Langheinrich, A.C. et al., *Invest Radiol*, 42, 263–73, 2007. With permission.)

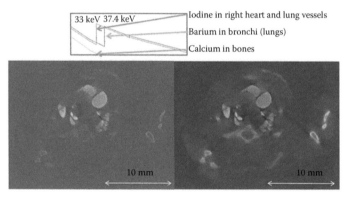

Figure 12.6 (See color insert.) (Left) CT image of a transaxial cross section of a mouse thorax. The bronchial tree had barium sulfate infused, and the pulmonary artery had an I-based contrast injected. These different materials and the skeletal features cannot be distinguished unambiguously on the basis of their CT gray scale values. (Right) Use of principal component analysis by virtue of the ability to extract the different x-ray photon energy components from the bremsstrahlung x-ray exposure allowed identification and quantitation of the three elements by virtue of their different attenuation-to-photon energy relationships, as illustrated in the top panel (From Anderson, N.G. et al., *Eur Radiol B*, 20, 2126–2134, 2010. With permission.)

trajectory. However, the refractive index of x-ray in water is very small, 7.4×10^{-7} (Lewis et al. 2003). Nonetheless, as shown in Figure 12.7, the phase-shift component (δ) of the refractive index is orders of magnitude greater than the attenuation component (β) of the compound refractive index n: $n = 1 - \delta - i.\beta$, where $i = \sqrt{-1}$. At 17.5 keV, there is a 180° phase shift caused by 50 μm of tissue, whereas the change in attenuation caused by 50 μm water is only 0.25%.

The deflection of the x-ray results from a shift in the phase of the x-ray that, in turn, is the result of the x-ray's interaction

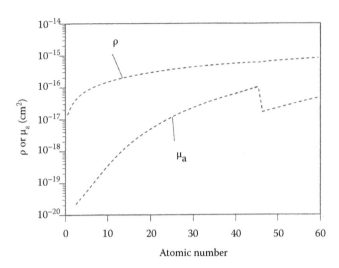

Figure 12.7 Atomic x-ray phase shift (ρ) and absorption (μ) of 24-keV x-ray as a function of atomic number. The x-ray refractive index of matter n = 1 − δ − i.β, where δ is the phase shift–related component and β the attenuation-related component. The step-change in the absorption curve corresponds to the Kedge effect. Note that the ρ value is orders of magnitude higher than the μ value at any one atomic number, indicating that either the x-ray refractive properties of matter can either be exploited to provide higher contrast resolution or reduced radiation exposure (From Momose, A. and Fukuda, J. *Med Phys*, 22, 375–9, 1995. With permission.)

with the material. Phase shift cannot be measured directly with current imaging systems because the frequency of approximately 10^{18}/s is much too fast. However, phase shift can be detected much more readily by virtue of interference patterns that can be generated by several means.

The most practical method (Figure 12.8), and the method that most readily can accommodate a broad-spectrum cone beam x-ray, involves use of multiple microscopic venetian blind–like gratings (for instance, consisting of micrometer-wide layers of Au alternating with layers of Si) placed between the source and specimen (to convert the full area beam into a series of parallel linear sources) and between the specimen and the detector array (to analyze the transmitted x-ray image). The slight deflection of the x-rays due to the refraction in the specimen can be quantitatively detected by moving the analyzer grating across the image in steps that are fractions of the interval between adjacent slats in the source grid, much like the function of a vernier micrometer (Nugent et al. 1996; Wilkins et al. 1996; Pfeiffer et al. 2006; Donnelly et al. 2007; Olivo and Speller 2007a,b; Zhou and Brahme 2008). Figure 12.9 is an example of the high contrast that is achievable with this methodology (Takeda et al. 2004).

The phase shift can be shown to be proportional to mass density for most biological materials, except when there is a high proportion of hydrogen present that has almost double the effect on phase due to its unique electron charge to Z ratio.

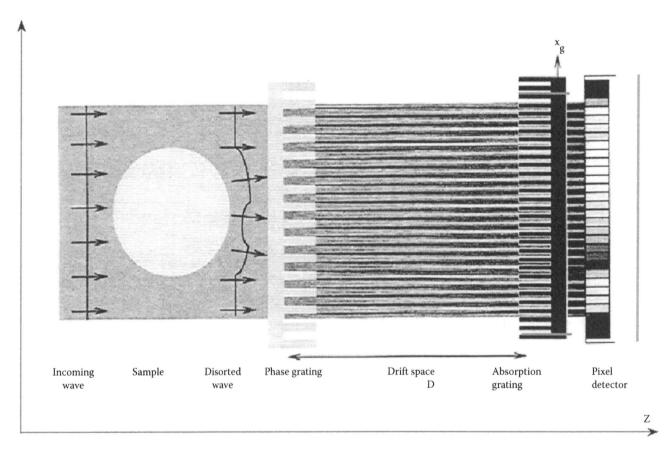

Figure 12.8 Four methods of using the very slight deviation of x-ray due to change in refractive index of material along the path of an x-ray beam. This effect also can be detected quantitatively; the phase shift of the x-ray shows how gratings can be used to generate "coded" x-ray images such that the distortion of that coded image by the phase shift can then be used to estimate the local refraction (From Weitkamp, T. et al., *Proc SPIE*, 5535, 138–42, 2004. With permission.)

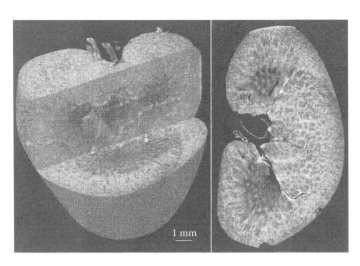

Figure 12.9 Phase contrast x-ray CT image of a rat kidney obtained at 35 keV. Whole structures of renal cortex, medulla, and pelvis were observed (From Takeda, T. et al., *Proc SPIE*, 5535, 380–91, 2004. With permission.)

12.4 TECHNICAL ISSUES

If the x-ray source–detector system is stationary and the specimen is rotated, the heavy components of the scanner can be rigidly and accurately positioned with great precision. This works very well for *in vitro* specimens but generally involves the use of a vertical rotation of the specimen (rather than horizontal) to minimize the gravity-induced movement or distortion of the specimen, relative to its axis of rotation. The living animal and its contents cannot be secured sufficiently rigidly to prevent motion of, and within, the body as it rotates about a horizontal axis. Although rotation of a living animal about a vertical axis minimizes this problem, maintenance of a vertical position over an extended period is not physiologically relevant position for larger quadrupeds and may interfere with cardiopulmonary function, although it is generally acceptable for small rodents.

Rotation of the x-ray source–detector system about a horizontal axis ensures that the animal can be positioned in its physiological horizontal position, and it will not distort with angle of view. A technical requirement for this arrangement is that the generally heavy source or detector components have to rotate so that deviation from the ideal trajectory about the axis of rotation is smaller than the detector pixel size.

The duration of a complete scan depends on the x-ray flux that can be generated by the x-ray source because this relationship governs the duration required to generate a projection image of sufficient quality (i.e., signal-to-noise and motion blurring) to be used for tomographic imaging and to a lesser extent the speed with which the necessary x-ray detection information can be recorded and transferred to an off-scanner memory. These factors vary greatly depending on the specific x-ray modality used to generate the tomographic image data.

12.5 RADIATION EXPOSURE

X-Ray exposure can result in disruption of chemical bonds and can generate super radicals that, in turn, damage nearby molecules, with DNA being of particular concern because this affects cell reproduction and its control (Bond and Robertson 1957; Ford et al. 2003; Boone et al. 2004). The number of photons absorbed in a region, represented by a voxel, determines the noise in the CT image (i.e., the variation in gray scale from voxel to voxel differs even though they represent the same material and hence attenuation coefficient). For a given exposure of the subject, the number of photons interacting within a voxel changes in direct relationship with the voxel volume. This, combined with some other consequences of the scanning process, results in the radiation exposure to the subject having to increase with approximately the fourth power of the voxel side dimension if the noise per voxel is to remain unchanged (Brooks and Di Chiro 1976). Consequently, the higher the spatial resolution, the higher the radiation exposure. The $LD_{50/30}$ lethal dose (i.e., dose after which 50% animals die within 30 days) for small animals is somewhat less than 8 Gray (Gy). A scan generating $(65-\mu m)^3$ voxels would involve a 5-Gy exposure (Carlson et al. 2007), tolerable in a terminal study but not in the first of several sequential scans of the same animal in a longitudinal study.

12.6 CONCLUSIONS

Molecular structure, in terms of elemental components (either as part of the molecule or as a synthetically labeled molecule) and certain chemical bonds (especially if they repeat along the length of long molecules), can be detected and somewhat characterized by x-ray micro-CT imaging methods. Micro-CT can enhance the utility of low-resolution, but high-sensitivity and -specificity, radionuclide tomography [single-photon emission computed tomography (SPECT) or positron emission tomography (PET)] because it provides the high spatial resolution confines of organs and physiological spaces in which molecules of interest are known to accumulate, be excluded, or washout from, and it provides the spatial distribution of x-ray attenuation that is needed to correct for attenuation of the gamma ray used in the tomographic image generated by the radionuclide within the body.

Although the attenuation aspect and the other imaging modalities such as radionuclide-based imaging can readily be individually integrated into a single micro-CT scanner so that time is saved and, more importantly, so that spatial registration of the two different images is greatly facilitated, no one multimodality combination is likely to meet all needs.

A major strength of small-animal-CT is that it provides clinically relevant image information on pathophysiology, at scale-equivalent of clinical CT scan resolution. Micro-CT can provide image data at resolutions much higher than achievable with clinical scanners so that greater insight into pathophysiological processes can be expected. Another strength of small-animal-CT is that it provides a test-bed for the development and evaluation of novel, clinically applicable x-ray imaging approaches.

ACKNOWLEDGMENTS

The micro-CT work was supported, in part, by National Institutes of Health grant EB000305.

REFERENCES

Anderson, N.G., Butler, A.P., Scott, N., et al. 2009. Medipix imaging – evaluation of data sets with PCA. *Eur Radiol B* 393: S228.

Anderson, N.G., Butter, A.P., Scott, N.J.A., Cook, N.J., et al. 2010. Spectroscopic (multi-energy) CT distinguishes iodine and borium contrast material in MICE. *Eur Radiol* 20, 2126–2134.

Badea, C.T., Drangova, M., Holdsworth, D.W., et al. 2008. In vivo small-animal imaging using micro-CT and digital subtraction angiography. *Phys Med Biol* 53: R319–50.

Badea, C.T., Hedlund, L.W., Johnson, G.A., et al. 2004. Micro-CT with respiratory and cardiac gating. *Med Phys* 31: 3324–9.

Badea, C.T., Hedlund, L.W., Mackel, J.F., et al. 2007. Cardiac micro-computed tomography for morphological and functional phenotyping of muscle LIM protein null mice. *Mol Imaging* 6: 261–8.

Bond, V.P. and Robertson, J.S. 1957. Vertebrate radiobiology (lethal actions and associated effects). *Annu Rev Nucl Sci* 7: 135–62.

Bonse, U., Johnson, Q., Nicols, M., et al. 1986. High resolution tomography with chemical specificity. *Nucl Instrum Methods Phys Res A* 246(1–3): 644–8.

Boone, J.M., Velazquez, O. and Cherry, S.R. 2004. Small-animal x-ray dose from micro-CT. *Mol Imaging* 3: 149–58.

Borah, B., Ritman, E.L., Dufresne, T.E., et al. 2005. The effect of risedronate on bone mineralization as measured by micro-computed tomography with synchrotron radiation: correlation to histomorphometric indices of turnover. *Bone* 37: 1–9.

Brooks, R.A. and Di Chiro, G. 1976. Statistical limitations in x-ray reconstructive tomography. *Med Phys* 3: 237–40.

Butzer, J.S., Butler, A.P.H., Butler, P.H., et al. 2008. Medipix imaging – evaluation of datasets with PCA. *Image and Vision Computing New Zealand.* 1–6. doi: 10.1109/IVCNZ.2008.4762080.

Cann, C.E., Gamsu, G., Birnbert, F.A., et al. 1982. Quantification of calcium in solitary pulmonary nodules using single and dual-energy CT. *Radiology* 145: 493–6.

Cao, G., Burk, L.M., Leo, Y.Z., et al. 2010. Prospective-gated cardiac micro-CT imaging of free-breathing mice using carbon nanotube field emission x-ray. *Med Phys* 37: 5306–12.

Carlson, S.K., Classic, K.L., Bender, C.E., et al. 2007. Small animal absorbed radiation dose from micro-computed tomography imaging. *Mol Imaging Biol* 9: 78–82.

Cowan, C.M., Aghaloo, T., Chou, Y.F., et al. 2007. Micro-CT evaluation of three dimensional mineralization in response to BMP-2 doses in vitro and in critical sized rat calvarial defects. *Tissue Eng* 13: 501–12.

Crone, C. 1963. The permeability of capillaries in various organs as determined by the use of the indicator diffusion method. *Acta Physiol Scand* 58: 292–305.

Dilmanian, F.A. 1992. Computed tomography with monochromatic x-rays. *Am J Physiol Imaging* 7(3–4): 175–9.

Donnelly, E.F., Price, R.R., Lewis, K.G., et al. 2007. Polychromatic phase-contrast computed tomography. *Med Phys* 34: 3165.

Drangova, M., Ford, N.L., Detombe, S.A., et al. 2007. Fast retrospectively gated quantitative four-dimensional (4D) cardiac microcomputed tomography imaging of free-breathing mice. *Invest Radiol* 42: 85–94.

Elliot, J.C. and Dover, S.D. 1982. x-ray tomography. *J Microsc* 162: 211–13.

Faulkner, K. and Moores, B.M. 1984. Noise and contrast detection in computed tomography images. *Phys Med Biol* 29: 329–39.

Feldkamp, L.A., Davis, L.C. and Kress, J.W. 1984. Practical cone-beam algorithm. *J Opt Soc Am A1*: 162–91.

Firsching, M., Butler, A.P., Scott, N., et al. 2009. Contrast agent recognition in small animal CT using the Medipix2 detector. *Nucl Inst Meth A* 607: 179–82.

Flannery, B.P., Deckman, H.W., Roberg, W.G., et al. 1987. Three dimensional x-ray microtomography. *Science* 237: 1439–44.

Ford, N.L., Thornton, M.M. and Holdsworth, D.W. 2003. Fundamental image quality limits for microcomputed tomography in small animals. *Med Phys* 30: 2869–98.

Gleason, S.S., Sari-Sarraf, H., Paulus, M.J., et al. 1999. Reconstruction of multi-energy x-ray computed tomography images of laboratory mice. *IEEE Trans Nucl Sci* 46: 1081–6.

Grodzins, L. 1983. Optimum energies for x-ray transmission tomography of small samples. *Nucl Instrum Methods* 206: 541–5.

Heismann, B.J., Schmidt, B.T. and Flohr, T. 2012. *Spectral Computed Tomography*. Bellingham, WA: SPIE.

Herman, G.T. 1980. *Image Reconstruction from Projections: The Fundamentals of Computerized Tomography*. New York: Academic Press.

Holdsworth, D.W. and Thornton, M. M. 2002. Micro-CT in small animal and specimen imaging. *Trends Biotech* 20(8): S34–9.

Kantor, B., Jorgensen, S.M., Lund, P.E., et al. 2002. Cryostatic micro-computed tomography imaging of arterial wall perfusion. *Scanning* 24: 186–90.

Kinney, J.H., Lane, N.E. and Haupt, D.L. 1995. In vivo three dimensional microscopy of trabecular bone. *J Bone Miner Res* 10: 264–70.

Langheinrich, A.C., Michniewicz, A., Sedding, D.G., et al. 2007. Quantitative x-ray imaging of intraplaque hemorrhage in aortas of apoE$^{-/-}$/LDL$^{-/-}$ double knockout mice. *Invest Radiol* 42: 263–73.

Lee, J., Beighley, P., Ritman, E., et al. 2007. Automatic segmentation of 3D micro-CT coronary vascular images. *Med Image Analysis* 11: 630–47.

Lewis, R.A., Hall, C.J., Hufton, A.P., et al. 2003. x-ray refraction effects: application to the imaging of biological tissues. *British J Radiol* 76: 301–8.

Momose, A. and Fukuda, J. 1995. Phase-contrast radiographs of nonstained rat cerebellar specimen. *Med Phys* 22: 375–9.

Nordsletten, D., Blackett, S., Bentley, M. D., et al. 2006. Structural morphology of renal vasculature. *Am J Physiol Heart Circ Physiol* 291: H296–309.

Nugent, K.A., Gureyev, T.E., Cookson, D.J., et al. 1996. Quantitative phase imaging using hard x-rays. *Phys Rev Letters* 77: 2961–4.

Olivo, A. and Speller, R. 2007a. A coded-aperture technique allowing x-ray phase contrast imaging with laboratory sources. *Appl Phys Lett* 91: 74106.

Olivo, A. and Speller, R. 2007b. Polychromatic phase contrast imaging as a basic step towards a widespread application of the technique. *Nucl Instrum Methods* A580: 1079–1082.

Op den Buijs, J., Bajzer, Z. and Ritman, E.L. 2006. Branching morphology of the rat hepatic portal vein tree: a micro-CT study. *Ann Biomed Eng* 34: 1420–8.

Panetta, D., Belcari, N., Baldazzi, G., et al. 2007. Characterization of a high-resolution CT scanner prototype for small animals. *Nuovo Cimento B* 122: 739–47.

Paulus, M.J., Geason, S.S., Kennel, S.J., et al. 2000. High resolution x-ray tomography: an emerging tool for small animal cancer research. *Neoplasia* 2: 36–45.

Pfeiffer, F., Weitkamp, T., Bunk, O., et al. 2006. Phase retrieval and differential phase-contrast imaging with low brilliance x-ray sources. *Nat Phys* 2: 256–61.

Ritman, E.L. 2011. Current status of developments and applications of Micro-CT. *Annual Rev Biomed Eng* 13: 531–52.

Ritman, E.L. 2013. Design considerations of small-animal CT systems. In: Zaidi, H. (ed.). *Multimodality Molecular Imaging of Small Animals: Instrumentation and Applications*. New York: Springer-Verlag, 182–201.

Ross, P.A. 1928 A new method of spectroscopy for faint x-radiations. *J Opt Soc Am* 16: 433–7.

Sasov, A. 1987. Non-destructive 3D imaging of the objects internal microstructure by microCT attachment for SEM. *J Microsc* 147: 169–92.

Schmermund, A., Bell, M.R., Lerman, L.O., et al. 1997. Quantitative evaluation of regional myocardial perfusion using fast x-ray computed tomography. *Herz* 22: 29–39.

Spanne, P. 1989. x-ray energy optimization in the computed tomography. *Phys Med Biol* 34: 679–90.

Takeda, T., Wu, J., Yoneyama, A., et al. 2004. SR biomedical imaging with phase-contrast and fluorescent x-ray CT. *Proc SPIE* 5535: 380–91.

Weitkamp, T., Diaz, A., Nöhammer, B., et al. 2004. Hard x-ray phase imaging and tomography with a grating interferometer. *Proc SPIE* 5535: 138–42.

Wilkins, S.W., Gureyev, T.E., Gao, D., et al. 1996. Phase-contrast imaging using polychromatic hard x-rays. *Nature* 384: 335–8.

Zhou, S.A. and Brahme, A. 2008. Development of phase-contrast x-ray imaging techniques and potential medical applications. *Phys Med* 24: 129–48.

13 Cardiac imaging

Katsuyuki (Ken) Taguchi and Elliot K. Fishman

Contents

13.1 INTRODUCTION

Cardiovascular disease remains the leading cause of death in the Western world, placing an ever-increasing burden on both private and public health services (CDCP 2004). The fiscal cost care for heart disease and stroke in the United States was U.S.\$194 billion in 2001 (Budoff et al. 2003) and is projected to triple from U.S.\$273 billion in 2010 to U.S.\$818 billion in 2030 (Heidenreich et al. 2011). Each year, more than 4 million patients undergo invasive coronary angiography (ICA) and catheterization in the United States, with 20%–40% of those examinations resulting in normal findings (Budoff et al. 2003). Patel et al. (2010) noted that "In this study slightly more than one third of patients without known disease who underwent elective cardiac catheterization had obstructive coronary artery disease. Better strategies for risk stratification are needed to inform decisions and to increase the diagnostic yield of cardiac catheterization in routine clinical practice." This diagnosis points to the potential value of coronary computed tomography angiography (CCTA) as a first study in select patients as Budoff et al. (2008) note that "Importantly the 99% negative predictive value at the patient and vessel level establishes CCTA as an effective noninvasive alternative to ICA to rule out obstructive coronary artery stenosis."

In addition, another disadvantageous outcome of catheter angiography is that it is a series of two-dimensional (2D) projection images obtained at almost a fixed viewing angle; thus, it is not the best approach to imaging a complex three-dimensional (3D) or four-dimensional (4D) object such as the heart. And, until recently, the use of computed tomography (CT) to obtain cross-sectional images of the heart and coronary arteries was not a practical solution because, as is discussed later, the combination of the CT scanner's slow rotation speed and long acquisition times as well as the changing shape of the heart during the scan due to the heart's beating resulted in severe motion artifacts in CT images.

Cardiac CT became feasible when gantry rotation time was improved to as short as 0.5 s per rotation in 1997 (McCollough and Zink 1999; Hu et al. 2000); the effective temporal resolution of 250 ms was just sufficient enough to capture a snapshot image at the middle of diastole when the heart rate is constant and is less than 60 beats per minute (bpm). Since 1997, cardiac applications have been a driving force behind improvements in CT imaging methods, including faster gantry rotation times, better spatial resolution, and more accurate image reconstruction. In this chapter, we outline some of the clinical benefits, technical issues, and current protocols for scanning and image reconstruction, and we outline future prospects for cardiac CT, including application to cone beam CT.

13.2 CLINICAL BENEFITS AND APPLICATIONS

Compared with other imaging modalities, the major strengths of cardiac CT imaging include the following: (1) it is quick, easily performed with properly trained personnel (e.g., radiologists and radiologic technologists), and noninvasive; (2) it has high isotropic spatial resolution (0.5^3 mm³) and full access to anatomical structure at any plane is provided; (3) images exhibit strong contrast resolution; and (4) the pixel value is physically meaningful (i.e., it is related to the x-ray attenuation coefficient). Thus, cardiac CT has strong potential for providing quantitative cardiac diagnoses.

Clinical applications of cardiac CT include detection and characterization of quantitative measurements of calcium deposits (i.e., calcium scoring) with good reproducibility, noncalcified plaque detection and quantification (i.e., atherosclerosis) (Gilard et al. 2006; Dey et al. 2010), various cardiac functions (e.g., wall motion, wall thickening, and global and regional ejection fraction), myocardial perfusion, infarcts, tumors, pericardial disease, postsurgical complications, and congenital malformations (Gilkeson et al. 2003; Desjardins and Kazerooni 2004; Lardo et al. 2006). In addition, evaluation of suspected aortic dissections and pulmonary embolization can be evaluated in a single exam often referred to as a triple rule-out study (Cury et al. 2013). However, it must be noted that some emerging applications require a higher heart rate (e.g., 90 bpm) than is currently recommended, which remains a significant challenge.

Among these uses, calcium scoring was one of the earliest and most widespread uses of cardiac CT in the early 2000s,

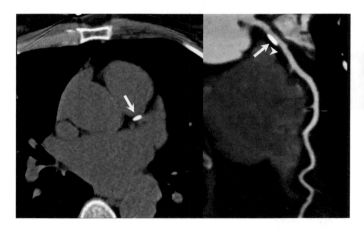

Figure 13.1 (a) A Ca deposit was found in non-contrast-enhanced CT scan (arrow). (b) A noncalcified plaque (arrowhead) was found next to the calcified plaque (arrow) by the following coronary CT angiography.

a method to scan the heart without contrast agent and quantify the amount of calcium deposits in coronary arteries (Figure 13.1). Studies showed that calcified plaques were in fact stable and would not trigger cardiac events; however, there is a high correlation between the total mass of calcium deposits and the possibility of future cardiac events (Jacobs et al. 2012; Petretta et al. 2012). Nasir and Clouse (2012) reported that "Coronary artery calcium (CAC) is an independent predictor of coronary artery disease (CAD) events and improves the ability to predict risk in vulnerable groups, adding information beyond current global risk assessment methods." They also noted that "A zero coronary calcium score stands as perhaps the most powerful negative risk factor for development of a coronary event." However, the presence of a zero calcium score does not necessarily mean the absence of coronary artery disease (Kelly et al. 2008).

The presence of vulnerable plaques is considered one of the causes that trigger myocardial ischemia, infarction, and stroke. Therefore, its detection, which in fact rules out the presence of soft plaques, is currently the primary purpose of many cardiac CT studies. It has been suggested that cardiac CT's high negative predictive value on a per-patient basis (>90%) would enable elimination of unnecessary diagnostic catheter coronary angiography procedures (Raff et al. 2005). Of late, a triple rule-out process has been one of the major clinical merits that cardiac CT is able to provide in the emergency room setting (Cury et al. 2011; Feuchtner et al. 2012; Gruettner et al. 2012; Henzler et al. 2012; Nance et al. 2012; Nielsen et al. 2012; Rich et al. 2012; Wallis et al. 2012). When a patient arrives at an emergency room suffering acute chest pain, emergency physicians must quickly and accurately diagnose whether a patient symptoms are an emergency condition involving CAD, aortic dissection, or pulmonary embolization that requires immediate treatment (Figure 13.2). Often, the accuracy of examinations such as chest x-ray and electrocardiography (ECG) as well as various blood tests is limited; thus, the patient must stay in the hospital for an extended time for precautious monitoring purposes. Cardiac CT scans have a high negative predictive value for all three of those emergency conditions

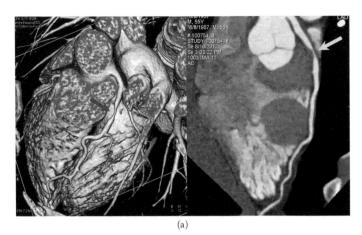

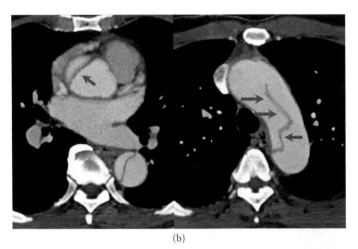

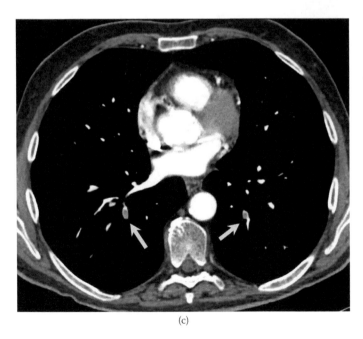

Figure 13.2 Typical images of triple rule-out examinations. (a) Noncalcified plaque. (b) Aortic dissection. (c) Pulmonary embolism.

(CAD, aortic dissection, or pulmonary embolization). Therefore, once a patient undergoes cardiac CT and all three of those conditions are ruled out, the patient can be discharged immediately with no further serious concern. This use of cardiac CT significantly changed the workflow of emergency departments.

13.3 TECHNICAL ISSUES UNIQUE TO CARDIAC IMAGING

13.3.1 CHALLENGES AND DEMANDS ON CARDIAC IMAGING

Cardiac CT imaging is a technically demanding application for CT because it requires the following:

1. High temporal resolution to "freeze" the cardiac motion
2. High spatial resolution to delineate the presence of plaque and quantify stenosis in the coronary arteries that are typically in the 4–5-mm size range
3. Low x-ray radiation dose with good low-contrast resolution and image noise
4. A region- or segment-specific optimal gating phase
5. Sufficient robustness against nonperiodic heart motion
6. Linearity of data
7. A short scan duration for a constant heart rate

These factors are organically connected to each other, and they affect the accuracy and the resolution of the final cardiac images. For example, if the temporal resolution is not sufficient for the motion of the target object (e.g., the coronary artery), the final image is blurred, even though the system has perfect spatial resolution (achieved only when the object does not move).

To meet the seven aforementioned demands, several methods, discussed in the next section, have been developed and implemented (Grass et al. 2003; Hsieh et al. 2006; Taguchi et al. 2006). Here, we review generic problems common to those methods.

13.3.2 GENERIC ISSUES ASSOCIATED WITH ECG-GATED SCANNING AND RECONSTRUCTION

Unlike the heart, most other organs do not deform and they can be imaged accurately from a set of 3D Radon transform data (calculated from a series of cone beam projections acquired by a multislice CT scanner or flat-panel C-arm CT scanner). Conversely, the heart is a challenging organ to image because it is a 4D object, constantly deforming over time due to the heartbeat. To image a 4D object, one needs to obtain a set of 4D Radon transform data and to invert the 4D transform. Current multislice cone beam CT scanners do not provide 4D Radon transform data, and it is therefore not possible to perform the inverse 4D transform. Thus, current cardiac CT uses a combined, prospectively or retrospectively ECG-gated scan and a 3D image reconstruction method (Figure 13.3). Regardless of specific schemes or protocols discussed in the next section, there are two inherent problems with this approach, intrawindow motion and interwindow (nonperiodic) motion, both of which result in problems pertaining to image artifacts and patient radiation doses.

Intrawindow motion: This approach aims to image the heart using cone beam projections that span just over one-half or one-quarter of a rotation acquired within the shortest reconstruction time window possible, ignoring the cardiac motion inside the time window (Hu et al. 2000; Kachelriess

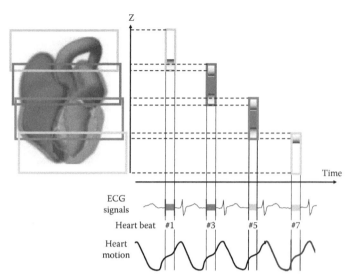

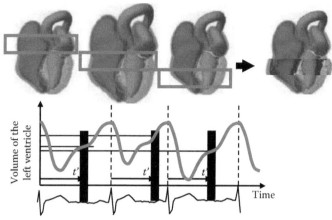

Figure 13.4 Banding artifacts appear when the heart motion is not periodic (i.e., when the interwindow motion is severe).

Figure 13.3 ECG-gated cardiac scan and reconstruction (single-sector method). Projection data acquired within time windows centering a cardiac phase of interest are used to reconstruct an image, ignoring the intrawindow motion of the heart. Several heartbeats are necessary when the detector coverage is smaller than the heart. It usually takes one heartbeat period to move the patient table to the next scan position; thus, the scan is performed every other heartbeat.

et al. 2000; Ohnesorge et al. 2000; Taguchi and Anno 2000; Grass et al. 2003; Taguchi 2003; Taguchi et al. 2006). To reduce motion artifacts and image blurriness, the gantry rotation time of CT scanners has been continually reduced to achieve the improved temporal resolution. At the same time, because the noise of reconstructed images is determined solely by the data acquired inside the reconstruction time window, to maintain the signal-to-noise ratio of images (i.e., to maintain the number of x-ray photons within the shortened reconstruction time window), x-ray tubes need to be more powerful to provide more intense x-ray beams.

Interwindow (nonperiodic) motion: For cardiac applications, a span of approximately 150 mm, from the aorta to the diaphragm, must be imaged. In contrast, the detector used in most CT systems is not large enough (e.g., 40 mm) to image the large range for one heartbeat; thus, a helical or a step-and-shoot scan is used over 5–10 heartbeats, wherein data from each heartbeat cover a limited range of the heart (Figure 13.4). An image reconstruction algorithm must "extract" data from the same ECG-phase with certain reconstruction time window widths from those 5–10 heartbeats and "connect" or "assemble" them to image the entire heart or coronary artery. Unfortunately, with the presence of nonperiodic heart motion, the shape of the heart may be different during subsequent heartbeats at the same ECG-phase (Figure 13.4). Therefore, each "connection" may generate banding artifacts (i.e., quasi-periodic horizontal shifts that make the heart wall discontinuous, the coronary arteries disjointed, and the soft plaques and Ca deposits smeared and distorted) (Figure 13.5) because most of the algorithms currently available assume a perfect, periodic motion.

Two cases that are exceptions (do not suffer from the interwindow motion problem) are a flash mode with a smaller detector (discussed in Section 13.4.1.4) and an axial scan mode

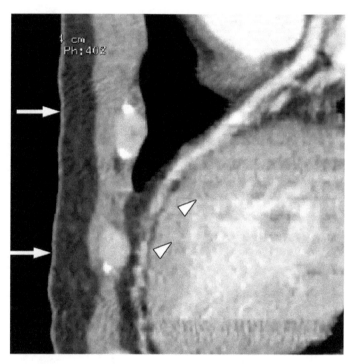

Figure 13.5 Sagittal maximum intensity projection image shows banding artifacts in the mediastinum (arrowheads) but not the chest wall (arrows), an effect caused by cardiac motion. (From Choi, H.S. et al., *Radiographics*, 24, 787–800, 2004. With permission.)

with a larger detector, such as 320 × 0.5 mm, covering 160 mm (discussed in Section 13.4.1.3).

13.3.3 SINGLE-SECTOR, DUAL-SECTOR, AND MULTISECTOR METHODS

In this section, using Figures 13.3, 13.6, and 13.7, we outline two typical methods, the single-sector method and the multisector method, used in cardiac CT. Opinions among manufacturers of CT systems have been divided as to which method is best. We relate the pros and cons of the two methods to the above-discussed intra- and interwindow motions. The temporal resolution of reconstructed images is limited by the *effective* reconstruction time window width for projection data over 180°.

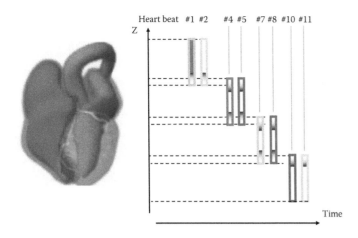

Figure 13.6 (See color insert.) A dual-sector method as opposed to a single-sector method shown in Figure 13.2. Two sectors are acquired at the same patient table position with about a half of the time window width for each sector.

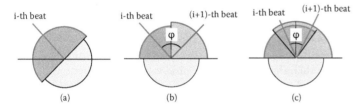

Figure 13.7 Single-sector and dual-sector methods. Polar angles denote projection angles. (a) The time window width for the single-sector method always is 180°. (b and c) The effective time window width for the dual-sector always is 90°; however, the two time windows may have an overlap depending on the relative projection angles at the phase of interest in adjacent heartbeats, φ. (b) There is no overlap when φ is 90°.

The projection data set extracted by one time window and one detector is referred to as a "sector" or "patch" (Flohr et al. 2003; Grass et al. 2003; Taguchi et al. 2006). The number of sectors used for reconstructing an image voxel or slice can be fixed at one or at two (Flohr et al. 2003), or it can be varied from one to five (Grass et al. 2003; Taguchi et al. 2006). Such strategies are referred to as the single-sector method, the dual-sector method, and the adaptive multisector method, respectively. See Figure 13.7 for a pictorial description of single- and dual-sector methods.

When the gantry rotation time is T_{rot}, the temporal resolution of images reconstructed by the single-sector method is the time it takes for 180° of gantry rotation, $T_{rot}/2$ (= $T_{rot} \times 180°/360°$). The single-sector method will thus provide images free of motion artifacts if, and only if, the intrawindow motion during $T_{rot}/2$ is reasonably small. In contrast, the temporal resolution of the multisector method is better with $T_{rot}/(2N)$, where N is the number of sectors. Thus, the multisector method is much less affected by intrawindow motion, resulting in much fewer motion artifacts. In practice, however, sectors from different heartbeats may contain different information due to nonperiodic heart motion, a change in contrast concentration, or patient motion such as breathing. Thus, if such interwindow motion is relatively large, and if a scheme to "connect" multiple sectors cannot mitigate the effect of the interwindow motion, images

reconstructed by the multisector method may exhibit blurred or double-contoured edges. Obviously, if the interwindow motion is small, or if an appropriate scheme to handle the interwindow motion is used, the multisector method will provide much sharper images than the single-sector method.

The effect of interwindow motion appears even with the single-sector method except for in the two cases that are exceptions (the flash mode or the large detector-CT). There may be severe banding artifacts between adjacent slices or voxels reconstructed from two different heartbeats and disconnected from each other (Figure 13.7). The multisector method with helical scanning may suffer from fewer banding artifacts, because among N heartbeats used to reconstruct such adjacent slices or voxels, the difference between them is only 1 (i.e., the effect of the discontinuity is limited to 1 out of N heartbeats [or $1/N$ of the single-sector method]).

Note, however, that the single-sector method requires roughly N times shorter scan duration than the N-sector method when the longitudinal detector height remains the same. To scan the same slice or voxel N times, the maximum table feed per heartbeat for the N-sector method is $1/N$ of the detector height. In contrast, the maximum table feed for the single-sector method can be the same as the detector height. In practice, a shorter scan duration (thus, a shorter breath-hold period) may result in fewer heart rate variations and less interwindow motion.

In short, neither of the two methods is ideal, and there is no clear winner in this comparison. When patient or phantom studies have negligible intra- and interwindow motion, both methods will provide excellent images. The radiation dose to patients can be comparable, if an appropriate tube current modulation technique is used to minimize the dose to out-of-phase projections (discussed in Section 13.3.5). When heart rates (thus, the intrawindow motion) are large, the multisector method with a better effective temporal resolution may be a better choice; however, heart rate variations (i.e., the interwindow motion) in such patients tend to be large as well, resulting in blurred axial images and some banding artifacts if used without an appropriate scheme such as (Taguchi et al. 2006). When the interwindow motion is too large, either method will fail to produce clear images.

13.3.4 DUAL-SOURCE CT SYSTEM

A dual-source CT system (Flohr et al. 2006) has two imaging chains, an x-ray tube and a detector, placed nearly perpendicular to each other. The major advantage of the system for cardiac CT is that it significantly decreases the intrawindow motion. The time duration needed to acquire projection data over 180°—the reconstruction time window width—is decreased to $T_{rot}/4$ (= $T_{rot} \times 180°/360°/N_i$, where N_i = 2 is the number of imaging chains). And unlike an N_i-sector method, these N_i sets of projections are acquired simultaneously from a single heartbeat, eliminating the possibility of misregistration between multiple heartbeats. This has proven to make a substantial difference in clinical practices, allowing the scanning of patients with larger heart rates with greater comfort (see Section 13.4.1.5).

One caveat is that interwindow motion remains the same or is potentially worse than it is with the single-source CT systems. When N_{row} detector rows are split into two imaging chains, the detector height is $N_{row}/2$, with two detectors covering the same longitudinal location. Let us discuss the following three scenarios: (1) a dual-source CT with *N*row/2 detector rows, use of the single-sector method, a table feed of *N*row/2 per heartbeat, and a scan duration of 2 × *T* seconds; (2) a single-source CT with *N*row detector rows, use of the single-sector method, a table feed of *N*row per heartbeat, and scan duration of *T* seconds; and (3) a single-source CT with *N*row detector rows, use of the dual-sector method, a table feed of *N*row/2 per heartbeat, and scan duration of 2 × *T* seconds. Compared with case 2, the dual-source CT (case 1) may suffer from a stronger effect of interwindow motion, because of the scan duration that is twice as long and may result in more severe heart rate variation. Compared with case 3, the dual-source CT also may suffer from a stronger effect of the interwindow motion, because each slice or voxel is reconstructed from projection data from one heartbeat (as opposed to two for case 3); thus, the discontinuity between adjacent slices or voxels is twice as strong as it is in case 3.

13.3.5 OUT-PHASE AND NEAR-PHASE DOSES

One of the concerns with cardiac CT has been the relatively large radiation dose to patients. Various prospective ECG-gated tube current modulation approaches have been implemented to decrease the two types of dose used to acquire projections that ultimately will not be used in the image reconstruction process: out-phase and near-phase dose. For explanations, see Figure 13.8. Before CT scans, a targeted cardiac phase and the corresponding ECG-phase can be determined using the heart rate. For example, the mid-diastole will be targeted when the heart rate is less than 70 bpm, and the end-systole will be target when the heart rate is more than 70 bpm. An ECG-phase that corresponds to the cardiac phase is patient-specific; however, in most patients, the mid-diastole can be found in the range of 60%–80% of the interval between the two adjacent R

waves (R-R interval) and the end-systole in 30%–50% of R-R intervals.

To decrease the out-phase dose, by prospectively synchronizing to ECG signals, an x-ray tube current is reduced or eliminated while ECG signals indicate that projections would not be used to reconstruct images of the cardiac phase of interest for the patient. The tube current for out-phase is decreased to 5%–20% of the full dose when functional information such as the ejection fraction or wall motion is necessary, and it is eliminated otherwise. To decrease the near-phase dose, it is desired to better predict an optimal phase for each patient, decreasing the range of the ECG-phases during which full-dose images can be obtained. For example, the range for the mid-diastole may be decreased from 20% R-R width (or 60%–80% R-R) to 10% R-R width (or 65%–75% R-R).

13.3.6 BLOOMING ARTIFACTS FROM CALCIUM DEPOSITS AND STENTS (BEAM-HARDENING, PARTIAL VOLUME, AND MOTION)

One of the major unsolved problems with coronary CT imaging is blooming artifacts caused by calcium deposits and stents. A typical image with such artifacts is shown in Figure 13.9. When dense calcium plaque or a stent is present in a coronary artery, the size of the calcium deposit or the stent depicted in the image is larger than the actual size; the pixel value is also inaccurate, and dark shadows or bright streaks are generated near the calcium or inside the stent. The last problem is the most significant, because the artifacts may mask or mimic soft, fatty vulnerable plaques that are often present near calcium plaques. When a stent is present, the radiologist must assess whether restenosis has occurred at the treated site or whether the site remains patent. Blooming artifacts make the assessment very challenging, if not impossible.

Blooming artifacts are caused by a combination of the beam-hardening effect in the x-ray spectrum, the partial volume effect, and cardiac motion. The extent of these causes varies depending on the CT scanner used; the patient imaged; and particular conditions, such as the projection angles, under which the objects are scanned. Alleviating artifacts due to

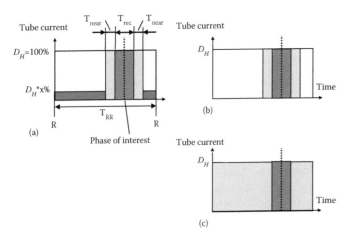

(a)

(b)

(c)

Figure 13.8 Tube current modulations for a patient dose reduction. T_{RR} refers to a time interval between two adjacent R waves of ECG signals. T_{rec} denotes a minimum time window width to reconstruct an image at one cardiac phase that corresponds to projection angles of 90° or 180°. T_{near} refers to an additional time period to reconstruct images at adjacent cardiac phases (e.g., 10% of T_{RR}). Blue area is the minimal dose required to image the heart at one phase. Yellow areas correspond to the near-phase dose, whereas the red areas corresponds to out-phase dose. The out-phase dose is reduced to x% (a) or eliminated (b). (c) Conventional scan with no tube current modulation.

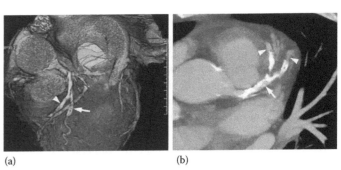

(a)   (b)

Figure 13.9 Blooming artifacts caused by coronary arterial calcifications. (a) Volume-rendered image shows high-attenuating artifacts caused by calcifications that prevent accurate evaluation of luminal patency in the left anterior descending artery (arrow) and the diagonal artery (arrowhead). (b) Multiplanar reformatted image shows variable contrast material filling in a patent but extensively calcified left anterior descending artery (arrow), as well as distal flow (arrowheads). (From Choi, H.S. et al., *Radiographics*, 24, 787–800, 2004, Figures 10a and 10b. With permission.)

these three causes through image reconstruction processes is not simple, especially using various data correction methods currently used in CT scanners. Thus, combatting one of the three problems might not improve the image quality in a straightforward manner.

Currently, the beam-hardening effect caused by bone (or Ca) is corrected by commercial CT scanners using an iterative (or two-pass) approach (Hsieh et al. 2000; Hsieh 2003). First, images are reconstructed assuming that x-ray beams are attenuated by water or soft tissue. The images are then segmented into bone, water, and air, mainly by using pixel values. Projections are synthesized through such segmented images, and the difference from the measured projections is calculated. Images are updated to decrease the difference in the second iteration (reconstruction) process. When Ca deposits are very dense or pixel values are inaccurate due to the other two causes of blooming artifacts, errors remain large after the second iteration. Metals require a third iteration or other processes for correction; thus, artifacts from metallic stents cannot be corrected for by using this second-pass beam-hardening correction.

Both Ca deposits and stents are tiny objects with sharp edges. When a part of an x-ray beam goes through the outside of such objects, and the rest of the beam goes through the inside, the attenuated x-ray intensity is nonlinearly related to the attenuation of such objects, because $\exp(A + B) \neq \exp(A) + \exp(B)$. This effect is called the partial volume effect, a major cause of artifacts in imaging with single-slice CT (or a single-detector row CT) when an acquisition detector collimator of CT scanners projected at the isocenter was as thick as 10 mm. Ca deposits and metallic stents are two objects that suffer from the partial volume effect even when the aperture of a detector pixel projected at the isocenter is reduced to 0.6 mm × 0.5–0.7 mm with multidetector row CT or multidetector CT (MDCT).

As discussed previously, the intrawindow motion is not accounted for in current cardiac imaging methods. Although the magnitude of such motion may be small and negligible for large objects, it may be significant and not negligible for small objects with large attenuation, such as Ca deposits and stents.

13.3.7 BEAM-HARDENING ARTIFACTS IN MYOCARDIUM

As discussed in the previous section, most beam-hardening artifact correction methods implemented in current CT scanners assume that large pixel values result from bones and small pixel values result from soft tissues. The I contrast agents present in a large quantity in the left ventricle, aorta, and other areas in cardiac scans have energy-dependent linear attenuation coefficients different from both bones and soft tissues; thus, beam-hardening artifacts remain severe and mimic ischemia, especially at the apex or the base of the heart (So et al. 2009; Kitagawa et al. 2010).

New beam-hardening correction methods have been developed that take into account the attenuation of I and soft tissues, or all of I, bones, and soft tissues (So et al. 2009). The performance of such methods seems promising, although more careful assessments are necessary, especially when quantitative accuracy of pixel values becomes more important for myocardial perfusion assessment.

13.3.8 HALF-SCAN ARTIFACTS FOR PERFUSION CT

It has been found that pixel values of reconstructed images fluctuate when the heart is imaged over multiple heartbeats (Primak et al. 2007; Stenner et al. 2009; Ramirez-Giraldo et al. 2012). This phenomenon is referred to as "half-scan artifacts," and it is critical that such artifacts are corrected for, especially when quantitative accuracy of pixel values are critical, such as in perfusion CT or computer-aided diagnosis.

The half-scan algorithm (Parker 1982; Silver 2000), as opposed to the full scan, has been used to reconstruct cardiac CT images to achieve better temporal resolution. The half-scan algorithm is a mathematically exact method when it is used for fan-beam axial scans. When computer simulations are performed using synthesized projections, the mean pixel values of reconstructed images are accurate and have no bias, although the amount of the noise depends on the projection angle range used in the reconstruction. That is, image pixels on the focal spot side have more noise than those on the detector side.

The mean pixel values, however, also seem to be projection-angle dependent, when the half scan is used with projections acquired using the multislice CT scanners. A specific mechanism that results in artifacts is not clear at this time; however, it is suspected to be caused by a combination of the scanner's projection angle-dependent inconsistencies that include the following issues: cross- or forward-scattered radiation (Petersilka et al. 2010), the approximate nature of cone beam half-scan algorithms (Taguchi 2003), inaccurate x-ray paths that result from ignored cone angles used in beam-hardening artifact corrections (Hsieh et al. 2000), the beam-hardening effect from I (see Section 13.3.7), and empirical data correction methods.

13.4 CURRENT PROTOCOLS

In this section, we present several clinical protocols that are currently being used. Two challenging and conflicting goals are to obtain the best image quality and to minimize patient radiation dose. Each year, new CT scanners improve the image quality and protocols for cardiac CT. Therefore, it is important to relate the current protocols to technical and clinical reasons that lead to these choices, so that it is possible to modify the protocols for the new scanners. For reference, the protocol choice for coronary CT angiography at our institution is shown in Tables 13.1 and 13.2 and is as follows: A prospective ECG-gated single heartbeat helical scan with a high pitch (flash mode, see Section 13.4.1.4) is used when a patient's heart rate is less than 60 bpm and a beta blocker is administered; a prospective ECG-gated sequential axial (step-and-shoot) scan is used otherwise.

13.4.1 SCAN AND IMAGE RECONSTRUCTION METHODS FOR SNAPSHOT 3D IMAGING

13.4.1.1 Helical scan with ECG-controlled (or "ECG-pulsing") tube current modulation

This protocol has been used since cardiac CT was first performed with four-detector row CT systems in the 1990s. A helical scan is performed while ECG signals are used to modulate a tube current (Figures 13.8a and 13.8b). The modulation is to use a higher tube current near the cardiac phase of interest and to use a lower

Table 13.1 Protocols for a prospective ECG-gated (or "ECG-triggered") sequential axial (step-and-shoot) scan

ACQUISITION PHASE	PHASE INCLUDED	SCAN DELAY	RESPIRATION PHASE	ANATOMICAL COVERAGE
Noncontrast	Yes	N/A	Inspiration	Carina through the apex of the heart
Arterial phase	Yes	Test bolus	Inspiration	Carina through the apex of the heart
Venous phase	N/A	N/A	N/A	N/A
Delayed phase	N/A	N/A	N/A	N/A

Scan comments: The adaptive sequence mode is very useful when imaging patients with irregular heart rates to avoid reconstruction artifacts. Cardiac Ca scoring is routinely preformed with a cardiac CTA to evaluate CAD. All scans are performed using the ALARA principle.

TECHNICAL PARAMETERS	PARAMETERS
kVp	100 or 120
Effective mAs	CareDose
CareDose reference mAs	270–320
Time (rotation)	0.28 s (flex)
Average acquisition time	HR dependent
Collimation	128 mm × 0.6 mm
Pitch value	HR dependent
Scan direction	Craniocaudal

Comments: Both the pitch and the scan time will vary based upon the patient's heart rate; 100 kVp is used for most average size patients. ECG dose modulation is always performed. When using the adaptive sequence, the radiation exposure can be limited to only the useful portion of the ECG wave that will significantly reduce the radiation dose.

RECONSTRUCTION PARAMETERS	SOFT TISSUE	THIN DATA	LUNG
Slice thickness	3 mm	0.75 mm	3 mm
Reconstruction spacing	3 mm	0.5 mm	3 mm
Reconstruction algorithm	B30f	B26 ASA	B80f
Window width and level	500/100	500/100	1600/−500

Reconstruction comments: Multiphase images are generated across the ECG pulsing range every 5%–10% of the R-R interval. Iterative reconstructions also are performed.

CONTRAST PARAMETERS	PARAMETERS
Contrast type	Nonionic
Contrast volume	80–100 mL
Saline flush	30 mL
Injection rate	6–7 mL/s
Oral contrast	N/A
Contrast volume	N/A

Comments: Test bolus is used to calculate the peak contrast enhancement to determine the correct scan delay. Bolus tracking can be used for patients with high heart rates (85 bpm).

Other comments: Selecting the correct range for ECG dose modulation is based primarily on the patient's heart rate and clinical indication. When multiphase images are not clinically indicated, the use of ECG dose modulation, specifically Siemens MinDose application, can significantly reduce the patient's exposure. See Cardiac CT flowchart in the following for more information on selecting the correct protocol based on the patient's heart rate and rhythm. Coronary CTA images should visualize the coronary arteries and left ventricle with adequate opacification of intravenous (IV) contrast. Images should limit venous contamination while flushing contrast out of the right ventricle to prevent image artifacts.

Note: N/A, not applicable.

tube current or turn off the x-ray outside the phase of interest. Typically, the higher tube current is used to acquire projection data that are needed to reconstruct images in multiple phases, not only at one possibly best cardiac phase, e.g., at 70% R-R, but also at some additional phases, such as 60%–80% R-R, because the optimal phase for each patient cannot be known before scanning. The range for the additional phases must be small to decrease the radiation dose, but it also must be large enough to not miss the optimal phase and result in an inconclusive examination. The range used has been decreasing as clinical studies continue to show

Table 13.2 Protocols for a prospective ECG-gated single heartbeat helical scan with a high pitch ("flash spiral" or "flash mode")

ACQUISITION PHASE	PHASE INCLUDED	SCAN DELAY	RESPIRATION PHASE	ANATOMICAL COVERAGE
Noncontrast	Yes	N/A	Inspiration	Carina through the apex of the heart
Arterial phase	Yes	Test bolus	Inspiration	Carina through the apex of the heart
Venous phase	N/A	N/A	N/A	N/A
Delayed phase	N/A	N/A	N/A	N/A

Scan comments: The flash spiral mode is most successful when imaging patients with heart rates 60 bpm or lower; using this mode will result in extremely low-dose exams. Cardiac Ca scoring is routinely preformed with a cardiac CTA to evaluate coronary artery disease. All scans are performed using the ALARA principle.

TECHNICAL PARAMETERS	PARAMETERS
kVp	100 or 120
Effective mAs	CareDose
CareDose reference mAs	320-360
Time (rotation)	0.28 s
Average acquisition time	0.4 s
Collimation	128 mm × 0.6 mm
Pitch value	3.4
Scan direction	Craniocaudal

Comments: Using the flash spiral mode allows for greatest dose savings potential. The entire data series is collected in one heartbeat, thus avoiding the need for a slow pitch with multiple images across the R-R interval of the ECG wave. 100 kVp is used for most average-sized patients. ECG dose modulation is not necessary for this protocol.

RECONSTRUCTION PARAMETERS	SOFT TISSUE	THIN DATA	LUNG
Slice thickness	3 mm	0.75 mm	3 mm
Reconstruction spacing	3 mm	0.5 mm	3 mm
Reconstruction algorithm	B30f	B26 ASA	B80f
Window width and level	500/100	500/100	1600/–500

Reconstruction comments: There is a limited reconstruction field of view when using the flash spiral mode 33.2 cm. Iterative reconstructions also are performed to improve image noise.

CONTRAST PARAMETERS	PARAMETERS
Contrast type	Nonionic
Contrast volume	80–100 mL
Saline flush	30 mL
Injection rate	6–7 mL/s
Oral contrast	N/A
Contrast volume	N/A

Comments: Test bolus is used to calculate the peak contrast enhancement to determine the correct scan delay.

Other comments: To successfully scan a patient using the flash spiral mode, the heart rate should be 60 bpm or less, and the patient should be no larger than average size to avoid increased image noise. See Cardiac CT flowchart in the following for more information on selecting the correct protocol based on the patient's heart rate and rhythm. Coronary CTA images should visualize the coronary arteries and left ventricle with adequate opacification of IV contrast. Images should limit venous contamination while flushing contrast out of the right ventricle to prevent image artifacts.

Note: N/A, not applicable.

good results with smaller ranges. Improved scanner performance has made image quality more robust against the chosen phase, and it also has helped to use a narrower range as well.

The maximum helical pitch is chosen to cover each voxel (or slice) for at least a little longer than one heartbeat (Figure 13.3) and thus is typically 0.2–0.3 of the detector height at the rotation center per rotation. The details on how the pitch is calculated can be found in Taguchi et al. (2006). The typical patient radiation dose range is 8–18 millisieverts (mSv). A helical, half-scan cone beam image reconstruction is then used to reconstruct the volume images at one phase or at multiple phases.

The strength of this protocol is that all cardiac phases can be imaged by one continuous helical scan, and these images are more consistent due to the continuous single scan. The weakness of this protocol is the relatively large radiation dose that is required. This protocol is now rarely used, because most current cardiac CT scans are performed to obtain a snapshot 3D image of the heart (and coronary arteries) at the most quiescent phase; thus, there is no need to acquire continuous data for the other cardiac phases.

13.4.1.2 Prospective ECG-gated (or "ECG-triggered") sequential axial (step-and-shoot) scans

This protocol has been used since 2005 with CT systems with 64 or more detector rows (or with >32-mm coverage). The current protocol used at our institution is shown in Table 13.1. ECG signals are used to trigger an axial scan several times. Usually, an x-ray is turned on while data sufficient to image the cardiac phase of interest only are acquired. Note that it is logically and technically possible to acquire data to image all cardiac phases (e.g., with parameters x of up and T_{near} of up in Figure 13.8a); however, such scans are rarely performed unless the cardiac functional analysis is desired. The patient table is then moved to the next scan location; and another axial scan is triggered to image the cardiac phase of interest. Typically, an axial scan is performed at every other heartbeat, and thanks to the large detector coverage, it only takes several heartbeats to image the entire heart, which is comparable to the earlier discussed helical scan. When a premature ventricular or atrial contraction (PVC and PAC, respectively) is detected on ECG signals, the scan is halted temporarily and resumed after a heartbeat or two. It is impossible to use this technique with a helical scan because the patient table will continue to move while the scan is halted, and some image slices will be missed. The typical patient radiation dose range is 1–5 mSv.

Axial, half-scan cone beam image reconstruction is then used to reconstruct a volume image at one cardiac phase (Taguchi 2003; Hsieh et al. 2006). Imaged areas with the two consecutive axial scans have some overlap, where a shift-variant-weighted summation is applied to the two reconstructed images at the same physical location to decrease the discontinuity between them (Hsieh et al. 2006).

The strengths of this protocol are the moderate radiation dose used and its ability to handle PVCs and PACs. The weakness of this protocol is the potential for discontinuity at the boundary of adjacent axial scans. This protocol is used when the heart rate is more than 60 bpm or when the patient has *not* been administered beta blocker.

13.4.1.3 Prospective ECG-gated single axial scan with 320-detector row CT

This protocol has been used with CT systems with a detector that is large enough (e.g., 320 detector rows for 160 mm) to image the entire heart with a single axial scan (Mori et al. 2004). ECG signals are acquired and trigger an axial scan that images the entire heart. An x-ray is turned on while data sufficient to image the cardiac phase of interest only are acquired out of one or two heartbeats. Note that it is logically and technically possible to acquire data to image all cardiac phases; however, such scans are not often performed. When a PVC or PAC is detected on ECG signals, the scan is postponed and triggered again after a heartbeat or two. The typical patient radiation dose range is less than 1–5 mSv.

The strengths of this protocol are that it uses the smallest radiation dose, it provides the ability to image the entire heart

simultaneously, it can handle PVCs and PACs, and it can improve the temporal resolution by multisector methods. The weaknesses of this protocol is a relatively large scattered radiation, which may decrease the contrast-to-noise ratio of images and a potential risk for cone beam artifacts.

13.4.1.4 Prospective ECG-gated single heartbeat helical scan with a high pitch (or "flash mode")

This protocol has been used with dual-source CT systems with two 64-row detectors, and it can be used with single-source CT systems with one 128-row detector. The current protocol used at our institution is shown in Table 13.2. A helical scan is used with a large pitch (or table feed) (e.g., 1.6 × 2 for dual-source CT or 1.6 for single-source CT) to image the entire heart with a single helical scan (Flohr et al. 2009). Note that the table feed per total detector coverage of the two systems remains the same: 1.6 × 2/rot × 64 rows × 0.6 mm/row = 123 mm/rot versus 1.6/rot × 128 rows × 0.6 mm/row = 123 mm/rot. The scan must be completed within a quiescent cardiac phase (e.g., 300-ms centering at mid-diastole). Thus, the table feed must be very fast (e.g., 440 mm per s).

ECG signals are used to trigger the helical scan with a large pitch that images the entire heart within 300 ms. Although the scan duration is quite short, the entire heart is not imaged simultaneously. One side of the volume belongs to the early part of the 300-ms window, whereas the other side of the volume belongs to the late part of the 300-ms window. The typical patient radiation dose range is less than 1–3 mSv.

The strengths of this protocol are that it uses the smallest radiation dose, and it provides the ability to image the entire heart during one heartbeat. The weakness of this protocol is the potential risk that one irregular heartbeat may jeopardize the entire examination. This protocol is used when the heart rate is less than 60 bpm and when the patient has been administered a beta blocker, which decreases the heart rate variation.

13.4.1.5 Dual-source CT

It is said that with a temporal resolution of 80 ms achieved by dual-source CT systems with a 300-ms gantry rotation time, 3D coronary images can be obtained with sufficient quality if heart rates are less than 70 bpm on a regular basis. Compared with single-source CT systems with 330 ms/rot (i.e., temporal resolution of 165 ms), dual-source CT systems are "more forgiving" to technologists, providing robust images with good quality. This factor cannot be overlooked, although it is difficult to quantify the benefit, to implement coronary CT angiography in clinical practices.

13.4.2 CONTRAST PROTOCOLS: BOLUS TRACKING VERSUS TEST INJECTION

Typically, 60–120 mL of I contrast agent is used with an injection rate of 4–7 mL/s, followed by a saline solution injection. The contrast agent ideally is kept in a warmer (37°C); warming is critical for reducing viscosity. We currently use either Visipaque-320 or Omnipaque-350 (General Electric Medical Solutions, Princeton, NJ), although some articles suggest higher concentrations of contrast should be used (370 concentration). With any scan and reconstruction protocol, either a bolus tracking method (Silverman et al. 1995; Kopka et al. 1996; Kirchner et al. 2000), a test bolus method (vanHoe et al. 1995;

Kaatee et al. 1998; Hittmair and Fleischmann 2001), or a test bolus tracking method (Yamaguchi and Takahashi 2009) is used. The choice between the three methods may be up to institution where the scan is performed; the discussion as to which method provides a better performance seems to be inconclusive (Cademartiri et al. 2004; Bae 2005, 2010). The bolus tracking method injects the contrast agent, monitors the pixel value of the ascending aorta, and triggers the scan 3–5 s after the value becomes higher than an established threshold, such as 150–200 Hounsfield units. The appropriate trigger value is dependent on both injection rate and the scanner type used. With the test bolus method, 10 mL of contrast agent is injected, and the pixel value of the ascending aorta is monitored. Once a time–density curve is acquired, an appropriate scan delay from the start of the injection is calculated and used in an actual scan with a full bolus. An advantage of test bolus is the ability to make certain the patient can cooperate with following the study breathing instructions before the main study is performed. Interested readers should visit the above-mentioned references. The test bolus tracking method is an emerging method that continuously performs the test bolus injection and the main bolus injection aiming at the best scan timing and the most efficient examination.

13.5 FUTURE PROSPECTS

In this section we discuss future prospects for cardiac CT, including and beyond applications in coronary CT angiography.

13.5.1 CORONARY CT ANGIOGRAPHY

Three major areas of research for coronary CT angiography are dose reduction, molecular imaging, and motion compensated image reconstruction.

Dose reduction: Efforts will continue to be made to further improve the detection and characterization of soft vulnerable plaques in coronary arteries and the applicability of cardiac CT in this process. Coronary imaging has been the driving force behind CT development over the past 10 years, and it appears that it will continue that way. Among all of the technical issues previously summarized for which improvements need made, it is safe to say that patient dose reduction is the most urgent topic for consideration (McCollough et al. 2012). Compared with the average dose of 10–20 mSv given in the early 2000s, significant dose reductions have been accomplished to achieve the current dose level (<1 mSv when the heart rate of a thin patient is below a threshold level, such as 60 bpm). Efforts on bow-tie filters, iterative reconstructions, and protocol optimizations also will be made to lower the radiation dose when the heart rate is above the threshold level (McCollough et al. 2012).

Molecular CT imaging for vulnerable plaque: It is the Holy Grail for those who are involved in coronary CT angiography to image vulnerable plaques specifically. Nanoparticle-based contrast agents have been developed (Hyafil et al. 2007; Bulte 2010; Cormode et al. 2010; Pan et al. 2010) that may allow imaging of vulnerable plaques specifically.

It is said that when soft plaque grows in size, it repeats multiple times minor ruptures of unstable caps and their healing processes, during which fibrins are heavily expressed, leading to minor thrombosis and agglutination of red blood cells. One type of nanoparticle-based contrast agent will be bound to the fibrin near the surface of such vulnerable plaques and will increase the x-ray attenuation due to a heavy atom, such as Bi, being attached to the particles (Caruthers et al. 2007; Pan et al. 2008, 2009, 2010). Thus, the contrast agents will only enhance the CT image pixel values where fibrins are present, indicating the high vulnerability of the soft plaque. Another type of nanoparticle-based contrast agent will be bound to macrophages of atherosclerotic lesions (Hyafil et al. 2007). The development of such nanoparticle contrast agents is still in the early phases; however, preliminary studies show promise and the ideas align well with one of the National Institutes of Health (NIH) visions for the future—personalized medicine. Improvement of imaging techniques such as iterative image reconstruction methods and photon-counting CT will allow further improvement of the sensitivity and specificity of the new contrast agents.

Motion compensated image reconstruction: As discussed, the current cardiac image reconstruction methods implemented in all CT scanners use only a part of the acquired data (blue areas in Figure 13.8) and leave the rest of data ("out-phase" in red areas and "near-phase" in yellow areas) unused, resulting in unnecessary radiation dose to the patient. Images are reconstructed neglecting the cardiac motion within the "in-phase" reconstruction time window for 70–150 ms, resulting in blurring and artifacts in the reconstructed images (Taguchi et al. 2006).

New approaches that allow for the following advantages *simultaneously* have been proposed (Taguchi 2008; Isola et al. 2010; Taguchi and Tang 2011; Bhagalia et al. 2012; Tang et al. 2012): to improve the quality of the images, reduce the patient radiation dose, and obtain an additional time-dependent 3D motion vector field. Fully 4D methods consist of a motion estimation (ME) method and a motion compensated image reconstruction (MCR) algorithm; and both are either processed sequentially or altered iteratively. The ME obtains the motion vector field from either images reconstructed at 10–50 cardiac phases or projection data. MCR then integrates the motion into reconstruction process and compensates for the effect of motion, providing better images. In case of iterative processes, the improved images are then used by the ME in the next iteration that will provide more accurate motion vector field to MCR. A larger reconstruction time window width may be used as the accuracy of ME improves with each iteration, improving the signal-to-noise of the images. Thus, the tube current (D_H in Figure 13.8) may be less than 100%. Motion artifacts of images may improve significantly, as shown in Figure 13.10.

13.5.2 COMPUTER-AIDED DIAGNOSIS, DETECTION, AND CHARACTERIZATION

Computer-aided diagnosis, detection, and characterization have been developed to detect soft plaques and to quantify the plaque burdens, the luminal opening, the ejection fraction, and wall motion and wall thickening. An example of such software output is shown in Figure 13.11. The performance of such software needs to be evaluated and possibly improved. Regardless, the software will become an invaluable tool in clinics, because reading coronary CT images requires specialized training and experience and is also a

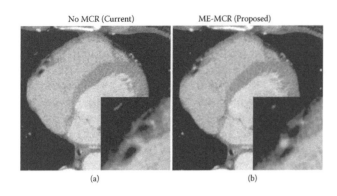

No MCR (Current)   ME-MCR (Proposed)

(a)   (b)

Figure 13.10 Images reconstructed by the current method (a) and the motion estimation-motion compensated image reconstruction (ME-MCR) method (b). Motion artifacts are reduced significantly by the ME-MCR method.

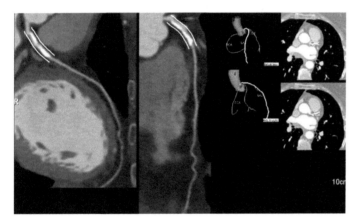

10cm

Figure 13.11 Automatic detection and analysis of coronary artery plaques.

time-consuming task of viewing images with various formats and angles such as curved multiplanar reformatting (MPR) images along three major coronary arteries, axial images, and slab maximum intensity projection (MIP) images. Quantifying the above-discussed factors will allow for assessment of therapeutic effects without interoperator variations. When an interscan variation of recent cardiac CT scans is sufficiently small—and it seems that most recent scanners have achieved this level of variation—the output of computer software is robust and stable and thus more valuable. Halpern et al. (2011) found that "Automated computer interpretation of cCTA with COR Analyzer provides high negative predictive value for the diagnosis of coronary disease in major coronary arteries as well as first-order arterial branches. False positive automated interpretations are related to anatomical and image quality considerations." "Among 207 cases evaluated by COR Analyzer, human expert interpretation identified 48 patients with stenosis. COR Analyzer identified 44/48 patients (sensitivity 92%) with a specificity of 70%, a negative predictive value of 97% and a positive predictive value of 48%. COR Analyzer agreed with the expert interpretation in 75% of patients." Arnoldi et al. (2010) found that "Compared with QCA (quantitative coronary angiogram) the automated detection algorithm evaluated had relatively high accuracy for diagnosing significant coronary

artery stenosis at cCta. If used as a second reader the high negative predictive value may further enhance the confidence of excluding significant stenosis based on a normal or near normal cCTA study."

13.5.3 PERFUSION CT

A recent study indicated that coronary CT angiography sent twice as many patients to a subsequent invasive cardiac procedure as stress myocardial perfusion scintigraphy (Shreibati et al. 2011). It is thus desired to combine coronary CT angiography and a stress test to decrease the number of false positives; however, it would increase the total cost, radiation dose, and time duration of exams if two scans were to be performed (Sato et al. 2008). The use of CT for the assessment of myocardial viability and the detection of myocardial infarction has been investigated since the 1970s in *ex vivo* studies. The improved image quality of recent CT scanners has shown promise in *in vivo* studies as well. A few approaches using CT scanners have been proposed: a first-pass stress test imaging for ischemia, delayed-enhancement imaging for infarct, or a true dynamic imaging for quantitative perfusion analyses. The first-pass stress imaging protocol performs the scan immediately after intravenous injection of the contrast agent, enabling imaging ischemic lesions as an adjunct to coronary CT angiography. The delayed-enhancement imaging protocol initiates the scan several minutes after the contrast injection. Because it is based on the same principle of delayed-enhanced magnetic resonance imaging (MRI), studies have demonstrated a strong correlation between the two examinations, with respect, for example, to infarct size. The dynamic perfusion protocol performs multiple axial scans over 30–90 s to capture the wash-in and wash-out of contrast-enhanced blood through myocardium. The patient table will be moved back and forth when the detector coverage is limited (e.g., 4–8 cm), a protocol called the "shuttle scan." This protocol provides quantitative perfusion indices such as maps of blood flow and mean transit time at the expense of radiation dose. Systematic animal and patient studies are being conducted to assess the accuracy and robustness of these approaches.

13.5.4 DUAL-ENERGY CT

Dual-energy CT can be applied to cardiac CT to better detect and characterize soft plaque, myocardial infarction, or both. When it is performed using the 2-kVp technique with dual-source CT, the temporal resolution of CT images will, in principal, be degraded to that of single-source CT. A unique image reconstruction method similar to those developed for half-scan artifacts will then be desirable for maintaining the effective temporal resolution of dual-energy images. When it is performed using kVp-switching technique, the temporal resolution will remain the same as the single-energy CT regardless of the scanner's type.

13.5.5 4D MOTION ANALYSIS AND ITS APPLICATION

An emerging 4D application has been developed (PhyzioDynamics) that uses image processing techniques such as estimating motion and warping images at different cardiac phases, and averaging them. Unlike the methods outlined in Section 13.5.1, this approach does not involve the image reconstruction process but uses images provided by the current

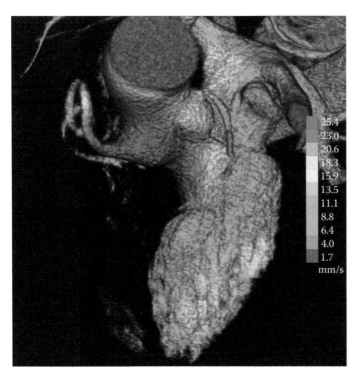

25.4
23.0
20.6
18.3
15.9
13.5
11.1
8.8
6.4
4.0
1.7
mm/s

Figure 13.12 (See color insert.) Velocity map fused on the surface of endocardium. (Courtesy of Ziosoft, Rodwood City, CA.)

ECG-gated reconstruction method. Thus, the image quality may be limited by the quality of the original images. Nonetheless, this approach shows promises in various applications such as detailed morphological dynamic imaging for the quadricuspid aortic valve (Karlsberg et al. 2012), noise reduction, and systolic velocity map (Figure 13.12). The idea of 4D imaging and motion analyses is intriguing (Sengupta et al. 2011), but its exact role in clinical practice needs to be determined through rigorous evaluations.

13.5.6 ITERATIVE IMAGE RECONSTRUCTION

Various iterative image reconstruction methods have been investigated in our community for many years. Of late, CT manufacturers have introduced some versions of such methods into clinic, attracting strong interest among radiologists and cardiologists. Because cardiac CT scans have been one of higher dose applications for the reasons outlined in Section 13.3, the use of iterative methods on cardiac image reconstruction makes sense. Thorough and careful, task-specific evaluations must be conducted on the accuracy of images, because the performance of nonlinear image reconstruction techniques such as iterative methods varies strongly with objects and conditions. Cardiac images involve a lot of conflicting, demanding components in a high spatial resolution in Ca plaques, stents, and pulmonary vessels; a good contrast resolution in soft plaques and perfusion defects in myocardium; a good linearity in Ca plaques and contrast agent.

13.5.7 CONE BEAM CT

Researchers have been trying to perform cardiac CT imaging using C-arm x-ray systems or cone beam CT in interventional suites. If successful, it will allow physicians to plan, monitor, and assess the progress before, during, and after the procedure,

respectively, qualitatively and quantitatively. Many technical developments discussed in this chapter, including perfusion CT and 4D motion analysis, will have great potential to be implemented to cone beam CT systems, once a problem with the slow gantry rotation speed discussed below is solved.

Cone beam CT provides a wide viewing range thanks to its large area detector with the in-plane spatial resolution as good as ~200 μm and the frame rate of as fast as 60 frames per s; however, because the gantry rotates slowly, it takes 4–6 s to cover the half-scan range. Images reconstructed using a standard cone beam method will present motion artifacts and blurred edges. Among many approaches to overcome these problems, two typical methods are (1) motion compensated image reconstruction methods similar to the method discussed in Section 13.5.1; and (2) highly regularized, iterative image reconstruction methods (e.g., total variation minimization) with an extremely narrow gating window. With the latter approach, the narrow gating window will minimize the intrawindow motion, but the number of projections from such a narrow gating window would be so few that such undersampled data would result in streaking artifacts with standard image reconstruction methods. Highly regularized methods could mitigate the sampling problem.

REFERENCES

Arnoldi, E., Gebregziabher, M. et al. 2010. Automated computer-aided stenosis detection at coronary CT angiography: initial experience. *Eur Radiol* 20(5): 1160–7.

Bae, K.T. 2005. Test-bolus versus bolus-tracking techniques for CT angiographic timing. *Radiology* 236(1): 369–70.

Bae, K.T. 2010. Intravenous contrast medium administration and scan timing at CT: considerations and approaches1. *Radiology* 256(1): 32–61.

Bhagalia, R., Pack, J.D. et al. 2012. Nonrigid registration-based coronary artery motion correction for cardiac computed tomography. *Med Phys* 39(7): 4245–54.

Budoff, M.J., Achenbach, S. et al. 2003. Clinical utility of computed tomography and magnetic resonance techniques for noninvasive coronary angiography. J Am Coll Cardiol 42(11): 1867–78.

Budoff, M.J., Dowe, D. et al. 2008. Diagnostic performance of 64-multidetector row coronary computed tomographic angiography for evaluation of coronary artery stenosis in individuals without known coronary artery disease: results from the prospective multicenter ACCURACY (Assessment by Coronary Computed Tomographic Angiography of Individuals Undergoing Invasive Coronary Angiography) trial. *J Am Coll Cardiol* 52(21): 1724–32.

Bulte, J.W.M. 2010. Science to practice: can CT be performed for multicolor molecular imaging? *Radiology* 256(3): 675–6.

Cademartiri, F., Nieman, K. et al. 2004. Intravenous contrast material administration at 16–detector row helical CT coronary angiography: test bolus versus bolus-tracking technique1. *Radiology* 233(3): 817–23.

Caruthers, S.D., Wickline, S.A. et al. 2007. Nanotechnological applications in medicine. *Curr Opin Biotechnol* 18(1): 26–30.

Cormode, D.P., Roessl, E. et al. 2010. Atherosclerotic plaque composition: analysis with multicolor CT and targeted gold nanoparticles. *Radiology* 256(3): 774–82.

Cury, R., Feuchtner, G. et al. 2011. Cardiac CT in the emergency department: convincing evidence, but cautious implementation. *J Nucl Cardiol* 18(2): 331–41.

Cury, R.C., Feuchtner, G.M. et al. 2013. Triage of patients presenting with chest pain to the emergency department: implementation of coronary CT agiography in a large urban health care system. *AJR Am J Roentgenol* 200(1): 57–65.

Department of Health and Human Services. 2004. *The Burden of Chronic Diseases and Their Risk Factors: National and State Perspectives.* Atlanta, GA: Centers for Disease Control and Prevention, National Center for Chronic Disease Prevetion and Health Promotion, Department of Health and Human Services.

Desjardins, B. and Kazerooni, E.A. 2004. ECG-gated cardiac CT. *AJR Am J Roentgenol* 182(4): 993–1010.

Dey, D., Schepis, T. et al. 2010. Automated three-dimensional quantification of noncalcified coronary plaque from coronary CT angiography: comparison with intravascular US. *Radiology* 257(2): 516–22.

Feuchtner, G.M., Plank, F. et al. 2012. Evaluation of myocardial CT perfusion in patients presenting with acute chest pain to the emergency department: comparison with SPECT-myocardial perfusion imaging. *Heart* 98: 1510–1517.

Flohr, T., McCollough, C. et al. 2006. First performance evaluation of a dual-source CT (DSCT) system. *Eur Radiol* 16(2): 256–68.

Flohr, T., Ohnesorge, B. et al. 2003. Image reconstruction and performance evaluation for ECG-gated spiral scanning with a 16-slice CT system. *Med Phys* 30(10): 2650–62.

Flohr, T.G., Leng, S. et al. 2009. Dual-source spiral CT with pitch up to 3.2 and 75 ms temporal resolution: image reconstruction and assessment of image quality. *Med Phys* 36(12): 5641–53.

Gilard, M., Cornily, J.-C. et al. 2006. Accuracy of multislice computed tomography in the preoperative assessment of coronary disease in patients with aortic valve stenosis. *J Am Coll Cardiol* 47(10): 2020–4.

Gilkeson, R.C., Ciancibello, L. et al. 2003. Pictorial essay. Multidetector CT evaluation of congenital heart disease in pediatric and adult patients. *AJR Am J Roentgenol* 180(4): 973–80.

Grass, M., Manzke, R. et al. 2003. Helical cardiac cone beam reconstruction using retrospective ECG gating. *Phys Med Biol* 48(18): 3069–84.

Gruettner, J., Fink, C. et al. 2012. Coronary computed tomography and triple rule out CT in patients with acute chest pain and an intermediate cardiac risk profile. Part 1: impact on patient management. *Eur J Radiol.* 82(1): 100–105.

Halpern, E.J. and Halpern, D.J. 2011. Diagnosis of coronary stenosis with CT angiography: comparison of automated computer diagnosis with expert readings. *Acad Radiol* 18(3): 324–33.

Heidenreich, P.A., Trogdon, J.G. et al. 2011. Forecasting the future of cardiovascular disease in the United States. *Circulation* 123(8): 933–44.

Henzler, T., Gruettner, J. et al. 2012. Coronary computed tomography and triple rule out CT in patients with acute chest pain and an intermediate cardiac risk for acute coronary syndrome: part 2: economic aspects. *Eur J Radiol.* 82(1): 106–111.

Hittmair, K. and Fleischmann, D. 2001. Accuracy of predicting and controlling time-dependent aortic enhancement from a test bolus injection. *J Comput Assist Tomogr* 25(2): 287–94.

Hsieh, J. 2003. *Computed Tomography: Principles, Design, Artifacts, and Recent Advances.* Bellingham, WA: SPIE.

Hsieh, J., Londt, J. et al. 2006. Step-and-shoot data acquisition and reconstruction for cardiac x-ray computed tomography. *Med Phys* 33(11): 4236–48.

Hsieh, J., Molthen, R.C. et al. 2000. An iterative approach to the beam hardening correction in cone beam CT. *Med Phys* 27(1): 23–9.

Hu, H., He, H.D. et al. 2000. Four multidetector-row helical CT: image quality and volume coverage speed. *Radiology* 215(1): 55–62.

Hu, H., Pan, T. et al. 2000. Multislice helical CT: image temporal resolution. *IEEE Trans Med Imaging* 19(5): 384–90.

Hyafil, F., Cornily, J.-C. et al. 2007. Noninvasive detection of macrophages using a nanoparticulate contrast agent for computed tomography. *Nat Med* 13(5): 636–41.

Isola, A.A., Grass, M. et al. 2010. Fully automatic nonrigid registration-based local motion estimation for motion-corrected iterative cardiac CT reconstruction. *Med Phys* 37(3): 1093–109.

Jacobs, P.C., Gondrie, M.J.A. et al. 2012. Coronary artery calcium can predict all-cause mortality and cardiovascular events on low-dose CT screening for lung cancer. *AJR Am J Roentgenol* 198(3): 505–11.

Kaatee, R., Van Leeuwen, M.S. et al. 1998. Spiral CT angiography of the renal arteries: should a scan delay based on a test bolus injection or a fixed scan delay be used to obtain maximum enhancement of the vessels? J Comput Assist Tomogr 22(4): 541–7.

Kachelriess, M., Ulzheimer, S. et al. 2000. ECG-correlated image reconstruction from subsecond multi-slice spiral CT scans of the heart. *Med Phys* 27(8): 1881–902.

Karlsberg, D.W., Elad, Y. et al. 2012. Quadricuspid aortic valve defined by echocardiography and cardiac computed tomography. *Clin Med Insights Cardiol* 6: 41–4.

Kelly, J.L., Thickman, D. et al. 2008. Coronary CT angiography findings in patients without coronary calcification. *AJR Am J Roentgenol* 191(1): 50–5.

Kirchner, J., Kickuth, R. et al. 2000. Optimized enhancement in helical CT: experiences with a real-time bolus tracking system in 628 patients. *Clin Radiol* 55(5): 368–73.

Kitagawa, K., George, R.T. et al. 2010. Characterization and correction of beam-hardening artifacts during dynamic volume CT assessment of myocardial perfusion. *Radiology* 256(1): 111–8.

Kopka, L., Rodenwaldt, J. et al. 1996. Dual-phase helical CT of the liver: effects of bolus tracking and different volumes of contrast material. *Radiology* 201(2): 321–6.

Lardo, A.C., Cordeiro, M.A.S. et al. 2006. Contrast-enhanced multidetector computed tomography viability imaging after myocardial infarction: characterization of myocyte death, microvascular obstruction, and chronic scar. *Circulation* 113(3): 394–404.

McCollough, C.H., Chen, G.H. et al. 2012. Achieving routine submillisievert CT scanning: report from the summit on management of radiation dose in CT. *Radiology* 264(2): 567–80.

McCollough, C.H. and Zink, F.E. 1999. Performance evaluation of a multi-slice CT system. *Med Phys* 26(11): 2223–30.

Mori, S., Endo, M. et al. 2004. Physical performance evaluation of a 256-slice CT-scanner for four-dimensional imaging. *Med Phys* 31(6): 1348–56.

Nance, J.W.J., Bamberg, F. et al. 2012. Coronary computed tomography angiography in patients with chronic chest pain: systematic review of evidence base and cost-effectiveness. *J Thorac Imaging* 27(5): 277–88.

Nasir, K. and Clouse, M. 2012. Role of nonenhanced multidetector CT coronary artery calcium testing in asymptomatic and symptomatic individuals. *Radiology* 264(3): 637–49.

Nielsen, L.H., Olsen, J. et al. 2012. Effects on costs of frontline diagnostic evaluation in patients suspected of angina: coronary computed tomography angiography vs. conventional ischaemia testing. *Eur Heart J Cardiovasc Imaging.* 14(5): 449–455.

Ohnesorge, B., Flohr, T. et al. 2000. Cardiac imaging by means of electrocardiographically gated multisection spiral CT: initial experience. *Radiology* 217(2): 564–71.

Pan, D., Caruthers, S.D. et al. 2008. Ligand-directed nanobialys as theranostic agent for drug delivery and manganese-based magnetic resonance imaging of vascular targets. *J Am Chem Soc* 130(29): 9186–7.

Pan, D., Roessl, E. et al. 2010. Computed tomography in color: nanoK-enhanced spectral CT molecular imaging. *Angew Chem Int Ed Engl* 49(50): 9635–9.

Pan, D., Williams, T.A. et al. 2009. Detecting vascular biosignatures with a colloidal, radio-opaque polymeric nanoparticle. *J Am Chem Soc* 131(42): 15522–7.

Parker, D.L. 1982. Optimal short scan convolution reconstruction for fanbeam CT. *Med Phys* 9(2): 254–7.

Patel, M.R., Peterson, E.D. et al. 2010. Low diagnostic yield of elective coronary angiography. *N Engl J Med* 362(10): 886–95.

Petersilka, M., Stierstorfer, K. et al. 2010. Strategies for scatter correction in dual source CT. *Med Phys* 37(11): 5971–92.

Petretta, M., Daniele, S. et al. Prognostic value of coronary artery calcium score and coronary CT angiography in patients with intermediate risk of coronary artery disease. *Int J Cardiovasc Imaging* 28(6): 1547–1556.

PhyzioDynamics. Redwood City, CA: Ziosoft Inc. Available from: http://www.ziosoftinc.com.

Primak, A.N., Dong, Y. et al. 2007. A technical solution to avoid partial scan artifacts in cardiac MDCT. *Med Phys* 34(12): 4726–37.

Raff, G.L., Gallagher, M.J. et al. 2005. Diagnostic accuracy of noninvasive coronary angiography using 64-slice spiral computed tomography. *J Am Coll Cardiol* 46(3): 552–7.

Ramirez-Giraldo, J.C., Yu, L. et al. 2012. A strategy to decrease partial scan reconstruction artifacts in myocardial perfusion CT: phantom and *in vivo* evaluation. *Med Phys* 39(1): 214–23.

Rich, M.E., Utsunomiya, D. et al. 2012. Prospective evaluation of the updated 2010 ACCF cardiac CT appropriate use criteria. *J Cardiovasc Comput Tomogr* 6(2): 108–12.

Sato, A., Hiroe, M. et al. 2008. Quantitative measures of coronary stenosis severity by 64-slice CT angiography and relation to physiologic significance of perfusion in nonobese patients: comparison with stress myocardial perfusion imaging. *J Nucl Med* 49(4): 564–72.

Sengupta, P.P., Marwick, T.H. et al. 2011. Adding dimensions to unimodal cardiac images. *JACC: Cardiovasc Imaging* 4(7): 816–8.

Shreibati, J.B., Baker, L.C. et al. 2011. Association of coronary CT angiography or stress testing with subsequent utilization and spending among medicare beneficiaries. *JAMA* 306(19): 2128–36.

Silver, M.D. 2000. A method for including redundant data in computed tomography. *Med Phys* 27(4): 773–4.

Silverman, P.M., Brown, B. et al. 1995. Optimal contrast enhancement of the liver using helical (spiral) CT: value of SmartPrep. *AJR Am J Roentgenol* 164(5): 1169–71.

So, A., Hsieh, J. et al. 2009. Beam hardening correction in CT myocardial perfusion measurement. *Phys Med Biol* 54(10): 3031–50.

Stenner, P., Schmidt, B. et al. 2009. Partial scan artifact reduction (PSAR) for the assessment of cardiac perfusion in dynamic phase-correlated CT. *Med Phys* 36(12): 5683–94.

Taguchi, K. 2003. Temporal resolution and the evaluation of candidate algorithms for four-dimensional CT. *Med Phys* 30(4): 640–50.

Taguchi, K. 2008. Toward motion compensated four-dimensional CT imaging. Medical Imaging and Information, Industrial Development Organization, Inc. 40: 1225–9.

Taguchi, K. and Anno, H. 2000. High temporal resolution for multislice helical computed tomography. *Med Phys* 27(5): 861–72.

Taguchi, K., Chiang, B.S. et al. 2006. Direct cone-beam cardiac reconstruction algorithm with cardiac banding artifact correction. *Med Phys* 33(2): 521–39.

Taguchi, K., Segars, W.P. et al. 2006. Toward time resolved 4D cardiac CT imaging with patient dose reduction: estimating the global heart motion. In: Flynn, M.J. and Hsieh, J., (eds.), in *Medical Imaging 2006: Physics of Medical Imaging2006*. San Diego, CA: SPIE, p. 61420J–9.

Taguchi, K. and Tang, Q. 2011. Four-dimensional cardiac imaging using x-ray CT. Medical Imaging and Information, Industrial Development Organization, Inc. 43: 1105–10.

Tang, Q., Cammin, J. et al. 2012. A fully four-dimensional, iterative motion estimation and compensation method for cardiac CT. *Med Phys* 39(7): 4291–305.

van Hoe, L., Marchal, G. et al. 1995. Determination of scan delay time in spiral CT-angiography: utility of a test bolus injection. *J Comput Assist Tomogr* 19(2): 216–20.

Wallis, A., Manghat, N. et al. 2012. The role of coronary CT in the assessment and diagnosis of patients with chest pain. *Clin Med* 12(3): 222–9.

Yamaguchi, T. and Takahashi, D. 2009. Development of test bolus tracking method and usefulness in coronary CT angiography. *Jap J Radiologic Technol* 65(8): 1032–40.

14

C-arm CT in the interventional suite: Current status and future directions

Rebecca Fahrig, Jared Starman, Erin Girard, Amin Al-Ahmad,
Hewei Gao, Nishita Kothary, and Arundhuti Ganguly

Contents

14.1 INTRODUCTION

Three-dimensional (3D) computed tomography (CT)–like reconstructions of patients based on image data acquired in the interventional suite *during an intervention* first became clinically available in the late 1990s. The projection image data were acquired using an x-ray image intensifier mounted on a C-arm gantry and then transferred to an off-line computer for processing, reconstruction, and visualization during the procedure. Maximum gantry rotation speeds of 20° per second, combined with the nonidealities of the image intensifier, including veiling glare and distortion, provided image quality sufficient only for high-contrast objects such as contrast-filled vessels and bone. The long acquisition times (on the order of 11 s for a rotation of 200° around the patient) limited the applications to those where patient motion was negligible. In spite of their limitations, these 3D images proved to be very useful, especially for guidance of minimally invasive intracranial procedures where anatomic motion is low.

Within the last decade, improvement to such C-arm systems replaced the XRII with a large-area amorphous-silicon (a-Si) digital flat-panel (FP) detector, providing a linear, wider dynamic range recording device that is free of distortion. When a sufficient number of projection images is acquired, and when the image acquisition geometry is ideal (i.e., no truncation

and a small irradiated volume so as to reduce scatter and cone beam artifact), low-contrast resolution in the order of 5 Hounsfield units (HU) can now be achieved in 3D C-arm CT reconstructions. In addition, robust mechanical designs for the C-arm gantry itself have significantly reduced the minimum image acquisition time, which can now be as low as 4 s. New applications for this 3D imaging system are quickly developing, as clinicians see the benefits to be gained from intraprocedural 3D image information.

In this chapter, we first outline the basic requirements for imaging in the interventional suite and for C-arm CT. We then describe the current state-of-the-art, the limitations of the current implementation, and some efforts that are underway to improve the image quality of C-arm CT reconstructions. We highlight several new applications that are under development that take advantage of fast C-arm rotation speeds and of new reconstruction algorithms. Finally, we outline new and future hardware developments that will further improve the image quality and utility of C-arm CT. A full discussion of detector technologies is beyond the scope of this chapter; an excellent pair of overview papers was recently provided by Cowen et al. (2008a,b). The remainder of this chapter focuses on developments using indirect detector cesium-iodide (Cs-I)–based FPs because these systems have been most often used for low-contrast CT imaging in the interventional suite.

14.2 IMAGING REQUIREMENTS: MINIMALLY INVASIVE INTERVENTIONS

The range of minimally invasive interventional procedures carried out in a catheterization laboratory (alternatively called an interventional suite) includes body interventions such as needle biopsy, stenting, and chemoembolization; cardiac interventions such as coronary balloon angioplasty and electrophysiology-guided ablation; and intracranial procedures such as mechanical thrombectomy and aneurysm repair. Although the imaging requirements differ somewhat for each body region, the general requirements are sufficiently similar that a single system can be used to guide all of them. The first requirement is for real-time low-dose x-ray fluoroscopy and 30 frames per s (fps) to guide wire, needle, and catheter placement. Dose rates incident on the detector are between 0.1 and 10 μR (Rowlands and Yorkston 2000). Higher quality diagnostic information is provided by higher dose radiographic sequences acquired at between 5 and 60 fps to allow visualization of small vessels, for example, in liver, brain, or in the beating heart. The range of exposures incident on the detector are between 30 μR and 3 mR. Detector-limiting resolution for fluoroscopy and radiography is on the order of 150 μm so that the small vessels of interest can be detected. The field of view (FOV) should be large enough to cover the region of interest (ROI) and is up to 40 cm × 40 cm for body and intracranial systems, and 20 cm × 20 cm for cardiac systems.

These imaging requirements have driven the design of "indirect conversion" digital FPs that must have a low noise floor (0.1 μR; Rowlands and Yorkston 2000), a wide dynamic range (12–14 bits), and high spatial resolution. Because much of the high-resolution imaging is done using I injections, the optimal operating point for the system is ~70–85 kilovoltage, peak (kVp), with the range depending mainly on the size of the patient. The CsI converter/scintillator material on the digital FPs has a columnar structure that maintains resolution while stopping as many incident x-rays as possible. A typical thickness is ~600 μm and provides a zero spatial frequency detective quantum efficiency [DQE(0)] of ~70%–75% at the kVps of interest (Busse et al. 2002, Granfors et al. 2003, Tognina et al. 2004).

The mechanical system should provide open, easy access to the patient, as well as a full range of angles and tilts so that projections with steep angulations can be acquired. The C-arm gantry, combined with a well-designed table for patient positioning, provides a good compromise between these two requirements. A typical geometry of such a system is shown in Figure 14.1a.

The aforementioned imaging requirements do not match well to those of clinical CT imaging. First, the standard CT detector is still a strip of ~2–5 cm wide by ~90 cm long, providing a large axial FOV but narrow coverage in the longitudinal (or z) direction. Pixel spacing at the detector is ~1.1 mm × 1.4 mm, with the larger dimension in the slice width direction, providing a minimum reconstructed slice width of ~0.6 mm. The individual detector elements used in CT are optimized to operate at 100–120 kVp, and the scintillator photodiode detectors provide a very fast response and a DQE(0) of ~78% at those energies

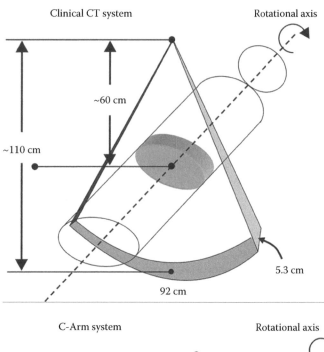

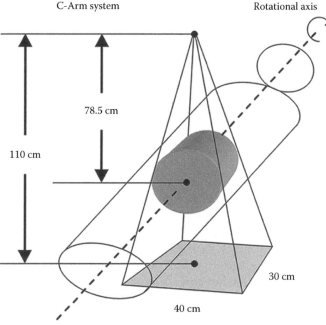

Figure 14.1 (Top) Typical geometric parameters for a C-arm system. One possible axis of rotation for 3D image acquisition is indicated. The C-arm also can angle in the craniocaudal direction. Focal-spot-to-detector distance can be varied by moving the detector closer to the patient. (Bottom) Typical geometric parameters for a clinical CT system. The detector shown here would have 64 rows. The geometry of the system is fixed; some gantries provide limited craniocaudal tilt.

[this DEQ(0) includes the reduction due to the geometric fill factor of the scintillator array, which is significantly reduced from the native quantum detection efficiency of the material by the need for intrapixel light and x-ray scatter barriers]. A full rotation of 360° can be acquired in under 0.5 s, and at least 1000 projections are acquired in this time. The x-ray tube has very high instantaneous and continuous heat loading specifications (typical values of 80–140 kV, 10–600 mA, and 20–100 kW) to support several rotations as is typical in helical mode when a large volume must be covered in a short time. For comparison purposes, a typical geometry of a clinical CT system is illustrated in Figure 14.1b.

14.3 CURRENT STATE-OF-THE-ART: C-ARM CT

C-arm CT imaging in the interventional suite is carried out using a floor- or ceiling-mounted gantry. A typical C-arm system, such as the system shown in Figure 14.2a, consists of an x-ray source and detector mounted on the ends of a C. The "C" is suspended by a second arm, the suspension, which is in turn attached to an L-arm mounted to the ceiling or the floor. The L-arm can rotate about a vertical axis, the C-arm suspension can rotate about a horizontal axis attached to the L-arm, and the C-arm itself can be rotated about the axis defined by the intersection of the planes containing the L-arm and the suspension axes of rotation. All three rotations meet at the isocenter of the system, thus motion is restricted to the surface of a single sphere. Linear motion can be provided by table motion or, for ceiling-mounted systems, linear tracks. To date, all clinical CT imaging uses the classic single-arc circular trajectory, sweeping through ~200°, or equivalently, sweeping through pi degrees plus the fan angle of the system. Two possible motions that can be used to provide the circular trajectory are shown in Figure 14.2 and are referred to as either propeller (Figure 14.2b, C-arm suspension) or sleeve (Figure 14.2c, C-arm itself). Neither axis permits a full rotation through 360° because high-voltage and signal cables impose limits; typical maximum ranges are 330° and 200° for propeller and sleeve modes, respectively. Maximum rotation speeds are between 40 and 60°/s, depending on the manufacturer, with acceleration to maximum speed requiring ~0.5 s.

The motion of the C-arm system through the minimum single arc is assumed to be reproducible and stable over time and is precalibrated to provide the required projection matrices for reconstruction. The circular trajectory provides sufficient angular coverage for reconstruction using the Feldkamp (FDK) algorithm (Feldkamp et al. 1984), with additional weights as suggested by Parker (1982) and Silver (2000). Correction and calibration of the projection data before reconstruction includes offset and gain of the FP, bad pixel correction, and calibration of the incident exposure (commonly called I_0) corresponding to each projection image. The calibration value for I_0 cannot be taken directly from an image due to the limited axial size of the flat panel, but it can be inferred from knowledge of the milliamperes-seconds (mA·s) per frame or can be measured for each frame using an external measurement device. If the automatic exposure control system is used during a C-arm CT acquisition, the mA·s per frame increases when the patient thickness is large (i.e., lateral projections) and decreases when the patient thickness is small (i.e., anteroposterior projections), providing an exposure profile as a function of angle that is similar to that of clinical milliampere (mA)-modulated CT.

For the a-Si detectors currently used in C-arm CT, trade-offs between detector size, pixel size, and frame rate are made (Roos et al. 2004, Tognina et al. 2004). The a-Si detectors have been optimized for fluoroscopy and so have resolutions better than 0.2 mm. Each row is addressed sequentially, with each column

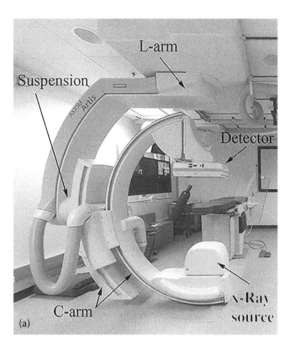

(a)

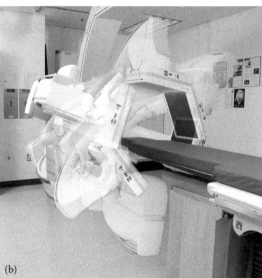

(b)

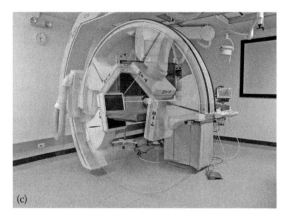

(c)

Figure 14.2 (a) Photograph of a C-arm system. (b) Composite photograph showing rotation in the propeller mode, or around the "suspension" axis, for which the maximum arc is 330°. This mode is used for CT imaging of the head and neck. (c) Composite photograph showing rotation in sleeve mode, using the double C-arm to provide a rotation through 200°. This mode is used for body CT imaging.

then connected to the readout electronics; the total time to read a single frame depends directly on the number of rows to be read out. A factor of two in readout rate is gained by addressing the top half of the detector up to one set of readout electronics, whereas the bottom half of the detector is addressed to a second set of readout electronics. The time necessary to read out each pixel is between 10–50 μs (Rowlands and Yorkston 2000), and this time is dominated by the pixel RC time constant due to the pixel and stray capacitance and data line and thin-film transistor (TFT) resistance. Given this readout time per pixel, between 20,000 and 100,000 lines can be read out per second. For current 40 × 30 FPs, frame rates of 7.5 fps are implemented at full resolution (~2048 rows total). Binning of pixels to 2 × 2 and 4 × 4 permits operation at 30 and 60 fps, with resolution reduced to 0.4 and 0.8 mm respectively; a frame rate of 120 fps, implemented with 4 × 4 binning, would require reading out 31,000 lines per s through each set of readout electronics to provide a 30 × 40 cm FOV with a matrix size of 512 × 512 pixels. Experimental modes and strip detectors with fewer lines (i.e., less z extent) have been operated as fast as 600–1000 fps (Colbeth et al. 2005, Edward et al. 2007). For C-arm CT applications, the pixels are often binned 4 × 4, providing pixels of ~0.8 mm at the detector, or isotropic voxels of ~0.5 mm in the reconstructed C-arm CT volume.

14.4 CLINICAL APPLICATIONS

A brief survey of the literature in the past 5 years for the specific area of C-arm CT generated more than 130 publications. Neurovascular applications accounted for more than 20% of the publications, and surprisingly, cardiac applications including imaging of the left atrium and coronary vessels accounted for another 20% of all publications. Other topics include numerical flow simulation, dose evaluation, physics of imaging system improvement and other high-contrast clinical applications such as guidance of spine interventions, and visualization of renal and other abdominal arteries. Because soft-tissue contrast has improved, applications have expanded to include quantitative cerebral blood volume (CBV) imaging and 3D imaging during minimally invasive therapies for hepatic cancers. The applications listed earlier acquire the data required for reconstruction using either one sweep (soft tissue or "native" reconstruction) or two sweeps (usually using subtracted projection images that show only I-filled vessels in the 3D reconstruction) around the patient. Applications requiring multiple sequential sweeps around the patient also are under investigation. Multiple sweeps in conjunction with recording of the electrocardiogram (ECG) signal can be used to provide ECG-gated images of the heart. If the time between sweeps is sufficiently short, multiple volumes acquired before, during, and after a short bolus of iodinated contrast can be used to obtain quantitative measurements of blood perfusion. In the following sections, we describe in more detail the newer applications of single/dual sweep C-arm CT, as well as two multisweep applications under development.

14.4.1 FUNCTIONAL IMAGING IN THE BRAIN

One of the potentially most useful applications of "native" reconstructions from a C-arm CT system would be the ability to distinguish between ischemic and hemorrhagic stroke and then to provide an evaluation of the severity of the ischemic insult and an estimate of the extent of salvageable tissue by measuring CBV and cerebral blood flow (CBF). Providing this information in the interventional suite would shorten the time to therapy while providing access to all of the available therapeutic techniques, for example, mechanical thrombectomy and arterially delivered tissue plasminogen activator (tPA). Preliminary studies have demonstrated the ability of C-arm CT to detect intracranial hemorrhages and have highlighted other applications for low-contrast intracranial imaging (Zellerhoff et al. 2005, Arakawa et al. 2008, Orth et al. 2008, Wallace et al. 2009). A study by Ahmed et al. (2009) has shown that clinically relevant measures of CBV can be obtained, with values that compare well with clinical CT measures of blood volume in a healthy canine model. The imaging protocol consists of two 10-s rotations: an initial mask run, followed by the power injection of iodinated contrast, and, after a delay to allow the contrast to reach steady state in the brain parenchyma, a second rotational acquisition. After appropriate image processing to calculate CBV from the two C-arm CT volumes, the maximum and mean deviations of CBV values between CT and C-arm CT were ~27% and 7%, respectively. Images from the same group in a canine model of stroke created by occlusion of the middle cerebral artery are shown in Figure 14.3. The region of increased CBV in C-arm CT correlates well with clinical CT and with the diffusion-weighted magnetic resonance image (MRI).

As pointed out, a single rotation currently requires ~4 s, with ~1 s for turnaround at the end of a sweep. Thus, a C-arm CT volume can be reconstructed every ~5 s; this temporal sampling is not sufficient to accurately measure the arterial input function, accurate knowledge of which is required for classic indicator dilution theory–based calculations of CBF. Neukirchen and Hohmann 2007) have proposed a dynamic iterative reconstruction algorithm to compensate for the effects of the slow C-arm rotational speed. A dynamic reconstruction approach for slow scanning using a conventional CT scanner based on filtered back-projection was presented by Montes and Lauritsch (2007). They have shown that the maximum rotation time for obtaining sufficient temporal sampling for neuroperfusion imaging is ~4 s. However, this assumes that the data were obtained over a full 2π arc. Also, for the initial human stroke case used for demonstrating this algorithm, the data were originally acquired at 0.5-s rotation every other rotation or one data set every 1 s. In postprocessing, the data at every 4 s were selected for perfusion calculations. Montes and Lauritsch (2007) developed a temporal interpolation and smoothing approach called partial block back-projection (PBB) technique that was used to improve losses in temporal resolution. The resultant perfusion maps were found to have a good qualitative match with the original data set. This can greatly help in reducing the number of scans and hence the overall dose to the patient during perfusion CT scans. An estimate of the noise in such images compared with the original scan is given by the following equation:

$$\sigma_K^2 = K\sigma_{Spl}^2 \approx 2.3\overline{\varepsilon^2}KT_{Smax} \qquad (14.1)$$

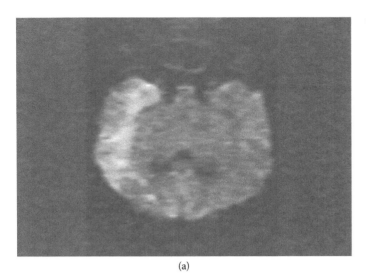

(a)

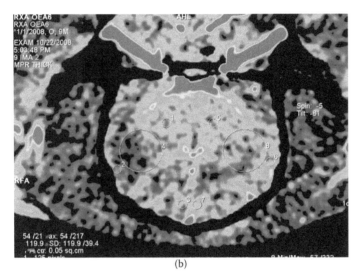

(b)

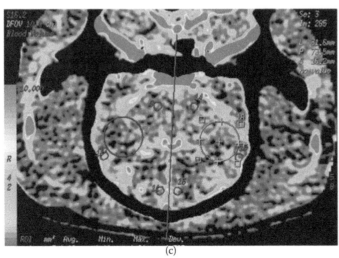

(c)

Figure 14.3 (See color insert (b) and (c).) (a) Diffusion-weighted MRI image acquired 3.5 hr after creation of stroke shows an infarct in the territory of the middle cerebral artery. (b) C-arm CT CBV image acquired during steady-state injection of iodinated contrast at 1.5 mL per s shows ~30% difference in CBV between right and left sides of the brain. (c) Clinical CT CBV image shows ~45% difference in CBV between right and left sides of the brain. (Images courtesy of Dr. Charles M. Strother, Department of Radiology, University of Wisconsin–Madison.)

where K indicates the downsampling factor, ε^2 is the variance in the original image, T_s is the time for single rotation, and ν_{max} is the maximum frequency of imaged flow signal that was estimated to be 0.0966 Hz for cerebral perfusion.

However, in C-arm CT, each dataset is acquired over the entire scan time of ~4 s followed by the turnaround time during which no data is acquired. Successive scans are obtained with the C-arm rotating in opposite direction with respect to the previous one. Also, the data set consists only of a short scan over a pi+fan angle. Thus, additional means to increase the data sampling rate are needed for imaging temporally varying signals (such as in brain perfusion) compared with the method suggested by Montes and Lauritsch (2007). One method is to acquire multiple data sets with appropriate offsets between start of image acquisition and start of a short bolus injection of contrast (assuming that the contrast dynamics are reproducible from injection to injection). For example, two interleaved six-sweep acquisitions with an offset of 5.55 s between acquisitions would provide volume images showing contrast dynamics with a sampling rate of 2.8 s over a total of ~33 s. Applying the method proposed by Montes and Lauritsch (2007) to these interleaved scans from multiple six-sweep acquisitions and using an interpolation scheme, CBF values that are comparable with those from corresponding clinical CT scans can be measured. This scheme is represented in the following equations and is illustrated in Figure 14.4.

Here, T_{rot} is the rotation time of the C-arm with view angle range $[0, \alpha_{max}]$ and T_w is the interval between two successive rotations. Given N_{rot} rotations of the C-arm, for each view angle the projection obtained has time varying signal given by

$$P_\infty(\gamma, \varphi, k) = P_\infty(\gamma, \varphi, t_\alpha(k)) \qquad (14.2)$$

where α is the fan angle and γ is the cone-angle at time points $t_\alpha(k)$ with $1 \le k \le N_{rot}$. Using an interpolation function in projection space $\varphi_{\alpha,k}(t)$ and assuming the contrast enhanced temporal signal is sufficiently smooth, the intermediate projections can be calculated. However, the projections from successive scans are not accurately reproducible with respect to the view angle. Taking this into account and also to increase computational speed, the interpolation is performed between partial blocks of reconstructed data. This follows from the method by Montes and Lauritsch (2007) but involves an additional interpolation step (Figure 14.4). The operator $PFBP_{\alpha_1}^{\alpha_2}(x)\{P_\alpha\}$ is defined for partial filtered back projection at position $\boldsymbol{x}$ using the block of projections P_α with $\alpha\varepsilon[\alpha_1, \alpha_2]$. For the filtered back-projection, the short-scan FDK-type algorithm can be used. To handle the temporal variation, the reconstruction is done as follows:

$$\mu(x,t) = \sum_{j=0}^{M-1} \sum_{k=1}^{N_{rot}} PFBP_{j.\triangle\alpha}^{(j+1)\cdot\triangle\alpha}(x)\{P_\alpha(k)\}\varphi_{(j+0.5)\cdot\triangle\alpha,k}(t - t_{(j+0.5)\cdot\triangle\alpha}(k))$$

$$(14.3)$$

Here, $\triangle\alpha = \alpha_{max}/M$ where M is the number of angular intervals. In general, the angular intervals $\triangle\alpha$ of the PBB also can be nonuniform. The interpolation function can be implemented as a linear or cubic spline function.

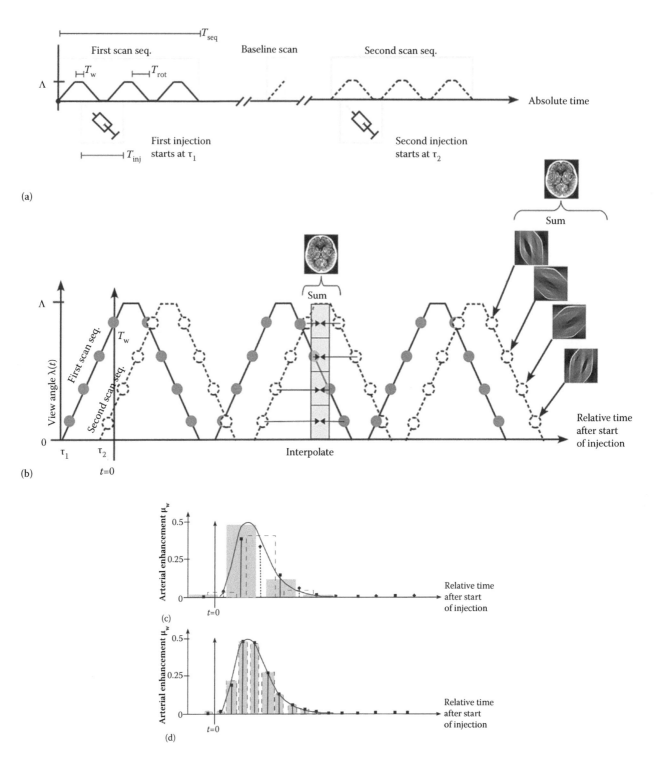

Figure 14.4 (a) Illustration of the data acquisition scheme using two interleaved scan sequences (six sweeps per sequence, first scan sequence solid, second scan sequence dashed) with different time intervals between start of injection and start of imaging. A single-sweep baseline scan is acquired between sequences to allow correction for residual contrast in the second scan resulting from the injection during the first scan sequence. (b) Distribution of acquired data relative to the time after the start of injection. Each circle represents a partial block reconstruction (shown here with $M = 4$). (c) Each reconstructed volume at time t can use all of the data from a single sweep (open circles, last sweep, temporal interval T_{rot}), resulting in overlapping sampling of the arterial enhancement curve. (d) Alternatively, partial block interpolation can be applied, resulting in a reduced temporal interval per reconstructed volume and providing the improved sampling.

To increase the temporal sampling, additional scans with different offsets to the start of the contrast injection are acquired and the data is interleaved (Figure 14.4). The temporal offset between the start of the sequences (normalized to the time of injection) can be given by

$$T_{offset} = T_p / N_{seq} \tag{14.4}$$

The data from the interleaved scans are then used in the dynamic reconstruction scheme.

Preliminary results comparing clinical perfusion CT with C-arm CT measurements of CBF in a healthy porcine model are promising.(Ganguly et al. 2011) Different injection rates and volumes (3 mL/s for 8 s at 67% and 100% I contrast concentration, and 6 mL/s for 8 s at 50% I contrast concentration) were tested for obtaining the CBF maps and compared with those acquired using a clinical CT scanner. For these preliminary studies, an intra-arterial approach was used for improved enhancement. Intra-arterial injections are conceivable in the interventional setting because the patients usually are catheterized for treatment.

Statistical analysis of the results showed that the correlation coefficient for the C-arm CT and clinical CT CBF values was best for the 6 mL/s 50% I concentration injection. The correlation was over 87% for scans with two, three, and six injections each followed by six bidirectional sweeps. Figure 14.5 shows the results for a two-injection scan sequence set.

Both of the techniques described earlier use the standard filtered back-projection algorithm to generate the volume images at sampled time points during the contrast injection. Neukirchen et al. (2010) have proposed an iterative method that uses a decomposition model-based approach, rather than interleaved scans, in an attempt to increase the temporal sampling of data acquired using a C-arm system. The goal of the algorithm is to estimate, on a voxel-by-voxel basis, the time–attenuation curves caused by the temporal variation of blood contrast concentration in vessels and tissue. The entire object attenuation is $\tilde{X} = \tilde{X}_d + \tilde{X}_s$, the sum of static anatomical contributions plus dynamic contrast. The first step of the approach is to subtract off the static line integrals in each projection using a baseline noncontrast acquisition to get $\tilde{p}_d = \tilde{p} + \tilde{p}_s$. Then, a parametric representation, $\tilde{X}_d(y)$, is used to describe the spatial and temporal characteristics of the time–attenuation curves in the object. The model parameters y should have a sufficient degree of freedom to reflect all possible variations in the spatial and temporal properties of the time-attenuation curves, while having limited complexity for computation reasons. This balance is achieved by using prior knowledge about expected features, defining a set of N basis functions $b_n(t)$ that cover the time interval of dynamic acquisitions. Temporal basis functions could be, for example, time-shifted Gaussian functions or gamma-variate functions. The time–attenuation model in the rth voxel

is then the superposition of basis functions $x_{dr}(t) = \sum_{n=1}^{N} y_{r,n} b_n(t)$,

where $y_{r,n}$ is the weighting factor for the nth basis function in the rth image voxel. Given a discrete time description of the imaging process for T acquired projection views, the contrast dynamics is represented by the $R \times T$ matrix X_d, where the tth column is the dynamic attenuation of all image voxels at time step t, and the rth row is the contrast time–attenuation curve for the rth image voxel. The decomposition model for the dynamic object can then be written as $X_d = Y \cdot B^{transpose}$, where Y is an $R \times N$ matrix of weight parameters $y_{r,n}$ and B is a $T \times N$ matrix of sampled function values created by adjoining all N vectors b_n.

The estimation procedure then proceeds as follows: The time–attenuation curve representation $X_d(y)$ is projected forward using nonsimultaneous projection F according to the temporal and geometric properties of the rotational movement of the imaging system to provide the parametric line integrals $\hat{p}_d(y) = FX_d(y)$. The optimal set of parameters $\hat{y}$ is obtained by minimizing the residual between the projected model and the acquired dynamic line integrals in a least-squares sense according to $\hat{y} = \text{argmin}_y \, p_d(y) - \tilde{p}_d^2$, where numerical minimization of the residual is achieved by an iterative, gradient-based optimization method.

The accuracy of the algorithm was investigated for a six-sweep, backward/forward C-arm acquisition using noisy data simulated from a CT perfusion data set. Normalized error in the arterial time–attenuation curve was between 12% and 16%, whereas error in normal and hypoperfused tissue was between 4% and 6%. Image maps of CBF, CBV, and mean transit time reproduced the main features of the clinical CT data well, although streak artifact due to the turning points in the trajectory led to a partial masking of the infarct area.

As described in Chapter 9, four-dimensional (4D) vascular blood flow information can be obtained from projection data acquired during a single C-arm rotation (Mistretta 2011). The technique uses a limited number of projections and constrained image reconstruction to obtain 3D volumes representing different time points throughout a contrast injection. The vascular flow information obtained could then be used to more accurately estimate the arterial input function in combination with other techniques (i.e., interleaved scanning or model-based reconstructions) that provide accurate estimates of tissue perfusion.

Constrained reconstruction using compressed sensing also has recently gained considerable attention in the imaging community (Fornasier and Rauhut 2011). For example, prior image constrained compressed sensing (PICCS) is a technique to reconstruct time-resolved images (Chen et al. 2008, 2009; Nett et al. 2010). In CT brain perfusion imaging, PICCS was investigated to reconstruct images using only a few projections (e.g., 4–20) distributed over the full view-angle range, thereby saving x-ray dose (Ramirez-Giraldo et al. 2011). Compared with standard filtered back-projection reconstruction, PICCS images have fewer streak artifacts; however, the accuracy of perfusion values using this technique has not yet been evaluated. A further application of PICCS is to improve temporal resolution by using limited sector reconstruction (TRI-PICCS) (Tang et al.

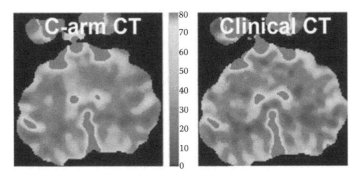

Figure 14.5 Comparison of brain perfusion maps in a healthy porcine model acquired using (left) two interleaved multisweep acquisitions on C-arm CT (right) standard acquisition parameters on clinical CT. Contrast was injected into the aortic arch for both cases.

2010, Mass and Kachelriess 2011) The projections are taken from a shorter view-angle range (e.g., 100°) than is necessary for traditional image reconstruction, thereby decreasing the temporal window associated with each reconstruction. This approach could be applied to an interleaved scanning protocol instead of the block interpolation technique described earlier, or it could potentially be used with a single injection protocol. The ability to accurately reconstruct low-contrast changes due to the contrast dynamics in tissue has not yet been verified, and computational times associated with this iterative method must be shortened to be practical during interventional stroke treatment.

14.4.2 DIRECTING MINIMALLY INVASIVE THERAPIES FOR HEPATIC CANCERS

A second application of growing importance for C-arm CT is guidance of abdominal interventions (Hirota et al. 2006, Sze et al. 2006, Kakeda et al. 2007, Meyer et al. 2007, Wallace et al. 2008). In particular, the ability to discern hypervascular lesions in multiple planes and the ability to reconstruct the 3D vascular geometry has proved to be of benefit during chemoembolization and other minimally invasive therapies to treat hepatic malignancies (Hirota et al. 2006, Virmani et al. 2007, Wallace 2007, Wallace et al. 2007). A typical acquisition acquires a total of 419 images during an 8-s rotational scan through 210° using a 50:50 mix of iodinated contrast and saline (rotation speed, 26°/s; image acquisition every 0.5°; 512 × 512 voxel matrix; 125 kVp; detector dose request of ~0.36 µGy per frame). Using this imaging protocol, C-arm CT was technically successful in 93 of 100 procedures (93%), provided information not available by DSA in 30 patients (35.7%) and resulted in a change in diagnosis, treatment planning or treatment delivery in 24 patients (28.6%) These included, amongst others, visualization of additional or angiographically occult tumors in 13 patients (15.5%) and identification of incomplete treatment in 6 patients (7.1%). An example of a tumor seen on neither pre-interventional 3D volume images (CT or MRI) nor on digital subtraction angiography (DSA) but that was identified in a C-arm CT volume is shown in Figure 14.6. Using the same interleaved multisweep approach described earlier for CBF, it also may be possible to get quantitative measures of blood perfusion in the liver during an intervention, providing additional information regarding distribution and impact of therapies.

14.4.3 NEW APPLICATIONS IN THE CARDIAC INTERVENTIONAL SUITE

The first cardiac application for C-arm-mounted digital FP detectors was rotational coronary angiography that was shown to use significantly less contrast (reported values between 19% and 40%) and less radiation (28%–59%) than standard coronary angiography using an image intensifier system (Kuon et al. 2003, Maddux et al. 2004, Raman et al. 2004, Akhtar et al. 2005). The single rotation provides for the visualization of the coronary arteries from multiple view angles, making the assessment more view-angle and operator-independent (Tommasini et al. 1998). Combining 3D reconstruction with cardiac gating (Rasche et al. 2006a,b), motion compensation (Schäfer et al. 2006, Hansis et al. 2008b, Bousse Ast et al. 2009), or vessel modeling

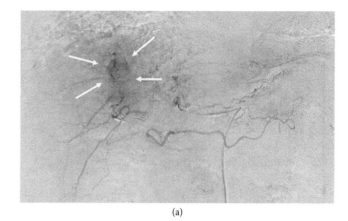

(a)

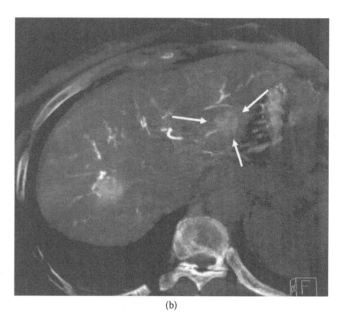

(b)

Figure 14.6 Comparison of 2D digital subtraction angiogram (a) and 3D C-arm CT image of the same patient (b) acquired during a chemoembolization procedure. A second lesion seen on C-arm CT was not visible in the angiogram, nor was it visible on *a prior* clinical CT. This case illustrates increased conspicuity due to local arterial injections of iodinated contrast during 3D imaging.

algorithms (Blondel et al. 2006, Jandt et al. 2009) further enhances visualization of the vascular tree (Garcia et al. 2007, Hansis et al. 2008a) and cardiac devices. Clinical applications include pacing lead placement (Mansour et al. 2005, Blendea et al. 2007, Tournoux et al. 2007, Knackstedt et al. 2008), improved quantification of artery and lesion dimensions, and cardiac device placement (Perrenot et al. 2007; Schoonenberg et al. 2008, 2009).

One important new application for single-sweep cardiac imaging is in the area of procedural planning and improved image guidance during radiofrequency ablation (RFA) procedures. RFA is becoming a first-line treatment for many cardiac arrhythmias, particularly for atrial fibrillation and ventricular tachycardia. The success of RFA depends on the extent of the lesion and the creation of lesions at specific anatomic locations, in continuous lines, or both. It is currently possible to register preprocedural CT or MR Is with electroanatomical mapping systems, but a more seamless integration and automatic registration of intraprocedural 3D C-arm CT cardiac images would be of benefit. Intraprocedural C-arm CT images more accurately represent the hemodynamic loading conditions and heart rhythm of the patient over preprocedural CT or MR images and also may help visualize surrounding structures, such as the esophagus (Orlov et al. 2007, Nölker et al. 2008), that are vulnerable to damage. Single-sweep C-arm CT images provide good image quality in 3D volumes of the left atrium and pulmonary vein anatomy, as well as the left ventricle (Thiagalingam et al. 2008). In addition, adenosine-induced asystole and rapid ventricular pacing have been investigated to improve the image quality of single-sweep acquisitions (Bartolac et al. 2009, Gerds-Li et al. 2009, Kriatselis et al. 2009, Tang et al. 2009). An example of three-way registration between an intraprocedural C-arm CT image of the left atrium, a segmented volume of the left atrium derived from a prior CT and the electroanatomical mapping system is shown in Figure 14.7.

ECG-gated C-arm CT, introduced by Lauritsch et al. (2006), provides more accurate images of the cardiac chamber, reconstruction at any time within the cardiac phase, and improved soft-tissue contrast due to decreased motion artifact (Al-Ahmad et al. 2008, Fahrig et al. 2008). A typical imaging protocol acquires images during four 5-s bidirectional sweeps and requires a single breath-hold of ~25 s. Iodinated contrast is injected into the inferior vena cava (6-f pigtail catheter; 75%–25% iodinated contrast-saline, at 5 cc/s; total volume injected, 185 mL) throughout the acquisition for cardiac chamber and pulmonary vein visualization. For myocardial visualization, contrast injection ends just before the start of imaging to reduce artifact from rapidly changing contrast concentration in the cardiac chambers.

The multisweep approach combines data from several cardiac cycles to reconstruct the desired cardiac phase. The challenge is to ensure that complementary coverage is obtained from each sweep so that a minimal pi+fan angle set of data is acquired. This approach is analogous to multisegment reconstruction in cardiac spiral CT. The angular position φ of the x-ray source for each projection can be plotted against the cardiac phase τ (ranging from 0% to 100%). Projection data are selected in a time window centered at the desired cardiac phase of the reconstruction τ_{recon}, and the resulting width $\Delta\tau$ required to obtain complete coverage (i.e., pi+fan angle) determines the temporal resolution of the resulting reconstruction, as illustrated in Figure 14.8.

The temporal width of a reconstruction can be minimized by ensuring that subsequent sweeps cover maximally disjoint intervals of the angular position φ of the x-ray source in the required time window. The time delay τ of the start of the jth sweep, *relative to the R-peak of the ECG*, depends on the direction of the sweep and the number of sweeps, N, as

$$\tau_j^f = \frac{j-1}{N}$$

with j = 1, 3, 5 … for forward sweeps and $\tau_j^b = 1 - \tau_{end} + 2 \cdot \tau_{recon} - (j-1)/N$ with j = 2, 4, 6 … for backward sweeps. Parameter τ_{end} is the heart phase at the end of the first forward run. Note that use of both forward and backward sweeps introduces a parameter

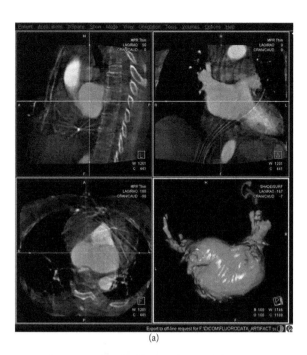

(a)

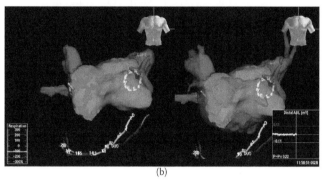

(b)

Figure 14.7 (See color insert.) (a) Segmentation of the volume of the left atrium from a single sweep C-arm CT acquisition during injection of iodinated contrast into the main pulmonary artery. (b) Overlay of C-arm CT-derived segmentation (red/gray), a segmentation from a prior clinical CT (blue), and points from the electromagnetic tracking system (red dots), indicating locations where radiofrequency ablation has been applied. Continuity of ablation points around the pulmonary veins can be monitored during the procedure.

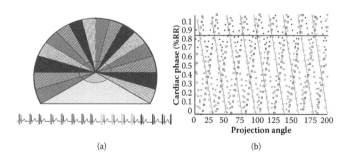

(a)   (b)

Figure 14.8 (a) Ideal angular sampling for a four-sweep ECG-gated acquisition; the timing of each sweep is adjusted to provide a narrow temporal window around 0.8 of the R-R interval. (b) Distribution of all acquired projections relative to cardiac phase; each symbol represents one projection; projections from a single backward sweep (open crosses) have been connected by a thin gray line for the purposes of illustration; the projections chosen from all sweeps for a reconstruction at 0.8 of the R-R interval are show as filled circles; the width of the window is approximately ±0.1 of the R-R interval.

that optimizes an acquisition to reconstruct a particular phase of the cardiac cycle. Because minimal heart movement occurs during end diastole, the acquisitions are typically optimized for 80% of the R-R interval. Other phases can be reconstructed using the same data, although the window widths associated with nonoptimal phases are larger, and the motion-related blurring in the reconstructed volumes is therefore more pronounced.

The most straightforward approach to select the set of projections used in a reconstruction is to use nearest neighbor interpolation to choose those projections closest to the chosen cardiac phase. The dynamic projection data for a particular cardiac phase can be described by $g(\varphi; x_1, x_2)$, where x_1 and x_2 are the detector coordinates. The index j of the acquisition run is included in the set of cardiac phases τ_j that have been acquired at a particular angular position φ. Let $\tilde{g}$ be the set of selected projection data that we choose for a given reconstruction. Nearest neighbor (subscript NN) interpolation, which provides the highest possible temporal resolution in the reconstructed volume, then requires

$$\tilde{g}_{NN}(\varphi; x_1, x_2) = g(\varphi; \tau_{NN}; x_1 x_2) \quad \text{with} \quad \tau_{NN} = \arg\min_{\tau_j} |\tau_j - \tau_{recon}|.$$

In some cases, additional smoothing may be advantageous to increase signal-to-noise ratio (SNR), in which case an interpolation which is normalized for nonuniform sampling should be used. An example proposed by Lauritsch et al. uses normalized convolution, which adapts automatically to the varying sample intervals. The interpolated image (subscript int) at the desired cardiac phase τ_{recon} is given by

$$\tilde{g}_{int}(\varphi; x_1, x_2) = \frac{\sum_{j=1}^{N} c(\tau_{recon} - \tau_j) g(\varphi, \tau_j; x_1 x_2)}{\sum_{j=1}^{N} c(\tau_{recon} - \tau_j)}.$$

The full width at half maximum characterizes the Gaussian interpolation kernel $c(\tau)$ according to

$$c(\tau) = e^{-\left[\frac{2\sqrt{\ln 2}}{FWHM}\tau\right]^2}$$

and thus for any FWHM, it is guaranteed that at least one sample will be available for every projection angle. Standard filtered back-projection as described earlier can then be used to reconstruct the data.

Note that the previous discussion regarding optimized time delays between sweeps to reconstruct a particular cardiac phase assumes that the heart rate is constant during data acquisition. It is known that heart rate varies during breathhold, and this trend is magnified by the length of the scan. The change in heart rate will produce some desynchronization between the acquisition and the heart phase, leading to a change in both the optimized phase of the reconstruction and the temporal, or window width. This effect could be reduced by optimizing the turnaround time based on real-time knowledge of the true sampling patterns of previous runs.

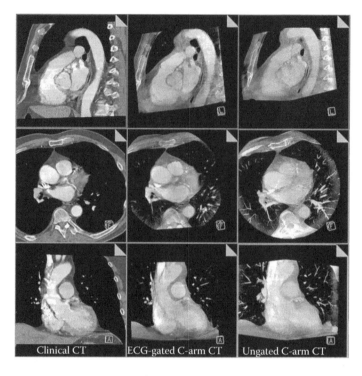

Figure 14.9 Comparison of clinical CT, ECG-gated C-arm CT, and ungated C-arm CT images obtained in the same patient. Image quality of ECG-gated C-arm CT permits delineation of the walls of the cardiac chamber that cannot be distinguished in the single-sweep reconstruction. Efforts to improve image quality include cardiac motion correction algorithms and reduction of total imaging time to reduce breathing-related motion.

An example of the image quality that can be achieved using the multi sweep protocol is shown in Figure 14.9. Our study comparing gated and nongated C-arm CT against clinical CT in seven patients showed that measurements of pulmonary vein dimensions more closely agreed with those of CT when gating was applied (correlation between 0.329 and 0.899 for ungated and between 0.732 and 0.962 for gated, with correlation coefficients that are highly dependent on which vessel was measured). Applications in low-contrast imaging beyond visualization of the cardiac chambers, such as intraprocedural visualization of RFA lesions (Girard-Hughes et al. 2009, Girard et al. 2011) and regions of myocardial infarction are now under investigation. Although ECG gating requires multiple sweeps of the C-arm and therefore a higher radiation dose, it is possible to further improve the SNR and low-contrast imaging within the myocardium using motion-corrected reconstructions (Prümmer et al. 2007). Cardiac C-arm CT is an imaging modality that brings together the current 3D anatomy of the patient with real-time fluoroscopic guidance in a system that permits open access to the patient (Ector et al. 2008, Kriatselis et al. 2009, Li et al. 2009). Future integration with a catheter guidance system will permit more minimally invasive procedures in the cardiac catheterization lab that would otherwise prove too complicated (Wallace et al. 2008).

In addition to the challenge of reconstructing the cardiac chambers and moving myocardium, a second imaging challenge of significance in the cardiac interventional suite is the visualization of coronary arteries in 3D. It is assumed that a 3D volume is easier to interpret that a series of 2D projection images and that the 3D volume could be used to select the best angle for 2D views to minimize vessel overlap and foreshortening. A 3D volume could be used to provide a roadmap during device placement and could therefore reduce the burden of iodinated contrast during a procedure. The ability to accurately quantify vessel properties such as patent diameter of the vessel, plaque burden, and geometry of the plaque (e.g., symmetric vs. asymmetric, composition, and heterogeneity) could affect choice of treatment, including stent location, length, and type. In combination with 2D flow evaluation, information regarding fractional flow reserve could be provided that could help to determine whether stenting is appropriate. Finally, after stent implantation, and 3D reconstruction of the stent and vessel could be used to verify stent position and to ensure good contact between vessel wall and stent.

Several approaches have been developed to permit 3D reconstruction of coronary vessels from a single-sweep C-arm CT acquisition. Because iodinated contrast is injected into the coronary arteries during imaging, the total time of the acquisition is limited to less than 15 s, and typical coronary injections are less than 5 s (Dodge et al. 1998). This limit is imposed by the fact that total replacement of coronary blood by contrast can produce oxygen starvation of cardiac tissues. Thus, C-arm CT reconstructions of the coronary arteries must be achieved using data from a single sweep of the C-arm, and advanced reconstruction techniques have been developed to improve the quality of the resulting reconstructions. Two such techniques highlighted here. One technique is based on iterative reconstruction and the second technique uses standard filtered back-projection; both illustrate the challenges and potential image quality that can be achieved.

A brief summary of the iterative reconstruction approach, based on the work by Hansis et al. (2010), is as follows:

Step 1: Acquire a rotational coronary angiography sequence, during selective injection of iodinated contrast into the ostium of the right or left coronary artery while the patient holds his or her breath. A typical acquisition takes 7.2 s, during a rotation through 180°, acquiring images 30 frames per s. An ECG is recorded simultaneously.

Step 2: Apply retrospective ECG gating to select angiograms corresponding to the desired heart phase, using the R peaks of the ECG as reference time points. Projection weights ω_ϕ for processing a cardiac phase at percentage Θ in the cardiac cycle are given by

$$\omega_\phi = \begin{cases} \cos^2\left(\dfrac{\tau_\phi - \Theta}{W}\pi\right), & \Theta - \dfrac{W}{2} \leq \tau_\phi \leq \Theta + \dfrac{W}{2}, \\ 0, & \text{else} \end{cases}$$

where W is the gating window width, given as percentage of the cardiac cycle, and τ_ϕ is the position of each acquired projection in the cardiac cycle in percent of the cardiac RR cycle. The simplest gating, with a gating window width of 0%, is NN for which the selected projections are given a weighting of 1, whereas all others receive weighting of 0.

Step 3: Apply a morphological top-hat filter (Bovic 2000) to flatten the background of the projection data.

Step 4: Perform an iterative reconstruction [simultaneous thresholding algebraic reconstruction technique (START)] (Hansis et al. 2008a) at a chosen phase, W_1, in the cardiac cycle with a narrow window width and few views (e.g., select the NN projections for a 0% width, therefore using between 5 and 10 projections in total, depending on the heart rate, that are closest to the selected cardiac phase) using L_1-norm minimization (Li et al. 2002, 2004) to regularize the reconstruction. Note that this initial reference reconstruction may suffer from blurring because coronary motion is not entirely reproducible, heart motion may be slightly aperiodic, and the breathhold may not be perfect.

Step 5: Compensate for residual coronary motion and improve the quality of the reconstruction by applying projection-based motion compensation (Hansis et al. 2008b). The motion compensation is carried out for each cardiac phase, using the reference, uncompensated reconstruction of step 4 at the phase of interest. Using a wider gating window, W_2, and applying the weighting of ω_ϕ, the projections for the motion-compensated reconstruction are chosen. Maximum-intensity projections are then generated from the reference volume of step 4 at angles corresponding to the projections in W_2. Centerlines are calculated in both forward projections (set M of N_m centerline points $\vec{d}'$ in the "model") and the projections in W_2 (set D of N_d centerline points $\vec{p}$ in the "data") by applying a multiscale curvature–based vesselness filter (Lorenz et al. 1997), directional maxima detection, and hysteresis thresholding (Canny 1986). Pairs of model-data points are determined by minimizing the Euclidean distance $R_M(\vec{p}) = \min_{\vec{d}' \in M} \|\vec{p} - \vec{d}'\|$ between each point $\vec{d}'$ and a corresponding closest point $\vec{p}$. Finally, an elastic deformation is applied to each projection pair such that the centerlines of the reference and the W_2 projections match as closely as possible. The elastic deformation parameters $\vec{q}$ are defined by dividing a projection into a grid of N_c square cells, with node centers u and four edges v, deforming the grid by translating its nodes, and interpolating new coordinates inside each grid cell with bilinear interpolation. The transformation parameters $\vec{q}$ are determined by minimizing the error function

$$E(\vec{q}) = \sum_{i=1}^{N_d} R_M^2 + \frac{\beta}{2}\frac{N_d}{4N_c}\sum_{u=1}^{N_c}\sum_{v=1}^{4}(l_{u,v} - l_0)^2,$$

where l_0 and $l_{u,v}$ are the lengths of cell edges before and after transformation. The second term regularizes the transformation by avoiding unrealistically large deformations, with the strength of the regularization controlled by parameter β, although rigid translational motion is not penalized. A nonlinear Levenberg–Marquardt optimization (Fitzgibbon 2001) is used to perform the minimization. After motion-compensating each projection pair, a new motion-compensated reconstruction is calculated from the motion-corrected projections again using the START algorithm. A typical value for W_2 is 20%. Note that multiple cycles of the projection-based motion compensation and START reconstruction can be performed to further improve image quality.

Representative image quality that can be achieved using this approach is shown in Figure 14.10.

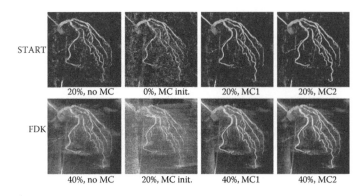

START

| 20%, no MC | 0%, MC init. | 20%, MC1 | 20%, MC2 |

FDK

| 40%, no MC | 20%, MC init. | 40%, MC1 | 40%, MC2 |

Figure 14.10 Different stages of the reconstruction process for a coronary artery using START or FDK. The value (in percentage) indicates the gating window width for each reconstruction. MC, motion compensation. MC1 and MC2 indicate the first and second motion compensation cycle, respectively. Images are maximum intensity projections. (Images courtesy of Dr. Eberhard Hansis, Nuclear Medicine, Philips Healthcare, San Jose, CA.)

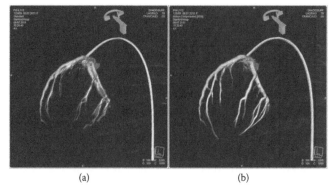

(a)   (b)

Figure 14.11 (See color insert.) Reconstructions using (a) standard ECG-gating and (b) FDK-based reconstruction coupling 3D motion estimation and reconstruction (5-s single sweep acquisition, 4 × 4 binning, 60 fps, coronary injection with a 3-s delay before start of acquisition). (Images courtesy of Dr. Christopher Rohkohl, Medical Image Reconstruction group, Pattern Recognition Lab, Friedrich-Alexander University Erlangen-Nuremberg, Germany.)

A brief summary of the filtered back-projection approach, based on the work by Rohkohl et al. (2010b), is as follows.

Step 1: See the preceding for typical acquisition parameters. Note that the approach described here is more robust to nonperiodic cardiac motion as well as residual breathing motion because the time-continuous motion field is parameterized by the acquisition time and not by the heart phase.

Step 2: See the preceding and generate an initial reference reconstruction using an ECG-gated reconstruction with a window width of 0%, $f_{\mathrm{REF}}(\boldsymbol{x})$.

Step 3: See the preceding.

Step 4: Apply an FDK-based reconstruction that couples motion estimation and reconstruction. The algorithm assumes a time-continuous motion model that maps a 3D voxel $\boldsymbol{x} = (x_0, x_1, x_2)^{\mathrm{T}}$ to a new voxel location x' for each time when a projection image is acquired. The motion model is defined via a set of control points placed uniformly in 3D space and time, and each control point is then assigned a displacement vector s. Cubic b-spline basis functions B_{j-t} are used to describe the motion, as given by $M(i,x,s) = x + \sum_{j,k,l,t} B_j(x_0)B_k(x_1)B_l(x_2)B_t(i)s_{jklt}$ at the time of the ith projection image.

The dynamic reconstruction algorithm is then given by

$$f(x,s) = \sum_i w(i, M(i,x,s)) \cdot p(i, A(i, M(i,x,s)))$$

where w is the distance weighting of the FDK reconstruction, and $p(i,u)$ returns the value of the ith projection image at the pixel u according to the 3D→2D perspective projection $A(i,\boldsymbol{x}) = \boldsymbol{u}$. Motion estimation is a multidimensional optimization problem where the $\hat{\boldsymbol{s}}$ are determined that minimize the joint intensity $L(\boldsymbol{s})$ of the sparse reference image $f_{\mathrm{REF}}(\boldsymbol{x})$ and the dense dynamic reconstruction $f(\boldsymbol{x},\boldsymbol{s})$, given by $L(\boldsymbol{s}) = -\sum_x f_{\mathrm{REF}}(\boldsymbol{x}) \cdot f(\boldsymbol{x},\boldsymbol{s})$. An iterative quasi-Newton algorithm designed for large-scale nonlinear optimization problems (L-BFGS-B; Byrd et al. 1995) is then applied to solve for the motion model parameters $\hat{\boldsymbol{s}}$.

Representative image quality that can be achieved using this approach is shown in Figure 14.11. Further work by the same group proposes addition of an affine motion model to the optimization so as to render the technique more robust to motion in the original reference reconstruction $f_{\mathrm{REF}}(\boldsymbol{x})$ (Rohkohl et al. 2010a)

14.5 IMPROVING C-ARM CT IMAGE QUALITY

Image quality in C-arm CT is, by most metrics of measurement, worse than that of clinical CT. The imaging geometry itself produces artifact due to axial image truncation, and the cone beam problem. Physical phenomena such as beam hardening and scatter produce artifacts in clinical CT imaging are worse in C-arm CT because compromises in system design are made to accommodate the real-time high-resolution large-area projection imaging requirements in the interventional suite. Some designs of FPs exhibit lag that produces streak artifact and other inaccuracies in the HU values. In addition, the limited dynamic range and frame rate of the FPs lead to saturation and streak artifact, respectively. Corrections for all of these nonidealities continue to be active areas of investigation. In addition, improvements to the FDK algorithm and alternative reconstruction strategies are under investigation as a means to improve image quality and lower radiation dose.

14.5.1 DETECTOR LAG

Lag is signal present in detector frames after the frame in which it was actually generated. The primary cause of detector lag is charge trapping in the a-Si layer of the detector (Street 1991, Siewerdsen and Jaffray 1999a, Overdick et al. 2001), and traps are inherent in a-Si because of the lack of long-range crystalline order and dangling bonds within the material. Scintillator afterglow and incomplete charge readout of the photodiode also contribute to lag. Trapping in the a-Si also leads to a gain change of the photodiode because when traps are filled, subsequently generated charge is collected by the readout electronics instead of falling into the trap states (Roberts et al. 2004). In fact, no well-defined edge to the band-gap exists; instead, the density of states function has exponential tails from the conduction and valence bands that extend far into the band-gap energies (Street 1991). Because of the broad spectrum of trap states that exist in the a-Si, lag effects can be seen at a wide range of panel frame rates (Siewerdsen and Jaffray 1999b, Edward et al. 2007). At higher frame rates, measured lag can appear to decrease because of shorter integration times, but the lag then persists for a longer number of frames.

Typically, the first frame lag after a significant length of irradiation is ~2%–10% of the irradiating signal (Siewerdsen and Jaffray 1999a), as shown in Figure 14.12. Although the lag signal decays exponentially, a significant amount of signal remains for many seconds and could easily equal the signal behind a large object. In CT reconstructions lag can lead to a range of image artifacts, such as streaks, comet-tails off of high-contrast objects, or blurring and shading artifacts for noncircular or off-center objects (Siewerdsen and Jaffray 1999a, Starman et al. 2006, Mail et al. 2008). The shading artifacts can be quite significant depending on the geometry and detector state, especially near

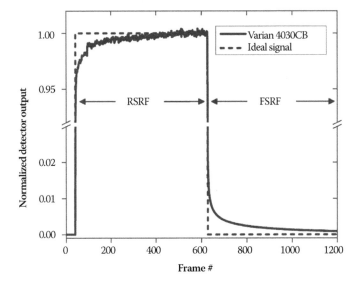

Figure 14.12 The measured and ideal responses of an a-Si FP to 600 frames of x-ray exposure. An increase of 4% over the course of irradiation is seen in the RSRF. First frame lag of the FSRF is 2.5% of the previous frame signal and slowly decays away with a multiexponential tail after the x-rays are turned off.

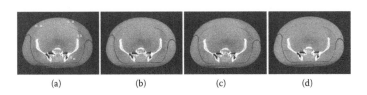

Figure 14.13 Comparison between reconstructions from a Varian 4030CB detector imaging a large pelvic phantom; errors are quoted for the ROI pairs across the artifact as (max error in HU | mean error in HU) (a) no lag correction, clearly showing a "radar" artifact (49 | 36); (b) the best possible correction using an linear time invariant (LTI) model (16 | 9); (c) the worst correction using an LTI model (26 | 21); and (d) correction using a non-LTI model (15 | 8). Window and level are 250 and 0 HU, respectively.

the object edge. Figure 14.13a shows a lag artifact of 49 HU for a Varian 4030CB panel used to image a large pelvic phantom.

Several software (Hsieh et al. 2000, Mail et al. 2008) and hardware methods to deal with lag in FP imagers exist. One effective hardware method uses light-emitting diodes (LEDs) built into the panels to backlight the photodiodes, which saturates the traps (Overdick et al. 2001). Forward biasing to push current through the diode can be used to achieve the same goal (Overdick et al. 2001, Mollov et al. 2008, Starman et al. 2008). Note that neither of these approaches correct for scintillator afterglow.

Current software methods typically model an impulse response function (IRF) for the panel by fitting a suitable model, such as a multiexponential (Yang et al. 2007) or power function (Weisfield et al. 1999), to the lag decay. A single IRF can be used to describe the entire panel, or an independent function can be fit for each individual pixel, allowing for variation in lag across the detector. A basic assumption of the IRF model is that the panel acts in a linear and time-invariant way. The correction is then provided by a temporal deconvolution of the detector output with the modeled IRF. Hsieh et al. (2000) model the IRF as a multiexponential signal for which the time constants and coefficients are known *a priori*. Expressing his approach in a discrete time version, we have

$$h(k) = b_0 \delta(k) + \sum_{n=1}^{N} b_n e^{-a_n k},$$

where k is the discrete time variable for x-ray frames, and N is the number of exponential terms in the IRF. The variable $\delta(k)$ is the impulse function of magnitude 1.0 and represents the portion of the input signal that is unaffected by lag. The coefficients b_n are the lag coefficients and the a_n are the exponential lag rates. Given the IRF h_k, the output of the detector is then

$$y(k) = x(k)^* h(k),$$

where $x(k)$ is the ideal signal in the absence of lag, and $y(k)$ is the actual measured signal, both in detector counts. Deconvolution is then applied recursively, such that only the current frame and a single state variable per pixel that contains the contribution from previous exposure must be stored. The recursive algorithm is as follows:

$$x(k) = x_k = \frac{y(k) - \sum_{n=1}^{N} b_n S_{n,k} e^{-a_n}}{\sum_{n=0}^{N} b_n}.$$

$$S_{n,k+1} = x_k + S_{n,k} e^{-a_n}$$

The state variable $S_{n,k}$ holds the contribution from previous image inputs, and is used to calculate the true signal x_k from the currently measured signal $S_{n,k+1}$. Note that b_n is normalized such that $h(0) = 1$. The IRF is determined by fitting either the falling (FSRF) or rising step response functions (RSFR) to N exponentials, where $u(k)$ is the discrete unit step function, according to $\mathrm{FSRF}(k) = \sum_{n=1}^{N} \frac{b_n}{1 - e^{-a_n}} e^{-a_n k} u(k)$

which, with $\tilde{b}_n = b_n/(1 - e^{-ti\theta})$, provides the simple exponential

$\mathrm{FSRF}(k) = \sum_{n=1}^{N} \tilde{b}_n e^{-a_n k} u(k)$ and similarly for the rising step response function

$$\mathrm{RSRF}(k) = \left(1 - \sum_{n=1}^{N} \tilde{b}_n e^{-a_n k}\right) u(k).$$

Fitting the N exponentials is a nonlinear problem, and a Gauss–Newton nonlinear least squares method can be used (Kelley 1999).

It has been shown that the accuracy of the linear time invariant model depends on the size of the object (i.e., on the dynamic range seen by individual pixels) and on the exposure at which the calibration of lag coefficients and lag rates is carried out. This implies that large signal variations across the panel, such as those near the boundary of an object, show nonlinear behavior (Starman et al. 2011).

14.5.2 BEAM HARDENING

Beam hardening is a well-known physical phenomenon that arises when a spectrum of x-rays passes through an object with variable attenuation. As the x-ray beam passes through the object, the average energy of the spectrum increases because low-energy photons are more likely to be absorbed than the high-energy photons. Even for a homogeneous object, the relationship between the measured value in the projection and the object thickness is nonlinear, violating one of the basic assumptions in CT reconstruction. This phenomenon causes cupping across the reconstructed image, and streaking between high-density objects. Current C-arm CT images are more likely to exhibit such artifacts than clinical CT systems because the C-arm system is optimized to operate at lower kVp and has a softer spectrum, causing the change in mean energy for a given change in object attenuation to be more significant. The magnitude of the problem is shown in Figure 14.14, where we have simulated beam hardening artifact caused by two inserts of cortical bone within a water cylinder, for a C-arm CT spectrum at 125 kVp (half-value layer = 4.4 mm Al; Boone et al. 2000) and for a clinical CT spectrum at 120 kVp (half-value layer = 8.2 mm Al; Fahrig et al. 2006). One easy solution to this problem is to add filtration to the x-ray beam, that is, to increase the mean energy of the beam before it hits the patient. Such an approach is limited, however, by the lower power ratings of C-arm x-ray tubes and generators, because filtration also reduces the overall number of quanta that hit the patient for a given mAs. Numerical approaches for correction use lookup tables to correct each pixel value based on an estimate of water equivalent thickness and work well for correction of cupping artifact although streak artifacts are still present (Zellerhoff et al. 2005, Star-Lack et al. 2009). This topic remains an active area of research, with particular emphasis placed on algorithms that do not require an initial reconstructed volume.

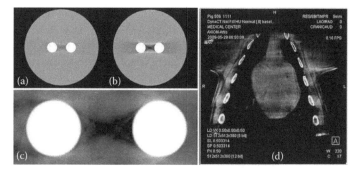

Figure 14.14 Reconstructions of a 25-cm circular water phantom with cortical bone inserts. A C-arm geometry was assumed for the simulations, except that 1000 projections covering a full 360° were used. The x-ray spectrum of a GE lightspeed CT scanner (a) and a Siemens C-arm system (b) were used to show the effects of beam hardening. Reconstructions are [windowed | leveled] at [500 HU | 0 HU]. (c) Reconstruction of two bone inserts in a soft-tissue equivalent phantom acquired at 90 kVp on a C-arm CT system. (d) The same artifact is seen as bright and dark bands in heart and lung regions corresponding to the locations of the ribs.

14.5.3 SCATTER

The importance of efficient and accurate correction for x-ray scatter is described in publications from the early days of CT (Glover 1982, Spital 1982) and has resurged with the development of volume scanning techniques such as C-arm CT that exhibit high scatter-to-primary ratios (SPRs). Several comprehensive investigations of SPR as a function of irradiated volume have recently been completed and are in good agreement given differences of acquisition geometry and measurement technique (Fox et al. 2001, Siewerdsen and Jaffray 2001, Endo et al. 2006). The SPR depends on such geometric factors as air gap between object and detector, focal-spot-to-object distance, and focal-spot-to-detector distance and is only weakly dependent on spectrum energy (kVp) over the range of diagnostic energies of interest. The magnitude of the SPR can be represented as a function of the irradiated volume, as was pointed out by Fox et al. (2001). Although this volume dependence is not obvious from physical principles, it is useful as an engineering approximation to predict scatter in large-field scanners and to allow comparison between measurements. Figure 14.11 illustrates this relationship, showing, for an uncollimated beam with no grid, approximately linear dependence on volume for low SPR. Estimated SPRs for geometries of interest to this discussion are also indicated, although differences in geometry (i.e., larger air gap due to long source-to-detector distance) have not been taken into account.

An intuitive explanation for the artifact caused by scatter is provided by referring to Figure 14.15a, where the scatter field is shown as a low-frequency background signal that does not vary from view to view, whereas the primary varies significantly depending on the angle. The signal in view 1 behind the two rods is significantly overestimated compared with the signal behind a single rod in view 2. The reconstructed image shows a dark streak connecting the rods as shown in the simulated images. Reconstructions as a function of SPR illustrate the magnitude of the streak and accuracy problem for even small SPRs (Figure 14.15b). Correction of the streak artifact requires

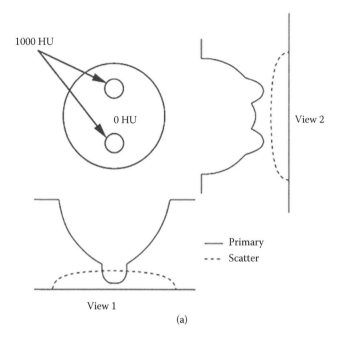

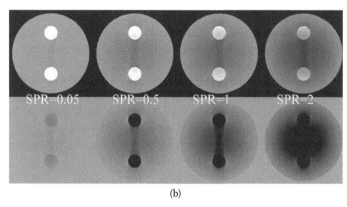

Figure 14.15 (a) Schematic illustrating the source of artifact due to scatter in CT reconstruction. (b) Magnitude of artifact in CT reconstructions of the object in panel a when scatter is not removed from the measured line integrals. The SPR is indicated for each column. (Top) CT images with display window [–500, 500] HU. (Bottom) Difference between ground truth and reconstructed images with display window [–300,300] HU. Even for small SPR, streak artifact is significant.

knowledge of the SPR for each primary ray position to achieve accurate scatter correction across the full FOV. An excellent summary of causes for scatter and strategies for scatter correction was recently presented by Ruhrnschopf and Klingenbeck (2011a,b) and Ruhrnschopf et al. (2011). These strategies include hardware-based approaches (Siewerdsen and Jaffray 2000, Endo et al. 2001a, Siewerdsen 2004, Wiegert et al. 2004, Siewerdsen et al. 2006) and algorithmic correction of x-ray scatter (Boone and Seibert 1988a,b; Spies et al. 2001; Ning et al. 2004; Bani-Hashemi et al. 2005; Bertram et al. 2005; Zhu et al. 2005; Jarry et al. 2006).

Two basic hardware approaches exist to reduce scatter in the recorded image: increase the air gap between the object and the detector (Siewerdsen and Jaffray, 2000) and use a grid or focused collimator (also between the object and detector) to preferentially remove higher angle photons while allowing most of the primary signal-carrying photons to be detected (Endo et al. 2001b, Siewerdsen 2004, Wiegert et al. 2004). Use of an air gap, although

very effective, increases system magnification, and reduces the solid angle subtended by the detector; thus, it is often limited by the focal spot and x-ray tube output of the imaging system. Conventional grids and focused collimators (Endo et al. 2001a, 2006) reject scatter; however, they may not provide an advantage compared with an air gap (Neitzel 1992, Siewerdsen and Jaffray 2000) because an increase in dose may be required to offset the loss of primary. For digital imagers, Neitzel (1992) concluded that air gaps and grids perform comparably under high-scatter conditions, but for low-scatter conditions an air gap is superior. This result is consistent with the work of Siewerdsen (2004).

An algorithmic approach requires an estimate of the scatter fluence in a projection image, and a variety of techniques have been developed, including analytical calculation (Boone 1986; Love and Kruger 1987; Boone and Seibert 1988a,b; Seibert and Boone 1988a,b; Honda et al. 1991; Kruger et al. 1994; Molloi et al. 1998; Bertram et al. 2005; Wiegert et al. 2005) Monte Carlo modeling (Boone and Seibert 1988b, Spies et al. 2001, Jarry et al. 2006) and "blocker-based" techniques (Ning et al. 2002, 2004; Zhu et al. 2005; Liu et al. 2006). Each of these techniques has merits and disadvantages in implementation. Analytical calculation of the scatter fluence using estimates based on the point spread function of single-scatter events is fast but is limited to simple models of patient shape and heterogeneity and was recently shown to be inadequate in the presence of variable anatomy (Tkaczyk et al. 2006). Monte Carlo modeling can be accurate but is computationally intense, requiring an initial 3D volume reconstruction and then several seconds per projection if the implementation is efficient. Blocker-based techniques (an array of Pb blockers placed between the x-ray source and the object) derive or measure the scatter fluence directly from the projection data. The Pb blockers produce holes of zero signal in the image; thus, either two projections per view, one view with the Pb blockers in place and one view without, are required, or artifacts are introduced because the missing data cannot be accurately recovered using interpolation. Our measurement-based scatter correction method uses a checkerboard pattern (primary modulator) of partially attenuating material placed between the x-ray source and the object (Zhu et al. 2006, 2007). This method provides an effective scatter correction using a single scan without loss of real-time imaging capabilities or increase in patient dose (Figure 14.16).

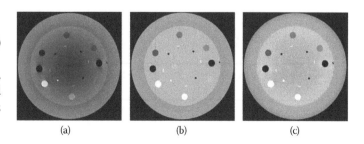

Figure 14.16 Image reconstructions of a Catphan@600 phantom from cone beam scans with and without scatter corrections. (a) Cone beam scan without scatter correction. (b) Fan-beam scan approximately scatter free as a reference. (c) Cone beam scan with scatter correction using the primary modulation method. Reconstruction size, 512 × 512 pixels. Display window, [–300, 300] HU. The maximum SPR in the cone beam scan is around 2. The contrast to noise ratio for images a, b, and c are 10.9, 23.4, and 19.2, respectively.

A fundamental difference between hardware and algorithmic techniques should be emphasized. Air gap and grids remove both scatter signal and the noise associated with the scattered photons from the recorded signal. Algorithmic techniques can be used to estimate and remove the mean scatter signal (on a per-pixel basis), thus eliminating the streak and cupping artifact, but they leave the noise associated with the scatter in the signal. Noise suppression schemes to deal with this challenge have been suggested previously (Zhu et al. 2009). It is likely that a final scatter-correction approach will include both hardware-based scatter removal, and an algorithm to remove the residual scatter signal from the recorded data to eliminate the remaining artifact.

14.5.4 LIMITING FRAME RATE

It is well known that streak artifacts from high-contrast edges, increasing with distance from their source, arise if the number of projections used in a CT reconstruction is not sufficient. The minimum number of views required for a given size of reconstruction is

$$N_{min} = (V_M \cdot 2 \cdot (\pi + 2\delta) \cdot R_s \cdot R)/(R_s - R),$$

where R_s is the focal spot to center-of-rotation distance, R is the maximum radius of the reconstructed volume, V_M is the maximum spatial frequency in the image, and δ is the half-fan angle (radians) of the system (Joseph and Schulz 1980). For example, if V_M is limited by the sampling interval of the detector to 2.0 mm^{-1}, the minimum number of views for a reconstructed FOV of 30 cm is ~700. For abdominal low-contrast imaging in the presence of peristaltic and cardiac motion, improved image quality is achieved for the shortest possible scan time, that is, for rotation rates of 60°/s in ~3.5 s. Acquiring 1000 frames (equivalent to a CT scanner) within only 4 s cannot be achieved with current FP technology.

Table 14.1 shows sample frame rates, binning and resulting pixel size for the Trixell Pixium 4700 detector and illustrates

the trade-offs between resolution, scan time, and frame rate that must be made. Correction approaches to alleviate the associated streak artifact (Galigekere et al. 1999) have been proposed, but resolution in the periphery of reconstructed images remains limited by the number of acquired projections.

14.5.5 TRUNCATION

The FOV of current C-arm CT systems is limited to 40 cm at the detector and is not likely to exceed this size in the near future. For abdominal and other large FOV imaging, truncation of projection images leads to severe artifacts around the edges of the reconstructed FOV and to inaccuracies in the reconstructed HU values due to the associated cupping effect in the image. Correction for truncation requires estimating data in such a way as to preserve the highest possible fidelity during the filtering step of the cone beam back-projection algorithm. In general, the different techniques attempt to estimate the missing data based on either *a priori* knowledge about the data, or by using redundancies in the data. These approaches include symmetric mirroring (Ohnesorge et al. 2000), use of redundant rays (Pan et al. 2005), polynomial extrapolation (Glover and Pelc 1985, Hsieh et al. 2004, Starman et al. 2005), use of *a priori* information (Ruchala et al. 2002), and iterative techniques as well as non-FDK reconstruction approaches such as projection onto PI-lines (Pan et al. 2005).

A comparison of an *in vivo* clinical CT image of the liver compared with that from an FD detector system with and without simple correction for truncation artifact is shown in Figure 14.17. Most of the artifact is confined to the periphery of the reconstructed slice, but HUs in the center of the reconstructed FOV are also inaccurate. These HU value inaccuracies for liver imaging with a C-arm CT system have been studied by Taguchi et al. (2009). In their study, an Allura Xper FD20 system (Philips Medical Systems, Best, The Netherlands) with the XperCT option (Release 1) was used for C-arm CT, with truncation between 3 and 10 cm depending on the view in clinical patients undergoing transcatheter

Table 14.1 Sample acquisition modes for different clinical applications

APPLICATION	SCAN TIME (S)	BINNING	FRAME RATE (FPS)	TOTAL PROJECTIONS	DETECTOR RESOLUTION (MM)	LIMITING RESOLUTION OF THE RECONSTRUCTION[a]
High-resolution, low inherent contrast, static object	20	1 × 1	7.5	496	148	4.0
High-contrast vascular applications	5	2 × 2	30	133	308	2.0
Single sweep of multisweep ECG-gated cardiac acquisition	5	4 × 4	60	249	616	1.0
Low-contrast body applications	8	4 × 4	60	397	616	1.0

Note that the total number of projections per reconstructed volume is less than half the number typically used for clinical CT, which is ~1000 projections.

[a] Limiting resolution of the reconstruction as measured using a line-pair spatial resolution phantom, Artis zee C-arm system (Siemens AG, Healthcare Sector, Forchheim, Germany) (Strobel et al. 2009). A summary of dose considerations for a "standard" Alderson phantom is provided in the same reference.

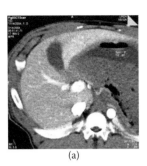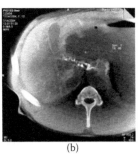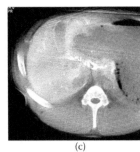

(a) (b) (c)

Figure 14.17 (a) Liver CT with venous bolus injection. (b) C-arm CT cone beam reconstruction of liver after subselective venous injection. (c) Same as in panel b with image truncation correction.

chemoembolization. Truncation correction was applied to all images in addition to other standard corrections such as beam hardening and scatter corrections. When I-contrast concentration was measured relative to local background, error in estimated contrast concentration was ~23%, and was independent of the level of truncation. Using a global estimate of background from a central ROI, error was ~40%. This illustrates the impact that truncation has on the ability to extract quantitative results from C-arm CT images.

14.6 HARDWARE IMPROVEMENTS FOR C-ARM CT

In the short term, two new mechanical implementations should provide shorter imaging times that will reduce patient-related motion in both standard and ECG-gated imaging as well as improve the temporal sampling for quantitative perfusion imaging. A brief discussion of each is provided in the following. To take full advantage of reduced scan times, improvements in detector performance also is required. The detector technology should have higher frame rates, lower noise floor, higher resolution, and higher detective quantum efficiency than current a-Si panels, while still providing a full area 30×40 cm detector. This significant design challenge will be met by alternative flat panel technologies, for example, complementary metal oxide semiconductor (CMOS), but a new detector platform is still several years from full integration into a clinical C-arm system.

Implementing C-arm CT on a biplane system for simultaneous data acquisition using both planes could reduce the time required to acquire a full pi+fan angle set of projections; each plane would rotate through a maximum of only ~110°, and even with a turnaround time of 1 second, volumes could be acquired every 3.5–4 s. This would lead to improved temporal sampling for applications such as perfusion imaging and also could significantly benefit ECG-gated cardiac imaging by permitting shorter breath-holds.

One further development in C-arm technology is the Siemens zeego® system, shown in Figure 14.18. The zeego system consists of the standard large FP detector (30×40 cm) and a rotating anode x-ray tube mounted on a C-arm, which is then attached to a six-axis KUKA® robot. The robot has six degrees of freedom, with an additional two degrees of freedom provided by rotation of the x-ray detector and x-ray tube to provide portrait and landscape acquisition modes (total angular range of 90°, tube

and detector move in synchrony). Unlike the L-arm mounted systems, the rotation *center* of the robot positioner can be placed *anywhere* in a sphere around the desired imaging volume (i.e., variable isocenter), can be moved during acquisition, and the imaging volume itself can be easily relocated or moved. The robot can move at a maximum speed of 120°/s with both acceleration to maximum speed and deceleration to 0 requiring approximately 0.5 s; motion accuracy and reproducibility has so far been verified for speeds up to 100°/s. This speed is twice that of the current Siemens Artis dTA and opens the door to a significant number of applications that require fast, serial volume CT images. The C-arm uses a standard C-arm geometry, with a focal-spot-to-detector distance of 120 cm and a focal-spot-to-isocenter of 78.5 cm if a propeller rotational acquisition is performed. High-voltage, control, and signal cables restrict total angular range of the system to 360° in propeller mode and to ~300° in the equivalent to "sleeve" mode. The flexibility of the robot motion can, in principle, be used to provide complex scan trajectories. One trajectory that cannot be achieved using L-arm systems is a "large-volume" data acquisition protocol; image data are acquired during a forward and backward sweep of the C-arm, with an offset in center of rotation between the two sweeps. Using new reconstruction algorithms, this effectively doubles the reconstructed FOV to 47 cm diameter with a z-extent of 18 cm. This reduces the impact of truncation (or missing projection data) for large objects that extend outside the standard 26-cm-diameter FOV and increases the accuracy of reconstructed HU numbers. Alternatively, by rotating the x-ray detector by 90°, the whole thoracic spine could be imaged with a 35-cm-diameter FOV and a z-extent of 25 cm. With the current factory-tested rotation speed of the zeego, such volumes can be acquired in ~7 s.

Design of and reconstruction for alternative trajectories for C-arm systems, even without the degrees of freedom provided by the robot system, has been an active area of investigation for many years. The main goal of this work is to permit the application of exact reconstruction algorithms to C-arm data (Kudo and Saito 1994, Khler 2001, Soimu et al. 2008). A simple way of increasing the ROI that satisfies Tuy's completeness condition is to use a nonplanar trajectory such as a saddle (Pack et al. 2004) or a circle plus line (Zeng and Gullberg 1992), and many alternatives have been proposed and developed (Clack et al. 1991; Danielsson 1992; Smith and Chen 1992; Weng et al. 1992; Yan and Leahy 1992; Axelsson and Danielsson 1994; Defrise and Clack 1994; Eriksson 1995; Kudo and

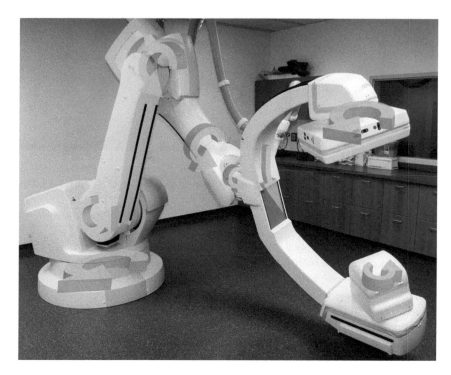

Figure 14.18 Photograph of the Siemens zeego system, with the possible degrees of freedom of motion indicated by the arrows. The increased flexibility of the system can be exploited to provide noncircular acquisition trajectories, for example, moving the center of rotation during a sweep to enable large FOV reconstructions.

Saito 1995, 1996, 1998; Noo 1997; Ramamurthi et al. 2004, 2005; Zambelli et al. 2007, 2008). To date, no manufacturer has offered C-arm CT using such trajectories, although one manufacturer has implemented a similar trajectory using the standard "L-arm"-mounting hardware described earlier. The XperSwing® (Philips Healthcare, Best, The Netherlands) provides optimized anatomical views of the coronary arteries during a single injection of contrast by rotating about the patient on a dual-axis trajectory moving in propeller and sleeve modes simultaneously (Horisaki et al. 2008, Klein and Garcia 2009). Although not intended for C-arm CT, a modified version of such a trajectory would match that described by Schomberg (2001) and could provide the data necessary to significantly reduce cone beam artifact in a spherical ROI with radius ~12.5 cm. The zeego system should provide additional benefits of realizing many such arbitrary 3D trajectories for a prescribed exact reconstruction ROI, allowing the appropriate trajectory to be matched to the imaging task. Note that trade-offs regarding SNR, dose, and artifact levels have to be explored in detail for each trajectory.

The increased speed of the robot system raises two new interesting challenges. The first area of investigation is providing real-time on-line feedback for patient-C-arm collision avoidance. Current C-arm systems have built-in hardware-based collision indicators (pressure sensors on x-ray tube and detector covers) that bring the C-arm to an immediate halt when triggered. The systems also have global collision avoidance software that makes assumptions about the location and size of the patient on the table and prevents the C-arm from entering a proscribed space. With the increased degrees of freedom of the robot system, these algorithmic solutions may prove too restrictive,

and research on real-time tracking of C-arm position relative to actual patient position is underway (Ladikos et al. 2008). A second challenge with such systems is the reproducibility of the position of the focal spot and detector during a high-speed acquisition. As stated earlier, the robot system has been shown to be reproducible for speeds up to 100°/s. If the highest-speed trajectories (i.e., 120° /s) cannot achieve the subpixel reproducibility required for accurate reconstruction then other approaches such as data-driven (iterative) misalignment correction may be necessary (Bodensteiner et al. 2007, Kyriakou et al. 2008).

14.7 SUMMARY

Innovation in both hardware and software will continue to provide improved image quality for C-arm CT volumes acquired in the interventional suite. Many of the iterative reconstruction algorithms currently under investigation for clinical CT will migrate to C-arm CT systems, leading to reduced noise, reduced dose, or both. The challenge for any of these increasingly complex algorithms is to provide the reconstructed volume within a clinically useful timeframe on the order of 2–5 min. However, if this goal can be achieved, then new dose reduction strategies may permit multiple 3D acquisitions during a single procedure, providing the ability to monitor progress during an intervention and to modify and improve the interventional approach on the fly. Closer integration of navigation tools, interventional instruments, and the imaging system will provide increased automation while improving the accuracy and diagnostic value of the 3D volume information acquired in the interventional suite.

REFERENCES

Ahmed, A.S., Zellerhoff, M., Strother, C.M., Pulfer, K.A., Redel, T., Deuerling-Zheng, Y., Royalty, K., Consigny, D. and Niemann, D.B. 2009. C-arm CT measurement of cerebral blood volume: an experimental study in canines. *AJNR Am J Neuroradiol* 30: 917–22.

Akhtar, M., Vakharia, K.T., Mishell, J., Gera, A., Ports, T.A., Yeghiazarians, Y. and Michaels, A.D. 2005. Randomized study of the safety and clinical utility of rotational vs. standard coronary angiography using a flat-panel detector. *Catheter Cardiovasc Interv* 66: 43–9.

Al-Ahmad, A., Wigström, L., Sandner-Porkristl, D., Wang, P.J., Zei, P.C., Boese, J., Lauritsch, G., Moore, T., Chan, F. and Fahrig, R. 2008. Time-resolved three-dimensional imaging of the left atrium and pulmonary veins in the interventional suite—a comparison between multisweep gated rotational three-dimensional reconstructed fluoroscopy and multislice computed tomography. *Heart Rhythm* 5: 513–9.

Arakawa, H., Marks, M.P., Do, H.M., Bouley, D.M., Strobel, N., Moore, T. and Fahrig, R. 2008. Experimental study of intracranial hematoma detection with flat panel detector C-arm CT. *AJNR Am J Neuroradiol* 29: 766–72.

Axelsson, C. and Danielsson, P. 1994. Three-dimensional reconstruction from cone-beam data in O(N/sup 3/logN) time. *Phys Med Biol* 39: 477–91.

Bani-Hashemi, A., Blanz, E., Maltz, J., Hristov, D. and Svatos, M. 2005. Cone beam x-ray scatter removal via image frequency modulation and filtering. *Med Phys* 32: 2093.

Bartolac, S., Clackdoyle, R., Noo, F., Siewerdsen, J., Moseley, D. and Jaffray, D. 2009. A local shift-variant fourier model and experimental validation of circular cone-beam computed tomography artifacts. *Med Phys* 36: 500–12.

Bertram, M., Wiegert, J. and Rose, G. 2005. Potential of software-based scatter corrections in cone-beam volume CT. *Proc SPIE* 5745: 259–70.

Blendea, D., Mansour, M., Shah, R.V., Chung, J., Nandigam, V., Heist, E.K., Mela, T., Reddy, V.Y., Manzke, R., Mcpherson, C.A., Ruskin, J.N. and Singh, J.P. 2007. Usefulness of high-speed rotational coronary venous angiography during cardiac resynchronization therapy. *Am J Cardiol* 100: 1561–5.

Blondel, C., Malandain, G., Vaillant, R. and Ayache, N. 2006. Reconstruction of coronary arteries from a single rotational x-ray projection sequence. *IEEE Trans Med Imaging* 25: 653–63.

Bodensteiner, C., Darolti, C., Schumacher, H., Matthaus, L. and Schweikard, A. 2007. Motion and positional error correction for cone beam 3D-reconstruction with mobile C-arms. *Med Image Comput Comput Assist Interv* 10: 177–85.

Boone, J.M. 1986. Scatter correction algorithm for digitally acquired radiographs: theory and results. *Med Phys* 13: 319–28.

Boone, J.M., Cooper, V.N., III, Nemzek, W.R., Mcgahan, J.P. and Seibert, J.A. 2000. Monte Carlo assessment of computed tomography dose to tissue adjacent to the scanned volume. *Med Phys* 27: 2393–407.

Boone, J.M. and Seibert, J.A. 1988a. An analytical model of the scattered radiation distribution in diagnostic radiology. *Med Phys* 15: 721–5.

Boone, J.M. and Seibert, J.A. 1988b. Monte Carlo simulation of the scattered radiation distribution in diagnostic radiology. *Med Phys* 15: 713–20.

Bousse Ast, A., Zhou, J., Yang, G., Bellanger, J.-J. and Toumoulin, C. 2009. Motion compensated tomography reconstruction of coronary arteries in rotational angiography. *IEEE Trans Biomed Eng* 56: 1254–7.

Bovic, A. 2000. *Handbook of Image and Video Processing*. London: Academic Press.

Busse, F., Rutten, W., Sandkamp, B., Alving, P.L., Bastiaens, R.J.M. and Ducourant, T. 2002. Design and performance of a high-quality cardiac flat detector. *Proc SPIE* 4682: 819–27.

Byrd, R.H., Lu, P.H., Nocedal, J. and Zhu, C.Y. 1995. A limited memory algorithm for bound constrained optimization. *SIAM J Sci Comput* 16: 1190–208.

Canny, J. 1986. A computational approach to edge-detection. *IEEE Trans Pattern Anal Mach Intell* 8: 679–98.

Chen, G.-H., Tang, J. and Hsieh, J. 2009. Temporal resolution improvement using PICCS in MDCT cardiac imaging. *Med Phys* 36: 2130–5.

Chen, G.-H., Tang, J. and Leng, S. 2008. Prior image constrained compressed sensing (PICCS): a method to accurately reconstruct dynamic CT images from highly undersampled projection data sets. *Med Phys* 35: 660–3.

Clack, R., Zeng, G.L., Weng, Y., Christian, P.E. and Gullberg, G.T. 1991. Cone beam single photon emission computed tomography using two orbits. *Information Processing in Medical Imaging Lecture Notes in Computer Science* 511: 45–54.

Colbeth, R.E., Mollov, I.P., Roos, P.G. and Shapiro, E.G. 2005. Flat panel CT detectors for sub-second volumetric scanning. *Proc SPIE* 5745: 387–98.

Cowen, A.R., Davies, A.G. and Sivananthan, M.U. 2008a. The design and imaging characteristics of dynamic, solid-state, flat-panel x-ray image detectors for digital fluoroscopy and fluorography. *Clin Radiol* 63: 1073–85.

Cowen, A.R., Kengyelics, S.M. and Davies, A.G. 2008b. Solid-state, flat-panel, digital radiography detectors and their physical imaging characteristics. *Clin Radiol* 63: 487–98.

Danielsson, P.-E. 1992. From cone-beam projections to 3D radon data in O(N3 longN) time. *IEEE Nucl Sci Symp Med Imag Conf.* 1: 1135–1137.

Defrise, M. and Clack, R. 1994. A cone-beam reconstruction algorithm using shift-variant filtering and cone-beam backprojection. *IEEE Trans Med Imaging* 13: 186–95.

Dodge, J.T., Nykiel, M., Altmann, J., Hobkirk, K., Brennan, M. and Gibson, C.M. 1998. Coronary artery injection technique: a quantitiative *in vivo* investigation using modern catheters. *Cathet Cardiovasc Diagn* 44: 34–9.

Ector, J., De Buck, S., Huybrechts, W., Nuyens, D., Dymarkowski, S., Bogaert, J., Maes, F. and Heidbüchel, H. 2008. Biplane three-dimensional augmented fluoroscopy as single navigation tool for ablation of atrial fibrillation: accuracy and clinical value. *Heart Rhythm* 5: 957–64.

Edward, G.S., Richard, E.C., Earl, T.D., Isaias, D.J., Ivan, P.M., Todor, I.M., John, M.P., Pieter, G.R., Josh, M.S.-L. and Carlo, A.T. 2007. Multidetector-row CT with a 64-row amorphous silicon flat panel detector. In: Jiang, H. and Michael, J.F. (eds.). *Medical Imaging 2007: Physics of Medical Imaging*. Bellingham, WA: SPIE, 65103X.

Endo, M., Mori, S., Tsunoo, T. and Miyazaki, H. 2006. Magnitude and effects of x-ray scatter in a 256-slice CT scanner. *Med Phys* 33: 3359–68.

Endo, M., Tsunoo, T., Nakamori, N. and Yoshida, K. 2001a. Effect of scattered radiation on image noise in cone beam CT. *Med Phys* 28: 469–74.

Endo, M., Tsunoo, T., Satoh, K., Matsusita, S., Kusakabe, M. and Fukuda, Y. 2001b. Performance of cone beam CT using a flat-panel imager. *Proc SPIE* 4320: 815–21.

Eriksson, J.A. and Danielsson, P-E. 1995. Helical scan 3D reconstruction using the linogram method. *Proceedings of the 1995 International Meeting on Fully Three-Dimensional Image Reconstruction in Radiology and Nuclear Medicine*, 287–90 (Aix-les-Bains, 1995).

Fahrig, R., Dixon, R., Payne, T., Morin, R.L., Ganguly, A. and Strobel, N. 2006. Dose and image quality for a cone-beam C-arm CT system. *Med Phys* 33: 4541–50.

Fahrig, R., Girard-Hughes, E., Mehdizadeh, A., Boese, J., Lauritsch, G., Wang, P., Rosenberg, J. and Al-Ahmad, A. 2008. Comparison of gated vs. non-gated cardiac C-arm CT (DYNA CT) of the left atrium and pulmonary veins in humans. *J Intervent Card Electrophysiol* 21: 169–174.

Feldkamp L.A., Davis L.C. and Kress J.W. 1984. Practical cone-beam algorithm. *J Opt Soc Am A* 1: 612–19.

Fitzgibbon, A. 2001. Robust registration of 2D and 3D point sets. In: Cootes, T. and Taylor, C. (eds.). *British Machine Vision Conference*. Manchester, UK: BMVC, 1145–1153.

Fornasier, M. and Rauhut, H. 2011. *Handbook of Mathematical Methods in Imaging*. Berlin, Germany: Springer.

Fox, T.R., Nisius, D.T., Aradate, H. and Saito, Y. 2001. Practical x-ray scatter measurements for volume CT detector design. *Proc SPIE* 4320: 808–14.

Galigekere R.R., Wiesent K., Holdsworth D.W. 1999.Techniques to alleviate the effects of view aliasing artifacts in computed tomography. *Med Phys* 26(6): 896–904.

Ganguly, A., Fieselmann, A., Marks, M., Rosenberg, J., Boese, J., Deuerling-Zheng, Y., Straka, M., Zaharchuk, G., Bammer, R. and Fahrig, R. 2011. Cerebral CT perfusion using an interventional C-arm imaging system: cerebral blood flow measurements. *AJNR Am J Neuroradiol* 32: 1525–31.

Garcia, J.A., Chen, S.-Y.J., Messenger, J.C., Casserly, I.P., Hansgen, A., Wink, O., Movassaghi, B., Klein, A.J. and Carroll, J.D. 2007. Initial clinical experience of selective coronary angiography using one prolonged injection and a 180 degrees rotational trajectory. *Catheter Cardiovasc Interv* 70: 190–6.

Gerds-Li, J., Tang, M., Kriatselis, C., Roser, M., Goetze, S., He, D. and Fleck, E. 2009. Rapid ventricular pacing to optimize rotational angiography in atrial fibrillation ablation. *J Interv Card Electrophysiol* 26: 101–7.

Girard, E.E., Al-Ahmad, A., Rosenberg, J., Luong, R., Moore, T., Lauritsch, G., Boese, J. and Fahrig, R. 2011. Contrast-enhanced C-arm CT evaluation of radiofrequency ablation lesions in the left ventricle. *JACC Cardiovasc Imaging* 4: 259–68.

Girard-Hughes, E., Al-Ahmad, A., Moore, T., Lauritsch, G., Boese, J. and Fahrig, R. 2009. Visualization and enhancement patterns of radiofrequency ablation lesions with iodine contrast-enhanced cardiac C-arm CT. *Proc SPIE* 7262, 72621N-1 -72621N-9.

Glover, G. and Pelc, N. 1985. *Method and apparatus for compensating CT images for truncated projections*. USA patent application U.S. Patent 4,550,371.

Glover, G.H. 1982. Compton scatter effects in CT reconstructions. *Med Phys* 9: 860–7.

Granfors, P.R., Aufrichtig, R., Possin, G.E., Giambattista, B.W., Huang, Z.S., Jianqiang, L. and Bing, M. 2003. Performance of a 41*41 cm/sup 2/amorphous silicon flat panel x-ray detector designed for angiographic and R&F imaging applications. *Med Phys* 30: 2715–26.

Hansis, E., Carroll, J.D., Schafer, D., Dossel, O. and Grass, M. 2010. High-quality 3-D coronary artery imaging on an interventional C-arm x-ray system. *Med Phys* 37: 1601–9.

Hansis, E., Schafer, D., Dossel, O. and Grass, M. 2008a. Evaluation of iterative sparse object reconstruction from few projections for 3-D rotational coronary angiography. *IEEE Trans Med Imaging* 27: 1548–55.

Hansis, E., Schäfer, D., Dössel, O. and Grass, M. 2008b. Projection-based motion compensation for gated coronary artery reconstruction from rotational x-ray angiograms. *Phys Med Biol* 53: 3807–20.

Hirota, S., Nakao, N., Yamamoto, S., Kobayashi, K., Maeda, H., Ishikura, R., Miura, K., Sakamoto, K., Ueda, K. and Baba, R. 2006. Cone-beam CT with flat-panel-detector digital angiography system: early experience in abdominal interventional procedures. *Cardiovasc Intervent Radiol* 29: 1034–8.

Honda, M., Kikuchi, K. and Komatsu, K.I. 1991. Method for estimating the intensity of scattered radiation using a scatter generation model. *Med Phys* 18: 219–26.

Horisaki, T., Katoh, O., Imai, S., Inada, T., Suzuki, T., Inimura, K., Sugiura, H., Bakker, N. and Melman, N. 2008. Feasibility evaluation of dual axis rotational angiography (XperSwing) in the diagnosis of coronary artery disease. *Medicamundi* 52: 11–15.

Hsieh, J., Chao, E., Thibault, J., Grekowicz, B., Horst, A., Mcolash, S. and Myers, T.J. 2004. A novel reconstruction algorithm to extend the CT scan field-of-view. *Med Phys* 31: 2385–91.

Hsieh, J., Gurmen, O.E. and King, K.F. 2000. Recursive correction algorithm for detector decay characteristics in CT. In: Dobbins, J.T., Boone, J.M. (eds.), *Medical Imaging 2000: Physics of Medical Imaging*. San Diego, CA: SPIE, 298–305.

Jandt, U., Schäfer, D., Grass, M. and Rasche, V. 2009. Automatic generation of 3D coronary artery centerlines using rotational x-ray angiography. *Med Image Anal* 13: 846–58.

Jarry, G., Graham, S.A., Moseley, D.J., Jaffray, D.J., Siewerdsen, J.H. and Verhaegen, F. 2006. Characterization of scattered radiation in kV CBCT images using Monte Carlo simulations. *Med Phys* 33: 4320–9.

Joseph, P.M. and Schulz, R.A. 1980. View sampling requirements in fan beam computed tomography. *Med Phys* 7: 692–702.

Joseph, P.M. and Spital, R.D. 1982. The effects of scatter in x-ray computed tomography. *Med Phys* 9: 464–72.

Kakeda, S., Korogi, Y., Ohnari, N., Moriya, J., Oda, N., Nishino, K. and Miyamoto, W. 2007. Usefulness of cone-beam volume CT with flat panel detectors in conjunction with catheter angiography for transcatheter arterial embolization. *J Vasc Interv Radiol* 18: 1508–16.

Kelley, C.T. 1999. *Iterative Methods for Optimization*. Philadelphia Society for Industrial Mathematics.

Khler, T.H., Proksa, R. and Grass, M. 2001. A fast and efficient method for sequential cone-beam tomography. *Med Phys* 28: 2318–27.

Klein, A.J. and Garcia, J.A. 2009. Rotational coronary angiography. *Cardiol Clin* 27: 395–405.

Knackstedt, C., Mühlenbruch, G., Mischke, K., Bruners, P., Schimpf, T., Frechen, D., Schummers, G., Mahnken, A.H., Günther, R.W., Kelm, M. and Schauerte, P. 2008. Imaging of the coronary venous system: validation of three-dimensional rotational venous angiography against dual-source computed tomography. *Cardiovasc Intervent Radiol* 31: 1150–8.

Kriatselis, C., Tang, M., Roser, M., Fleck, E. and Gerds-Li, H. 2009. A new approach for contrast-enhanced x-ray imaging of the left atrium and pulmonary veins for atrial fibrillation ablation: rotational angiography during adenosine-induced asystole. *Europace* 11: 35–41.

Kruger, D.G., Zink, F., Peppler, W.W., Ergun, D.L. and Mistretta, C.A. 1994. A regional convolution kernel algorithm for scatter correction in dual-energy images: comparison to single-kernel algorithms. *Med Phys* 21: 175–84.

Kudo, H. and Saito, T. 1994. Derivation and implementation of a cone-beam reconstruction algorithm for nonplanar orbits. *IEEE Trans Med Imaging* 13: 196–211.

Kudo, H. and Saito, T. 1995. An extended completeness condition for exact cone-beam reconstruction and its application. *Nuclear Science Symposium and Medical Imaging Conference. 1994 IEEE Conference Record (Cat. No.94CH35762)*, 1710–14, vol. 4.

Kudo, H. and Saito, T. 1996. Extended cone-beam reconstruction using Radon transform. *1996 IEEE Nuclear Science Symposium Conference Record (Cat. No.96CH35974)*, 1693–7 vol. 3.

Kudo, H. and Saito, T. 1998. Fast and stable cone-beam filtered backprojection method for non-planar orbits. *Phys Med Biol* 43: 747–60.

Applications

Kuon, E., Glaser, C. and Dahm, J.B. 2003. Effective techniques for reduction of radiation dosage to patients undergoing invasive cardiac procedures. *Br J Radiol* 76: 406–13.

Kyriakou, Y., Lapp, R.M., Hillebrand, L., Ertel, D. and Kalender, W.A. 2008. Simultaneous misalignment correction for approximate circular cone-beam computed tomography. *Phys Med Biol* 53: 6267–89.

Ladikos, A., Benhimane, S. and Navab, N. 2008. Real-time 3D reconstruction for collision avoidance in interventional environments. *Med Image Comput Comput Assist Interv* 11: 526–34.

Lauritsch, G., Boese, J., Wigstrom, L., Kemeth, H. and Fahrig, R. 2006. Towards cardiac C-arm computed tomography. *IEEE Trans Med Imaging* 25: 922–34.

Li, J.H., Haim, M., Movassaghi, B., Mendel, J.B., Chaudhry, G.M., Haffajee, C.I. and Orlov, M.V. 2009. Segmentation and registration of three-dimensional rotational angiogram on live fluoroscopy to guide atrial fibrillation ablation: a new online imaging tool. *Heart Rhythm* 6: 231–7.

Li, M., Yang, H. and Kudo, H. 2002. An accurate iterative reconstruction algorithm for sparse objects: application to 3D blood vessel reconstruction from a limited number of projections. *Phys Med Biol* 47: 2599–609.

Li, M.H., Kudo, H., Hu, J.C. and Johnson, R.H. 2004. Improved iterative algorithm for sparse object reconstruction and its performance evaluation with micro-CT data. *IEEE Trans Nucl Sci* 51: 659–66.

Liu, X., Shaw, C.C., Wang, T., Chen, L., Altunbas, M.C. and Kappadath, S.C. 2006. An accurate scatter measurement and correction technique for cone beam breast CT imaging using scanning sampled measurement (SSM) technique. *Prog Biomed Opt and Imaging Proc SPIE*, 6142 II.

Lorenz, C., Carlsen, I.-C., Buzug, T.M., Fassnacht, C. and Weese, J. 1997. Multi-scale line segmentation with automatic estimation of width, contrast and tangential direction in 2D and 3D medical images. *Lect Notes Comput Sci* 1205: 233–42.

Love, L.A. and Kruger, R.A. 1987. Scatter estimation for a digital radiographic system using convolution filtering. *Med Phys* 14: 178–85.

Maddux, J.T., Wink, O., Messenger, J.C., Groves, B.M., Liao, R., Strzelczyk, J., Chen, S.-Y. and Carroll, J.D. 2004. Randomized study of the safety and clinical utility of rotational angiography versus standard angiography in the diagnosis of coronary artery disease. *Catheter Cardiovasc Interv* 62: 167–74

Mail, N., Moseley, D.J., Siewerdsen, J.H. and Jaffray, D.A. 2008. An empirical method for lag correction in cone-beam CT. *Med Phys* 35: 5187–96.

Mansour, M., Reddy, V.Y., Singh, J., Mela, T., Rasche, V. and Ruskin, J. 2005. Three-dimensional reconstruction of the coronary sinus using rotational angiography. *J Cardiovasc Electrophysiol* 16: 675–6.

Mass, C. and Kachelriess, M. 2011. Quantification of temporal resolution and its reliability in the context of TRI-PICCS and dual source CT. In: Pelc, N.J., Ehsan S., Nishikawa, R.M. (eds.), *SPIE Medical Imaging 2011: Physics of Medical Imaging.* Lake Buena Vista, FL: SPIE 79611M1-7.

Meyer, B.C., Frericks, B.B., Albrecht, T., Wolf, K.J. and Wacker, F.K. 2007. Contrast-enhanced abdominal angiographic CT for intra-abdominal tumor embolization: a new tool for vessel and soft tissue visualization. *Cardiovasc Intervent Radiol* 30: 743–9.

Mistretta, C.A. 2011. Sub-Nyquist acquisition and constrained reconstruction in time resolved angiography. *Med Phys* 38: 2975–85.

Molloi, S., Zhou, Y.F. and Wamsely, G. 1998. Scatter-glare estimation for digital radiographic systems: comparison of digital filtration and sampling techniques. *IEEE Trans Med Imaging* 17: 881–8.

Mollov, I., Tognina, C. and Colbeth, R. 2008. Photodiode forward bias to reduce temporal effects in a-Si based flat panel detectors. *Proceedings of the SPIE - The International Society for Optical Engineering*, 6913, 69133S-1-69133S-69133S-9.

Montes, P. and Lauritsch, G. 2007. A temporal interpolation approach for dynamic reconstruction in perfusion CT. *Med Phys* 34: 3077–92.

Neitzel, U. 1992. Grids or air gaps for scatter reduction in digital radiography: a model calculation. *Med Phys* 19: 475–81.

Nett, B.E., Brauweiler, R., Kalender, W., Rowley, H. and Chen, G.-H. 2010. Perfusion measurements by micro-CT using prior image constrained compressed sensing (PICCS): initial phantom results. *Phys Med Biol* 55: 2333–50.

Neukirchen, C. and Hohman, S. 2007. An iterative approach for model-based tomographic perfusion estimation. In: Proc Fully 3-D. 104–7.

Neukirchen, C., Giordano, M. and Wiesner, S. 2010. An iterative method for tomographic x-ray perfusion estimation in a decomposition model-based approach. *Med Phys* 37: 6125–41.

Ning, R., Tang, X. and Conover, D. 2004. X-ray scatter correction algorithm for cone beam CT imaging. *Med Phys* 31: 1195–202.

Ning, R., Tang, X. and Conover, D.L. 2002. X-ray scatter suppression algorithm for cone-beam volume CT. *Proc SPIE* 4682: 774–81.

Nölker, G., Gutleben, K.J., Marschang, H., Ritscher, G., Asbach, S., Marrouche, N., Brachmann, J. and Sinha, A.M. 2008. Three-dimensional left atrial and esophagus reconstruction using cardiac C-arm computed tomography with image integration into fluoroscopic views for ablation of atrial fibrillation: accuracy of a novel modality in comparison with multislice computed tomography. *Heart Rhythm* 5: 1651–7.

Noo, F. 1997. Cone-beam reconstruction from general discrete vertex sets using radon rebinning algorithms. *IEEE Trans Nucl Sci* 44: 1309.

Ohnesorge, B., Flohr, T., Schwarz, K., Heiken, J.P. and Bae, K.T. 2000. Efficient correction for CT image artifacts caused by objects extending outside the scan field of view. *Med Phys* 27: 39–46.

Orlov, M.V., Hoffmeister, P., Chaudhry, G.M., Almasry, I., Gijsbers, G.H.M., Swack, T. and Haffajee, C.I. 2007. Three-dimensional rotational angiography of the left atrium and esophagus—a virtual computed tomography scan in the electrophysiology lab? *Heart Rhythm* 4: 37–43.

Orth, R.C., Wallace, M.J. and Kuo, M.D. 2008. C-arm cone-beam CT: general principles and technical considerations for use in interventional radiology. *J Vasc Interv Radiol* 19: 814–20.

Overdick, M., Solf, T. and Wischmann, H.A. 2001. Temporal artefacts in flat dynamic x-ray detectors. *Proc SPIE* 4320: 47–58.

Pack, J.D., Noo, F. and Kudo, H. 2004. Investigation of saddle trajectories for cardiac CT imaging in cone-beam geometry. *Phys Med Biol* 49: 2317–36.

Pan, X., Zou, Y. and Xia, D. 2005. Image reconstruction in peripheral and central regions-of-interest and data redundancy. *Med Phys* 32: 673–84.

Parker, D.L. 1982. Optimal short scan convolution reconstruction for fanbeam CT. *Med Phys* 9: 254–7.

Perrenot, B., Vaillant, R., Prost, R., Finet, G., Douek, P. and Peyrin, F. 2007. Motion correction for coronary stent reconstruction from rotational x-ray projection sequences. *IEEE Trans Med Imaging* 26: 1412–23.

Prümmer, M., Fahrig, R., Wigström, L. and Boese, J. 2007. Cardiac C-arm CT: 4D non-model based heart motion estimation and its application. *Proc SPIE* 6510, 651015-1-651015-12.

Ramamurthi, K., Prince, J.L. and Strobel, N. 2004. Exact 3D cone-beam reconstruction from projections obtained over a wobble trajectory on a C-arm. *IEEE Int Symp Biomed Imaging* 5674: 932–5.

Ramamurthi, K., Strobel, N. and Prince, J.L. 2005. Exact 3D cone-beam reconstruction from two short scans using a C-arm imaging system. *Proc SPIE* 5674: 87–98.

Raman, S.V., Morford, R., Neff, M., Attar, T.T., Kukielka, G., Magorien, R.D. and Bush, C.A. 2004. Rotational x-ray coronary angiography. *Catheter Cardiovasc Interv* 63: 201–7.

Applications

Ramirez-Giraldo, J.C., Trzasko, J., Leng, S., Yu, L., Manduca, A. and Mccollough, C.H. 2011. Nonconvex prior image constrained compressed sensing (NCPICCS): theory and simulations on perfusion CT. *Med Phys* 38: 2157–67.

Rasche, V., Movassaghi, B., Grass, M., Schafer, D. and Buecker, A. 2006a. Automatic selection of the optimal cardiac phase for gated three-dimensional coronary x-ray angiography. *Acad Radiol* 13: 630–40.

Rasche, V., Movassaghi, B., Grass, M., Schafer, D., Kuhl, H.P., Gunther, R.W. and Bucker, A. 2006b. Three-dimensional X-ray coronary angiography in the porcine model: a feasibility study. *Acad Radiol* 13: 644–51.

Roberts, D.A., Moran, J.M., Antonuk, L.E., El-Mohri, Y. and Fraass, B.A. 2004. Charge trapping at high doses in an active matrix flat panel dosimeter. *IEEE Trans Nucl Sci* 51: 1427–33.

Rohkohl, C., Lauritsch, G., Biller, L. and Hornegger, J. 2010a. ECG-gated interventional cardiac reconstruction for non-periodic motion. *Med Image Comput Comput Assist Interv* 13: 151–8.

Rohkohl, C., Lauritsch, G., Biller, L., Prummer, M., Boese, J. and Hornegger, J. 2010b. Interventional 4D motion estimation and reconstruction of cardiac vasculature without motion periodicity assumption. *Med Image Anal* 14: 687–94.

Roos, P.G., Colbeth, R.E., Mollov, I., Munro, P., Pavkovich, J., Seppi, E.J., Shapiro, E.G., Tognina, C., Virshup, G.F., Yu, J.M., Zentai, G., Kaissl, W., Matsinos, E., Richters, J. and Riem, H. 2004. Multiple gain ranging readout method to extend the dynamic range of amorphous silicon flat panel imagers. *Proc SPIE* 5368: 139–49.

Rowlands, J.A. and Yorkston, J. 2000. *Flat Panel Detectors for Digital Ragiography*. Bellingham, WA: SPIE.

Ruchala, K.J., Olivera, G.H., Kapatoes, J.M., Reckwerdt, P.J. and Mackie, T.R. 2002. Methods for improving limited field-of-view radiotherapy reconstructions using imperfect a priori images. *Med Phys* 29: 2590–605.

Ruhrnschopf, E.P. and Klingenbeck, K. 2011. A general framework and review of scatter correction methods in cone beam CT. Part 2: scatter estimation approaches. *Med Phys* 38: 5186–99.

Ruhrnschopf, E.P. and Klingenbeck, K. 2011a. Erratum: a general framework and review of scatter correction methods in x-ray cone beam CT. Part 1: scatter compensation approaches. *Med Phys* 38(7): 4296–4311.

Ruhrnschopf, E.P. and Klingenbeck, K. 2011b. A general framework and review of scatter correction methods in x-ray cone-beam computerized tomography. Part 1: scatter compensation approaches. *Med Phys* 38: 4296–311.

Schäfer, D., Borgert, J., Rasche, V. and Grass, M. 2006. Motion-compensated and gated cone beam filtered back-projection for 3-D rotational x-ray angiography. *IEEE Trans Med Imaging* 25: 898–906.

Schomberg, H. 2001. Complete source trajectories for C-arm systems and a method for coping with truncated cone-beam projections. *Proceedings VIth International Meeting on Fully Three-dimensional Image Reconstruction in Radiology and Nuclear Medicine*, 221–4.

Schoonenberg, G., Florent, R., Lelong, P., Wink, O., Ruijters, D., Carroll, J. and Ter Haar Romeny, B. 2009. Projection-based motion compensation and reconstruction of coronary segments and cardiac implantable devices using rotational x-ray angiography. *Med Image Anal* 13: 785–92.

Schoonenberg, G., Lelong, P., Florent, R., Wink, O. and Ter Haar Romeny, B. 2008. The effect of automated marker detection on *in vivo* volumetric stent reconstruction. *Med Image Comput Comput Assist Interv* 11(2): 87–94.

Seibert, J.A. and Boone, J.M. 1988a. Medical image scatter suppression by inverse filtering. *Proc SPIE* 914: 742.

Seibert, J.A. and Boone, J.M. 1988b. X-ray scatter removal by deconvolution. *Med Phys* 15: 567–75.

Siewerdsen, J.H., Daly, M.J., Bakhtiar, B., Moseley, D.J., Richard, S., Keller, H. and Jaffray, D.A. 2006. A simple, direct method for x-ray scatter estimation and correction in digital radiography and cone-beam CT. *Med Phys* 33: 187–97.

Siewerdsen, J.H. and Jaffray, D.A. 1999a. Cone-beam computed tomography with a flat-panel imager: effects of image lag. *Med Phys* 26: 2635–47.

Siewerdsen, J.H. and Jaffray, D.A. 1999b. A ghost story: spatio-temporal response characteristics of an indirect-detection flat-panel imager. *Med Phys* 26: 1624–41.

Siewerdsen, J.H. and Jaffray, D.A. 2000. Optimization of x-ray imaging geometry (with specific application to flat-panel cone-beam computed tomography). *Med Phys* 27: 1903–14.

Siewerdsen, J.H. and Jaffray, D.A. 2001. Cone-beam computed tomography with a flat-panel imager: magnitude and effects of x-ray scatter. *Med Phys* 28: 220–31.

Siewerdsen, J.H., Moseley, D.J., Bakhtiar, B., Richard, S. and Jaffray, D. 2004. The influence of antiscatter grids on soft-tissue detectability in cone-beam computed tomography with flat-panel detectors. *Med Phys* 31: 3506–20.

Silver, M.D. 2000. A method for including redundant data in computed tomography. *Med Phys* 27: 773–4.

Smith, B.D. and Chen, J. 1992. Implementation, investigation, and improvement of a novel cone-beam reconstruction method [SPECT]. *IEEE Trans Med Imaging* 11: 260–6.

Soimu, D., Buliev, I. and Pallikarakis, N. 2008. Studies on circular isocentric cone-beam trajectories for 3D image reconstructions using FDK algorithm. *Comput Med Imaging Graph* 32: 210–20.

Spies, L., Ebert, M., Groh, B.A., Hesse, B.M. and Bortfeld, T. 2001. Correction of scatter in megavoltage cone-beam CT. *Phys Med Biol* 46: 821–33.

Star-Lack, J., Mingshan, S., Kaestner, A., Hassanein, R., Virshup, G., Berkus, T. and Oelhafen, M. 2009. Efficient scatter correction using asymmetric kernels. *Proceedings of the SPIE - The International Society for Optical Engineering*, 7258, 72581Z (12 pp.)-72581Z (12 pp.).

Starman, J., Pelc, N., Strobel, N. and Fahrig, R. 2005. Estimating 0th and 1st moments in C-arm CT data for extrapolating truncated projections. In: Fitzpatrick, J.M. (ed.). *SPIE Medical Imaging 2005: Image Processing*. San Diego, CA: SPIE, 378–387.

Starman, J., Star-Lack, J., Virshup, G., Shapiro, E. and Fahrig, R. 2011. Investigation into the optimal linear time-invariant lag correction for radar artifact removal. *Med Phys* 38: 2398–411.

Starman, J., Tognina, C., Virshup, G., Star-Lack, J., Mollov, I. and Fahrig, R. 2008. Parameter investigation and first results from a digital flat panel detector with forward bias capability. *Proceedings of the SPIE - The International Society for Optical Engineering*, 6913, 69130L-1-69130L-69130L-9.

Starman, J., Virshup, G., Bandy, S., Star-Lack, J. and Fahrig, R. 2006. TH-E-330A-04: investigation into the cause of a new artifact in cone beam CT reconstructions on a flat panel imager. *Med Phys* 33: 2288.

Street, R.A. 1991. *Hydrogenated Amorphous Silicon*. Cambridge: Cambridge University Press.

Strobel, N., Meissner, O., Boese, J., Brunner, T., Heigl, B., Hoheisel, M., Lauritsch, G., Nagel, M., Pfister, M., Rührnschopf, E.P., Scholz, B., Schreiber, B., Spahn, M., Zellerhoff, M. and Klingenbeck-Regn, K. 2009. 3D Imaging with flat-detector C-arm systems. In: Reiser, M.F.E.A. (ed.). *Multislice CT*, 3rd edn. Berlin: Springer, 33–51.

Sze, D.Y., Strobel, N., Fahrig, R., Moore, T., Busque, S. and Frisoli, J.K. 2006. Transjugular intrahepatic portosystemic shunt creation in a polycystic liver facilitated by hybrid cross-sectional/angiographic imaging. *J Vasc Interv Radiol* 17: 711–15.

Taguchi, K., Funama, Y., Zhang, M., Fishman, E.K. and Geschwind, J.F. 2009. Quantitative measurement of iodine concentration in the liver using abdominal C-arm computed tomography. *Acad Radiol* 16: 200–8.

Tang, J., Hsieh, J. and Chen, G.-H. 2010. Temporal resolution improvement in cardiac CT using PICCS (TRI-PICCS): performance studies. *Med Phys* 37: 4377–88.

Tang, M., Kriatselis, C., Ye, G., Nedios, S., Roser, M., Solowjowa, N., Fleck, E. and Gerds-Li, J. 2009. Reconstructing and registering three-dimensional rotational angiogram of left atrium during ablation of atrial fibrillation. *Pacing Clin Electrophysiol* 32: 1407–16.

Thiagalingam, A., Manzke, R., D'Avila, A., Ho, I., Locke, A.H., Ruskin, J.N., Chan, R.C. and Reddy, V.Y. 2008. Intraprocedural volume imaging of the left atrium and pulmonary veins with rotational x-ray angiography: implications for catheter ablation of atrial fibrillation. *J Cardiovasc Electrophysiol* 19: 293–300.

Tkaczyk, J.E., Trousset, Y., Walter, D., Du, Y., Thompson, R.A. and Harrison, D. 2006. Novel features of the x-ray scatter profile not modeled by convolution of the primary. *Prog Biomed Opt Imaging Proc SPIE*, 6142 II.

Tognina, C., Mollov, I., Yu, J.M., Y., Webb, C., Roos, P.G., Batts, M., Trinh, D., Fong, R., Taie-Nobarie, N., Nepo, B., Job, I.D., Gray, K., Boyce, S. and Colbeth, R. 2004. Design and performance of a new a-Si flat panel imager for usein cardiovascular and mobile C-arm imaging systems. *Proc SPIE* 5368: 648–56.

Tommasini, G., Camerini, A., Gatti, A., Derchi, G., Bruzzone, A. and Vecchio, C. 1998. Panoramic coronary angiography. *J Am Coll Cardiol* 31: 871–7.

Tournoux, F.B., Manzke, R., Chan, R.C., Solis, J., Chen-Tournoux, A.A., Gérard, O., Nandigam, V., Allain, P., Reddy, V., Ruskin, J.N., Weyman, A.E., Picard, M.H. and Singh, J.P. 2007. Integrating functional and anatomical information to facilitate cardiac resynchronization therapy. *Pacing Clin Electrophysiol* 30: 1021–2.

Virmani, S., Ryu, R.K., Sato, K.T., Lewandowski, R.J., Kulik, L., Mulcahy, M.F., Larson, A.C., Salem, R. and Omary, R.A. 2007. Effect of C-arm angiographic CT on transcatheter arterial chemoembolization of liver tumors. *J Vasc Interv Radiol* 18: 1305–9.

Wallace, M.J. 2007. C-arm computed tomography for guiding hepatic vascular interventions. *Tech Vasc Interv Radiol* 10: 79–86.

Wallace, M.J., Kuo, M.D., Glaiberman, C., Binkert, C.A., Orth, R.C. and Soulez, G. 2008. Three-dimensional C-arm cone-beam CT: applications in the interventional suite. *J Vasc Interv Radiol* 19: 799–813.

Wallace, M.J., Kuo, M.D., Glaiberman, C., Binkert, C.A., Orth, R.C. and Soulez, G. 2009. Three-dimensional C-arm cone-beam CT: applications in the interventional suite. *J Vasc Interv Radiol* 20: S523–37.

Wallace, M.J., Murthy, R., Kamat, P.P., Moore, T., Rao, S.H., Ensor, J., Gupta, S., Ahrar, K., Madoff, D.C., Mcrae, S.E. and Hicks, M.E. 2007. Impact of C-arm CT on hepatic arterial interventions for hepatic malignancies. *J Vasc Interv Radiol* 18: 1500–7.

Weisfield, R.L., Hartney, M.A., Schneider, R., Aflatooni, K. and Lujan, R. 1999. High-performance amorphous silicon image sensor for x-ray diagnostic medical imaging applications. *Conference on Physics of Medical Imaging*. San Diego, CA: SPIE.

Weng, Y., Zeng, G.L. and Gullberg, G.T. 1992. A reconstruction algorithm for helical cone beam SPECT. In: *Nuclear Science Symposium and Medical Imaging Conference*, 1992., Conference Record of the 1992 IEEE, 1992. 1077-1079 vol. 2.

Wiegert, J., Bertram, M., Rose, G. and Aach, T. 2005. Model based scatter correction for cone-beam computed tomography. *Proc SPIE* 5745: 271–82.

Wiegert, J., Bertram, M., Schafer, D., Conrads, N., Timmer, J., Aach, T. and Rose, G. 2004. Performance of standard fluoroscopy anti-scatter grids in flat detector based cone beam CT. *Proc SPIE* 5368: 67–78.

Yan, X. and Leahy, R.M. 1992. Cone beam tomography with circular, elliptical and spiral orbits. *Phys Med Biol* 37(3): 493.

Yang, K., Kwan, A.L. and Boone, J.M. 2007. Computer modeling of the spatial resolution properties of a dedicated breast CT system. *Med Phys* 34: 2059–69.

Zambelli, J., Nett, B.E., Leng, S., Riddell, C., Belanger, B. and Chen, G.-H. 2007. Novel C-arm based cone-beam CT using a source trajectory of two concentric arcs. In: Jiang, H. and Michael, J.F. (eds.). *Medical Imaging 2007: Physics of Medical Imaging*. Bellingham, WA: SPIE, 65101Q.

Zambelli, J., Zhuang, T., Nett, B.E., Riddell, C., Belanger, B. and Chen, G.-H. 2008. C-arm based cone-beam CT using a two-concentric-arc source trajectory: system evaluation. In: Jiang, H. and Ehsan, S. (eds.) *Medical Imaging 2008: Physics of Medical Imaging*. Bellingham, WA: SPIE, 69134U.

Zellerhoff, M., Scholz, B., Ruehrnschopf, E.P. and Brunner, T. 2005. Low contrast 3D reconstruction from C-arm data. *Proc SPIE* 5745: 646–55.

Zeng, G.L. and Gullberg, G.T. 1992. A cone-beam tomography algorithm for orthogonal circle-and-line orbit. *Phys Med Biol* 37: 563–77.

Zhu, L., Bennett, N.R. and Fahrig, R. 2006. Scatter correction method for x-ray CT using primary modulation: theory and preliminary results. *IEEE Trans Med Imaging* 25: 1573–87.

Zhu, L., Starman, J., Bennett, N.R. and Fahrig, R. 2007. MTF measurement and a phantom study for scatter correction in CBCT using primary modulation. *2006 IEEE Nuclear Science Symposium Conference Record (IEEE Cat. No.06CH37832)*, 5 pp.-5 pp.

Zhu, L., Strobel, N. and Fahrig, R. 2005. X-ray scatter correction for cone-beam CT using moving blocker array. In: Flynn, M.J. (ed.). *Medical Imaging 2005: Physics of Medical Imaging*. Bellingham, WA: SPIE, 251–8.

Zhu, L., Wang, J. and Xing, L. 2009. Noise suppression in scatter correction for cone-beam CT. *Med Phys* 36: 741–52.

Cone beam CT: Transforming radiation treatment guidance, planning, and monitoring

John W. Wong, David A. Jaffray, Jeffrey H. Siewerdsen, and Di Yan

Contents

15.1 INTRODUCTION

In radiation therapy (RT), ionizing radiation is directed to eradicate disease in a clinical target volume internal to the patient. Accuracy in beam placement is paramount to avoiding the double detriments of missing the target and damaging surrounding structures. The highly localized nature of RT predisposes its close tie to advances in imaging. The community has long relied on some form of imaging to verify patient setup and beam placement. But as treatment methodologies evolve with technological advances, cone beam computed tomography (CBCT) has rapidly become the main enabling technology for image guidance since its commercial introduction in 2005. The advent of CBCT heralded our present era of image-guided radiation therapy (IGRT), and perhaps more appropriately, the era of volumetric or three-dimensional (3D) IGRT.

15.2 HISTORICAL DEVELOPMENT: THE EVOLUTION OF TREATMENT VERIFICATION

Table 15.1 provides a summary of the milestones in evolution of x-ray imaging technologies to verify treatment setup. Beginning with the introduction of ready pack films for RT in the 1970s, projection imaging with megavoltage (MV) x-rays from the linear accelerator was the mainstay of treatment verification into the 1990s. Portal images were acquired to ensure the correct placement of the beam port or aperture in the body. The lack of conspicuous anatomic information (i.e., visibility of high-contrast bone structures or airways only) in a portal image was compensated with the use of an open (nonshaped) beam to localize the surrounding anatomy. Double exposure imaging, combining both localization and portal image, was thus devised to verify the correct placement of the treatment aperture with respect to the patient anatomy. Weekly double exposure MV imaging remains a standard practice for those treatments that do not require kilovoltage (kV) volumetric image guidance.

During the 1980s, electronic portal imaging devices (EPIDs) mounted onboard the linear accelerators became available. These systems were quickly embraced by the community (van Herk et al. 1988; Shalev et al. 1989; Munro et al. 1990; Wong et al. 1990; Herman et al. 2001) with the exciting anticipation that they would improve setup accuracy by allowing more frequent portal imaging and improved image quality. Using the first-generation matrix ionization chamber system (van Herk et al. 1988) and the camera-based systems (Shalev et al. 1989; Munro et al. 1990; De Neve et al. 1992), many studies were performed that greatly increased our understanding of setup uncertainties. Portal imaging with large-area flat-panel detectors (i.e., active matrix arrays of amorphous Si thin film transistors and photodiodes coupled to a scintillator) offered further improvement in image quality (Antonuk et al. 1992, 1993). It was soon recognized that patient setup was an inherent and substantial 3D problem. More anatomic information was needed from two or more distinct projections for deriving more robust setup correction (Pisani et al. 2000). Such online interventions required a separate imaging and evaluation session before treatment and deterred routine daily application then. As a result, these procedures were reserved for those treatments that required a high degree of setup accuracy. Contrary to expectation, EPIDs did not have significant impact on improving setup accuracy in the community. They were mostly used as a convenient film replacement, and their digital format was also desirable for

Table 15.1 Summary of milestones in the evolution of x-ray imaging technologies to verify treatment setup

TIME	1970S	1980S	1990S	2000S	2010
Technologies	Film ————————————▶ CR			————————————————▶	
		EPID ——————————————————————————▶			
			In-room kV imaging; CT-on-the-rail	————————————————▶	
				CBCT ——————————▶	
Applications	Weekly imaging ———————————————————————————————————————▶				
			Off-line adaptive RT	————————————————▶	
				Daily online correction ————▶	

improving visualization (Shalev et al. 1989). Perhaps the most important benefits provided by EPIDs were the increased number of portal images made available for analysis and the speed at which they were made available to the clinician while the patient was on the treatment table. New insight about systematic and random setup errors led to the development of various off-line and online correction strategies (Bel et al. 1993). In most off-line approaches, the systematic error is estimated from some number of initial treatment sessions and corrected. In the online approach, setup errors are determined at the time of treatment and corrected, thus accounting for both systematic and random errors. Because of the need to invest in a significant workflow infrastructure to manage the data and implement corrections, only a handful of clinics have implemented the off-line approach, such as the program of adaptive radiation therapy (ART) at William Beaumont Hospital in Michigan (Yan et al. 1998) or the No Action Level (NAL) strategies in the Netherlands (de Boer et al. 2001).

Quantitative analysis of setup uncertainty measured with EPIDs raised several important questions. The low quality of portal images acquired with MV x-rays (namely, a lack of contrast and spatial resolution, even for visualization of bone edges) brought into question whether setup accuracy would be compromised. The desire to acquire more projection images also would impart additional MV dose. A logical approach to address these questions was to compare MV with kV portal imaging. The use of kV x-rays to verify treatment setup was not new and indeed was introduced as early as the 1950s (Johns et al. 1959). Various embodiments as room-mounted or gantry-mounted systems were deployed over the years for radiographic and fluoroscopic applications. Unfortunately, these early gantry-mounted systems with offset kV sources used standard film and screen detectors and were not conducive for routine clinical use.

It was in mid-1990s that investigators at William Beaumont Hospital (Beaumont) used a camera-based EPID to capture open field images from a gantry-mounted kV source (Pisani et al. 2000). The markedly improved portal image quality motivated the installation of a kV imaging system (i.e., an x-ray tube and EPID camera-based imager orthogonal to the MV imager) onboard an Elekta medical accelerator already equipped with its own MV EPID system. The effort was significant, and necessarily

(a)   (b)

Figure 15.1 (a) A member of the design team, Mr. Gabriel Blosser, testing the modification of the steel drum to receive the retractable arm. (b) Members of the team that installed the orthogonal kV imaging system on board the Elekta medical accelerator. From left to right: M. Moreau, D. Drake, D. Jaffray of William Beaumont Hospital, United States; and R. Cooke of Elekta, United Kingdom.

in-house, because it required major mechanical modification, as well as the rearrangement of electromechanical components, of a functioning clinical treatment machine. In 1997, the kV imaging system was instrumented orthogonal to the MV system with the kV x-ray source supported by struts that could be retracted into the drum gantry of the accelerator (Jaffray et al. 1997). The kV imager was identical to the camera-based system used for MV imaging. To alleviate the risk of damage to the clinical system during construction, a drum that had completed life-testing for mechanical stability in the United Kingdom was shipped to Beaumont as an engineering test-rig in 1997. Figure 15.1a shows a picture of Mr. Gabriel Blosser, a member of the design team, testing the modification of the steel drum to receive the retractable arm. After completing testing of other mechanical assembly and loading on the clinical system, mechanical modifications were made to the clinical gantry drum over the course of a weekend. Installation of the entire kV imaging system occurred over another weekend 3 months later (Jaffray et al. 1997). Figure 15.1b shows members of the team in front of the modified accelerator with the orthogonal kV imaging system.

With onboard kV imaging capabilities, studies were made to address questions raised about setup uncertainties using MV imaging (Figure 15.2). In patients with lung, prostate, and head-and-neck cancers, orthogonal MV and kV portal images were taken over the entire course of treatment. Corrections were made based on

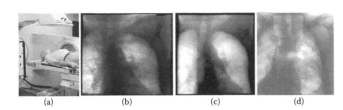

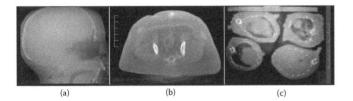

Figure 15.2 (a) Acquisition of a setup image of the thorax using the onboard kV imaging system. (b) A kV open field image. (c) An MV open field image. (d) A dual beam image consisting of a kV open field image and an MV portal image.

Figure 15.3 CBCT images acquired with the camera-based kV imaging system of a head phantom (a), small dog (b), and euthanized rats in a plastic container (c).

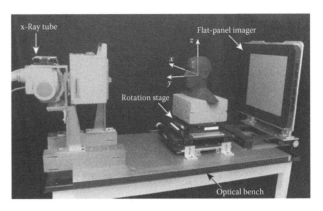

Figure 15.4 Bench-top system where rotate–translate stages were used to facilitate object motion for CBCT imaging research.

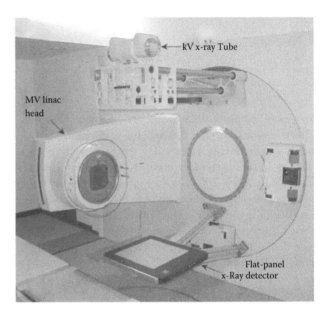

Figure 15.5 First flat-panel detector rigidly mounted to the face of the gantry drum.

MV and kV images alone on alternating weeks. A most surprising finding revealed that the measured setup variations and corrections were similar using either beam energy (Pisani et al. 2000). It was hypothesized that with adequate anatomic information for evaluation, differences in image quality due to beam energies would not be the decisive factor. In contrast, more projections and more frequent imaging would be possible with kV x-rays as the imaging dose was significantly lower than that with MV x-rays.

The time of the comparative kV and MV studies coincided with heightened concerns over the impact of organ motion on dose escalation with intensity modulated radiation treatment (IMRT) (van Herk et al. 1995). Because projection portal images did not contain soft-tissue information, alternative verification methods were exploited, such as imaging of implanted radio-opaque or electromagnetic markers (Kitamura et al. 2002), ultrasound imaging (Lattanzi et al. 2000), and in-room CT-on-rails (Uematsu et al. 1996); even an integrated magnetic resonance imaging (MRI)-accelerator had been proposed (Raaymakers et al. 2004). With the kV imaging system onboard a treatment gantry, extension to CBCT for soft-tissue imaging was most logical. The approach was also highly feasible given the seminal work on reconstruction by Felkamp et al. (1984) and other feasibility studies for medical applications (Cho et al. 1995). The first CBCT using the in-house kV imaging system was demonstrated using phantoms and small animals in 1997 (Jaffray et al. 1997). Figure 15.3 shows CBCT images of a head phantom (Figure 15.3a), small dog (Figure 15.3b), and euthanized rats (Figure 15.3c) in a plastic container acquired with the camera-based kV imaging system. These first CBCT images were most encouraging as proof of principle. However, with an unacceptably high CBCT imaging dose, limited imaging field of view, influence of veiling glare, and geometric instabilities of the optical collection path, the camera-based portal imaging system was never used for clinical studies.

Another important milestone in the evolution of CBCT for IGRT was the advent of large-area flat-panel detectors based

on amorphous silicon (a-Si:H) technology. The pioneering work at the University of Michigan was the beginning of a technological revolution for radiological imaging (Antonuk et al. 1996). Anticipating its availability, research and development of CBCT at Beaumont continued in the laboratory setting with a smaller 21 cm × 21 cm a-Si:H panel; 0.4 mm × 0.4 mm pixels manufactured by EG&G (currently PerkinElmer) and provided by Elekta (Courtesy of K. Brown). Figure 15.4 shows the bench-top system constructed in 1999 where rotate–translate stages were used to facilitate object motion for imaging (Jaffray et al. 2000). Numerous basic imaging performance studies were performed on this platform to determine the limits of soft-tissue imaging and spatial resolution of flat-panel CBCT (Siewerdsen et al. 1999a,b, 2000, 2001, 2004; Jaffray et al. 2000, 2002).

In 2000, a prototype 41 cm × 41 cm flat-panel a-Si:H detector for kV imaging was made available by Elekta, replacing the camera-based EPID at Beaumont (Jaffray et al. 2002) as shown in Figure 15.5. The flat-panel detector was rigidly mounted to the face of the gantry drum and was not intended for clinical use. Nevertheless, both the bench-top and

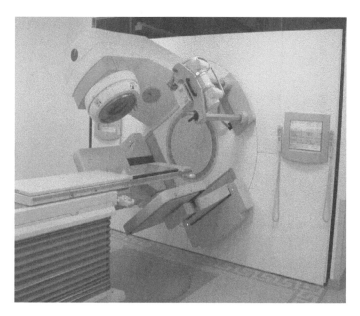

Figure 15.6 First prototype Synergy accelerator with onboard CBCT at Christie Hospital (2001).

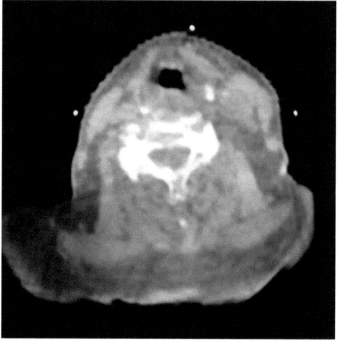

Figure 15.7 CBCT image of the first head-and-neck patient acquired with the early system at Beaumont in 2002. (Courtesy of D. Letourneau. With permission.)

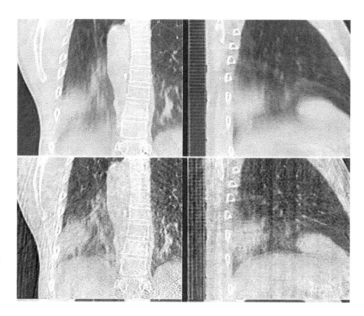

Figure 15.8 Sagittal and coronal images from free-breathing CBCT (top) and 4D-CBCT (bottom). (Courtesy of J. Sonke. With permission.)

gantry-mounted prototype systems made possible important studies to characterize and optimize the performance of flat-panel detectors for CBCT, including spatiotemporal response, image lag, imaging geometry, and effects of scatter (Siewerdsen et al. 1999a,b, 2000, 2001). The preparatory work that began in 1997 cumulated into deployment of four CBCT prototypes by Elekta beginning in 2001 at the Christie Hospital (Manchester, United Kingdom), the Netherlands Cancer Institute (NKI, Amsterdam), Beaumont, and Princess Margaret Hospital (Toronto, ON, Canada). Figure 15.6 shows the first prototype Synergy accelerator with onboard CBCT at Christie Hospital. These systems were constructed on site as the infrastructure for production was lacking then.

These early systems provided the platforms for further technical refinement, and most importantly, they allowed the development of clinical hardware and software tools to support quality assurance (QA) techniques, system calibration, and IGRT workflow for IGRT. Rapid progress was made, enabling clinical feasibility studies within 2 years. Figure 15.7 shows a CBCT image of the first head-and-neck patient acquired with the system at Beaumont in 2002 under an institutional review board (IRB)–approved protocol (Letourneau et al. 2005). In 2003, the investigators at the NKI developed the four-dimensional (4D) CBCT method (Sonke et al. 2005). Figure 15.8 (top) shows the sagittal and coronal images of free-breathing CBCT in comparison with 4D-CBCT (bottom) where motion blurring is significantly reduced.

In 2004, Elekta officially launched Synergy, their commercial accelerator product with onboard CBCT. In the same year, Varian implemented CBCT capability on their kV onboard imaging (OBI) system at Henry Ford Hospital (Detroit, MI). Siemens has made available a similar product based on MV CBCT, and more recently, Artiste, the kV CBCT system in line with the MV treatment beamline. Today, CBCT has become the central technology that drives IGRT. Almost all accelerators delivered to the North American community today come equipped with

CBCT. To date, the application of flat-panel CBCT in RT represents one of its largest deployments in medicine.

15.3 IGRT PROCESS WITH CBCT

Many of the technical aspects to optimize, calibrate, and maintain CBCT image quality to guide interventions are covered in other chapters of this book. There are, however, considerations and tasks that are specific to using CBCT to guide the delivery of radiation treatment. The picture in Figure 15.6 shows the various

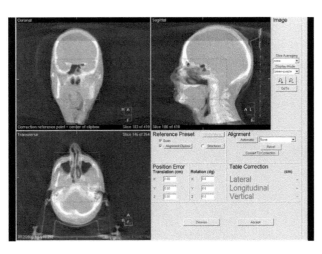

Figure 15.9 Elekta XVI software utility used to evaluate CBCT data for patient adjustment.

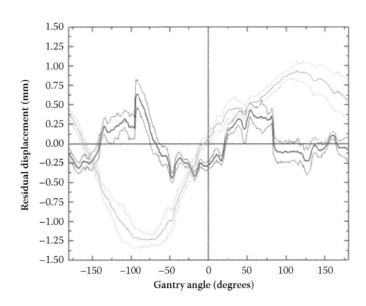

Figure 15.10 Flex maps of an Elekta accelerator that were acquired over 2 years.

components of a medical accelerator equipped for IGRT. Other than the MV head assembly, the other parts are the kV CBCT subsystem and the patient couch; both can be retracted or moved remotely from outside the treatment room. Central to the delivery of radiation treatment is the concept of a treatment isocenter. The isocenter is, ideally, a point in space where the axes of all rotations involving MV beam, the kV beam, and couch intersect. In reality, because perfect alignment of mechanical motion is difficult to achieve, the isocenter is a not a singular point, but a 3D region in space that is kept as small as possible. At treatment planning, a point typically within the target volume inside the patient is identified for placement at isocenter (Jaffray et al. 2002).

Before the advent of CBCT, patients were commonly set up for treatment by aligning skin marks, determined at planning or the first treatment session, with optical beams from wall-mounted lasers that intersect at the isocenter. The accuracy of the setup was then verified by comparing a double exposure portal image with a digitally reconstructed radiograph to ensure correct placement of the patient (bony) anatomy within the treatment beam aperture. Correct positioning of the patient at isocenter was thus implied. In an IGRT session with CBCT, placement of the patient at isocenter was explicit. A patient on the couch is again first positioned by aligning skin marks with lasers. A CBCT is then acquired and explicitly compared with the planning computed tomography (CT) scan in 3D to determine the internal target location with respect to the machine isocenter. Once done, the patient is repositioned, often via remote adjustment of the treatment couch. Figure 15.9 shows the XVI software utility on the Elekta accelerator that is used to evaluate the CBCT data for patient adjustment. Similar utilities are available from the Varian and Siemens systems.

The IGRT process brings into focus that the mechanical integrity of CBCT imaging subsystem and MV treatment system are inter-related. Both must be accurately implemented and properly maintained to achieve the necessary accuracy with IGRT. With the many retractable components, the rigidity of the accelerator is a significant consideration. The resultant flex and sag of the system must be corrected for proper CBCT reconstruction and also for proper setup adjustment (Jaffray et al. 2002). In the

Varian system, an active servo mechanism continuously adjusts the OBI system to maintain its proper alignment with the isocenter. The accelerator flex as a function of gantry rotation also can be determined from projection images of a ball bearing phantom acquired as the gantry rotates. The resultant flex map that relates the motion of the beam central axis with that of the flat-panel detector can then be used to adjust the kV projection images for reconstruction. Because the flex map is highly reproducible, it is measured on the Elekta and Siemens accelerators at time of installation and maintained as part of the QA program. Figure 15.10 shows the flex maps of an Elekta accelerator that were acquired over a period of 2 years. Varian has recently released their IsoCal feature that addresses this issue and is used in combination with the servo scheme.

It should be noted that because of the variation in accelerator flex, the MV and kV beam axes will not be colinear at the same gantry position with respect to the room; that is, a kV image acquired in the anterior–posterior direction of a patient will be slightly shifted in comparison with that acquired with the MV beam. These differences need to be accounted for if the kV projection image is used to verify or guide the placement of the MV beam. Finally, because patient setup is reliant on automated couch motion, accurate and precise motion of the couch is also paramount for successful IGRT. Calibration procedures and phantoms to establish and maintain the mechanical integrity of an IGRT system based on CBCT are available from the manufacturers or other vendors (Bissonnette et al. 2012). These procedures must be performed periodically to maintain the CBCT system for proper IGRT operation. It is also advisable to carry out a simple end-to-end test on a frequent basis to verify the correct positioning of a phantom to the machine isocenter based on CBCT.

15.4 CLINICAL APPLICATION OF CBCT

As discussed, the lack of definitive soft-tissue information in portal images (even with the highest quality EPID) spurred the development of various imaging methods to guide high-precision treatment. Patient setup with CT appeared most attractive, because it would reproduce the anatomic information used for planning. Before CBCT, this reproduction was feasible only with the installation of a conventional CT inside the treatment room. With the so-call CT-on-rail systems, the CT scanner is translated to scan the patient on a stationary couch that has been rotated to fit through the bore of the CT scanner. These high-precision systems provide excellent image quality to manage interfraction patient setup and organ motion. However, the cost, complexity of patient movement between imaging and treatment, and concerns about stability after correction were major impediments to clinical uptake. Nevertheless, these systems continue to be used today with interest in applying them in the context of adaptive RT (Barker et al. 2004; Schwartz 2012).

The advent of flat-panel a-Si:H detectors made possible onboard CBCT to provide 3D anatomic information of the patient in the treatment position at time of treatment. Image quality is expected to be inferior to that acquired with the more mature technology of helical CT; however, it is nevertheless adequate for guiding patient setup, particularly when a high-quality planning CT is available as reference. The onboard embodiment offers significant advantages for convenient intrafraction verification with repeat CBCT, or radiographic and fluoroscopic imaging throughout the treatment procedure.

15.5 CBCT-GUIDED INTERVENTION

CBCT makes available an enormous amount of patient treatment data, including the daily positions of bony and soft-tissue structures. A course of treatment with daily CBCT imaging would readily generate 3–5 GB of data. The wealth of data can be used to improve the accuracy and quality of individual patient treatment. However, analysis of the large volume of data also will consume tremendous amounts of resources. It is thus important that the individual clinician as well as the department properly defines the treatment objectives and matches them with the most efficient and efficacious strategies. As introduced previously, there are two general categories of patient setup position correction strategies, off-line and online, both with the goal of setting up the patient according to the treatment plan. Any correction strategy is intricately related with the prescription of an appropriate margin for the planning target volume (PTV). Figure 15.11 shows their relationship with the PTV margin in a cartoon. A cross-hair is shown on each of the panels, representing the treatment room coordinates. On each panel, dots of the same color represent the distribution of treatment position of a particular patient. The ideal position of each dot would be at the cross-hair center. Figure 15.11a depicts the customary observations of the variability of individual patient treatments. The mean setup positions of all patients do not coincide with the cross-hair center, indicating that there is a systematic error in the setup of each patient. The spread of the setup distribution also varies for each patient, indicating that there is a random variation for each patient's setup, which may be different. Note that the position of each dot is unknown and can be measured only after initiation of treatment. In the case of Figure 15.11a, the gray circle represents a large institutional margin or PTV that is prescribed to accommodate the treatment setup of all patients if there is no image-guided

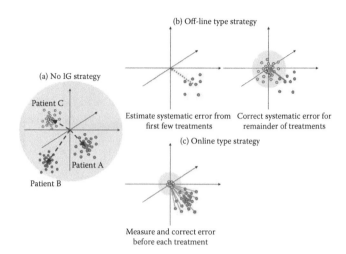

Figure 15.11 Schematic illustrations of image guidance strategies where no image guidance (IG) is used (a), the PTV is modified after off-line analysis of limited number of daily IG data (b), and a small PTV margin is used where daily image guidance corrections are made (c).

position error management. Much of this institutional margin is spent accounting for interpatient variation.

An individual patient can be treated with a smaller margin, if the distribution of the setup variation is known. In that case, as shown in Figure 15.11b, the systematic error of the patient setup can be corrected with a move to the cross-hair center, and the residual setup error can be accommodated with a smaller margin. Thus, in an off-line position correction strategy, imaging information is acquired for a limited number of early treatment sessions to provide an adequate estimation of the systematic and residual errors. The systematic error is then corrected. Because the random nature of residual errors will result mostly in blurring the delivered dose distribution, it has been recommended by several margin recipes that a much reduced margin for PTV will suffice after correction of the systematic error. The derivation of these recipes has been enlightening on the relative importance of systematic and random errors. However, they are based on various assumptions that should not be applied indiscriminately for all cases. Specifically, they assume a degree of stability in the patient position (mean) over the subsequent fractions. This has been proven to be robust in sites dominated by setup error or organ motion, but it may not be the case in situations wherein the tumor is shrinking over the course of therapy.

In the online position correction model, represented in Figure 15.11c, imaging information acquired at the time of, or during, treatment is examined to facilitate immediate adjustment. Online correction reduces both systematic and random errors and is typically applied via 3° translate couch motion. More aggressive treatments, such as single fraction stereotactic radiosurgery (SRS) or short course stereotactic body radiation therapy (SBRT), are almost always delivered using online image guidance where corrections can be made with 6° translate–rotate couch motion, usually with a threshold for allowable rotations. More comprehensive corrections using daily CBCT information for online plan reoptimization have been attempted but remain rare.

It is worth noting that off-line position correction strategies, such as those implemented based on decision rules with Action Level (Bel et al. 1993) or NAL (de Boer et al. 2001), were developed before the advent of CBCT where elucidation of setup error based on

MV projection images could be ambiguous and time-consuming. The volumetric information in CBCT has significantly reduced the ambiguity. At present, corrections of setup based on CBCT are usually conducted entirely online. In addition to the perception that correction of both systematic and random position errors would result in more accurate treatment, the preference for online correction is also due to its simpler workflow. It follows that off-line position correction strategy has become unnecessary with CBCT.

15.6 ONLINE POSITION CORRECTION STRATEGY

Despite its simplicity, several technical and workflow inter-related issues impact the effectiveness of online setup correction using CBCT.

Time constraints are a major factor for online correction. It must be noted that CBCT produces a snapshot of information whose richness has often been misinterpreted as complete. Without invasive immobilization, the integrity of the CBCT representation of the patient setup becomes questionable with time (Purdie et al. 2006), such as when excessive time is spent on manipulating the display parameters, determining the shifts to be applied, or deriving a new treatment plan to be delivered. All such intratreatment variation factors need to be considered in the determination of residuals and target margin. Effective online corrections need to be applied efficiently, and it is best that the parameters for CBCT acquisition and display as well as the action level for corrections be predetermined to minimize the duration of an IGRT session. Repeat CBCT or other forms of monitoring need to be acquired and evaluated to ensure that the initial setup correction remains valid. The development of remote couch control has altered online practices to move away from verification imaging to confirm the shift, except in the case of hypofractionation schedules (Li et al. 2009).

Imaging surrogates are often used to augment the snapshot nature of CBCT, its less-than-ideal image quality, and to provide intrafraction monitoring during extended IGRT session. The surrogates include room-mounted kV fluoroscopy systems, optical tracking of surface markers or patient's topography, and implanted electromagnetic beacons or radio-opaque markers. The surrogate signals are often less complex and easier to interpret than the fully 3D CBCT data. Because clinics are encouraged to adopt consistent practices to minimize operator variability, they have often chosen to use surrogates for the less aggressive treatment (Yin et al. 2009). In reality, very few studies have been conducted to evaluate the inter- and intrafraction validity of these surrogates to represent the intended target or to establish the appropriate PTV margin using surrogates. The question as to what is the most appropriate technology—radiographic or CBCT—for each clinical application remains unaddressed.

Determination of setup error with CBCT is central to accurate IGRT. It has become apparent that with CBCT, setup corrections based on 3D rigid bony structures can be made accurately and efficiently, certainly minimizing the ambiguity with setup rotation using two orthogonal projection images. However, aligning on soft-tissue targets (and avoiding soft-tissue organs at risk) is a new paradigm for radiation therapists and may cause some confusion in the evaluation process. It is thus important for the physician who

orders IGRT for a specific clinical site, or even for an individual patient, to communicate the objectives of the guidance with the physicist and the treatment personnel (Yin et al. 2009). The correction for cranial sites, understandably, will be primarily based on bony landmarks. For pelvic disease such as prostate cancer, where the central location of the target volume is less sensitive to dosimetric variation due to minor changes in the patient torso, the correction may be primarily based on the soft-tissue target itself.

For the thoracic and upper abdominal sites, respiratory motion will result in blurred CBCT image quality. Nevertheless, CBCT can be a very powerful tool in the context of stereotactic body radiation therapy (SBRT) of lung tumors where contrasts are intrinsically higher. Figure 15.12 shows the axial, sagittal, and coronal planes of the tumor in the planning CT (a, red panel) compared with the same slices from the daily CBCT (b, yellow) for a case of lung SBRT. Despite the motion artifacts in the CBCT, it provides the important information for accurately positioning the tumor with respect to the PTV (green).

Definitive correction directives, however, are often not possible to execute for many sites. For the head-and-neck sites, the patient setup variation may involve flexing of the neck, rendering rigid body evaluation ineffective, and where nonrigid (deformable) registration solutions are difficult to apply. Often, compromise is made by emphasizing the importance of certain vertebral bodies or bony landmarks, or deriving an average correction from rigid body registration of several 3D regions of interest. Figure 15.13 shows an example where alignment of the nasopharyngeal structures

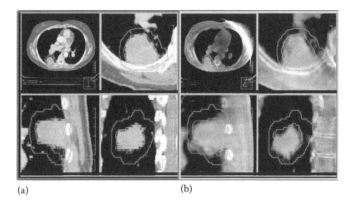

(a)   (b)

Figure 15.12 (See color insert.) Axial and coronal planes of the tumor for a case of lung SBRT in the planning CT (a, red) and from the daily CBCT (b, yellow). Despite the motion artifacts in the CBCT, the information can be used to for accurately positioning the tumor with respect to the PTV (green).

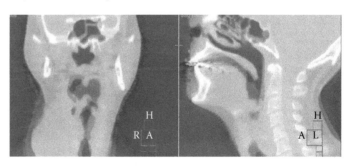

Figure 15.13 (See color insert.) Example where alignment of the nasopharyngeal structures results in misalignment of the cervical spine and variations in the skin contour in a head-and-neck cancer treatment.

results in misalignment of the cervical spine and variations in the skin contour in the head-and-neck region. The example points out that the nonrigid and mobile nature of the soft-tissue targets or adjacent structures often render perfect setup or correction infeasible. The identification of an appropriate PTV margin is paramount to the optimal execution of corrective actions and successful IGRT. There will also be those situations that evade any correction, and situations where restarting the patient setup become necessary. Examples of these situations include large rotations in the context of paraspinal SBRT wherein the patient needs to be simply repositioned on the couch, or evidence of excessive flex in the c-spine for IMRT in the head and neck.

In addition to geometric analysis of the guidance images, there is growing support for the integration of dosimetric data at the treatment machine to guide the interpretation of the guidance images. For example, the creation of a "limiting dose structure" to compare with the location of organs at risk during hypofractionated treatments, such as stereotactic lung RT. Furthermore, the use of "guidance structures" to link the expectations of the planning process to what is happening at the treatment machine is recommended. For example, visualization of the spinal cord organ at risk (OAR) while aligning to a target in the head and neck can assist in the decision-making process. The development of alignment tools that accommodate different regions of registration (often defined by the oncologist) also has emerged from the clinical need to interpret these images in a consistent manner across the radiation treatment team (oncologists, physicists, dosimetrists, therapists).

Dose recalculation based on treatment CBCT provides important information to refine a particular IGRT technique, but it is rarely performed. It needs to be recognized that irrespective of whether a rigid or nonrigid model is used to determine the setup error, the correction of patient setup errors is entirely based on rigid body adjustment. The 3° or 6° motion correction is applied to the entire patient via the motion of the mechanical components of the machine. As such, the effects of the applied correction on the delivered dose to the patient are unclear in the online position correction strategy. Similarly, analysis of the shift information can be useful to determine a more appropriate population margin. However, the revised margin contraction or expansion would be uniform and does not reflect the dependence of margin expansion on beam direction. The dosimetric effects of the setup errors and corrections for more optimal margin design would require the off-line analysis of the online IGRT data involving dose recalculation and replanning.

As a first-order approximation, the isocenter shifts from IGRT can be applied to planning CT to facilitate dose recalculations. A more comprehensive analysis is to perform dose calculation on the CBCT acquired at each treatment session. The conversion of CBCT voxel values to calibrated CT numbers (Hounsfield units) for purposes of dose calculation can be achieved by proper calibration. Commercial products that support dose calculation based on CBCT are available from Varian and Nucletron. Given that the patient anatomy on CBCT at the time of treatment would not be identical to that at treatment planning, the more demanding challenge would be the application of deformable registration to consolidate the cumulative dosimetry data of all treatment sessions on a common CT. Figure 15.14 shows a panel of treatment plans in the sagittal plane representing the prescribed

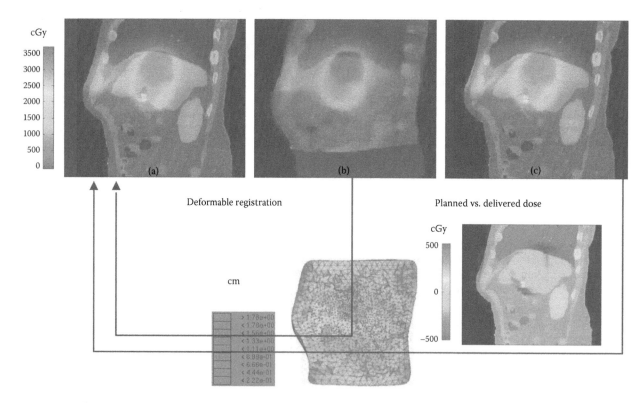

Figure 15.14 (See color insert.) Ability to deform anatomic structures is critical for accurate dose estimation during adaptation of treatment plans. Shown is a panel of treatment plans in the sagittal plane representing the prescribed dose distribution on the planning CT (a), the delivered dose distribution based on the CBCT of the day (b), and the delivered dose distribution deformed onto the original planning CT (c) (Courtesy of K. Brock, M. Velec, and L. Dawson, Princess Margaret Hospital).

dose distribution on the planning CT (Figure 15.14a), the delivered dose distribution based on the CBCT of the day (Figure 15.14b), and the delivered dose distribution deformed onto the original planning CT (Figure 15.14c) (Velec et al. 2011).

Finally, implementation of online correction strategy requires a thorough understanding of the uncertainties associated with the process. All CBCT and correction data at the initial phase of implementation, regardless of online or off-line approaches, should be analyzed systematically. The information is needed to determine action level for setup corrections that balance efficiency and effectiveness or to derive an appropriate PTV margin for planning a treatment using a particular IGRT technique. Correction residuals that cannot be eliminated should be quantified and compensated. Often overlooked is the analysis of user variability in the determination of setup error. The subjective component of applied correction is unavoidable given that an exact solution often does not exist. Such off-line analysis of the CBCT correction data will clarify the relative preponderance of the translational and rotational components of the setup error, or the uncertainty of the perceived setup error, so as to develop guidelines for more robust online setup evaluation by the user. This is often referred to as "rational margin design" and is clearly a prerequisite for online strategies.

15.7 ADAPTIVE TREATMENT STRATEGY

ART is a correction strategy that addresses more than positioning errors. Unlike the online–off-line position correction strategies, the goal of ART is not to maintain the patient at an intended treatment position. Instead of correction to achieve the objectives specified at the initial plan, the adaptive treatment strategy aims to proactively analyze the variations measured during the course of treatment and to apply the information to reoptimize the individual treatment when necessary. The improvement in ART is patient-specific.

ART was introduced by Yan and colleagues at William Beaumont Hospital (Yan et al. 1997, 1998) as a closed-loop feedback approach to correct the systematic variation in patient setup and organ variation, and to include the random variation to modify the treatment plan. Daily CT and MV portal imaging was acquired during the early phase of the treatment course to derive the feedback for reoptimizing the treatment of prostate cancer. Figure 15.15 shows an example where the target volumes and PTVs are modified during the course of prostate cancer treatment.

The advent of CBCT provides a more effective and efficient means to provide the necessary data for ART. With CBCT, the principles of ART have since been applied and expanded in the treatment of various clinical sites. The wide variety of approaches to implement adaptive treatment deal with rigid or nonrigid variations of patient position and anatomy; tumor shrinkage; normal organ volume and shape changes; and the frequency of the treatment plan modifications, such as online or off-line, single or multiple modifications. For head-and-neck treatments, ART focuses on the timing and frequency of replanning as the tumor shrinks. Figure 15.16 shows the resultant improved sparing of various critical structures for a group of head-and-neck patients

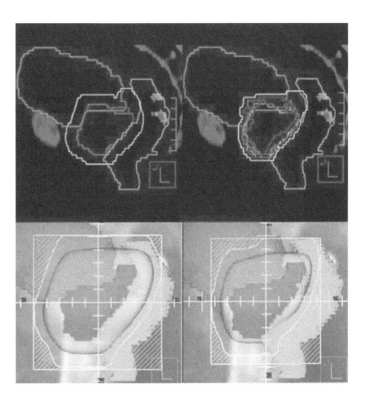

Figure 15.15 (See color insert.) PTV prescribed at treatment planning (left) is modified (right) during the course of prostate cancer treatment.

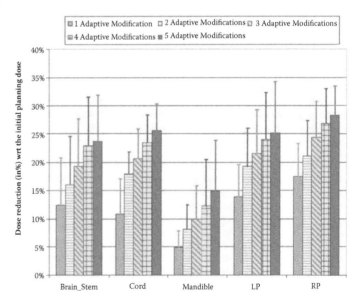

Figure 15.16 Sparing of various critical structures is improved with increasing frequency of modifications for a group of head-and-neck patients treated with the ART process.

treated with ART process. For lung cancer treatments, 4D CT is analyzed to optimize a motion weighted PTV for free-breathing treatment with significantly reduced margin than that of the internal target volume (ITV) method encompassing the entire range of tumor motion. 4D-CBCT also is used to perform 4D adaptive inverse planning optimization during the course of treatment. An excellent review of ART for various clinical sites has recently been published by Yan (2010).

Despite the synergistic potential of CBCT-based guidance and ART, its implementation imparts a significant resource challenge to the clinic. The integration of software tools to analyze the initial set of CBCT data; evaluate the impact of the patient anatomy variations, perhaps involving dose recalculation and replanning; and carry out the correction for the individual patient are daunting. At present, the few institutions that deploy ART successfully are those that have invested significant in-house efforts and, in most cases, they are limited to specific disease sites (Nuver et al. 2007). Given that the simplest software tools to analyze the trends and statistics of daily correction data of a patient's treatment are not available with commercial IGRT products, the majority of the community adopts online setup correction strategies with CBCT. Regardless of the current challenges, the methods for dose accumulation and integration of IGRT data will progress. It is likely that patients will be treated with a hybrid of online position correction and off-line adaptation methods that ensure each patient's anatomic changes are accommodated and exploited to the individual's benefit. The online adaptation method, in contrast, could be eventually possible for hypofractionation and SRS cases, but it fully depends on the computation speed and the reliability of clinical QA method.

15.8 CHALLENGES AND FUTURE DEVELOPMENT

It is likely that CBCT, the subject matter of this book, will continue to improve with the advances of x-ray imaging technologies. Future advances in computer technologies also will enhance the implementation of more powerful image processing and reconstruction algorithms. The anticipated improvement in CBCT image quality and reduction in imaging dose will benefit RT, allowing more frequent imaging. For RT, emphasis also needs to be placed on improving the application of CBCT for IGRT in the future.

A major shortcoming of CBCT is the snapshot nature of its acquisition. The integrity of the CBCT information will deteriorate with time. In those cases where the treatment session is inherently long, such as with high-dose short-course SBRT or single-course SRS, additional means of intrafraction monitoring is necessary to complement CBCT or to advise on the need to acquire a new CBCT for the IGRT session. Novel x-ray imaging techniques with reduced projection images, with the incorporation of imaging priors, or both, are being investigated to support fast, local reconstruction (Lee 2012; Stayman 2012). These approaches are promising and leverage the advances in iterative 3D image reconstruction.

The use of surrogates, such as implanted or surface markers that can be localized with x-ray or nonionization detection techniques, are commonly used to complement CBCT. There are a variety of systems that are gantry-mounted, room-mounted, or mobile. In addition to providing continuous or more frequent monitoring of the patient setup, surrogates (particularly implanted makers) also are being used to address the present deficiency of CBCT to confidently delineate soft-tissue targets in low-contrast contexts. The use of surrogates can be highly effective although validation studies are rare (Moseley et al. 2006). There are

nonnegligible differences between correction information derived from implanted markers and CBCT, although their relationship has not been resolved (Barney et al. 2011).

There is, however, some resistance to the use of implanted markers due to the invasiveness of the procedure, and in some cases, the additional x-ray imaging dose. The ideal image guidance method would be nonionizing and noninvasive. To that end, various embodiments of integrated MRI-medical linear accelerators are being developed to supplant CBCT for IGRT in a variety of treatment sites. The onboard system is particularly powerful to allow real-time imaging of tumors influenced by respiratory motion. These novel systems are becoming available commercially and may well represent the next major advancement in IGRT. However, their widespread dissemination in the community awaits reports from early adopters, given the cost of each system and the significant workflow adjustment needed to accommodate an MRI unit in the radiation treatment environment (Jaffray et al. 2010).

Another complementary development is the integration of 3D ultrasound imaging onboard a medical accelerator to enhance soft-tissue localization with CBCT (Roland 2012; Zhong 2013). The ultrasound device is placed under robotic control on the patient for transabdominal or transperineal scanning at time of CT simulation and CBCT treatment. It is CT-scanned with the patient, and different than its current use, the ultrasound probe will be left in place on the patient to provide continuous soft-tissue monitoring during treatment. Robotic control ensures reproducible probe placement and tissue deformation, overcoming the concern of operator dependence with ultrasound imaging. The use of a 3D wobbler ultrasound probe minimizes the need of mechanical adjustment of the probe. Given that IMRT delivery involves highly degenerate solutions in the optimal beam placements, beam arrangements will be found to avoid probe irradiation. Figure 15.17 shows a sagittal view of the CBCT of a prostate patient (Figure 15.17a), a transperineal ultrasound scan acquired of the patient's pelvic region (Figure 15.17b), and the overlay of the ultrasound and CBCT scan (Figure 15.17c) to provide improved soft-tissue delineation on the CT anatomy. The integrated system offers a low-cost approach for continuous intrafraction soft-tissue monitoring to enhance CBCT, albeit not applicable for lung tumors.

Arguably and as previously discussed, the fundamental obstacle to improving the use of CBCT for IGRT is the lack of integrated software tools to help the clinic analyze its correction

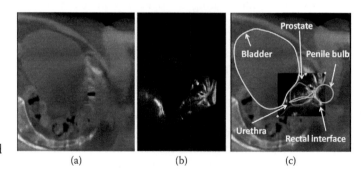

(a)   (b)   (c)

Figure 15.17 (a) Sagittal view of the CBCT of a prostate patient. (b) Transperineal ultrasound scan acquired of the patient's pelvic region. (c) Overlay of the ultrasound and CBCT scan.

data. Many of the necessary software tools to perform deformable registration of daily CBCT and (if pursuing ART) to perform dose recalculation and deformable registration of the delivered dose distributions are now available as commercial products. They are, unfortunately, not well integrated. Such an integrated software package, whether available commercially or as open source, will be of tremendous value to the entire community (Brock et al. 2008). Analysis of the CBCT correction data establishes the baseline to improve clinical practice, validates the quality of the clinical data to support cooperative clinical trials, enhances the development of cost-effective IGRT, and provides many other benefits. The present use of CBCT for IGRT in the community is the tip of the iceberg with the wealth of information yet to be exploited.

REFERENCES

Antonuk, L.E., Boudry, J., Huang, W., et al. 1992. Demonstration of megavoltage and diagnostic x-ray imaging with hydrogenated amorphous silicon arrays. *Med Phys* 19: 1455–66.

Antonuk, L.E., Yorkston, J., Huang, W., et al. 1996. Megavoltage imaging with a large-area, flat-panel, amorphous silicon imager. *Int J Radiat Oncol Biol Phys* 36: 661–72.

Barker, Jr., J.L., Garden, A.S., Ang, K.K., et al. 2004. Quantification of volumetric and geometric changes occurring during fractionated radiotherapy for head-and-neck cancer using an integrated CT/linear accelerator system. *Int J Radiat Oncol Biol Phys* 59(4): 960–70.

Barney, B.M., Lee, R.J., Handrahan, D., et al. 2011. Image-guided radiotherapy (IGRT) for prostate cancer comparing kV imaging of fiducial markers with cone beam computed tomography (CBCT). *Int J Radiat Oncol Biol Phys* 80(1): 301–5.

Bel, A., van Herk, M., Bartelink, H., et al. 1993. A verification procedure to improve patient set-up accuracy using portal images. *Radiother Oncol* 29(2): 253–60.

Bissonnette, J.P., Balter P.A., Dong L. et al. 2012. Quality assurance for image-guided radiation therapy utilizing CT-based technologies: A report of the AAPM TG-179. *Med Phys* 39(4): 1946–63.

Brock, K.K., Hawkins, M., Eccles, C., et al. 2008. Improving image-guided target localization through deformable registration. *Acta Oncol* 47(7): 1279–85.

Cho, P.S., Johnson, R.H., Griffin T.W. 1995. Cone-beam CT for radiotherapy applications. *Phys Med Biol* 40(11): 1863–83.

de Boer, H.C.J. and Heijmen, B.J.M. 2001. A protocol for the reduction of systematic patient setup errors with minimal portal imaging workload. *Int J Radiat Oncol Biol Phys* 50: 1350–65.

De Neve, W., Van den Heuvel, F., Coghe, M., et al. 1993. Interactive use of on-line portal imaging in pelvic radiation. *Int J Radiat Oncol Biol Phys* 25: 517–24.

De Neve, W., Van den Heuvel, F., De Beukeleer, M., et al. 1992. Routine clinical on-line portal imaging followed by immediate field adjustment using a tele-controlled patient couch. *Radiother Oncol* 24: 45–54.

Feldkamp, L.A., Davis, L.C., Kress, J.W. 1984. Practical cone-beam algorithm. *J Opt Soc Am* 1: 612–9.

Herman, M.G., Balter, J.M., Jaffray, D.A., et al. 2001. Clinical use of electronic portal imaging: report of AAPM Radiation Therapy Committee Task Group 58. *Med Phys* 28(5): 712–37.

Jaffray, D. and Carlone, M., et al. 2010. *Image-guided Radiation Therapy: Emergence of MR-Guided Radiation Treatment (MRgRT) Systems. Society of Photo Optical Instrumentation and Engineering.* San Deigo, CA: Springer.

Jaffray, D.A. and Siewerdsen, J.H. 2000. Cone-beam computed tomography with a flat-panel imager: initial performance characterization. *Med Phys* 27(6): 1311–23.

Jaffray, D.A., Siewerdsen, J.H., Wong, J.W., et al. 2002. Flat-panel cone-beam computed tomography for image-guided radiation therapy. *Int J Radiat Oncol Biol Phys* 53(5): 1337–49.

Jaffray, D.A. and Wong, J.W. 1997. Exploring target of the day strategies for a medical linear accelerator with conebeam-CT scanning capability. In: Leavitt, D. and Starkschall, G. (eds.). *Proceedings of the International Conference on the Use of Computers in Radiotherapy.* Salt Lake City, Utah: Medical Physics, 172–5.

Johns, H.E. and Cunningham, J.R. 1959. A precision cobalt 60 unit for fixed field and rotation therapy. *Am J Roentgenol* 81: 4–12.

Kitamura, K., Shirato, H., Shimizu, S., et al. 2002. Registration accuracy and possible migration of internal fiducial gold marker implanted in prostate and liver treated with real-time tumor-tracking radiation therapy (RTRT). *Radiother Oncol* 62(3): 275–81.

Lattanzi, J., McNeeley, S., Hanlon, A., et al. 2000. Ultrasound-based stereotactic guidance of precision conformal external beam radiation therapy in clinically localized prostate cancer. *Urology* 55: 73–8.

Lee, J., Stayman, J.W., Otake, Y., et al. 2012. Volume-of-change cone-beam CT for image-guided surgery. *Phys Med Biol* 57: 4969–89.

Letourneau, D., Wong, J.W., Oldham, M., et al. 2005. Cone-beam-CT guided radiation therapy: technical implementation. *Radiother Oncol* 75(3): 279–86.

Li, W., Moseley, D.J., Manfredi, T., et al. 2009. Accuracy of automatic couch corrections with on-line volumetric imaging. *J Appl Clin Med Phys* 10(4): 3056.

Moseley, D.J., White, E.A., Wiltshire, K.L., et al. 2007. Comparison of localization performance with implanted fiducial markers and cone-beam computed tomography for on-line image-guided radiotherapy of the prostate. *Int J Radiat Oncol Biol Phys* 67(3): 942–53.

Munro, P., Rawlinson, J.A. and Fenster, A. 1990. A digital fluoroscopic imaging device for radiotherapy localization. *Int J Radiat Oncol Biol Phys* 18: 641–9.

Nuver, T.T., Hoogeman, M.S., Remeijer, P., et al. 2007. An adaptive off-line procedure for radiotherapy of prostate cancer. *Int J Radiat Oncol Biol Phys* 67(5): 1559–67.

Pisani, L., Lockman, D., Jaffray, D., et al. 2000. Setup error in radiotherapy: on-line correction using electronic kilovoltage and megavoltage radiographs. *Int J Radiat Oncol Biol Phys* 47(3): 825–39.

Purdie, T.G., Moseley, D.J., Bissonnette, J.P., et al. 2006. Respiration correlated cone-beam computed tomography and 4DCT for evaluating target motion in Stereotactic Lung Radiation Therapy. *Acta Oncol* 45(7): 915–22.

Raaymakers, B.W., Raaijmakers, A.J., Kotte, A.N., et al. 2004. Integrating an MRI scanner with a 6 MV radiotherapy accelerator: dose deposition in a transverse magnetic field. *Phys Med Biol* 49(17): 4109–18.

Roland, T., Herman, J., Iordachita, I., et al. 2012. Real time image guided radiotherapy for pancreatic tumors – the concept of dual modality monitoring using kV-CBCT and Robot assisted ultrasound imaging. *Proceedings of the first joined meeting of the International Conference on Translational Research in Radiation Oncology and Physics for Health in Europe*, Geneva, Switzerland, Feb 27–Mar 2, Abstract 55.

Schwartz, D.L. 2012. Current progress in adaptive radiation therapy for head and neck cancer. *Curr Oncol Rep* 14(2): 139–47.

Shalev, S., Lee, T., Leszczynski, K., et al. 1989. Video techniques for on-line portal imaging. *Comput Med Imag Graph* 13: 217–26.

Siewerdsen, J.H. and Jaffray, D.A. 1999a. A ghost story: spatio-temporal response characteristics of an indirect-detection flat-panel imager. *Med Phys* 26(8): 1624–41.

Siewerdsen, J.H. and Jaffray D.A. 1999b. Cone-beam computed tomography with a flat-panel imager: effects of image lag. *Med Phys* 26(12): 2635–47.

Siewerdsen, J.H. and Jaffray, D.A. 2000. Optimization of x-ray imaging geometry (with specific application to flat-panel cone-beam computed tomography). *Med Phys* 27(8): 1903–14.

Siewerdsen, J.H. and Jaffray, D.A. 2001. Cone-beam computed tomography with a flat-panel imager: magnitude and effects of x-ray scatter. *Med Phys* 28(2): 220–31.

Siewerdsen, J.H., Moseley, D.J., Bakhtiar, B., et al. 2004. The influence of antiscatter grids on soft-tissue detectability in cone-beam computed tomography with flat-panel detectors. *Med Phys* 31(12): 3506–20.

Sonke, J.J., Zijp, L., Remeijer, P., et al. 2005. Respiratory correlated cone beam CT. *Med Phys* 32(4): 1176–86.

Stayman, J.W., Prince, J.L. and Siewerdsen, J.H. 2012. Information propagation in prior-image-based reconstruction. *Proceedings of the 2nd International Conference on Image Formation in X-ray CT*, Salt Lake City, Utah, June 24–27; 334–338.

Uematsu, M., Fukui, T., Shioda, A., et al. 1996. A dual computed tomography linear accelerator unit for stereotactic radiation therapy: a new approach without cranially fixated stereotactic frames. *Int J Radiat Oncol Biol Phys* 35(3): 587–92.

van Herk, M., Bruce, A., Kroes, A.P., et al. 1995. Quantification of organ motion during conformal radiotherapy of the prostate by three dimensional image registration. *Int J Radiat Oncol Biol Phys* 33(5): 1311–20.

van Herk, M. and Meertens, H. 1988. A matrix ionization chamber imaging device for on-line patient set-up verification during radiotherapy. *Radiother Oncol* 11: 369–78.

Velec, M., Moseley, J.L., Eccles, C.L., et al. 2011. Effect of breathing motion on radiotherapy dose accumulation in the abdomen using deformable registration. *Int J Radiat Oncol Biol Phys* 80(1): 265–72.

Wong, J.W., Binns, W.R., Cheng, A.Y., et al. 1990. On-line radiotherapy imaging with an array of fiber-optic image reducers. *Int J Radiat Oncol Biol Phys* 18: 1477–84.

Yan, D. (ed.). 2010. Seminars in Radiation Therapy Special Issues. Adaptive radiation therapy. *Semin Radiat Oncol* 20(2): 79–146.

Yan, D., Vicini, F., Wong, J., et al. 1997. Adaptive radiation therapy. *Phys Med Biol* 42: 123–32.

Yan, D., Ziaja, E., Jaffray, D., et al. 1998. The use of adaptive radiation therapy to reduce setup error: a prospective clinical study. *Int J Radiat Oncol Biol Phys* 41(3): 715–20.

Yin, F.F., Wong, J.W., Balter, J., et al. 2009. *TG-104: The Role of In-Room kV X-Ray Imaging for Patient Setup and Target Localization*. College Park, MD: AAPM.

Zhong, Y., Stephans, K., Qi, P., et al. 2013. Assessing feasibility of real-time ultrasound monitoring in stereotactic body radiotherapy of liver tumors. *Technol Cancer Res Treat* 12(30): 243–50

16 Breast CT

Stephen J. Glick

Contents

16.1 INTRODUCTION

Breast cancer is the second most common cancer among women, accounting for nearly 25% of new cancers diagnosed in U.S. women. In 2011, an estimated 230,480 new cases of invasive breast cancer were diagnosed, along with 57,600 new cases of *in situ* breast cancer. It is estimated that one out of every eight women will be diagnosed with breast cancer in their lifetime, and approximately 40,000 women will die from the disease every year. Unfortunately, there is no known cure for breast cancer, and the most successful means for decreasing breast cancer mortality is early detection and improved therapeutic methods.

Many randomized controlled trials have indicated that routine screen-film mammography can significantly reduce breast cancer mortality. Mammographic screening began in the mid-1980s. Since then, it has been observed that approximately 39.6% fewer women die every year due to breast cancer (range of mortality reduction over six models was 29.4%–54%) (Hendrick and Helvie 2011). For the 50-year period previous to this, the mortality rate due to breast cancer was unchanged. In the past decade, the development of full-field digital mammography, using either amorphous-Si or amorphous-Se detectors, has come to fruition. Recently, a multi-institutional clinical study funded by the American College of Radiology Imaging Network (ACRIN) involving 50,000 subjects was conducted to compare performance with screen-film and digital mammography (Pisano et al. 2005). The results of this study suggested that digital mammography can provide improved performance in certain subgroups of women, such as women aged less than 50 years, at pre- or perimenopausal stages, and with dense breast tissue. Another benefit in the development of digital mammography was that the detectors developed for digital mammography could be used in dedicated breast computer tomography (CT) systems with some modification.

Although screening mammography has been very successful in reducing breast cancer mortality, it is far from perfect. Sensitivity values for mammography that are reported in the literature range from 74% to 80% and may be lower for women with dense breast tissue (Kerlikowske et al. 1996; Rosenberg et al. 1996, 1998; Kolb et al. 1998). It is likely that performance is substantially worse for more difficult-to-image women, such as women with dense breast tissue. It was hopeful that digital mammography would substantially improve the sensitivity of screening mammography; however, a recent ACRIN-funded study designed to compare screen-film and digital mammography reported that 30% of cancers were not detected on either modality (Pisano et al. 2005). In addition to the problem of missing breast cancer lesions, studies have suggested that 55%–85% of biopsies performed to examine suspicious breast lesions are negative (Meyer et al. 1990; Parker et al. 1991; Elvecrog et al. 1993; Opie et al. 1993). This low positive predictive value (PPV3) of findings sent to biopsy is problematic because it ultimately drives up healthcare costs, as well as causes unnecessary patient anxiety. In fact, one of the driving forces in the recent controversial recommendations from the 2009 U.S. Preventative Services Task Force (USPSTF) to limit mammograms for women in their 40s was that younger women face increased risks of harm from screening such as anxiety, false-positive findings, and unnecessary biopsies (Nelson et al. 2009).

In addition to somewhat limited sensitivity and specificity, screening mammography also is limited with respect to the tumor size that can be reliably detected. Michealson et al. (2002) showed that there exists a correlation between tumor size and lethality, and this correlation can be expressed by a simple exponential equation. For example, Figure 16.1 shows this relationship based on a population study reported by Tabar et al. (2000). This relationship clearly demonstrates the importance of early detection for screening of breast cancer. For example, the average-sized breast tumor detected with conventional mammography is on the order of 10 mm. From Figure 16.1, it can be ascertained that a new imaging modality that could reliably detect breast tumors of size 3 mm would be able to increase survival rates by approximately 10%.

One of the major limiting problems with mammography is that it attempts to portray a three-dimensional (3D) object (i.e., the breast) using a two-dimensional (2D) image. Thus, the recorded mammographic image represents the 3D breast superimposed onto a 2D plane, with normal anatomical breast structure combining with important diagnostic information (such as breast lesions) in such a way as to make the extraction of diagnostic information difficult for radiologists. Using statistical decision theory, Burgess et al. (1999, 2001) have shown that the detection of lesions in such a structured background is limited by both quantum measurement noise and anatomical noise (due to the structured background of the breast). It has been shown that for mammography, the effect of anatomical noise on lesion detection is 30–60 times more important than the effect of x-ray quantum noise. Metheany et al. (2008) studied the anatomical noise in breast CT and showed that similarly to mammography, it can be characterized as having a power law spectrum, but with lower exponent. From this observation, one would expect improved breast tumor detection with breast CT.

One recently developed approach to obviating the problem of structured overlap is digital breast tomosynthesis (DBT). DBT uses a limited angle tomography geometry to generate tomographic slices through the compressed breast. The principle of tomosynthesis is not new, and in fact it was first discussed in the 1930s by Ziedesdes Plantes (1932). Niklason et al. (1997) reported the use of a breast tomosynthesis system with a stationary amorphous-Si flat-panel detector to image breast phantoms and breast specimens. Since then, DBT has been commercialized by many companies and has received approval by the U.S. Food and Drug Administration (FDA). The geometry of DBT is very similar to conventional mammography and thus can be implemented by a relatively simple upgrade. However, because DBT exhibits a rather large blurring along the direction perpendicular to the detector, it does not provide truly isotropic 3D breast images. Nevertheless, preliminary clinical studies have suggested that DBT can provide improved visualization of masses and areas of architectural distortion compared with conventional mammography. However, there have been some reports that visualization of microcalcifications is inferior on DBT (Poplack et al. 2007).

In this chapter, the focus is on the imaging of the breast with x-ray CT, where CT is defined as tomographic imaging using projection data acquired with complete angular sampling. There are two avenues of research using CT of the breast. The first avenue is to use a conventional, whole-body CT scanner similar to that found in most radiology clinics. Imaging of the breast with a whole-body CT scanner has focused on diagnostic applications (i.e., analyzing suspicious lesions visualized on mammography) and typically uses iodinated contrast agent to improve tumor contrast. The second avenue of research is the use of smaller CT scanners dedicated for imaging of the breast. Dedicated breast CT has several advantages, including lower radiation dose (very little radiation dose is given to the body), higher spatial resolution, and higher cost efficiency.

Dedicated breast CT is currently being researched by many investigators (Boone et al. 2001; Chen and Ning 2002; Boone 2004; McKinley et al. 2004; Tornai et al. 2005; Gong et al. 2006; Kalender et al. 2011). A few experimental prototypes have been fabricated and preliminary clinical studies are ongoing (Partain et al. 2007; Lindfors et al. 2008; Prionas et al. 2010;

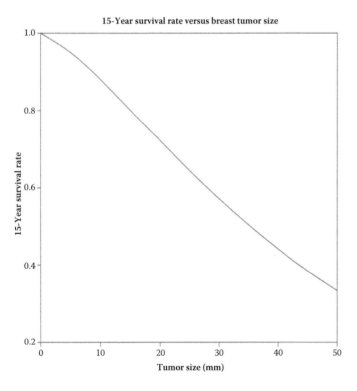

Figure 16.1 Graph showing the 15-year survival rate as a function of the size of breast tumor upon diagnosis based on data reported by Tabar, L. et al. (*Surg Oncol Clin North Am*, 9, 33–77, 2000).

O'Connell et al. 2010). These studies are discussed later in the chapter. It is currently unclear as to how dedicated breast CT will be used for patient care; more clinical studies are needed to investigate this question. In addition to the potential of dedicated breast CT as a screening tool for imaging asymptomatic women, there are several diagnostic clinical applications that might prove to be helpful in the diagnosis and treatment of breast cancer. One important potential application for breast CT might be for lesion analysis to further investigate suspicious lesions found on mammography. The current approach for lesion analysis is the use of diagnostic mammography including spot compression, modified projection views, and magnification mammography. In addition, other modalities such as ultrasound and breast magnetic resonance imaging (MRI) are used. Dedicated breast CT could be produced at low enough cost such as that it might be feasible to have a breast CT system located in the breast imaging clinic. In this situation, breast CT could be used as an adjunct or possibly as a replacement for diagnostic mammography. In particular, breast CT should be efficient at confirming suspected summation artifacts seen on screening mammography.

Another potential application of dedicated breast CT would be for the assessment of tumor response to neoadjuvant chemotherapy, typically used to shrink a tumor that is inoperable in its current state. It is desirable to predict the response of neoadjuvant chemotherapy to discontinue the administration of possibly harmful drugs that are not effective. Breast CT is a high-resolution modality and could be successfully used to monitor early response (or nonresponse) to chemotherapeutic treatment, as well as to assess residual disease after completion of therapy.

Dedicated breast CT also could be used for better staging of known breast cancer. Many patients diagnosed with early stage breast cancer choose to undergo breast conservation surgery (BCS), otherwise known as lumpectomy. During the BCS procedure, the surgeon attempts to remove the malignant tissue with a surrounding margin of healthy tissue. After the surgical excision of the tumor, the specimen is submitted to pathology for evaluation to determine whether tumor margins are negative. Surprisingly, positive margins indicated by pathology and surgical re-excision procedures are currently too commonplace. It has been estimated that approximately 20%–70% of BCS procedures need to be repeated due to positive margins. (Fleming et al. 2004; Dillon et al. 2007; Jacobs 2008). It has been hypothesized that staging information obtained from high-resolution breast CT might be able to reduce the number of positive margins in BCS procedures.

Other potential applications for dedicated breast CT include using it to guide interventional procedures such as robotic biopsy or radiofrequency ablation.

16.2 PREVIOUS STUDIES WITH BREAST CT

The first documented report describing CT imaging of breast mastectomy specimens was published in 1976 by Reese and colleagues (Reese et al. 1976; Gisvold et al. 1979). Results from these studies motivated General Electric (GE) to fabricate an experimental prototype breast CT scanner called CT/M. This breast CT scanner was installed at both the Mayo Clinic in

Rochester, MN, and at the University of Kansas College of Health Sciences in Lawrence. This GE prototype used a fan beam geometry to acquire 1-cm-thick CT slices in approximately 10 s. The breast was imaged in the prone position by having the woman lie on a canvas table with a hole for the breast. To make the x-ray fluence more uniform across the detector, the breast was placed in a holder containing continuously flowing heated water. Projection data were reconstructed into 127 × 127 voxel matrices, with each square voxel equal to 1.56 mm and with slice thickness equal to 1 cm. Patients were imaged before and after the administration of iodinated contrast agent. Two studies were conducted characterizing the performance of the GE CT/M scanner. Chang et al. (1978, 1979) studied 655 patients separated into groups of asymptomatic women and women with suspicious lesions. Gisvold et al. (1979) imaged 724 patients with suspicious lesions on mammography. Both studies presented promising results, suggesting that CT breast imaging (CTBI) has high sensitivity for detecting breast cancer. Lesions as small as 6–8 mm were visualized. The reported specificity of breast CT was less impressive with different types of benign tissue appearing to have similar uptake properties as malignant tissue. Ultimately, GE decided against further development owing to concerns about high dose to the breast, limited spatial resolution (1.56-mm voxels), and high cost.

In the subsequent years after the development of the GE CT/M scanner, many clinical studies were conducted to evaluate the use of conventional whole-body multidetector CT as a diagnostic tool to evaluate breast lesions (Hagay et al. 1996; Sardanelli et al. 1998; Akashi-Tanaka et al. 2001; Uematsu et al. 2001; Nakahara et al. 2002; Inoue et al. 2003; Kim and Park 2003; Nishino et al. 2003; Miyake et al. 2005; Shimauchi et al. 2006; Yamamoto et al. 2006). All of these studies administered iodinated contrast agent to the patient before imaging. The primary application studied was for staging breast cancer and determining its extent before breast conservation surgery. Whether CT could be used to differentiate malignant and benign lesions also was studied. The literature seems to suggest that whole-body, contrast-enhanced CT imaging of the breast for these applications provides excellent sensitivity; however, specificity is somewhat limited. Because multidetector, whole-body CT scanners are present in most radiology clinics, there is some appeal to using them to image the breast. However, because the entire thorax is exposed, the radiation dose to the patient is substantially higher than in mammography. In addition, spatial resolution is somewhat less than desired. Within the past decade, high-resolution flat-panel detectors have been developed for mammography and radiographic applications. This development has motivated investigators to look at the feasibility of incorporating these flat-panel detectors into cone beam CT scanners dedicated to imaging of the breast. In addition to the lower radiation dose to the thorax, dedicated cone beam breast CT using flat-panel detectors have several advantages over imaging the breast with conventional whole-body CT. Flat-panel detectors currently used for dedicated breast CT have small pixel size; subsequently, dedicated breast CT systems have substantially better spatial resolution than whole-body CT. However, current breast CT systems have inferior spatial resolution compared with mammography, and it is uncertain whether future breast CT systems will be able to match the high resolution of mammography. For a few reasons, high spatial resolution is important in imaging of the breast. First, very small microcalcifications visualized in the breast provide important diagnostic indication for ductal carcinoma *in situ* (DCIS). In fact, it has been estimated that 29%–48% of nonpalpable carcinomas are visible based on microcalcifications alone (Wolfe 1974; Feig et al. 1977; Bjurstam 1978; Frankl and Ackerman 1983). Thus, one can conclude that high spatial resolution is required to maximize detection of DCIS. Another reason why high spatial resolution is desired in imaging of the breast is to clearly examine tumor margins. Irregular-shaped breast masses with spiculated margins indicate a high probability of malignancy, and visualization of these very thin spiculations is typically a task requiring high spatial resolution.

In the next sections, the designs of current dedicated, cone beam breast CT systems are discussed in detail.

16.3 DEDICATED CONE BEAM BREAST CT SYSTEM DESIGN

A dedicated cone beam breast CT design is illustrated in Figure 16.2. The patient lies on a table and the breast hangs in the pendant position through a hole in the table. Unlike mammography, breast compression is not needed, although it is possible to use some support to minimize breast motion. For example, to reduce nonuniform x-ray fluence on the detector, a slight compression upward to the nipple could be applied. Beneath the table, the x-ray tube and flat-panel detector rotate around the breast, collecting truncated cone beam projection images at many views. These projections can then be reconstructed to obtain an isotropic 3D representation of the breast. A handful of academic investigators, and a few companies, are currently investigating similar dedicated breast CT systems as described here. One group at the University of California–Davis has built two experimental prototypes and is currently performing clinical testing (Boone et al. 2001; Boone 2004).

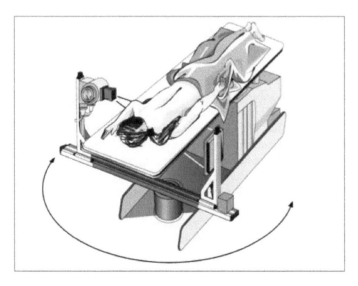

Figure 16.2 Illustration of a dedicated breast CT system. The patient lies on a table, and the breast hangs in the pendant position through a hole in the table.

Another group developing breast CT is at the University of Rochester in New York (Chen and Ning 2002). This group also is conducting clinical testing and is currently commercializing the technology (Koning Inc., Rochester, NY). A group at Duke University in Durham, NC, is working on a combined single-photon emission computed tomography (SPECT)/CT breast scanner using an innovative gantry that can potentially maximize the coverage of breast tissue near the chest wall (McKinley et al. 2004a; Tornai et al. 2005). Another group at The University of Texas MD Anderson Cancer Center in Houston, TX, is investigating the use of magnification breast CT that can display lesions of interest at very high resolution (Chen et al. 2009a). Other groups investigating breast CT include University of Massachusetts (UMASS, Worcester), University of Erlangen-Nuremberg (Germany), and Emory University (Atlanta, GA). The group at UMASS has developed a bench-top prototype breast CT system to optimize imaging parameters (Gong et al. 2006; Glick et al. 2007), and they are exploring other physics-based breast CT-related issues (Vedantham et al. 2013). This group is also in the process of conducting clinical trials to investigate the performance of breast CT. Although there are many groups investigating the feasibility of dedicated breast CT, there are still many uncertainties as to the optimal design and acquisition parameters. Some of these issues are discussed in the following.

16.3.1 PATIENT TABLE AND CT GANTRY

One of the important advantages of dedicated breast CT over conventional mammography is that breast compression is not required. Two of the primary reasons for compression in mammography are to reduce breast superposition and improve tumor contrast. Of course, image reconstruction inherently eliminates the superposition problem and improves contrast, thus compression is not needed. Dullum et al. (2000) have reported that more than 50% of women experience moderate or greater discomfort from compression. In a study by Linfors et al. (2008), women undergoing dedicated breast CT were asked to compare their comfort on breast CT to that of mammography using a continuous 10-point scale. It was reported that breast CT was significantly more comfortable than conventional mammography ($P < .001$).

Breast tissue can occasionally extend into the chest wall and the axilla, as well as laterally past the anterior axillary line and down into the upper abdomen. It is for this reason that breast compression is used to slightly pull breast tissue away from the chest wall. In breast CT, gravity is used to help pull the pendant breast away from the chest wall; however, it is still unclear whether an adequate amount of breast tissue near the chest wall can be imaged. All of the prototype breast CT scanners developed have table designs that allow for easier imaging of tissue near the chest wall. Boone et al. (2004) have used a flexible neoprene hammock to create a swale in the table, whereas Crotty et al. (2007) have discussed various innovative table designs for maximizing chest wall coverage. In addition to the table design, another important aspect to maximizing chest wall coverage is the minimization of the distance between the bottom of the table and the x-ray tube focal spot. This requires an x-ray tube with focal spot positioned near the physical end of the x-ray tube housing. Various approaches for x-ray tube selection are discussed in the following.

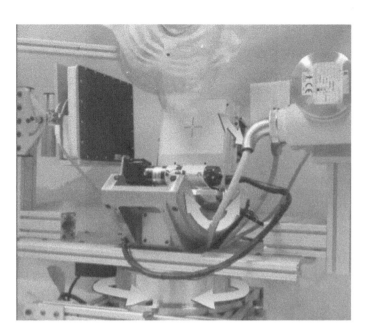

Figure 16.3 The gantry system being developed at Duke University for dedicated breast SPECT/CT imaging. This system features a goniometer for acquiring SPECT and CT scans with different orbits.

The various prototype breast CT scanners developed to date have different CT gantry systems. The primary design goals for a breast CT gantry are to (1) provide good sampling coverage to minimize cone beam CT artifacts; (2) maximize coverage of breast tissue near the chest wall; and (3) allow fast imaging, preferably within a breath hold to minimize motion artifacts. The investigators at Duke University have been developing a specialized gantry (used for both SPECT and CT imaging) using a goniometer that allows acquisition of cone beam projection images using complex noncircular type orbits (Tornai et al. 2005; Figure 16.3). To minimize patient motion, it is desired to perform CT scanning in less than the time required for a patient to hold her breath. This requires both a gantry that moves very fast (i.e., 360° rotation in 10–15 s) and a detector with fast readout. Thus, cable management of breast CT systems is challenging. The early prototypes at University of California–Davis accommodated cables using commercial chain housing positioned on a low-friction, high-density polyethylene platform. Another CT gantry option is a slip-ring, a technique for forming an electrical connection to the rotating x-ray tube and detector assembly. The prototype scanners developed at University of Rochester have used a very fast, slip-ring gantry.

16.3.2 X-RAY TUBE AND X-RAY FILTER

The primary design goals for the x-ray tube in dedicated breast CT are (1) compact size to fit underneath the patient table, (2) focal spot positioned near the end of the x-ray tube housing, (3) powerful enough to collect many projections (possibly many CT scans for dynamic protocols) in a short period, and (4) options for operating in pulsed mode. The requirement of compact size and focal spot placement are needed to maximize coverage of the chest wall by allowing the focal spot to be very close to the bottom of the table. The University of California–Davis prototype breast CT scanner uses an industrial, water-cooled,

stationary anode end-windowed x-ray tube with the physical end of the tube only 47.5 mm from the focal spot. The University of Rochester uses a mammographic x-ray tube (maximum 49 kVp) that also has a relatively small distance from the focal spot to the end of the tube housing. Zhang et al. (2005) are investigating carbon nanotube x-ray sources that could potentially be very useful for dedicated breast CT. These x-ray sources are small, have a high heat-load capacity, and can be used in multisource arrays with multiplexing.

In addition to the inherent filtering of the x-ray beam due to the x-ray housing, external filters are often applied with the goal of shaping the x-ray spectrum. McKinley et al. (2004) have shown that x-ray filters can be applied to obtain a more optimal "quasi-monochromatic x-ray beam." They concluded that using a tungsten (W) anode x-ray tube with kVp setting between 50 and 70, along with an x-ray filter consisting of material with $Z = 57$–63, produced optimal spectra in terms of signal-to-noise ratio (SNR) per dose. Glick et al. (2007) also have studied the impact of x-ray spectral shape on the ideal observer SNR in breast CT, exploring such factors as kVp settings, filter material types, and filter thickness.

16.3.3 BREAST CT DETECTORS

An ideal breast CT detector should have (1) high resolution to allow visualization of small microcalcifications and fine tumor margins, (2) low image lag to allow fast acquisition speeds, and (3) low electronic noise so as to allow imaging at low exposure per projection view. Most current prototypes use indirect conversion flat-panel detectors, consisting of cesium iodine (CsI) phosphor coupled to a pixelized array of thin-film transistors and photodiodes on an amorphous-Si substrate. Both the University of California–Davis and University of Rochester experimental prototype breast CT scanners use a 30 cm × 40 cm amorphous-Si flat-panel with 600-µm-thick CsI, and an effective pixel dimension of 194 µm × 194 µm. However, to read out the data at 30 frames per s (fps) and perform the CT acquisition in less than 15 s, 2 × 2 binning has to be implemented that effectively increases the pixel dimension to 388 µm × 388 µm. Due to magnification, the reconstructed voxel size can be somewhat smaller than this value.

Because it is desired to keep the radiation dose in breast CT approximately equivalent to that of mammography, CT acquisitions can have a low x-ray fluence incident on the detector. Thus, the detector's performance at low exposure can be important. One feature that is available on breast CT detectors is a dynamic gain option (Roos et al. 2004). With this option, pixel gain can vary across the flat-panel detector, so that it will be higher in areas of low x-ray fluence (e.g., directly behind the center of the breast) and lower in areas of high x-ray fluence (e.g., toward the periphery of the breast).

16.3.4 RADIATION DOSE FROM BREAST CT

Studies using Monte Carlo simulation software to analyze the mean glandular dose from uncompressed breast CT have been reported previously (Boone et al. 2004; Thacker and Glick 2004; Sechopoulos et al. 2008). These studies have provided the glandular dose coefficients needed to compute radiation dose for a range of breast composition and size, as well as kVp settings for the truncated cone beam geometry. The dose distribution within the breast was also reported, and it was observed that the dose in CT is more uniformly distributed throughout the breast compared with mammography. This suggests that the maximum dose in the breast can be lower in CT than mammography. In addition, because CT uses more energetic photons, the biologic effect should be lower than in mammography. Boone et al. (2005) have provided a comprehensive summary of technique factors and their relationship to radiation dose determined from both physical measurements and computer modeling. Sechopoulos et al. (2008) have simulated the dose to other organs (besides the breast) in breast CT. Their study reported that the lungs, heart, and thymus received the highest dose besides the breast and skin; however, this dose was very low compared with the glandular dose. An early clinical study performed at University of California–Davis reported an initial clinical experience by imaging 79 women, with mean glandular dose reported in the range of 2.5–10.3 mGy (Lindfors et al. 2008). Another early clinical study at University of Rochester reported imaging 23 women with mean glandular dose in the range of 4–12.8 mGy (O'Connell et al. 2010).

16.3.5 IMAGE RECONSTRUCTION FOR BREAST CT

Current experimental breast CT prototypes use a truncated cone beam geometry that requires cone beam reconstruction algorithms. The most common cone beam reconstruction method is filtered back-projection (FBP) as described by Feldkamp et al. (1984). The Feldkamp FBP algorithm is not an exact solution and becomes more problematic in reconstructing regions farther away from the central plane (i.e., plane at the center of the cone angle) (Vedantham et al. 2011). Thus, it is possible that reconstruction artifacts could arise more toward the anterior region of the pendant breast. Wang and Ning (1999) have shown that cone beam acquisition orbits such as the "circle plus line orbit" that satisfy Tuy's cone beam sampling requirements can reduce these cone beam reconstruction artifacts.

Another approach to breast CT reconstruction is the use of iterative reconstruction methods. These methods are based on formulating the reconstruction problem as a system of linear equations and then developing iterative algorithms that solve this system of equations using either deterministic or statistical criterion (Fessler 2000). Iterative reconstruction methods have several advantages over FBP methods. They typically model the noise in the projection measurements more accurately and thus can have lower noise for the same image resolution. This advantage can be translated into performing breast CT at lower dose. Another advantage of iterative reconstructions is that they provide a framework for the correction of degradations that occur in the imaging process. For example, modeling of the focal spot blur, or detector blur can be included into the iterative reconstruction algorithm, thereby correcting for these blurring effects.

Tornai et al. (2005) have reported the use of an ordered subsets transmission iterative algorithm (OSTR) for the multimodality breast CT/SPECT scanner being developed at Duke University. Studies by Makeev and colleagues (Makeev et al. 2012; Makeev and Glick 2013) investigated benefits of a penalized maximum-likelihood iterative reconstruction method for breast CT.

16.4 PERFORMANCE OF BREAST CT SYSTEMS

16.4.1 SPATIAL RESOLUTION

High spatial resolution is important in breast CT for detection of small microcalcifications and for the accurate visualization of spiculations and irregular-shaped malignant tumor boundaries. One of the primary advantages of dedicated breast CT systems using flat-panel detectors is their higher resolution compared with imaging of the breast with whole-body conventional CT. Spatial resolution in flat-panel, cone beam breast CT is limited by the x-ray tube focal spot size, the inherent detector blurring properties, and the reconstruction filter and voxel size. Determining the optimal spatial resolution desired for breast CT is a complicated task and involves determining the required trade-off between resolution and noise. Most current prototype breast CT scanners use detectors with pixel size 194 μm; however, 2 × 2 binning also is performed for purposes of faster readout, thus making the effective pixel size 388 μm. Taking into account magnification, reconstructed voxel sizes can be somewhat lower, and one prototype system (Koning, West Henrietta, NY) uses a 273-μm voxel size. A few studies have investigated spatial resolution of breast CT using both simulations and experimental measurements. Kwan et al. (2007) have reported on spatial resolution using a breast CT system with continuous x-ray tube operation during the acquisition. They reported little variation in spatial resolution with reconstructed matrix size or cone angle; however, resolution did degrade radially from the axis of rotation. Yang et al. (2007) conducted further computer simulation studies and suggested that this worsening of resolution toward the periphery could be corrected by using a pulsed x-ray source, or by increasing the frame rate and collecting more projection views during the gantry rotation. The latter solution could be problematic in that the exposure per view would reduce, thereby increasing the impact of electronic noise.

16.4.2 DEGRADATION OF IMAGE QUALITY DUE TO SCATTER

One of the primary concerns in current breast CT prototypes is how to reduce the degrading effects of the rather large scattered radiation component in the resulting images. Siewerdsen et al. (2001) have reported that the measured detected scattered radiation with flat-panel cone beam CT imaging increases with increasing cone angle, leading to image artifacts, a reduction in image contrast, and quantitative errors in measured CT numbers. Kwan et al. (2005) have measured scatter properties on a prototype CT breast imaging scanner and reported scatter-to-primary ratios of up to 100% (depending on breast size). They concluded that scattered radiation in CTBI would likely affect image quality. Chen et al. (2009c) also have evaluated the characteristics of scatter using both experimental measurements and Monte Carlo simulation studies of a dedicated CTBI system and confirmed the large scatter component reported by Kwan et al. (2005). Based on these characterization studies, it appears as if scattered radiation can have important effects on image quality in flat-panel CTBI. If this is the case, it will be necessary to develop methods for reducing the scatter in CTBI.

Several approaches for minimizing the effect of scatter have been proposed. A simple tactic is to modify the measurement geometry, for example, using a larger air gap between the object and detector (Neitzel 1992) or limiting the field of view. Ning et al. (2004) has proposed a beam-stop array (BSA) method where preliminary scout projection views are acquired with an array of small circular lead disks between the x-ray source and object. Although this approach can provide for an estimate of the patient-dependent scatter distribution, it does have some disadvantages. Because the BSA method requires additional acquisitions, it results in patients having to hold their breath for longer time. In addition, the BSA method requires the administration of a higher dose and can result in a noisy correction for large breasts.

Another approach for reducing scatter is the use of an antiscatter grid (Wiegert et al. 2004). Antiscatter grids are routinely used in conventional mammography to reduce scatter. The most common mammography grids consist of one-dimensional (1D) linear focused arrays of Pb lamellae, with C fiber interspace material. Siewerdsen et al. (2004) have explored the influence of antiscatter grids on image quality for cone beam CT of various anatomical sites. Experimental measurements using various linear grids were performed on a flat-panel cone beam CT system, and the influence of these grids on soft-tissue detectability was explored. This study concluded that although linear antiscatter grids did reduce scatter artifacts and improve subject contrast, minimal improvement (if any) in image quality was observed for most circumstances. Siewerdsen et al. (2004) suggested that although the antiscatter grids did increase contrast due to the reduction of scatter, they also substantially decreased primary radiation (primary transmission factors were 60%–70%), thereby resulting in increased noise. That is, the antiscatter grid imparted a trade-off between improved contrast and increased image noise. This study used 1D linear grids and was focused on cone beam CT for guidance of radiation therapy; thus, the conclusion could have been different for the use of focused 2D antiscatter grids designed specifically for breast CT.

16.4.3 OPTIMAL OPERATING CONDITIONS

Dedicated cone beam breast CT systems have many acquisition and design parameters that can affect image quality. Some of these parameters include kVp setting, x-ray tube filter, scintillator thickness, electronic noise, detector pixel size, reconstructed voxel size, reconstruction filter, and imaging geometry. Many studies have been conducted to investigate optimal imaging conditions of breast CT. Boone et al. (2001) have compared contrast-to-noise ratio (CNR) at equivalent dose for 80-, 100-, and 120-kVp spectra and proposed the use of 80 kVp. Chen and Ning (2002) have used theoretical studies to compute dose efficiency (SNR/dose) versus keV. They showed that 30–40 keV provides the optimal energy range for dose efficiency. The Koning prototype system developed by Chen and Ning uses a 49-kVp W anode spectra. Weigel et al. (2011) have used a contrast-to-noise figure of merit to conclude that breast CT should be performed at tube voltages of 50 kVp and higher. McKinley et al. (2004) have proposed the use of a quasi-monochromatic x-ray spectra by using x-ray filters with atomic numbers in the range of Z = 51–63. Glick et al. (2007) have proposed use of a theoretical framework for optimization of breast CT that uses the ideal observer SNR as the figure of merit. This framework was used to evaluate performance using kVp spectra ranging from 30 to 100 kVp with x-ray filters using various materials, with Z ranging from 10 to 70. Results showed that the optimal kVp setting for various tasks ranges from 40 to 70 kVp depending on the selection of filter and breast size. This ideal observer SNR methodology was proposed for optimization of other breast CT parameters.

16.4.4 DETECTION OF MICROCALCIFICATIONS

The visualization of microcalcifications is very important for the accurate detection of DCIS. With conventional mammography, 90% of DCIS is identified on the basis of suspicious microcalcifications (Dershaw et al. 1989). Because 14%–50% of all DCIS eventually becomes invasive (Kopans 1998), detection of DCIS is important and can contribute to a decreased breast cancer mortality rate.

Lai et al. (2007) have conducted an experimental study to evaluate visibility of microcalcifications on breast CT. They used a bench top breast CT system with a similar detector system to other prototype breast CT systems currently being evaluated (Varian PaxScan 4030CB, 388-μm effective pixel size after 2 × 2 binning). Experimental phantoms were constructed and filled with gelatin to simulate uncompressed breasts with composition of 100% glandular tissue. Eight different sizes of calcium carbonate grains were used to emulate microcalcifications, ranging in size from 180 to 200 μm and 355 to 425 μm in diameter. The results suggested that visibility of microcalcifications increased with dose but decreased with breast size. With a 50% detection threshold, the minimum detectable sizes in the 14.5-cm diameter phantom were 355, 307, and 275 μm for 6, 12, and 24 mGy mean glandular dose, respectively.

Gong et al. (2004) have conducted an observer study using receiver operating characteristic (ROC) analysis with computer-simulated breast CT images to evaluate microcalcification detection accuracy. They used a 50% glandular, 50% adipose tissue breast composition model and a mean glandular dose of 4 mGy. Various detector pixel sizes were studied. The results suggested that microcalcifications with a diameter more than 175 μm can be detected by using a detector with 100- or 200-μm pixel size. One reason for the discrepancy between the conclusions of these two studies could be that the background composition was modeled differently. This suggests that detection of microcalcifications that are embedded within adipose tissue is an easier task than detection of microcalcifications embedded in dense fibroglandular tissue. Although the Gong study did suggest that 100-μm pixel size would perform well, this would require a large increase in the amount of detector data needed to be read out and would be problematic if acquisition times were lengthened because patient motion also could have a blurring effect on detection of small microcalcifications.

16.4.5 CLINICAL TRIALS

To date, a few clinical trials to evaluate breast CT have been conducted, and more are being planned. The group at the University of California–Davis has reported on a pilot study with 10 healthy volunteers and a study with 69 women in the BIRADS 4 and 5 category (Lindfors et al. 2008). In the latter study, breast CT images and corresponding screen-film mammograms were

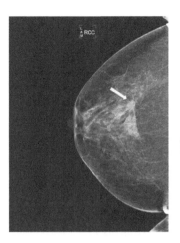
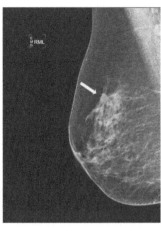

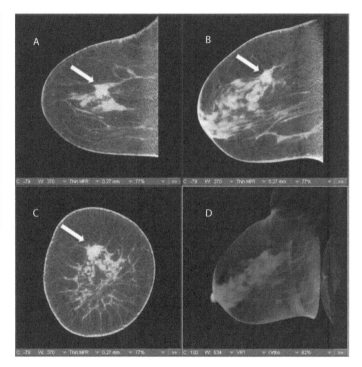

Figure 16.4 A subject imaged at the University of Rochester's Highland Breast Imaging Center with both mammography and breast CT. CC and MLO view mammograms of the right breast showing an irregular high-density mass with indistinct margins. (Courtesy of Avice O'Connell, University of Rochester Medical Center.)

read by an experienced mammographer and subjectively rated for lesion visualization. In addition, comfort level was rated and compared between the two modalities. The results indicated that masses were significantly more conspicuous on breast CT images, whereas mammography outperformed breast CT for the visualization of microcalcification lesions. Women found breast CT to be significantly more comfortable than mammography, most likely due to the lack of breast compression. However, the breast CT table required women to arch their back into the scanner, which some felt was uncomfortable. One potential clinical application of breast CT is as a diagnostic tool to reduce recalls due to summation artifacts present in mammography. In this study, two lesions that were visualized on mammography but not breast CT turned out to be summation artifacts. The authors reported limitations in visualizing the pectoralis musculature and axillary tail on breast CT but suggested that refinements in the CT table to improve visualization of the chest wall are being worked on.

Another prospective breast CT patient study was performed at the University of Rochester's Highland Breast Imaging Center (O'Connell et al. 2010). In this study, 23 BIRADS 1 and 2 patients were imaged using a breast CT experimental prototype. Breast coverage, radiation dose, and subjective image quality were evaluated and compared with mammography. Figures 16.4 and 16.5 are images from a representative patient in this cohort. Figure 16.4 illustrates the mammograms of the right breast [craniocaudal (CC) and mediolateral oblique (MLO) views], showing an irregular high-density mass with indistinct margins. Figure 16.5 shows sagittal, axial, and coronal noncontrast breast CT slices through the center of the mass, showing an irregular mass with indistinct margins. Also shown is a 3D volume rendering showing the whole-breast view. Histopathologic analysis showed invasive ductal carcinoma.

From these 23 patients, 90% (60/67) of benign findings (defined as masses, calcifications, clips, or saline implants) that were observed on mammography also were found on breast CT. All findings that were observed on mammography but not CT were microcalcifications. In total, 86.5% (45/52) of

Figure 16.5 Breast CT images acquired with the experimental Koning system of the same subject as shown in Figure 16.4. Sagittal (A), axial (B), and coronal (C) CT slices through the index mass (as indicated by the arrows) show an irregular mass with indistinct margins. Also shown is a 3D volume-rendering display (D). (Images courtesy of Avice O'Connell, University of Rochester Medical Center.)

microcalcification findings were observed on both mammography and CT. It is important to note that unlike the University of California–Davis breast CT system, the x-ray source was not continuously on, but rather on for 8-ms pulses. Thus, it is likely that spatial resolution near the breast periphery is higher. The average glandular dose in the patient cohort for mammography ranged from 2.2 to 15 mGy, whereas for breast CT the average glandular dose ranged from 4 to 12.8 mGy. The FDA limits the average glandular dose for two-view screening mammography to 6 mGy; however, there is no limitation for diagnostic mammography, which typically involves more than two views. It is likely that breast CT will initially be used for diagnostic workup; thus, the radiation dose for CT should be similar or lower than diagnostic mammography. It was observed that mammography had better coverage of the axilla and axillary tail than breast CT, but coverage with CT was significantly better in the lateral, medial, and posterior aspects of the breast. The chest wall in CT was consistently visualized; in fact, ribs were sometimes observed.

16.5 FUTURE IMPROVEMENTS TO DEDICATED BREAST CT

16.5.1 IODINATED CONTRAST-ENHANCED BREAST CT (CE-BCT)

Over the last decade, the use of magnetic resonance (MR) breast imaging with intravenously administered gadolinium diethylene triamine pentaacetic acid (Gd-DTPA) contrast agent has played an important role in the diagnosis of breast cancer. Although not promoted as a primary screening tool for the general population, CE-MR breast imaging has been suggested for use in a number of indications. Clinical studies also have been conducted to evaluate breast imaging using conventional whole-body CT for many of these same indications. Some of these studies used modern multidetector CT scanners, and all used a protocol involving the intravenous infusion of nonionic iodinated contrast media. Reports indicate similar results to CE-MR breast imaging, that is, high sensitivity but somewhat variable specificity. Perhaps this is not surprising, because both CE-CT and MR use vascular contrast agents that take advantage of the same mechanism: leaky blood vessels formed through tumor angiogenesis.

A noted shortcoming of CE-MR breast imaging is limited specificity, a shortcoming that is primarily attributed to the observation that both benign and malignant lesions can show contrast enhancement. Most reports suggest that CE-MR breast imaging has very high sensitivity for the detection of invasive carcinoma. However, it might be that MR is too sensitive, making it sometimes hard to distinguish between malignant and benign lesions. The problem in distinguishing between malignant and benign lesions might be exacerbated by the nonlinear contrast (with Gd) sensitivity observed with MR (Partain et al. 2007). It is possible that the linear contrast sensitivity of CT might allow a better differentiation between malignant and benign lesions

Prionas et al. (2010) have published the first dedicated breast CT study to evaluate contrast enhancement of malignant and benign lesions and to compare their conspicuity. In this study, 46 women of mean age 53.2 years with BIRADS 4 and 5 lesions were imaged with CE-BCT. The protocol consisted of four sequential breast CT scans imaging one breast at a time before and after administration of contrast agent. The average delay between injection of iodinated contrast agent and postcontrast scanning was 96 s. Figure 16.6 is an example of the imagery obtained in this study. Figure 16.6A shows an implant-displaced

craniocaudal mammogram with an obscured mass in the midbreast. Figure 16.6B shows the precontrast sagittal breast CT image, and Figure 16.6C shows the postcontrast CT image with iodine uptake in the invasive ductal carcinoma.

Two radiologists subjectively scored conspicuity using a continuous scale of 1 to 10. In addition, quantitative analysis was conducted by computing the mean and standard deviation of volumes of interest (VOIs) within the lesion and normalizing to adipose tissue intensity. Results suggested that malignant lesions were visualized better on CE-CBT than on noncontrast BCT ($P < .001$) or mammography ($P < .001$). Malignant lesions manifested on mammography as only microcalcifications were observed better on CE-CBT than on noncontrast BCT ($P < .001$) and were observed similarly to that of mammography. It was observed that 29 malignant lesions enhanced with 55.9 Hounsfield units (HU) ± 4.0 (standard error), whereas 23 benign lesions had mean HU of 17.6 ± 2.8. An ROC analysis of lesion enhancement provided an area under the ROC curve of 0.876.

This study with a small patient cohort showed that iodinated CE-BCT shows promise as a tool for differentiating malignant and benign tissue. Further studies are needed, especially to compare performance of CE-CBT to that of CE breast MR.

16.5.2 BREAST CT WITH PHOTON COUNTING DETECTORS

One particular promising technology for breast CT is photon counting detectors (Le et al. 2010; Kalender et al. 2011; Shikhaliev and Fritz 2011; Ding et al. 2012; Ding and Molloi 2012). It is expected that the next generation of x-ray detectors for digital radiography and CT will have the capability of counting individually measured photons and recording their energy. Unlike x-ray detectors operating in an energy-integrating mode, photon counting detectors can record and analyze each individual x-ray interacting within the detector. However, due to the high-count rate (i.e., x-ray flux incident on the detector) typically present in x-ray CT, it has historically been impossible

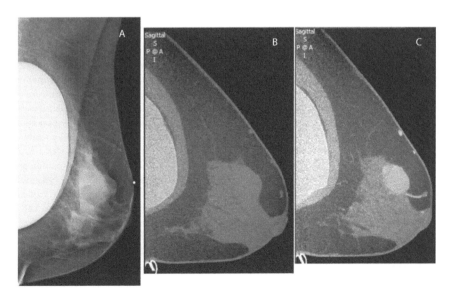

Figure 16.6 A subject imaged at the University of California–Davis with iodinated CE breast CT. (A) Implant-displaced MLO mammogram illustrates an indistinct lesion. Precontrast (B) and postcontrast (C) sagittal breast CT slices show an invasive ductal carcinoma. (Images courtesy of John M. Boone, University of California–Davis.)

to operate CT detectors in a photon counting mode. Due to recent technological improvements in x-ray detectors and associated electronics, it has now become a lot more feasible for photon counting detectors to be used for CT applications with lower x-ray fluence requirements. Given that the dose to the breast is typically constrained to approximately that of the dose given for mammography, one of these applications is CT breast imaging. Breast CT with a photon counting detector promises to provide several advantages over current prototypes, including (1) improved spatial resolution; (2) improved tumor contrast; (3) minimization of detector electronic noise; (4) minimization of Swank noise; (5) minimization of image lag and ghosting effects; (6) increased dynamic range; (7) very fast detector readout, thereby minimizing motion blurring; (8) improved SNR through x-ray energy weighting; and (9) the potential for using single-exposure multiple-energy imaging to improve quantitative accuracy for CE-CT breast imaging.

16.5.3 VOI BREAST CT

Chen et al. (2009b) have proposed a methodology for obtaining high-resolution images of a VOI within the breast. The basic approach described is to perform an initial breast CT scan using a full-field flat-panel detector to localize the suspicious index lesion (e.g., found on mammography). The radiologist then reads this initial breast CT scan, selecting an appropriate VOI to view in high resolution. The initial CT scan will then be followed with a VOI CT scan using a high-resolution smaller detector. In this scan, a lead VOI mask with a rectangular opening would be placed between the source and the breast to deliver x-rays to the VOI while eliminating x-rays to the rest of the breast. Chen et al. (2009b) have conducted Monte Carlo studies to compare radiation dose with the VOI mask in place. Two scenarios are compared: (1) moving the VOI to the center of the CT rotation and (2) moving the lead VOI mask at each projection acquisition. Using the VOI-centered approach, the dose reduction factor increased from 2.0 to 11.3 as the distance from the VOI increases. Similarly, for the moving VOI mask approach, the dose reduction factor increased from 2.8 to 21.1. The study showed marginal improvement in visibility of calcifications in a phantom. Clinical studies are needed to investigate the use of this methodology for examining lesions and microcalcifications *in vivo*.

ACKNOWLEDGMENTS

The writing of this chapter was supported, in part, by National Institutes of Health National Cancer Institute grants CA140400, CA102758, and CA134128 and a research grant from the Toshiba Medical Research Institute.

REFERENCES

Akashi-Tanaka, S., Fukutomi, T., Miyakawa, K., Nanasawa, T., Matsuo, K., Hasegawa, T. and Tsuda, H. 2001. Contrast-enhanced computed tomography for diagnosing the intraductal component and small invasive foci of breast cancer. *Breast Cancer* 8: 10–15.

Bjurstam, N. 1978. Radiology of the female breast and axilla. *Acta Radiol Suppl* 357.

Boone, J.M. 2004. Breast CT: its prospect for breast cancer screening and diagnosis. In: Giger, A.K.A.M. (ed.). *RSNA Categorical Courses in Diagnostic Physics: Advances in Breast Imaging – Physics, Technology, and Clinical Applications*. Chicago, IL: RSNA.

Boone, J.M., Kwan, A.L., Seibert, J.A., Shah, N., Lindfors, K.K. and Nelson, T.R. 2005. Technique factors and their relationship to radiation dose in pendant geometry breast CT. *Med Phys* 32: 3767–76.

Boone, J.M., Nelson, T.R., Lindfors, K.K. and Seibert, J.A. 2001. Dedicated breast CT: radiation dose and image quality evaluation. *Radiology* 221: 657–77.

Boone, J.M., Shah, N. and Nelson, T.R. 2004. A comprehensive analysis of DgN_{CT} coefficients for pendant-geometry cone-beam breast computed tomography. *Med Phys* 31: 226–35.

breastcancer.org [Online]. Available from: http://www.breastcancer.org/symptoms/understand_bc/statistics.jsp.

Burgess, A.E., Jacobson, F.L. and Judy, P.F. 1999. On the detection of lesions in mammographic structure. *Proc SPIE* 3663: 304–15.

Burgess, A.E., Jacobson, F.L. and Judy, P.F. 2001. Human observer detection experiments with mammograms and power-law noise. *Med Phys* 28: 419–37.

Chang, C.H., Sibala, J.L. and Fritz, S.L. 1979. Specific value of computed tomographic breast scanner (CT/M) in diagnosis of breast disease. *Radiology* 132: 647–52.

Chang, C.H., Sibala, J.L., Fritz, S.L., Gallagher, J.H., Dwyer, S.J. and Templeton, A.W. 1978. Computed tomographic evaluation of the breast. *Am J Roentgenol* 131: 459–64.

Chen, B. and Ning, R. 2002. Cone-beam volume CT breast imaging: feasibility study. *Med Phys* 29: 755–70.

Chen, L., Shen, Y., Lai, C.-J., Han, T., Zhong, Y., Ge, S., Liu, X., Wang, T., Yang, W.T. and Whitman, G.J. 2009a. Dual resolution cone beam breast CT: a feasibility study. *Med Phys* 36: 4007.

Chen, L., Shen, Y., Lai, C.-J., Han, T., Zhong, Y., Ge, S., Liu, X., Wang, T., Yang, W.T., Whitman, G.J. and Shaw, C.C. 2009b. Dual resolution cone beam breast CT: a feasibility study. *Med Phys* 36: 4007–14.

Chen, Y., Liu, B., O'Connor, J.M., Didier, C.S. and Glick, S.J. 2009c. Characterization of scatter in cone-beam CT breast imaging: comparison of experimental measurements and Monte Carlo simulation. *Med Phys* 36: 857–69.

Crotty, D.J., Madhav, P., Mckinley, R.L. and Tornai, M.P. 2007. Investigating novel patient bed designs for use in a hybrid dual-modality dedicated 3D breast imaging system. *SPIE Proc* 6510, Medical Imaging 2007: Physics of Med Imag, 65104, doi: 10.1117/12.713764.

Dershaw, D., Abramson, A. and Kinne, D.W. 1989. Ductal carcinoma in situ: mammographic findings and clinical implications. *Radiology* 170: 411–5.

Dillon, M.F., Mc Dermott, E.W., O'Doherty, A., Quinn, C.M., Hill, A.D. and O'Higgins, N. 2007. Factors affecting successful breast conservation for ductal carcinoma in situ. *Ann Surg Oncol* 14: 1618–28.

Ding, H., Ducote, J.L. and Molloi, S. 2012. Breast composition measurement with a cadmium-zinc-telluride based spectral computed tomography system. *Med Phys* 39: 1289–97.

Ding, H. and Molloi, S. 2012. Quantification of breast density with spectral mammography based on a scanned multi-slit photon-counting detector: a feasibility study. *Phys Med Biol* 57: 4719–38.

Dullum, J.R., Lewis, E.C. and Mayer, J.A. 2000. Rates and correlates of discomfort associated with mammography. *Radiology* 214: 546–52.

Elvecrog, E.L., Lechner, M.C. and Nelson, M.T. 1993. Nonpalpable breast lesions: correlation of stereotaxic large-core needle biopsy and surgical biopsy results. *Radiology* 188: 453–5.

Feig, S.A., Shaber, G.S. and Patchefsky, A. 1977. Analysis of clinically occult and mammographically occult breast tumors. *AJR Am J Roentgenol* 128: 403–8.

Feldkamp, L.A., Davis, L.C. and Kress, J.W. 1984. Practical cone-beam algorithm. *J Opt Soc Am* 1: 612–9.

Fessler, J.A. 2000. Statistical image reconstruction methods for transmission tomography. In: Sonka, M. and Fitzpatrick, J.M. (eds.). *Handbook of Medical Imaging, Volume 2. Medical Imaging Processing and Analysis*. Bellingham, WA: SPIE, 3–58.

Fleming, F.J., Hill, A.D., Mc Dermott, E.W., O'Doherty, A., O'Higgins, N.J. and Quinn, C.M. 2004. Intraoperative margin assessment and re-excision rate in breast conserving surgery. *Eur J Surg Oncol* 30: 233–7.

Frankl, G. and Ackerman, M. 1983. Xeromammography and 1200 breast cancers. *Radiol Clin North Am* 21: 81–91.

Gisvold, J.J., Reese, D.F. and Karsell, P.R. 1979. Computed tomographic mammography (CTM). *AJR Am J Roentgenol* 133: 1143–9.

Glick, S.J., Thacker, S., Gong, X. and Liu, B. 2007. Evaluating the impact of x-ray spectral shape on image quality in flat-panel CT breast imaging. *Med Phys* 34: 5–24.

Gong, X., Glick, S.J., Liu, B., Vedula, A.A. and Thacker, S. 2006. A computer simulation study comparing lesion detection accuracy with digital mammography, breast tomosynthesis, and cone-beam CT breast imaging. *Med Phys* 33: 1041–52.

Gong, X., Vedula, A.A. and Glick, S.J. 2004. Microcalcification detection using cone-beam CT mammography with a flat-panel imager. *Phys Med Biol* 49: 2183–95.

Hagay, C., Cherel, P.J., De Maulmont, C.E., Plantet, M.M., Giles, R., Floiras, J.L., Garbay, J.R. and Pallud, C.M. 1996. Contrast-enhanced CT: value for diagnosing local breast cancer recurrence after conservative treatment. *Radiology* 200: 631–8.

Hendrick, R.E. and Helvie, M.A. 2011. United States Preventive Services Task Force screening mammography recommendations: science ignored. *AJR Am J Roentgenol* 196: W112–6.

Inoue, M., Sano, T., Watai, R., Ashikaga, R., Ueda, K., Watatani, M. and Nishimura, Y. 2003. Dynamic multidetector CT of breast tumors: diagnostic features and comparison with conventional techniques. *AJR Am J Roentgenol* 181: 679–86.

Jacobs, L. 2008. Positive margins: the challenge continues for breast surgeons. *Ann Surg Oncol* 15: 1271–2.

Kalender, W.A., Beister, M., Boone, J.M., Kolditz, D., Vollmar, S.V. and Weigel, M.C. 2012. High-resolution spiral CT of the breast at very low dose: concept and feasibility considerations. *Eur Radiol,* 22: 1–8.

Kerlikowske, K., Grady, D., Barclay, J., Sickles, E.A. and Ernster, V. 1996. Effect of age, breast density, and family history on the sensitivity of first screening mammography. *JAMA* 276: 33–8.

Kim, S.M. and Park, J.M. 2003. Computed tomography of the breast: abnormal findings with mammographic and sonographic correlation. *J Comp Assisted Tomo* 27: 761–70.

Kolb, T.M., Lichy, J. and Newhouse, J.H. 1998. Occult cancer in women with dense breasts: detection with screening US— diagnostic yield and tumor characteristics. *Radiology* 207: 191–9.

Kopans, D.B. 1998. *Breast Imaging.* Philadelphia, PA: Lippincott, Williams & Wilkins.

Kwan, A.L., Boone, J.M., Yang, K. and Huang, S.Y. 2007. Evaluation of the spatial resolution characteristics of a cone-beam breast CT scanner. *Med Phys* 34: 275–81.

Kwan, A.L.C., Boone, J.M. and Shah, N. 2005. Evaluation of x-ray scatter properties in a dedicated cone-beam breast CT scanner. *Med Phys* 32: 2967–75.

Lai, C.J., Shaw, C.C., Chen, L., Altunbas, M.C., Liu, X., Han, T., Wang, T., Yang, W.T., Whitman, G.J. and Tu, S. 2007. Visibility of microcalcification in cone beam breast CT: effects of x-ray tube voltage and radiation dose. *Med Phys* 34: 2995–3004.

Le, H., Ducote, J. and Molloi, S. 2010. Radiation dose reduction using a CdZnTe-based computed tomography system: comparison to flat-panel detectors. *Med Phys* 37: 1225–36.

Lindfors, K.K., Boone, J.M., Nelson, T.R., Yang, K., Kwan, A.L.C. and Miller, D.F. 2008. dedicated breast CT: initial clinical experience. *Radiology* 246: 725–33.

Makeev, A., Das, M. and Glick, S.J. 2012. Investigation of statistical iterative reconstruction for dedicated breast CT. *Proc SPIE* 8313, Medical Imaging 2012 Physics of Medical Imaging, 8313W, doi: 10.1117/12.912786.

Makeev, A. and Glick, S.J. 2013. Investigation of statistical iterative reconstruction for dedicated breast CT. *Med Phys* 40: 081904.

McKinley, R.L., Tornai, M.P., Samei, E. and Bradshaw, M.L. 2004. Simulation study of a quasi-monochromatic beam for x-ray computed mammotomography. *Med Phys* 31: 800–13.

Metheany, K.G., Abbey, C.K., Packard, N. and Boone, J.M. 2008. Characterizing anatomical variability in breast CT images. *Med Phys* 35: 4685–94.

Meyer, J.E., Eberlein, T.J., Stomper, P.C. and Sonnenfeld, M.R. 1990. Biopsy of occult breast lesions. Analysis of 1261 abnormalities. *JAMA* 263: 2341–3.

Michaelson, J.S., Silverstein, M., Wyatt, J., Weber, G., Moore, R., Halpern, E., Kopans, D.B. and Hughes, K. 2002. Predicting the survival of patients with breast carcinoma using tumor size. *Cancer* 95: 713–23.

Miyake, K., Hayakawa, K., Nishino, M., Nakamura, Y. and Morimoto, T. 2005. Benign or malignant? Differentiating breast lesions with computed tomography attenuation values on dynamic computed tomography mammography. *J Comp Assisted Tomo* 29: 772–9.

Nakahara, H., Namba, K., Wakamatsu, H., Watanabe, R., Furusawa, H., Shirouzu, M. and Matsu, T. 2002. Extension of breast cancer: comparison of CT and MRI. *Radiat Med* 20: 17–23.

Neitzel, U. 1992. Grids or air-gaps for scatter reduction in digital radiography: a model calculation. *Med Phys* 11: 475–81.

Nelson, H.D., Tyne, K., Naik, A., Bougatsos, C., Chan, B., Nygren, P. and Humphrey, L. 2009. *Screening for Breast Cancer: Systematic Evidence Review Update for the US Preventive Services Task Force.* Rockville, MD Ann Intern Med 15:727–737.

Niklason, L.T., Christian, B.T., Niklason, L.E., Kopans, D.B., Opsahl-ong, B.H. and Landberg, C.E. 1997. Digital tomosynthesis in breast imaging. *Radiology* 205: 399–406.

Ning, R., Tang, X. and Conover, D. 2004. X-ray scatter correction algorithm for cone beam CT imaging. *Med Phys* 31: 1195–202.

Nishino, M., Hayakawa, K., Yamamoto, A., Nakamura, Y., Morimoto, T., Mukaihara, S. and Urata, Y. 2003. Multiple enhancing lesions detected on dynamic helical computed tomography mammography. *J Comp Assisted Tomo* 27: 771–8.

O'Connell, A., Conover, D., Zhanh, Y., Seifert, P., Logan-Young, W., Lin, C.L., Sahler, L. and Ning, R. 2010. Cone-beam CT for breast imaging: radiation dose, breast coverage, and image quality. *AJR Am J Roentgenol* 195: 496–509.

Opie, H., Estes, N.C., Jewell, W.R., Chang, C.H., Thomas, J.A. and Estes, M.A. 1993. Breast biopsy for nonpalpable lesions: a worthwhile endeavor? *Am Surg* 59: 490–3; discussion 493–4.

Parker, S.H., Lovin, J.D., Jobe, W.E., Burke, B.J., Hopper, K.D. and Yakes, W.F. 1991. Nonpalpable breast lesions: stereotactic automated large-core biopsies. *Radiology* 180: 403–7.

Partain, L., Prionas, S., Seppi, E., Virshup, G., Roos, G., Sutherland, R. and Boone, J.M. 2007. Iodine contrast cone-beam CT imaging of breast cancer. *Proc SPIE* 6510 Medical Imaging 2007: Physics of Med Imag 65102; doi: 10.1117/12.712637.

Pisano, E.D., Gatsonis, C., Hendrick, E., Yaffe, M., Baum, J.K., Acharyya, S., Conant, E.F., Fajardo, L.L., Bassett, L., D'Orsi, C., Jong, R. and Rebner, M. 2005. Diagnostic performance of digital versus film mammography for breast-cancer screening. *N Engl J Med* 353: 1773–83.

Poplack, S.P., Tosteson, T.D., Kogel, C.A. and Nagy, H.M. 2007. Digital breast tomosynthesis: initial experience in 98 women with abnormal digital screening mammography. *AJR Am J Roentgenol* 189: 616–23.

Prionas, N.D., Lindfors, K.K., Ray, S., Huang, S.Y., Beckett, L.A., Monsky, W.L. and Boone, J.M. 2010. Contrast-enhanced dedicated breast CT: initial clinical experience. *Radiology* 256: 714–23.

Reese, D.F., Carney, J.A., Gisvold, J.J., Karsell, P.R. and Kollins, S.A. 1976. Computerized reconstructive tomography applied to breast pathology. *AJR Am J Roentgenol* 126: 406–12.

Roos, P.G., Colbeth, R.E., Mollov, I., Munro, P., Pavkovich, J., Seppi, E.J., Shapiro, E.G. and Tognina, C.A. 2004. Multiple-gain ranging readout method to extend the dynamic range of amorphous silicon flat-panel imagers. *Proc SPIE* 5368, Medical Imaging 2004: Physics of Med Imag, 139; doi:10.1117/12.535471.

Rosenberg, R.D., Hunt, W.C., Williamson, M.R., Gilliland, F.D., Wiest, P.W., Kelsey, C.A., Key, C.R. and Linver, M.N. 1998. Effects of age, breast density, ethnicity, and estrogen replacement therapy on screening mammographic sensitivity and cancer stage at diagnosis: review of 183,134 screening mammograms in Albuquerque, New Mexico. *Radiology* 209: 511–8.

Rosenberg, R.D., Lando, J.F., Hunt, W.C., Darling, R.R., Williamson, M.R., Linver, M.N., Gilliland, F.D. and Key, C.R. 1996. The New Mexico Mammography Project. Screening mammography performance in Albuquerque, New Mexico, 1991 to 1993. *Cancer* 78: 1731–9.

Sardanelli, F., Calabrese, M., Zandrino, F., Melani, E. and Parodi, R.C. 1998. Dynamic helical CT of breast tumors. *J Comp Assisted Tomo* 22: 398–407.

Sechopoulos, I., Vedantham, S., Suryanarayanan, S., D'Orsi, C.J. and Karellas, A. 2008. Monte Carlo and phantom study of the radiation dose to the body from dedicated CT of the breast. *Radiology* 247: 98–105.

Shikhaliev, P.M. and Fritz, S.G. 2011. Photon counting spectral CT versus conventional CT: comparative evaluation for breast imaging application. *Phys Med Biol* 56: 1905–30.

Shimauchi, A., Yamada, T., Sato, A., Takase, K., Usami, S., Ishida, T. and Moriya, T. 2006. Comparison of MDCT and MRI for evaluating the intraductal component of breast cancer. *AJR Am J Roentgenol* 187: 322–9.

Siewerdsen, J.H. and Jaffray, D.A. 2001. Cone-beam computed tomography with a flat-panel imager: magnitude and effects of x-ray scatter. *Med Phys* 2001: 220–31.

Siewerdsen, J.H., Moseley, D.J., Bakhtiar, B., Richard, S. and Jaffray, D.A. 2004. The influence of antiscatter grids on soft-tissue detectability in cone-beam computed tomography with flat-panel detectors. *Med Phys* 31: 3506–20.

Tabar, L., Dean, P.B., Duffy, S.W. and Chen, H.H. 2000. A new era in the diagnosis of breast cancer. *Surg Oncol Clin North Am* 9: 33–77.

Thacker, S. and Glick, S.J. 2004. Normalized glandular dose (DgN) coefficients for flat-panel CT breast imaging. *Phys Med Biol* 49: 5433–44.

Tornai, M.P., Mckinley, R.L., Bryzmialkiewicz, C.N., Madhav, P., Cutler S.J., Crotty, D.J., Bowsher, J.E., Samei, E. and Floyd, C.E., Jr. 2005. Design and development of a fully-3D dedicated x-ray computed mammotomography system. *Proc SPIE* 5745 Medical Imaging 2005: Physics of Med Imag. 189, doi: 10.1117/12595636.

Uematsu, T., Sano, M., Homma, K., Shiina, M. and Kobayashi, S. 2001. Three-dimensional helical CT of the breast: accuracy for measuring extent of breast cancer candidates for breast conserving surgery. *Breast Cancer Res Treat* 65: 249–57.

Vedantham, S., Shi, L., Glick, S.J. and Karellas, A. 2013. Scaling-law for the energy dependence of anatomic power spectrum in dedicated breast CT. *Med Phys* 40: 011901–8.

Vedantham, S., Shi, L., Noo, F., Glick, S.J. and Karellas, A. 2011. Cone-beam artifacts in dedicated breast CT. *AAPM Annual Meeting* 38(6): 3430.

Wang, X. and Ning, R. 1999. A cone-beam reconstruction algorithm for circle-plus-arc data acquisition geometry. *IEEE Trans Med Imaging* 18: 815–24.

Weigel, M., Vollmar, S.V. and Kalender, W.A. 2011. Spectral optimization for dedicated breast CT. *Med Phys* 38: 114–24.

Wiegert, J., Bertram, M., Schafer, D., Conrads, N., Timmer, J., Aach, T. and Rose, G. 2004. Performance of standard fluoroscopy anti-scatter grids in flat detector based cone beam CT. *Proc SPIE* 5368: 67–8.

Wolfe, J.N. 1974. Analysis of 462 breast carcinomas. *AJR Am J Roentgenol* 121: 846–53.

Yamamoto, A., Fukushima, H., Okamura, R., Nakamura, Y. and Morimoto, T. 2006. Dynamic helical CT mammography of breast cancer. *Radiat Med* 24: 35–40.

Yang, K., Kwan, A.L. and Boone, J.M. 2007. Computer modeling of the spatial resolution properties of a dedicated breast CT system. *Med Phys* 34: 2059–69.

Zhang, J., Yang, G., Cheng, Y., Gao, B. and Qui, Q. 2005. Stationary scanning x-ray source based on carbon nanotube field emitters. *Appl Phys Lett* 86: 184104.

Ziedes Des Plantes, B.G. 1932. Eine neue methode zur diffenzierung in der rontgenographie (planigraphie). *Acta Radiol* 13: 182–92. [German]

Index

T - #0250 - 111024 - C0 - 280/210/13 - PB - 9780367576189 - Gloss Lamination